By the same author:

(With G. Frere Cook) *The Guinness History of Sea Warfare* (1975)

(With D. Brown and C. Shores) *The Guinness History of Air Warfare* (1976)
The Guinness History of Land Warfare (1976)
Technology in War (1987)
For Want of a Nail (1989)

(With William Woodhouse) *The Penguin Encyclopedia of Modern Warfare* (1991)

Kenneth Macksey

The Penguin Encyclopedia of

Weapons and Military Technology

Prehistory to the Present Day

VIKING

VIKING

Published by the Penguin Group
Penguin Books Ltd, 27 Wrights Lane, London W8 5TZ, England
Penguin Books USA Inc., 375 Hudson Street, New York, New York 10014, USA
Penguin Books Australia Ltd, Ringwood, Victoria, Australia
Penguin Books Canada Ltd, 10 Alcorn Avenue, Toronto, Ontario, Canada M4V 3B2
Penguin Books (NZ) Ltd, 182–190 Wairau Road, Auckland 10, New Zealand

Penguin Books Ltd, Registered Offices: Harmondsworth, Middlesex, England

First published 1993
10 9 8 7 6 5 4 3 2 1
First edition

Typeset by Datix International Limited, Bungay, Suffolk
Set in 8½/10½ pt Lasercomp Photina
Printed in England by Clays Ltd, St Ives plc

A CIP catalogue record for this book is available from the British Library

ISBN 0-670-84411-X

Contents

List of Maps and Figures

Acknowledgements for Maps and Figures
Michael Haine
Gordon E. MacKay MSAI

Introduction

It is among the paradoxes of history – military history in particular – that the role of technology tends to be relegated to a lowly place within the unfolding story of events. As a result scientists and technologists are often demoted to insignificance, and even obscurity, while statesmen, politicians, admirals, generals and air marshals enjoy the prominence and glory. As Winston **Churchill** wrote in one of his dispatches as a war correspondent in 1897: 'To be technical is a grave offence . . .', but then defended his temerity by describing technology's military impact – long before himself becoming a symbol of technological, innovative insight as First Lord of the Admiralty, Minister for Munitions and War and Minister of Defence in two world wars.

Arguably, the developments of **strategy** and **tactics** in war – and indeed as the causes of war – have far more often than not been inspired by the interaction of personalities and technology. When politicians and military men have opted to take up the sword for aggrandisement, ambition, greed and glory, among many vices, the temptations have frequently been generated by scientists, technologists and industrialists with their specious visions and promises of war-winning, deadly weapon systems in combination with new organizations and doctrines.

So, the aim of this Encyclopedia is to present in compact form the essential and vital elements creating the interaction of technology and weapons upon the evolution of warfare. It does so with entries about key scientists, inventors and industrialists, by describing the materials they worked with, the products of their labours and the techniques which evolved. And it also records the innovatory military **philosophers** and commanders whose thoughts and concepts influenced important changes in organization and doctrine; or who made decisive strategic and tactical use of the latest weapons and methods.

It poses a chicken-and-egg question to ask whether technology changes attitudes and methods, or the other way about. Let the student judge. It is suggested, however, that a unifying motif of this Encyclopedia is its Chronology; one which not only illustrates in time the advent of new technology and weapons in relation to wars (and therefore, by implication, their effect on commanders, campaigns and battles) but also shows how events triggered new technology in response to fluctuating moods of bellicosity.

When war-making, as so often in the past, became virtually an industry for the acquisition of wealth, as well as power, by politicians and magnates, and when humanitarianism was unknown or cast aside, technology and **armaments industries** became vital. No doubt **bronze** and **iron** were discovered accidentally (round about, if not before, 4000 BC), because nobody at that time could possibly have been invited to discover them through **research and development**, as we understand it today. But their outstanding utility must have been recognized quickly and eagerly adapted for **edged**

weapons and **armour** in combat. Nevertheless, only as social intercourse ripened by way of better **communications**, improved **education** and a burgeoning inquisitiveness, did the momentum of discovery swell.

Comparison of the increasing pace of technology and weapons is salutary. Take, for example, the time elapsing between the discovery of bronze, *c.* 4000 BC, and cast iron, *c.* 1000 BC; between the invention of the maritime **navigation** log *c.* 600 BC and mail armour *c* AD 200; and between the invention of the crossbow *c.* 1100, gunpowder *c.* 1200 and firearms *c.* 1330. Compare this very slow evolution with the rate of progress after James **Watt** invented a practical steam engine in 1769 (to launch the Industrial Revolution by powering the mass-production, flow-line armaments firms, such as Boulton and Watt) and to supply the forces fighting in the **French revolutionary** and Napoleonic **wars**. All of this was tortuously slow by comparison with what happened almost immediately after Michael **Faraday** explained **electromagnetic effects**; after Gottlieb **Daimler** invented the internal-combustion engine; the **Wright brothers** flew in the first man-lifting, engine-powered, heavier-than-air machine; Lise Meitner and others uncovered the secrets of **nuclear** power; and Alan **Turing** employed binary mathematics to make feasible the fast, **electronic computer**.

How to use the Encyclopedia. Cross-references in running text, printed in **bold** type, to entries in the Encyclopedia, are intended to highlight selected subjects to help the student recognize related subjects and thus make research in depth easier. For example, the entry for **Artillery** draws attention to numerous entries such as **anti-aircraft** and **anti-tank weapons**, **bows and arrows**, **rockets**, **siege engines**; vital **wars**; the battles of **Sluys**, **Crécy** and **Jutland**; the techniques of **gunnery**, **siege warfare**, **communication** and **surveillance**; gunners such as Jean **Bureau** and Noel **Birch**; inventors such as Henry **Shrapnel** and Henry **Bessemer**; and companies like **Armstrong's**, **Whitworth's** and **Schneider's**.

Maps, diagrams and pictures are included for amplification and clarification; and are placed where they will be most effective. There is a simplified Chronology following this Introduction, setting out the principal wars, technologies and weapons. The Bibliography on page 373 is restricted to those encyclopedias, bibliographies and reference works judged most helpful for further study. To provide an exhaustive bibliography of the innumerable works on war and technology would be utterly self-defeating in any general work, let alone a compact one such as this. However, the entry entitled **Bibliography of war** may be of assistance to researchers.

It is perhaps worth repeating what William Woodhouse and I wrote in *The Penguin Encyclopedia of Modern Warfare*, that the sum total of knowledge acquired is in ratio to the application and quality of reading and thought. That, in military parlance, encyclopedias simply offer a firm base for reconnaissance followed by attack upon a selected objective of knowledge, and to indicate lines of exploitation to where further information can be found and new ideas developed.

In effect, the student is being invited to consult this Encyclopedia for the sort of voyage of discovery which experience has shown to be highly rewarding and satisfying.

Kenneth Macksey

Chronology

Important Wars, New Technology and Weapons

Year	War	Technology/Weapon
? 4000 BC		Wheel Chariot
c. 4000 BC		Bronze
c. 3500 BC		Wrought iron
c. 3000 BC		Steel
c. 3000 BC		Galleys
c. 1000 BC		Cast iron
c. 600 BC		Abacus
500 BC to AD 182	Roman expansion	
499–326 BC	Greek–Persian	
c. 350 BC		Maritime navigation log
265–146 BC	Punic	
c. 150 BC		Piston
c. 100 BC		Water-wheel
58–51 BC	Gallic	

Year	War	Technology/Weapon
50–30 BC	Roman Civil	
c. AD 100		Reaction **turbine**
c. 200		Mail **armour**
370	Huns invade Europe	
406	Vandals invade Gaul	
410	Goths invade Italy	
632	**Muslim** expansion begins	
672		**Flame**-thrower
1066	Normans in Britain	
1096–1444	**Crusades**	
c. 1100		Crossbow
c. 1160		Longbow
1180		Magnetic compass
c. 1200		Gunpowder
1206–79	**Mongol** in Asia	
1236–43	Mongol in Europe	
c. 1250		**Rocket missile**
1258–60	Mongol in Arabia	
1297–1327	Mongol in India	
c. 1300		**Clock**
1337–1457	**Hundred Years**	
1340		Cannon at sea
1345		Cannon at Crécy

Year	War	Technology/Weapon
c. 1350		Small arms
1361–1923	**Ottoman Empire**	
1380		Blast furnace
c. 1390		Small-arms match ignition
1419–36	Hussite	
c. 1425		Matchlock small arms
c. 1450		**Printing** press
1451		**Mortar**
c. 1470		Caravel Seaman's log
1484		Maritime astrolabe
c. 1485		**Pistol**
1495–1598	**Spanish**	
c. 1500		Watch mainspring Bastion fortifications
c. 1515		Hourglass Wheel-lock small arms
c. 1556		Flintlock small arms
c. 1560		**Carrack**
1563–1629	Polish–Swedish (sporadic)	
1570		Galleon
1589		Lenses
1592–8	Korean–Japanese	**Iron-armoured** warship
c. 1600		Musket **Camera** obscura

Year	War	Technology/Weapon
1608		Telescope
1609–61	**Thirty Years**	
1609–18	Russo-Polish	
1620		Leather light gun
1629		Impulse steam turbine
1630		One-piece artillery **ammunition**
1635–1783	**French**	
1642–51	English Civil	
1652–74	Anglo-Dutch	
1653–9	Franco-Spanish	
c. 1660		**Vauban** fortifications
1667–88	Franco-Spanish	
1672–8	Franco-Dutch	
1688–97	Great French	
1698		Military academy
1701–14	Spanish Succession	
1718		Puckle **machine-gun**
1726		Gridiron pendulum
1740–48	Austrian Succession	
1741		Robin's ballistic pendulum
c. 1750		Breech-loading **rifle** in service
1756–63	**Seven Years**	
1757		Sextant

Year	War	Technology/Weapon
1761		Chronometer
1766		Hydrogen isolated
1769		Steam-powered road vehicle
1775–83	**American of Independence**	
1776		Man-powered **submarine** Sea-**mine**
1779		Carronade in action
1783		Hot-air **balloon** Steam paddle-boat
1786		Shrapnel **ammunition**
1789		**Uranium** discovered
1791		**Titanium** discovered Portland-type cement concrete
1792		Semaphore network
1792–1815	**French revolutionary**	
1796		Flow-line mass production
1797		**Parachute** Precision lathe
1798		Interchangeable manufacture
1800		Chemical-electric primary cell
1805		Percussion cap
1809		Bottled food
1813		Steam-powered armoured warship
1820		**Electromagnetic** rotation
1821		Natural-gas well

Year	War	Technology/Weapon
1822		Photographic plate
1827		Water-**turbine** engine
1831		Steam-turbine engine Electromagnetic induction
1833		**Plastics**
1834		Electric motor **Computer**
1837		Screw-propelled boat Practical **telegraph** system
1839–42	First **Afghan** First Opium	
1839		Fuel cell Canned food
1845–9	Anglo-Sikh	
1845		Pneumatic tyre patented
1846		Nitrocellulose Rodman gun barrel Anaesthetics Arc light, Carbon
1849		**Bombs** from unmanned balloons
1850		Morse code with telegraph 'sounding key'
1850–64	Taiping Rebellion	
1851		All-steel artillery piece
1852		Repeater **rifle** Man-carrying, non-rigid **airship**
1854–7	**Crimean**	
1854		**Periscope**

Year	War	Technology/Weapon
1855		Cowen's **armoured fighting vehicle** Refrigeration Rodman powder
1856–60	Second Opium	
1856		Bessemer steel
1857–8	Indian Mutiny	
1858		**Heliograph** Aerial **photography**
1859	Austro-Piedmont	*La Gloire* armoured ship Breech-loading artillery Hand-rotated **machine-gun** First successful **oil** well Lead-acid storage **battery**
1860		Magazine rifle
1861–5	**American Civil**	Turreted ironclad warship
1861–7	Franco-Mexican	Pasteurized food
1864	Danish–Prussian	
1865		Antiseptics
1866	Austro-Prussian Austro-Italian	Locomotive **torpedo**
1870–71	**Franco-Prussian**	**Anti-aircraft** gun
1870		Xylonite (Celluloid)
1873		Spar **torpedo-boat** Practical **typewriter**
1874		**Barbed wire**
1875		Ballistite smokeless powder Cordite

Year	War	Technology/Weapon
1876		**Telephone** Four-stroke petrol engine Locomotive torpedo-boat
1877–8	Russo-Turkish	
1878–80	Second Afghan	
1879		Electric-light bulb **Dynamo**
1880–81	First Boer	
1881		**Searchlight**
1882		Armoured **steel**
1883–5	Sino-French	
1884		**Destroyer** Steam-turbine dynamo
1885		Four-wheel motor carriage Smokeless powder Semi-automatic machine-gun
1886		Electrically powered submarine
1888		Portable roll-film **camera** Pneumatic tyre
1889		Cine-camera **Radio** waves Staged steam turbine
1890		Electromechanical computer
1892		Detected radio signal
1894–5	Sino-Japanese	
1895		Radio transmission and reception **Gyroscope**

Year	War	Technology/Weapon
1897		Steam-turbine-engine ship French 75mm field gun Cathode-ray-tube oscilloscope
1898	Spanish–American	Wire recorder (forerunner of tape recorder)
1899–1901	Second Boer	**Armoured car**
1899		Plastic explosive (later RDX)
1900–1901	Boxer Rebellion	
1900		Petrol–electric submarine Rigid **airship**
1901		Transatlantic radio link
1902		Thermionic diode valve TNT
1903		Powered, heavier-than-air flying machine
1904–5	**Russo-Japanese**	Radio direction-finding (DF)
1905		Dreadnought **battleship** Scott naval fire-control system
1906		Radio crystal detector
1908		Voice radio
1909		Duralumin Bakelite
1910		Aerial bombing Armed aircraft **Nuclear** theory Shipborne aircraft
1911–50	Chinese revolutionary	
1911–12	Italo-Turkish	Seaplane Airborne torpedo
1912–13	Balkan	Aircraft catapult

Year	War	Technology/Weapon
1914–18	**World War I**	
1915		Poison gas **Depth charge** Unmanned air vehicle (UAV)
1916		**Tank** Predicted artillery fire
1917		**Aircraft-carrier** Unmanned ground vehicle (UGV)
1918		**Sonar** (ASDIC)
1918–22	Russian revolutionary	
1919	Third Afghan	
1920	Russo-Polish	
1920–22	Greek–Turkish	
1921		Teleprinter
1923		Autogiro **Television** (patented)
1925		Short-wave, crystal-controlled radio
1926		Enigma encoding machine
1927		Radio-beacon **navigation** Parachute troops
1929		**Penicillin** Electro-welded **cruiser**
1930		Catalyst
1933		**Radar**
1935–6	Italo-Ethiopian	
1936–9	**Spanish Civil**	

Year	War	Technology/Weapon
1937		**Helicopter** Radio-beam navigation
1937–45	Sino-Japanese **World War II**	Mechanized **landing-craft**
1939	Russo-Japanese	**Jet** aircraft Electromechanical computer
1942		Nerve gas
1943		Infrared night vision instrument Guided-rocket missile
1944		Programmable, electronic digital computer **Cruise rocket missile**
1945–91	**Cold**	**Atom bomb**
1945–9	Indonesian of Independence	
1945–54	French Indo-China	
1946		**Titanium** processed
1946–9	Greek Civil	
1948–	**Arab–Israeli**	
1948–60	Malayan of Independence	
1949		Transistor
1950–53	**Korean**	
1953		**Hydrogen bomb** **Ground-effect machine** (GEM)
1954–	African of Independence	Inertial navigation
1955		**Nuclear**-powered **submarine**
1956	Suez	Unmanned underwater vehicle (UUV)
1957		**Space vehicle**

Year	War	Technology/Weapon
1959–74	**Vietnam**	
1960		**Laser** Microchip
1965	Indo-Pakistan	
1967	Six-Day	
1971	Indo-Pakistan (Bangladesh)	
1973	Yom Kippur	
1979–88	**Iran–Iraq**	
1979–89	Russo-Afghan	
1982	**Falkland Islands**	
1990–91	**UN–Iraq**	Anti-ballistic missile in action MLRS in action

A

Adrianople, Battles of. Adrianople (Edirne) stands on the River Maritsa at a strategic gateway between Bulgaria and Constantinople (Istanbul). In AD 323 it was the scene of the battle in the civil war which founded the Byzantine empire. The Emperor Constantine I's army of roughly 130,000 men outflanked and assaulted the fortified camp of his rival, Valerius Licinius, to rout an army of similar size.

In 378 the Byzantine Emperor Valens attempted with 60,000 dispirited men to storm the **wagon laager** established by invading 'barbaric' Goths, Huns, Samartians and Alans under Frigidern. Sometimes mistakenly called the start of the **cavalry** era, this battle was certainly modern. It began with a clash between cavalry forces, in which the Roman mobile arm was swept away, and came to a decision when the unsupported Roman legion was destroyed by Frigidern's cavalry and **infantry** supported by arrows fired from the laager. In the ensuing rout, Valens and his commanders were slain.

There would be many more battles at Adrianople in 718, 914, 1003, 1094, 1205, 1224, 1255, 1225, 1365, 1829 and 1913, but none so important strategically, tactically and technically as that of 378.

Afghanistan, Wars of. This mountainous region has been the target of Arab, Maratha, Mogul and Persian armies, besides intensive internal strife, from the beginning of time. The British, using the technology of the day (including **aircraft** and **armoured cars** in 1919), became involved in the so-called *First* (1839–42), *Second* (1878) and *Third* (1919) *Afghan Wars*, but eventually were forced to withdraw. As also were the Russians after they moved in in 1979 and, with 100,000 mechanized troops with **tanks**, jet aircraft and **helicopters** sought to subjugate the people, only to stir up an internal hornets' nest as well as aggravating world-wide feeling. So long as the Russians managed to control the country's communications and were able to make life extremely unpleasant for the hard core of tribesmen who endured in the mountains and among the less accessible fertile valleys, they had their way at bearable cost in a war which became increasingly unpopular at home. Until weapons from friendly nations, including the USA and Britain, began to make their way via numerous frontier byways into the hands of the guerrillas (by no means a unified force), the tribesmen were at a disadvantage, chiefly because low-flying jet aircraft and armed helicopters could strike much as they pleased. But when hand-held surface-to-air (SAM) missiles reached the guerrillas the tide turned because the Russians and the disheartened Afghan army were no longer able to harass their victims from out of range. Thereupon the guerrillas were able to move with greater safety, bring in supplies with some ease and multiply their ambushes to the extent that sometimes, for months on end, enemy garrisons were besieged. In these circumstances against a background of world-wide indignation, and with the need to recast her internal affairs, the Russians had little option but to copy an old British precedent and depart (in 1989) with as good a grace as possible, leaving the Afghans to their traditional devices. Since then, strife has continued between factions.

Airborne forces. The first flight of a hot-air **balloon** in 1783 and the invention of the **parachute** in 1797 inspired numerous prophets and propagandists to flights of imagination about airborne armies. But not until powered **aircraft** were developed and Colonel William **Mitchell** proposed their use in 1919 were they deemed feasible.

In 1927 the Italians dropped a 'stick' of nine equipped parachutists. But it was the Russians who first formed an airborne division, in 1934, and made massed drops of **infantry** and a light **tank** along with **glider**-borne troops. Meanwhile, in the 1920s, the British delivered troops by **transport aircraft** to threatened places in their empire. Then, at the start of the **Spanish Civil War**, the Germans (with Italian help) airlifted 13,523 troops from Morocco to Spain; and in 1938 took the lead in airborne warfare by spearheading the bloodless invasion of Austria with parachutists and air-landed artillery.

Subsequently in **World War II**, German airborne troops successfully spearheaded the invasions of **Norway**, Holland and France in 1940, and Greece in April 1941. However, their attempt to invade **Crete** in May 1941, though eventually successful, was so costly that **Hitler** drew the conclusion that their day was done. Henceforward they were relegated to small-scale raiding as the British and Americans enthusiastically developed parachute and glider-borne forces for prominent roles in the invasions of **North Africa**, Sicily, France, in Holland (**Arnhem**) and in **Germany**. Airborne forces were also used at divisional level in the Burma **jungle**, though with only limited success.

The Japanese and Russians, meanwhile, made only limited use of airborne troops, the former during their expansion into South-East Asia in 1941 and 1942, the latter mainly in support of **guerrilla** forces behind the enemy lines.

Since World War II major parachute operations have taken place in **Korea** and the **Arab–Israeli wars**, but owing to vulnerability and inflexibility their opportunities have become fewer and farther between. Since the **Vietnam War** the **helicopter** has dominated airborne operations.

Aircraft. Almost immediately after the first **balloon** went aloft in 1783 concepts of lighter-than-air flying machines with a military capability abounded. Within weeks of François de Rozier's ascent an article appeared in Amsterdam envisaging the capture of Gibraltar by 'flying globes'. A year later the American Benjamin Franklin was envisaging 'ten thousand men descending from the clouds'. And soon there were visions of 'galleons of the sky' with 200 guns apiece engaged in **air warfare**. But the first action from the air occurred at the Battle of Fleurus when a French tethered balloon went aloft to reconnoitre the Austrian positions.

It was not until 1849 that an attempt was made by the Austrians to drop bombs from unmanned free-flight balloons on the Venetians. In the same year a 10-year-old boy flew in a **glider** designed by Sir George Cayley. And three years before, in France, Henri Giffard flew, under slight control, in a manned, steam-powered, steerable (so-called dirigible) balloon. It was followed in 1888 by the first **airship**, a dirigible powered by a **Daimler** petrol engine. All such machines, including Count Ferdinand **Zeppelin's** rigid airships, were initially suggested to the military as **reconnaissance aircraft**, although the carrying of bombs was always envisaged.

Flights by man-lifting kites, manned gliders and powered models, along with a few short 'hops' by powered machines, became ever more frequent towards the end of the 19th century. But it was the three flights (one of 59 seconds' duration) on 17 December 1903 by the *Flyer*, the petrol-engined biplane designed by the **Wright brothers** which proved the feasibility of controllable, load-carrying aircraft. This led in 1907 to an American military specification calling for a machine with a speed of 40mph, a duration of one hour and a crew of two, met in 1909 by the delivery for trials of the first warplane.

Henceforward progress was substantial and rapid. Before the outbreak of **World War I** in 1914, the feasibility of **bombing**, aerial firing of small arms and cannon as well as operating from ships had been demonstrated, as also had **photography** and **radio** communications to and from the ground. And in 1911 the Italians had used nine aeroplanes and two non-rigid airships against the Turks in Tripolitania during the Italo-Turkish War.

Nevertheless, only the rudiments of future **air warfare** were practised during the opening campaigns of World War I. But, as aircrew took to fitting **machine-guns** and **bomb** dropping became more prevalent, it was shown that the performance of these functions demanded specialized aircraft. And no sooner were the combat limitations of airships exposed, than improvisations to existing **reconnaissance aircraft** were followed by the design and production of **fighter** and **bomber aircraft** (including seaplanes and flying boats) incorporating special weapon systems and navigation aids. Naturally, evolving **strategy** and **tactics** demanded ever-improving technology to satisfy requirements for increased speed, ceiling, rate of climb, manoeuvrability, range, armament, payload, **armour** protection, reliability, ease of maintenance, **navigation**, night flying and structural strength. The result was that an infant **aircraft construction industry** had to grow up extremely rapidly, with particular emphasis upon the design and development of more powerful engines (*see* **propulsion**) and structural systems, at the same time as ways of saving weight also became paramount. Towards the end of World War I the **wood**, cloth and wire used for airframes and wings gave way to **aluminium** alloys; engine-power output had more than trebled; and multi-engined machines, with clear potential as **transport aircraft**, were in operation.

Thus the dynamic processes which continue to drive air technologists were established early on, along with the industrial base which would respond to the strategic concepts of leaders such as Air Marshal Sir Hugh **Trenchard** and General Giulio **Douhet**. In the 1920s and 1930s, when biplanes were being supplanted by aerodynamically cleaner monoplanes and the search for higher performance seemed to place emphasis on greater specialization, the reactive forces of operational as well as financial economy sometimes called for multi-role aircraft. Such were transport machines, like the Junkers Ju52, which could operate as a bomber; and Junkers Ju87 dive-bombers, which could function as fighter-bombers.

World War II witnessed dramatic advances in performance, notably through **jet engines**; coupled with the impact of burgeoning **electronic** and **radar** systems, without which the latest operational procedures could hardly function. The introduction of **rockets and guided missiles**, smart munitions and **space vehicles** also had a profound effect, made possible by new fuels, as well as **metallurgy** and synthetic materials to strengthen and further reduce weight. Paradoxically, machines with automated command and control facilities not only vastly boosted operational effectiveness but, owing to their complexity, vastly multiplied unit costs, though reducing the need for numbers of machines and aircrew. In the 1990s a few sophisticated, multi-role aircraft, crewed by two, can, with precision, do the work of several squadrons of World War II hit-or-miss bombers crewed by eight or ten men.

Aircraft-carrier. In 1910 the American Eugene Ely took off by plane from the deck of a US cruiser. In 1911 the first take-offs and landings from water took place and in 1912 catapult launchings from ships. During **World War I** the British used seaplane carriers which lowered these aircraft into the water, but in 1917 built a deck, suitable for take-off only, on to the cruiser HMS *Furious*, which carried three seaplanes and five scouts in its hangar. In 1918 the first 'true' carrier, HMS *Argus*, with a full, unobstructed flight deck was ready but the war ended before it could be used in action.

After the war the British, Americans and Japanese experimented extensively with carriers, with particular attention paid to flight-deck arrester gear and, led by the British, the introduction of a bridge, the so-called 'island', to one side of the deck. The problem of smoke obscuration was also tackled along with the type of aircraft most desirable – **fighters**, dive-bombers and torpedo-bombers were all developed with varying degrees of enthusiasm by each of the committed navies. In 1922 the Washington Naval Treaty, which limited each nation's number of battleships, left a loophole permitting the construction of carriers. As a result the main signatories, with the exception of Italy, at once converted a number of uncompleted **battleship** and **battle-cruiser** hulls into large carriers of up to about 27,000 tonnes, capable of taking, in the American case, up to 72 aircraft, the Japanese 60 and the British 48. These were all, of necessity, fast ships in order to handle their aircraft, the stalling speeds of which gradually increased. Lifts were developed to the hangars below. Armament was chiefly **anti-aircraft** since it was envisaged that carriers would be prime targets for enemy air attacks. In Britain's later carriers an armoured flight deck was fitted.

Japan was first to use carriers, against the Chinese during the 1932 'incident' at Shanghai. Shore targets were bombed and aerial combat took place. At the start of World War II in 1937, when Japan fought an undeclared war against China, carriers were frequently employed, sometimes three at a time.

At the beginning of the war in Europe in 1939, British carriers initially were used for convoy escort and suffered their first loss to a German submarine. Subsequently they were heavily involved in the Battle of **Norway**, the Battles of the **Mediterranean Sea** (including the significant sinking by torpedo-bombers of Italian battleships at **Taranto** and in support of convoys to break the siege of **Malta**) and the sinking of the battleship *Bismarck*. These operations showed that the aircraft-carrier was in process of assuming the battleship's role as a fleet capital ship, as was made plain when the Japanese carrier aircraft executed their devastating blow against **Pearl Harbor** at the commencement of a series of campaigns in the Pacific and Indian Oceans which were

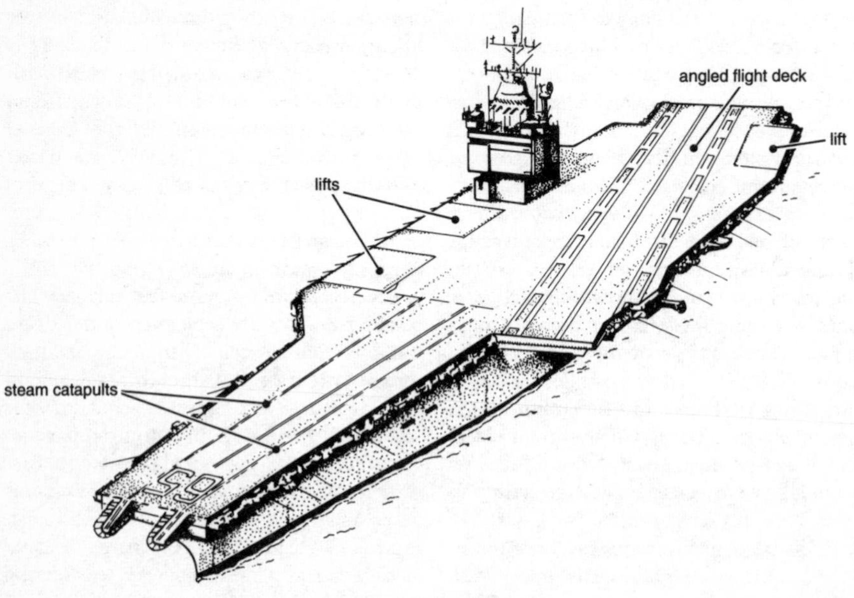

Aircraft-carrier, c. *1960*

dominated by the rival carrier squadrons.

A turning-point of the war occurred at the Battle of the **Coral Sea** and **Midway** in 1942 when carrier fought carrier through their aircraft, without coming in sight of each other, battles which cost Japan 6 of her original 10 carriers and America 4 of her original 8. But whereas the Japanese lost the élite of their pre-war crews, the Americans not only went from strength to strength in numbers but also in prowess and technical capability until, in the Battles of the Philippines and at **Leyte Gulf**, they largely destroyed the Japanese Navy.

Meanwhile, especially to impose air power against submarines in the Battle of the **Atlantic**; but also for universal employment, light carriers had been built, some based on merchant-ship hulls, some with catapults to launch fighters against bombers in mid-Atlantic. As convoy escorts these smaller vessels were economically effective and, as was shown in the Pacific, useful in the midst even of fleet actions.

Post-1945, despite the **nuclear** threat, carriers were developed apace as battleships were being scrapped. British inventions such as the armoured flight deck, the steam-powered catapult, the angled flight deck and the mirror-landing device all contributed to greater survivability and efficiency and were incorporated in carriers which attended most of the major wars of the times. Carriers were intensively involved, for example, off Korea (**Korean wars**), Suez in 1956, **Vietnam**, and in the **Falklands** and **UN–Iraq** wars. Their potential has been considerably increased by the arrival of British Harrier VSTOL aircraft which can operate off specially built light carriers or from the decks of merchant ships and, at a pinch, small warships.

Aircraft construction industry. From the outset the design, development and construction of **aircraft** have been at the frontier of technology, driven largely by military requirements. For example, **balloons** required gas-proof fabric and the large-scale manufacture of hydrogen gas. And **airships** were not practicable until suitable means of **propulsion** were developed; and at the same time they stimulated the demand by Count **Zeppelin** for and use of **aluminium** and, post-1909, Alfred Wilm's far harder and stronger duralumin.

Generally, at first, the constructors of heavier-than-air machines, including **gliders**, were their enthusiastic inventors – like the **Wright brothers**, who were bicycle-makers, and Louis Blériot, who made automobile lights. While others such as Thomas **Sopwith** were pilots who wanted better racing aeroplanes. They built workshops and recruited cadres of craftsmen, many from furniture manufacturing, to make machines which were mainly of wood and fabric. Most would profit by the sudden, enormous demand for aircraft of all kinds generated by **World War I**.

Of necessity, owing to the environment in which aircraft operated, standards of design, workmanship and inspection had to be higher even than with **shipbuilding**. But not all were as thorough as the Wright brothers in calculating engineering stresses and in wind-tunnel testing of their models. In response to pressure for better performance as well as large numbers, numerous faulty designs and practices were adopted. Inferior quality was permitted along with lax or non-existent inspection, exacerbated by difficulties in the setting up of mass production, the recruitment of labour and the acquisition of materials, along with unexpected demands for more powerful petrol engines and numerous accessories such as bomb-sights and bomb-racks, lightweight **machine-guns** and **radio** equipment.

Gradually unfit companies collapsed or were absorbed; in Britain, for example, by the government-owned Royal Aircraft Factory and the private Vickers Company, while entrepreneurial firms such as Handley Page, set up in 1909 and with sound designs, capital and organization, flourished and designers of the calibre of Geoffrey **de Havilland** made their mark. Similarly in France the Voisin, Caudron and Hanriot companies were thriving, while, elsewhere, the Wright and the Rolls-Royce companies were busy developing and selling engines. In Germany too, alongside the pre-war Zeppelin and Deutsche Flugzeug-Werk (DFW) companies, there grew in strength the Albatross (Chief designer Dr Ernst **Heinkel**) and Gotha firms in competition with Holland's Anthony **Fokker**'s works building his fighters with a machine-gun synchronized to fire through the arc of the propeller.

By 1919 the various enterprises in Germany and Austria–Hungary had built 51,135 aircraft, and in France, Britain, Italy and the USA approximately 154,300. Most would be swept away by peacetime economies and meagre interest in civil aviation. The lean survivors, however, hung on with sporadic government **research and development** projects and uneconomically small civil and military production runs. Prior to rearmament in the early 1930s, the emphasis, through racing (notably the Schneider Trophy competition for seaplanes) was on speed plus long-distance or height records attempts. All of this stimulated engine design and the shift to all-metal construction (led by the Junkers company); and from biplanes to monoplanes by means of engineering improvements. Meanwhile firms such as Vickers, in order to increase orders and maintain their manufacturing base, played significant roles in setting up the firms of Nakajima and **Mitsubishi** to create a military-aircraft industry in Japan; and also, along with American and German companies, greatly assisted the weak Russian state aircraft industry.

Rearmament found most aircraft constructors unready for mass production, even in the predatory German, Italian and Japanese nations. Yet the vast and hastily initiated flight-testing and factory-building programmes gave a mighty boost not only in numbers of machines but also to economic growth in harness (if not always harmony) with burgeoning technology. This, throughout **World War II** and into the **Cold War**, was a continuing theme, in company sometimes with the conflicting interests of improvement or modifications of design, on the one hand, and expanded production on the other. For the increasing complexity, and lengthening time from inception of a design on the frontiers of technology to completion, produced machines which were obsolescent as they entered service. Nevertheless by 1946, Britain, the USA and Russia had built approximately 559,000 aircraft and the Axis powers 198,000.

In paying this price for lead-time over the opposition in the so-called military 'high-technology race', particularly in the construction of guided weapons and **space vehicles**, the aircraft industry contributed to an incalculable extent to improve, by its frenetic energy, the quality of life and means of communication. The vast majority of all projects to reduce weight through miniaturization and in the use of lighter materials (notably **plastics**) are attributable to aircraft design. Similarly, the interrelation of government programmes for bombers indirectly subsidizing firms like **Boeing** in the production of their famous airliners at competitive prices was very important in helping draw peoples and nations together.

Air forces, Organizations of. At the beginning air forces, whether naval or army, tended to copy army organizations, but with new nomenclature soon appearing. Thus the basic squadron or company was subdivided into flights; and in due course squadrons became parts of groups, battalions or wings, which were placed within brigades or groups commanded by divisions, corps, armies or, later on, air forces. Naturally, as specialist **aircraft** evolved for specific tasks, units and formations assumed titles befitting their role. By November 1914 the French had formed a Bomber Group and the Germans a so-called corps to bomb London. Fighter, reconnaissance, artillery observation, balloon, flying-boat and other units, some specializing in night operations, soon followed, along with the **communications**, **logistics** and administrative organizations required. In several air forces, too, **anti-aircraft** units were included.

The development of **air warfare** and, especially, **strategic bombing** produced demands, notably in Britain, for reprisals in kind as well as improved defence. This led to the formation in 1918 of the Royal Air Force (RAF) by the amalgamation of the Royal Naval Air Service (RNAS) with the Royal Flying Corps (RFC), a move towards centralization of the functions of air power which was copied by several other air forces prior to **World War II** – though not, significantly, by the USA until 1947. No sooner had the RAF come into being than, under public and political pressure to counter the effects of German strategic bombing of Britain, an **Independent Air Force** under General Sir Hugh Trenchard was formed to bomb German industry by day and night.

After 1918, under the impact of rationalization policies, General **Douhet**'s theory in 1921 that wars could be decided by air power almost unaided, and the belief that 'air power was indivisible', opinions among the arbiters of air power became divided. Whereas the Americans and Japanese continued to let their navies and armies retain control of separate air arms (with largely beneficial results) and the Russians and Germans, who formed composite air fleets (*see diagram*), tended to place everything in support of land forces, with only a passing nod to their navies' requirements, the British placed everything under the RAF – to the detriment of their navy and army during World War II. The belated return of the Fleet Air Arm to the Royal Navy came far too late to compensate for the inferior aircraft which the RAF had provided.

Be that as it may, the functional commands – fighter, bomber, coastal, army co-operation, balloon, training, and maintenance – had much to recommend them and over the years have been copied by other nations. They complemented the formation, as operations and the demands of land forces required, of tactical air forces, which

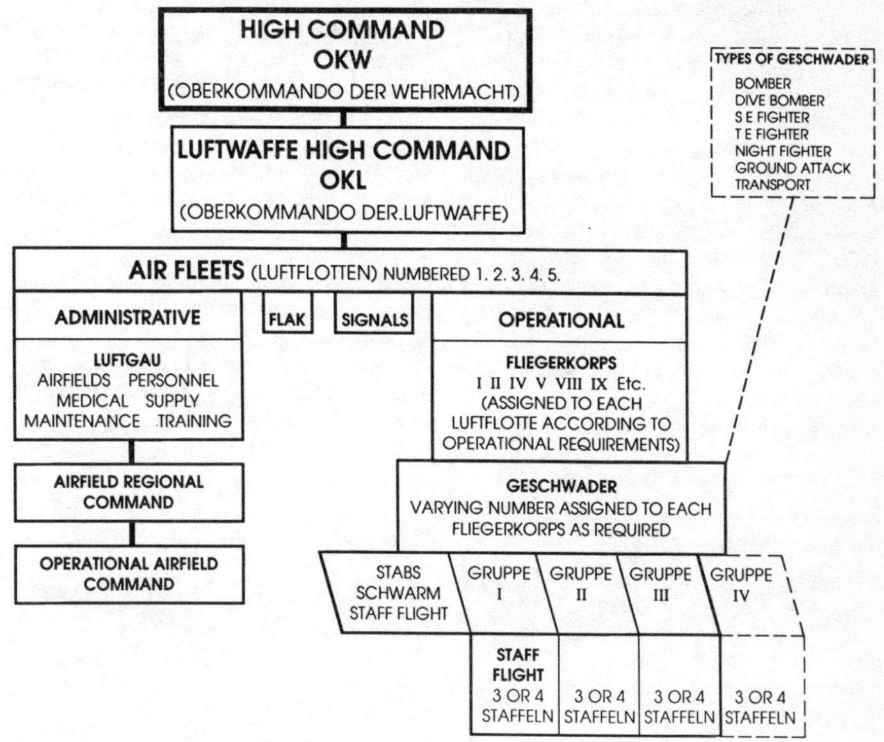

Operational chain of command in the German Air Force, 1939

comprised reconnaissance, fighter and medium-bomber units, whose task was to work closely with armies – formations that, like commands, exist to the present day.

Indeed, despite technological, strategic and tactical changes, the post-1945 framework remains similar to what had evolved earlier. There has been tinkering and, of course, with space warfare looming ahead, immense increases in both missile and transport commands. But as the vastly increased performance and interchangeability of modern aircraft becomes plainer, the old divisions between commands become less distinct. At the same time as navies benefit vastly from control of their own aircraft, armies struggle persistently for possession of close-support aircraft. It was a notable event when, in 1966 during the **Vietnam War**, the US Army did a deal with the US Air Force to exchange its fixed-wing machines for **helicopters**.

Airship. No sooner had the first man-carrying **Montgolfier brothers' balloon** flown in 1783 than researchers examined the feasibility of a powered, steerable airship. This was not achieved before 1852 when Henri Giffard went aloft in a balloon with a 3hp, coke-fired, steam engine driving a three-bladed propeller – a hazardous business with 88,000ft^3 of highly inflammable hydrogen in the envelope. Safer, therefore, was the French Army's *La France* of 1884 with its electric engine, and in 1888 the slightly better ship of Dr Karl Wolfert's with a Daimler petrol engine – although the Daimler's hot-tube ignition was also a fire risk.

It was, however, a Hungarian, David Schwarz, who in the 1880s built the first 'rigid' airship, comprising a hydrogen-filled envelope made of aluminium girders, covered by aluminium sheets and powered by a 16hp Daimler engine. This impractical design crashed on first flight. But Schwarz's sound concept was vastly improved upon by Colonel Count von **Zeppelin** when, in 1900, his LZ1 'rigid' (built for the Army) put Germany ahead of existing French non-rigid airships.

With its silk- or cloth-covered aluminium frame housing gas cells hung from the girders, the twin-engine, somewhat defective LZ1 acted as the prototype of 173 rigids to come, of which all but 20 were German-built. Prior to **World War I** the Germans carried out extensive trials over land and sea, in which their Navy was more active than the Army. With their long endurance and range, high payloads (in the region of 20,000lb) and operational ceilings which, in due course, rose to 23,000ft, German airships could at first outclimb existing heavier-than-air **aircraft**. Initially they were used for **reconnaissance**, with a very few tactical bombing raids for good measure – until several were lost or damaged at low level when seeking targets within range of **anti-aircraft** guns. But on 6 June 1915 one was destroyed over land, in daylight at 11,000ft, by six 20lb **bombs** dropped from a single-seater monoplane.

Thereafter airships usually flew only by night over enemy territory, (with inevitable **navigation** difficulties) when engaged in the strategic **bombing** terror missions deep into France and Britain. But because the Germans were denied possession of the rare inert gas

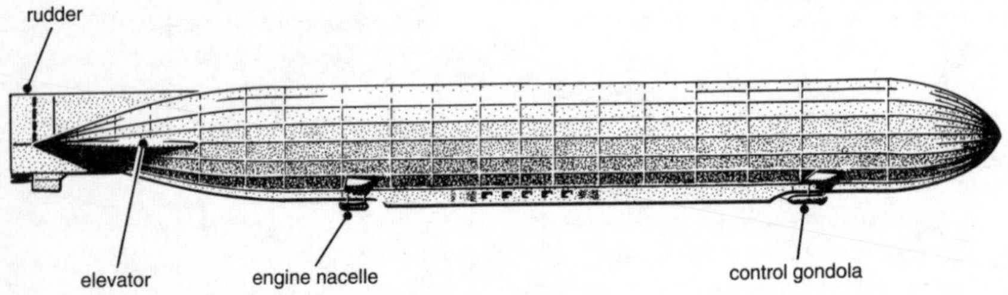

Airship

helium, and the latest **fighter aircraft**, using incendiary **ammunition**, could outclimb them, this had fatal consequences. Many were shot down in flames or crashed after suffering damage from the enemy or rough weather. Even when in relative safety scouting for the **North Sea battles**, they usually failed to produce worthwhile information. Even the imaginative **logistic** mission to fly 16 tonnes of supplies in L59 from Bulgaria to General von Lettow-Vorbeck's **guerrilla** army in East Africa was aborted following the loss of landing facilities at the destination. By 1918 the airship's effective operational role was doomed, although until the 1930s the British and Americans persisted with expensive military projects.

Meanwhile the British and French operated small, non-rigid airships for **convoy** escort and **anti-submarine** tasks – a role still envisaged for small, present-day helium-filled, non-rigid and rigid ships with **aluminium** skin when little opposition is expected.

Air warfare. In 1794 an observer in a tethered French **balloon** provided information about the Austrian enemy at the Battle of Fleurus; and in 1810 Julius von Voss, a Prussian officer, prophesied combat between **airships**. But not until 1849 was **bomb**-dropping from unmanned balloons tried over Venice; and 1911 before Italian aircraft carried out reconnaissance and bombing missions during the Italian–Turkish War in Tripolitania. In 1912 a paper by Captain Murray Sueter RN claimed, rather prematurely, that 'The fight for the supremacy of the air in future wars will be of the first and greatest importance,' going on to imply that the loser would have to terminate the struggle.

Be that as it may, within a year of the outbreak of **World War I** the essentials of air warfare had been identified and to some extent practised. Once the great value of air reconnaissance and the difficulty of preventing it by ground **anti-aircraft weapons** fire had been established, the need for specialized **fighter aircraft** to win air supremacy was recognized. This was underlined by the ineffectiveness of improvised armaments on existing reconnaissance machines, added to the need to counter the first strategic bombing (**Bombing, Strategic**) when German aircraft and airships raided Paris, London and other cities, and the French and British retaliated with successful attacks by **bomber aircraft** against airship hangars and less profitable raids against industrial and communications targets in Germany.

Aerial combat grew in proportion to the quantity and effectiveness of aircraft available in relation to the strategies and tactics evolved under the pressure of events. With the appearance in 1915 of fighters with fixed, forward-firing **machine-guns** aimed by pointing the machine at its target, aerial jousting by individual pilots became crucial to the winning of air superiority and, thereby, the freedom to reconnoitre and bomb by day. Paramount in technology was the striving for greater engine power (*see* **Propulsion**) as the means to increase speed and ceiling. Performance was rapidly improved, numbers increased and techniques developed. Machines were grouped into formations for mutual protection and numerical advantage in so-called dog-fights. But usually it was their technical qualities and pilots' skill which were decisive.

Lacking communication facilities (such as **radio**), missions were handicapped by rigid drills and inflexible pre-flight instructions, supplemented by hand-signals, prior to clashes with the enemy. Depending for **navigation** on map-reading and rudimentary dead-reckoning, the finding and hitting of targets deep in enemy territory was haphazard, particularly in bad weather and at night. Nevertheless, by 1918 night bombing by single aircraft against long-range so-called strategic targets was practised, with meagre success; while by day fighter-escorted bomber formations attempted to fight their way to targets at shorter range. But bombs were small and of low power. Therefore material damage was slight and the impact on civilian morale ephemeral.

Nor did aircraft achieve much success when used in support of naval operations. Reconnaissance at sea by airships was poor;

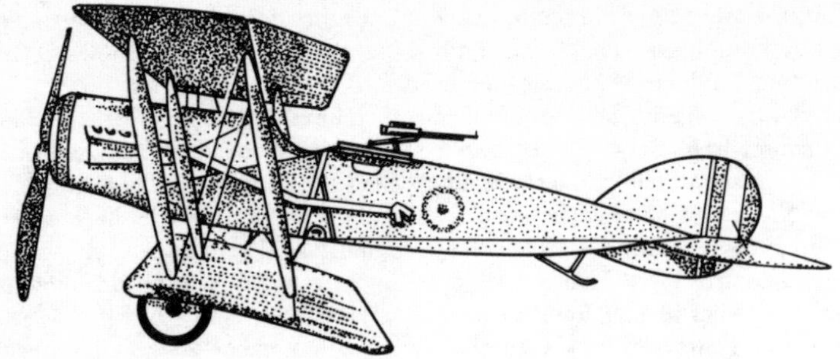

British Bristol fighter, 1917

a few bombing raids by seaplanes launched from ships indicated great potential. With armies, however, very significant results were achieved. Usually, if weather conditions were good, vital strategic and tactical information could often be obtained by aerial **photography** and low-level scouting; while fighters regularly machine-gunned and, later, bombed front lines, **artillery** positions and supply vehicles in the forward areas. **Bombers** were left to go more for **logistic** and communication facilities in the rear, including airfields as part of the fight for air superiority. Also a few transport missions were being flown, notably in the Middle East, and to insert agents behind enemy lines.

Between **World Wars I** and **II** prevalent techniques were enhanced by dive-bombing, designed to destroy ships and pinpoint ground targets, and the practitioners of strategic bombing (egged on by visionaries such as **Douhet** and **Trenchard**, who postulated victory through the paralysis of industry and the terrified populace by saturation with **explosives**, incendiary and poison-gas weapons). But with unreliable astral navigation by night and inaccurate ocular bombsights for high-level bombing, results were so poor that it was decided that daylight attacks by close formations of heavily armed, multi-engine bombers would have the better chance of success. Beams (*see* **Radio**) and other **electronic** aids offered promising ways to find targets, but in 1939 were in their infancy, with Germany in the lead. Area bombing was therefore an unavoidable option.

But while the British and, to a lesser extent the Americans, embraced strategic bombing as a means to a decision by so-called **independent air forces**, most nations (above all the Germans and Japanese) preferred mainly to concentrate air power in support of ground or sea operations. German use of air power in **blitzkrieg** and Japanese employment of **aircraft-carrier**-borne dive-bombers and torpedo-bombers in the **Pacific Ocean War** epitomized these methods. And also under development were **airborne** forces carried by **transport aircraft** and **gliders**.

Meanwhile it was fighters with **electronic** assistance which really held the key to the achievement of air superiority, as shown during the defeat of the Germans in the **Battle of Britain**. For anti-aircraft guns and balloon barrages had only minor defensive effect. Despite the introduction of faster, well-armed bombers and, in due course, technically sophisticated **rockets and guided missiles**, the fighter's ability to outclimb, overtake and overpower other machines, remained central to the exercise of air power – provided **C³I** systems and, especially, **radar**, enabled them to engage their opponents advantageously. Yet, as fighters grew bigger and more powerful, they naturally evolved into multi-role machines fitted with reconnaissance aids as well as bombs and guided missiles. The introduction of **jet engines** merely amplified the status quo in 1944, while imposing demands for vastly more complex technology and ground-support services which were themselves high-value targets for attack.

Yet, despite the enormous outlays on independent strategic bombing as well as in support of navies, armies and **guerrilla** forces, air power failed to win World War II in the manner foretold. Its effect remained only one vital, complementary element in the conduct of war; partly because of its inability to occupy and hold territory, but also because it was unable to destroy pinpoint targets with assurance at a time when there were indications (linked to repugnance) that indiscriminate destruction of area targets often was counterproductive and not necessarily decisive. Eventually those who banked on aircraft-delivered **nuclear weapons** as a means to a strategic decision, if only as a **deterrence** to war, were frustrated when alternative means of waging non-nuclear war were developed and the nuclear option devalued out of dread and revulsion, if not for moral reasons.

In the **Cold War** and **limited wars**, such as the **Korean**, Arab–Israeli, **Vietnam** and **UN–Iraqi**, air warfare was conducted by further enhanced, traditional methods under the nuclear shadow, though widened in scope and power by the introduction of large transport aircraft, **helicopters**, guided missiles (*see* **Rockets**), smart munitions, **robots** and **space vehicles**. Underlying very many of the technical developments of these wars was an urge to acquire the capability to find and destroy discrete and vital targets with economic precision, while improving the mobility and effectiveness of surface forces. Because of the vulnerability of large, slow bombers to missile-armed fighters and surface-to-air missiles (SAM), smaller, more agile multi-role aircraft came into their own – particularly when, in Vietnam, they first were armed with smart munitions. Helicopters began to play a vital role in all kinds of operation, notably at sea against **submarines** and surface vessels; as well as overland in reconnaissance, combat and logistic roles. Meanwhile the huge transport fleets, utilizing the enormous number of airfields constructed world-wide, time and again demonstrated their mobility by rapidly deploying and supplying considerable forces of all arms at great distances from their starting-points.

The **UN–Iraq War** in 1991 gave evidence of a significant extension of air power's ability to dominate. Because, with very low losses, aircraft hit small strategic targets with devastating effect, it was claimed that a new, economical factor in the execution of deterrence had been attained. And yet air power did not and could not win that war on its own. Many vital targets escaped **surveillance** by space vehicles or were missed by even the best smart weapons. Even though Iraq's air defences were obliterated and her economy and infrastructure ruined, she might yet have continued resistance if there had been no invasion of her territory by land forces.

Alamein, Battles of El. The turning-point in the **North African** campaigns was reached in the aftermath of the fall of Tobruk and the pursuit by Field Marshal Rommel's Axis forces of the routed British Eighth Army. His opponent, General Sir Claude Auchinleck, withdrew to the 40-mile-wide bottleneck at El Alamein between the sea and the almost impassable Qattara Depression. From hastily constructed defences he stopped Rommel in a series of hard-fought clinches in the *First Battle* (1–27 July) which exhausted both sides.

The *Second Battle*, sometimes known as Alam Halfa after the ridge of that name, which was the key to the British defensive position, began on 31 August. Rommel attempted once more to smash the Eighth Army (now under General Bernard **Montgomery**) and reach the Suez Canal. Battered by **bombing** and intense **artillery** fire, while enmeshed in minefields and difficult going, Rommel was repulsed by well-coordinated anti-tank fire. His withdrawal, prior to preparing deep, intricate defences of his own, was only tentatively followed up by Montgomery.

The *Third Battle* was at first led on 23 October by British infantry and **mine**-clearing teams, after intensive bombing of Axis lines of communication, airfields and **logistics**. By seizing vital ground within the enemy defences, Montgomery drew the Axis into an attritional struggle in which his numerically superior **tanks** and **anti-tank weapons** systematically destroyed or neutralized 500 tanks and 400 guns. Rommel withdrew and, as news of Allied landings in Algeria and Tunisia was received, turned it into a 700 miles non-stop run for the Mareth Line, the next safe refuge.

Alanbrooke, Field Marshal Lord. *See* **Brooke**.

Alexander III (the Great) (356–323 BC). As the son of Philip II of Macedonia and the pupil of Aristotle, the great philosopher and scientist, Alexander had already demonstrated considerable military prowess by defeating Greek states in battle before succeeding to the throne at the age of 20. Throughout his energetic lifetime his interests in research and statecraft were to complement his destruction of the Persian threat and the introduction of Greek culture to the east.

Alexander inherited a trend-setting army along with his father's **strategic** and **tactical** doctrines, plus the ambition to crush Persia. (*see* **Greek–Persian wars**). The army which Alexander used to bring the Grecian states under his hegemony was controlled by a talented military **staff** and was founded on well-drilled armoured **infantry**, which fought shoulder-to-shoulder in a massed phalanx, plus some light infantry; a large force of heavy and light **cavalry** as the arm of decision; and a train of **siege engines** manned by engineers (*see* **Engineering, Military**).

In 334 BC the Greeks crossed the Dardanelles and sought battle with the larger but inferior army of King Darius III. In numerous sieges and at the battles of Granicus (334) and Issus (333) Alexander defeated the Persians and occupied Egypt. After a Persian-financed Spartan revolt in Greece had been put down he caught and routed Darius at **Arbela** (Gaugamela) (331) before invading Persia and India to hunt down and destroy Darius's murderers and their allies at the culminating Battle of Hydaspes (Jhelum) (326). But although he endeavoured to create a unified confederation of his conquests under Greek hegemony, a mutiny within the army enforced a bungled retreat to Egypt, and taxed his diplomatic talents to the limit. He died of malaria at Alexandria.

Allenby, Field Marshal Sir Edmund (Lord) (1861–1936). Service in a cavalry regiment during small wars in South Africa in the 1880s, and later throughout the Boer War, taught Allenby the value of surprise and manoeuvre and the vital importance of logistics, lessons he never forgot during **World War I**. In 1914 he shrewdly commanded the British Expeditionary Force's Cavalry Division during the retreat from Mons to the **Marne** and in the first Battle of Ypres. Appointed to command a corps and then Third Army, his conduct of the Battle of **Arras** in April 1917, though denied full success, demonstrated a determination to break, by many technical innovations, the stalemate of trench warfare. Sent to command the Egyptian Expeditionary Force in the aftermath of its second defeat at Gaza in 1917, he soon belied, by subtle strategy and tactics, his nickname, earned by brusqueness of manner, of 'The Bull'. Victories over the Turks by deception, surprise, manoeuvre and sound logistics at Gaza, in the pursuit to Jerusalem and during the subsequent campaigns of 1918 which, at Megiddo and in the capture of Damascus and Aleppo, forced Turkey to sue for peace, implied more than battlefield triumphs. For his handling of **Lawrence**'s Arab irregulars and of his own **cavalry** and **armoured cars** hinted strongly at the shape of war to come, misinterpreted as some of the lessons later were.

Aluminium (US **aluminum**) was first isolated by Hans Oersted in 1825 and introduced to the public as a metal rarer than gold in 1855. Not until the 1880s was this lightweight, corrosion-resistant, adaptable metal produced in large quantities by electrolytic means by Charles Hall in the USA and Paul Héroult in France. Almost at once it was used in the construction of an **airship** (the *Schwarz*). In 1909 the first of the many important, strong alloys – Duralumin, developed by Alfred Wilm by the addition of copper and manganese – was available and by the 1920s was becoming the principal material for **aircraft** construction. Its casting together with hardened alloys as **armour** was also developed. Indeed, aluminium frequently supplanted **steel** as it acquired universal civil and military uses that included house construction and electric-cable manufacture.

American Civil War. Although a principal reason for the war's outbreak in 1861 was failure to resolve long-standing differences over slavery, the humanitarian issue was paralleled by rivalry between the northern industrial and the southern agrarian economies. Therefore, when the militarily educated President Jefferson Davis severed relations between his southern Confederacy and the North and opened hostilities against the northern Federals at Fort Sumter on 12 April 1861, it was in the knowledge that his army, strong as it soon would be with 112,000 men and under some of the best-trained officers, was only marginally superior to the North's 150,000, and that it might not be self-sufficient in a prolonged struggle. Yet Davis, for political reasons, adopted a defensive strategy in the hope that the North under its lawyer President Abraham **Lincoln** would tire of war. This was Davis's greatest mistake, if only because he overlooked the danger of a naval **blockade** which, as time went by, the 90 obsolete warships initially at Federal disposal would impose in compliance with Lincoln's order of 19 April. Countervailing every future Confederate success on land had to be a strangulation which made it largely impossible to export the South's staple cotton crop to help purchase and smuggle in sufficient munitions.

Both sides were engrossed by the importance of their capital cities. Whenever, as occurred from the outset in July 1861, the untrained Federal army moved south from Washington to capture Richmond, the Confederate capital, the Confederacy would both defend stoutly and distract the Federals by threats against Washington. The essence of the many campaigns in the Shenandoah Valley and against the **fortifications** of the two cities were politically orientated thrusts which tied large armies to this relatively small arena while smaller forces manoeuvred in the open spaces of the West. At the root and heart of manoeuvre and coordination of widely separated fronts lay the extensive use of **railways** and the **telegraph**.

A pattern which was to outlast the war emerged after skirmishes in the Valley culminated in a Federal defeat at the first Battle of Bull Run on 21 July. At the same time the Federals began to gain a hold in the West while implementing the naval blockade. This process developed in 1862, as the armies grew in size and skill, when Federal forces under General George McClellan landed on the Yorktown Peninsula and attempted to seize Richmond from the East. His defeat by General Robert E. **Lee** in the **Seven Days Battle** was, however, no less important than the ***Merrimack* versus *Monitor*** fight in Hampton Roads, in March, and General Ulysses Grant's bloody victory at Shiloh in April. For the one eliminated any Confederate pretensions to sea power and the other tightened the Federal grasp in the West, a few days before Commodore David Farragut won the naval battle of New Orleans, as a prelude to his advance up the Mississippi to **Vicksburg** to make junction there with Grant and effectively cut off the Confederacy from the West.

Belatedly in 1863, Davis realized the peril into which his defensive strategy had led the South. Grant, ably supported by General William **Sherman**, closed in on Vicksburg with an improvised **amphibious** operation and General Joseph Hooker, now commanding in the East, led the Federal forces to defeat at Chancellorsville (where the best Confederate general, Thomas Jackson, was killed). Davis let Lee strike northwards into Federal territory by way of exploitation but also in the hope of preventing the Federals from reinforcing their menacing operations in the West. It all went wrong for Davis. On 4 July, as Lee admitted defeat at Gettysburg, Grant brilliantly captured Vicksburg after an approach march of 200 miles in 19 days, living off the

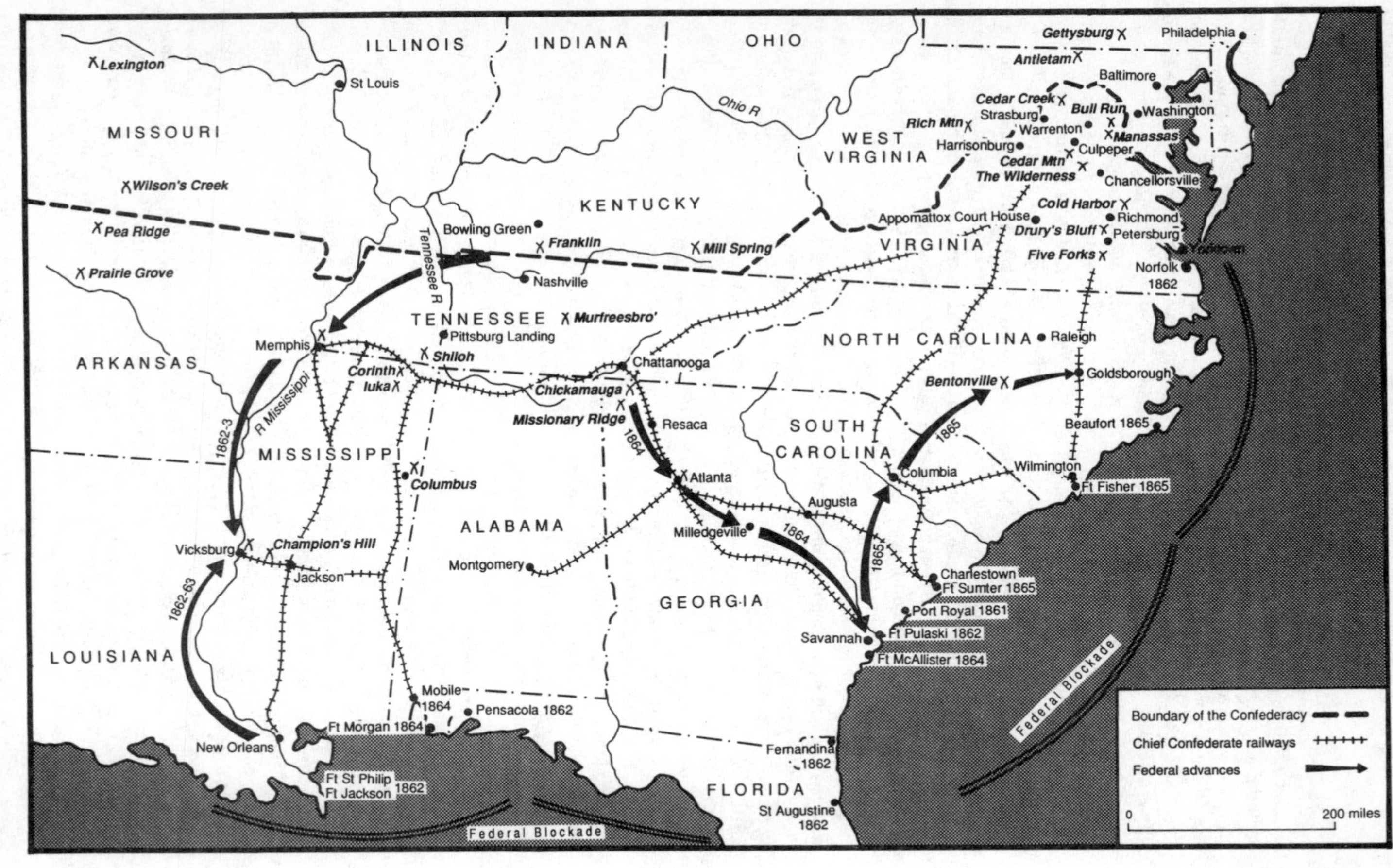

The American Civil War

country. But he judged the **logistic** problem nicely: the siege of Vicksburg, screened by Sherman's mobile troops and supported by ironclads, lasted over six weeks but was always itself sufficiently supplied despite difficult conditions.

The Federal Army now turned its attention to the capture of the vital route centre of Chattanooga. But after a series of battles attendant upon a siege of the outmanoeuvred Federal forces penned in the town, and which lasted from 19 September to 27 November, it took the appointment of Grant to overall command in the West to break the Confederate grip. At this point Lincoln recognized in Grant and Sherman the generals who would win the war for him and circumvent the challenge by the Democrat presidential candidate, the rejected General McClellan, whose party spoke of negotiating peace. Grant was appointed General-in-Chief in the field and moved to Washington to conduct operations in the East. Sherman took over in the West. Their opponents, meanwhile, were in distress. Cut off from the outside world apart from supplies by a few blockade runners and smuggling from Mexico via Texas, their manpower was depleted by heavy casualties; their armouries, ill-supplied by an ever-inadequate **armament industry**, were all but empty; and their communication systems, in particular their railways, were foundering from lack of spares and maintenance.

So when Grant attacked in the direction of Richmond in April 1864 with 105,000 men against Lee's 61,000, and initiated what evolved into the vicious Wilderness campaign and the Siege of **Petersburg**, there was little Davis or Lee could spare for the West when Sherman lunged from Chattanooga to Atlanta prior to launching his celebrated **March to the Sea**. Although by brilliant generalship and the valour of his armies Lee managed to check Grant in the Wilderness, the Confederacy was on its last legs as Sherman ripped through its heartlands. Only scattered Confederate armies kept the field when Sherman turned northwards in 1865 in a scythe-like sweep to capture the remaining Atlantic ports and, in conjunction with Grant, close the ring upon Lee at Appomattox on 9 April – just prior to Lincoln's assassination on the 14th and the final laying down of Confederate arms on 26 May 1865.

The Civil War occurred just beyond the point of change between the first and second industrial revolutions – between the substitution of **iron** by **steel** for many uses, of cart tracks by railways, visual, hand signalling by electrical transmissions (*see* **Communications, Signal**), wooden walls by ironclads, **balloons** by **airships** for **surveillance** and **bombing**, and smooth-bore, muzzle-loading weapons by rifled, breech-loading small arms and **artillery**. It was not quite in time to register the shift of emphasis in **naval warfare** to underwater attack by locomotive **torpedoes**, but in plenty of time for **machine-guns**. So heavy were casualties in the face of intense fire that fortifications, particularly the hastily constructed field variety, became extremely popular with **infantry** bent on survival – and to the detriment of **cavalry** as the arm of decisive mobility. It was also distinguished by the unusually good care taken of the wounded (*see* **Medical services**) in campaigns which, logistically, were reasonably well provided for by comparison with those of the past elsewhere. It seriously retarded the economic growth of the South while vastly enhancing the industrial development of the North. And it would be many years before the **strategic**, **tactical** and technical lessons were digested and applied.

American War of Independence. Although the causes of this war in 1775 were complex and the natural result of conflicts of loyalty and interest between Britain and the American colonies, it has to be viewed also in connection with the continuing sequence of **French wars** commencing in 1635 and France's loss of Canada after the Battle of Quebec in 1759. Moreover, despite the first stirrings of the industrial revolution in Europe, the initial fighting in America was largely **guerrilla warfare**, and there were few technical or tactical innovations. Strategically, however, it was dominated by sea power and French opportunism in reinforcing and assisting the poorly trained and armed dissident Americans.

As at Bunker Hill in June 1775, when American irregulars challenged disciplined British troops, the rebels usually were defeated by superior firepower. That same year an American invasion of Canada was routed. And in 1776, against a greatly reinforced British army, they again were thrown back from Canada prior, in October, to losing control of Lake Champlain to a flotilla of five small British warships, which had been moved there in sections and assembled on the lake side. Yet in December, following defeat in the approaches to New York, the Americans made a remarkable recovery with a surprise advance under General George Washington to win small victories at Trenton on 26 December and at Princeton eight days later. In September, however, the British **blockade** of the mouth of the Hudson River had been weakened by a withdrawal of ships to seaward following an abortive underwater attack by David **Bushnell**'s **submarine** *Turtle*.

In 1777 British failure to coordinate superior, converging sea and land approaches against the rebels in the Hudson valley culminated in the surrender of General Burgoyne's logistically starved army at Saratoga. This defeat encouraged the French to join with the Americans to break the blockade, aiming to build up and convert Washington's guerrilla force into a disciplined, so-called 'Continental Army' capable of facing the British in conventional action.

In 1778 the training of the Continental Army was put under General Baron Friedrich von Steuben, a very experienced Prussian who had fought in the **Seven Years War**. He drilled the irregulars in volley-firing of **muskets** and the new Pennsylvania **rifle**, the scientific use of **artillery** he had learnt under King **Frederick II**, and the construction of European-type **fortifications**. His training contributed to a costly victory over the British at Monmouth in June, but the crucial moves were made at sea when, to counter two French fleets, the British were compelled to relax the coastal blockade in order to defend home waters and also bring the French to action. But the engagements off Newport, Rhode Island and Ushant in July, were indecisive.

In 1779, however, an American **amphibious** attack, with 40 vessels carrying 200 guns and 3,000 troops from Boston against the British at Castine, was routed by a British squadron from New York. And in May the following year not only were the French unable to prevent British landings in the Carolinas and the overrunning of Georgia, but Count Jean de Rochambeau declined to take his army from Boston to reinforce the hard-pressed Americans until the arrival of a French fleet with reinforcements.

In 1781, to exploit their tactical successes, the British under General Lord Charles Cornwallis advanced into Virginia and in August occupied the Yorktown Peninsula at the entrance to the strategically vital Chesapeake Bay. The arrival of a French fleet and 3,000 troops, under Admiral François de Grasse, persuaded Washington and Rochambeau that Cornwallis could be destroyed if they concentrated American and French forces in Chesapeake Bay. But on 5 September poor signal communications (*see* **Communications, Signal**) allied to faulty orders lost the British Admiral Thomas Gates an opportunity to defeat piecemeal a panicky de Grasse. After a damaging but inconclusive exchange of fire, Gates retired to New York, leaving Cornwallis to his fate after Washington was further reinforced with Frenchmen and his siege train.

In 1782, de Grasse would lose five ships and suffer severe damage and many casualties (much from the new carronade – *see* **Artillery**) to Admiral Sir George Rodney at the Battle of the Saintes in the West Indies. But Britain had lost the war – and America too in 1783.

Amiens, Battles of. When this vital commercial and communications centre became a main **railway** junction prior to 1850, its long-standing importance as a bastion of northern France was augmented. In November 1870 it was occupied by the Germans after a sharp engagement at Villers-Bretonneux, but was lost and then recaptured in December as the **Franco-Prussian War** drew to a close.

In 1914 it was temporarily occupied by the Germans in their sweep to the Battle of the **Marne**, but strongly held by the Allies for the rest of **World War I**. Seriously threatened as a principal objective of **Hindenburg**'s 1918 offensive, the line, once again, was held at Villers-Bretonneux, which became a starting-point for the surprise Allied massed tank attack on 8 August. Including reserve and supply tanks 604 British heavy and light **tanks** and 12 **armoured cars** plus 90 French light tanks attacked through mist, in conjunction with infantry and cavalry, and with massive artillery support. The poorly organized German defences were smashed over a frontage of 20 miles to a depth of six miles in the first 24 hours. A few armoured cars and light tanks cut loose in the enemy rear where panic ensued. German troops gave up in droves and some called on their comrades to cease prolonging the war. As a result General **Ludendorff** concluded that the war had to be brought to an end. The Allied attack, however, was handicapped by the inability of tanks, **cavalry** and **infantry** to co-operate, owing to communication problems and their differences of speed and protection which made them tactically incompatible. After five days the offensive had lost momentum here and was transferred elsewhere.

Ammunition. The English word 'ammunition' is a *c.* 16th–17th-century army corruption of the French word *munition* and in this entry refers to ballistic projectiles which have been hurled or fired. No doubt even the first **stones** thrown by primitive men were selected for their ease of throwing as well as impact on target – as most certainly were the rocks flung by catapults and other **siege engines**. Design decided selection, however, when **spears**, javelins and arrows (*see* **Bows and arrows**) were being acquired with careful attention paid to the balance between shaft, feathers and heads, be the latter of stone or metal (*see* **Metallurgy**).

With the invention of gunpowder (*see* **Explosives**) ammunition ceased (for the time being) to be made, like arrows, in one piece. For use with the first **artillery** and hand guns, the powdered *propellant* (the function of which was to burn very quickly and produce expanding gases resulting in high pressure to give acceleration to the *projectile*) was loaded separately into the barrel. The burning was initiated by a *detonator* – in early days 'touched-off' through a hole in the barrel by a gunner with a flaming match. The projectile was sometimes a metal 'feathered' dart, but usually a stone sphere, although, until the 15th century, **bronze**, **iron** or lead were occasionally used.

At first, in order to avoid separation of its components, the gunpowder's ingredients were mixed on the gun site shortly before use. Transport problems, demands for safer, more reliable and powerful charges, along with ways of keeping powder dry, led in the 17th century to ready-mixed and carefully selected powder being issued in bags and, later still, in metal containers which could be rammed either into the breech or down from the muzzle.

Gunpowder remained the standard propellant until the 1860s, but there were endless versions of projectiles, even for small arms for which the sphere (ball or round shot) was gradually superseded after the mid 19th century by elongated, conical, streamlined bullets for improved velocity and accuracy. The limitations of solid artillery shot, however, as a smasher of ships or stone alone were regarded as unacceptable as early as 1410 when *canister* rounds were developed: metal containers filled with small projectiles arranged to spread for anti-personnel effect. And at about the same time *shells* were invented: hollowed out rounds filled with gunpowder and ignited by a crude, pre-ignited fuse of some sort. Much later, in 1784, Lieutenant Henry **Shrapnel** would merge these methods into a shell fused to burst in mid-air and spray bullets over the target.

It was for use against ships that projectiles were most highly developed to cope with a variety of special roles. Heated shot, despite the danger of it causing a premature explosion in the barrel, was popular until iron

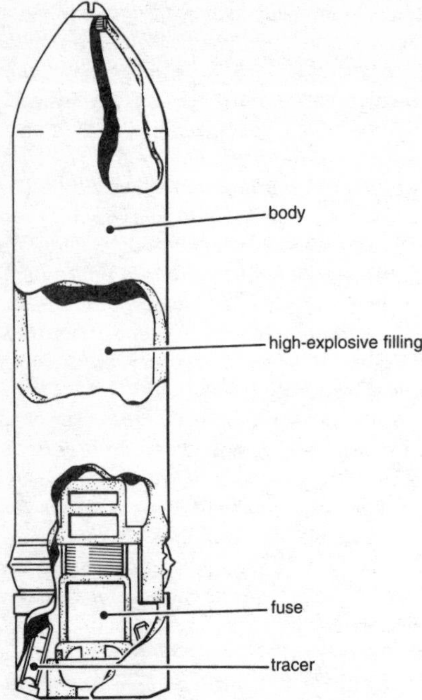

Ammunition: High-Explosive Squash Head (HESH) shell

replaced wooden ships in the 19th century. Bar shot (two balls linked by a bar) and chain (with a chain instead of a bar) were favoured for attack on spars and rigging until sail propulsion gave way to steam engines.

The most crucial changes to ammunition were brought about, however, by the further development of breech-loading guns, made feasible (a) by the latest copper cartridge cases which came about as a result of the invention of the percussion cap in 1805 by the Reverend Alexander Forsyth, using fulminate of mercury, which had been discovered by Edward Howard in 1799; and (b) by the use of stronger gun barrels capable of withstanding higher pressures. Early in the early 1840s, Daniel Treadwell invented barrels built up by iron or **steel** hoops shrunk around a central cylinder; and Captain Thomas Rodman, US Army, discovered how to cast very big, hollow gun barrels around a removable water-cooled core. Rodman followed this in the next decade by creating prismatic and perforated-cake gunpowder which burned more evenly and slowly with controlled expansion of gases.

These inventions were overtaken by all-steel barrels and the introduction into service in 1860 of Christian Schönbein's somewhat unreliable, but much more powerful, nitro-cellulose (gun-cotton) explosive. But in 1875 Alfred **Nobel**'s smokeless ballistite, with its vast tactical implications (including making feasible the semi-automatic Maxim **machine-gun**) overtook them all.

Subsequently, to satisfy the demands by navies and armies for longer-range, more accurate and rapid-firing artillery which could penetrate armour, kill men and animals in the open, destroy earthworks and dispense poison gas or smoke, the ammunition designers have devised a host of solutions.

For most pieces of ordnance over 120mm in calibre, they chose separated ammunition, usually with a bagged propellant but sometimes, for smaller guns, metal, so-called quick-fire, one-piece rounds with the projectile fixed to the cartridge case. At the same time as the gun-designers overcame the problems of sealing the breech, ammunition was developed to fit snugly into the bore, and thus prevent loss of power by escaping gases. Cordite propellants, discovered in Britain by Frederick Abel and James Dewar, have been developed for very flexible use, varying from small grains for small arms to large cylinders for some **rockets**. Liquid propellants, on the other hand, have not so far proved very satisfactory for artillery.

Of necessity projectiles have received enormous attention. The simplest is the *small-arms bullet*, where the important parameter is stopping-power. This requires about 80 joules of energy and the bullet must be given sufficient weight and velocity to achieve this over the desired range. Nickel is often used for the rifle bullet's outer casing, being sufficiently ductile to engage the rifling yet strong enough to withstand firing pressures. Lead and antimony are commonly used for the core, to achieve penetration combined with stability in flight. The trend is to have a lighter bullet fired at a much higher velocity so as to produce the required stopping-power.

Artillery projectiles tend to be more complex in design and can be broadly classified, whether for use on land, at sea or in the air, either as shells filled with high explosive (HE), with all their variants, or as anti-armour projectiles.

For maximum effect conventional HE shell should break up into small fragments when detonated on impact; special fusing may be fitted, for example in air-defence weapons, to enable the shell to burst close to a target instead of actually having to hit it. The introduction in 1944 of the **electronic** variable time (VT) fuse, which was activated by a radio sensor when in close proximity to a shell or **bomb**'s objective, was a revolution in the engagement of air and surface targets.

Anti-armour ammunition is either kinetic energy (KE), relying on the velocity of a dense core to penetrate the target, or chemical energy (CE), which relies on the explosive filling to do the work. To achieve high velocity a large base area needs to be presented to the propellant gases but for KE penetration a small dense slug is required. The complex Armour-Piercing Discarding Sabot (APDS) has been developed with the broad-based 'sabot' being discarded at the muzzle, leaving the core to fly on to the target. A further refinement is the 'long rod' penetrator which has to be fin-stabilized, resulting in the Armour-Piercing Fin-Stabilized Discarding Sabot (APFSDS) round.

The CE round most widely used is High-Explosive Anti-Tank (HEAT) which, on striking the target, penetrates by means of a semi-molten jet of explosive and armour fragments. A more effective jet is formed if the projectile does not spin, so fin stabilization is normally used and the round is well suited to the smooth-bore tank gun. An alternative is High-Explosive Squash Head (HESH), which relies on the explosive charge 'pancaking' on the outside of the target and, when detonated, sending a shock wave through the armour. This in turn detaches a large scab from the inner face, causing widespread damage within the vehicle.

Tactical developments on the modern battlefield have called for engagement of targets at ever-increasing ranges, with a consequent demand for increased accuracy, in particular the need to engage large armoured formations outside the range of tank guns. This has led to the development of Terminally Guided Munitions (TGM), giving the projectile the ability to seek out its target on arrival in the target area. Such carrier shells typically release a large number of small bomblets which provide lethal coverage of the target spread out below. It is in the development of sophisticated fusing and means of terminal guidance that the future of the gun-fired projectile lies, including nuclear weapons.

Amphibious warfare is operations by forces landed from the sea. Hence it always has been integrated with **naval warfare** and since the invention of **aircraft**, vitally influenced by **air warfare**. Even since the advent of steam-powered vessels, it has been hostage in some degree to weather, water currents, wind and muscle power in the delivery of seaborne forces ashore. Amphibious operations are, therefore, among the most hazardous of all, especially if undertaken against opposition.

It may be assumed that ancient river communities waged amphibious warfare against their upstream or downstream neighbours and that, by *c.* 3000 BC, at least, Mediterranean peoples with seagoing vessels were coming into conflict with each other (*see* **Shipbuilding**). Certainly in *c.* 1200 BC so-called Sea People were embarking in Asia Minor to invade Cyprus and Egypt. Migratory movements which initiated the everlasting surge of maritime mobility reached a famous climax in 490 BC when **Darius the Great** attacked Greece (*see* **Greek–Persian wars**) with some 50,000 men carried across the Aegean Sea and landed at **Marathon** where they met their doom.

The **Roman wars** are abundant with amphibious operations, not least Julius **Caesar**'s first, heavily opposed landing by **infantry**, supported by seaborne catapults, against the Britons in 55 BC. It was followed in 54 BC by his more successful, unopposed attempt with **cavalry** as well as infantry, which was hampered by subsequent storm damage to many of his ships. These events foreshadowed more than 1,000 years of amphibious raiding and invading along the coasts of Europe, most notably during the period of **Viking expansion**.

Regardless of the gradual replacement of muscle-powered rowing vessels by the sailing caravel, amphibious **tactics** altered but little, though their **strategic** importance became increasingly vital in company with the outward thrusts of the great maritime nations: Portugal, Spain, France, Holland and Britain. Their wars teemed with raids, and major landings were often connected with invasions and directed against strategic and economic objectives such as ports and **dockyards**. For example, in 1588 there was the Spanish attempt to invade England, which was thwarted by the defeat of their Armada; then in 1667, during the second Anglo-Dutch war, the successful Dutch raid against the Medway ports; and, in 1798, **Napoleon**'s unopposed invasion of Egypt, which was neutralized by Horatio **Nelson**'s destruction of his fleet at the Battle of the Nile; followed in 1808 by the British unopposed landings in Portugal as the start of the five-year struggle to drive the French from Spain.

The advent of steamships made amphibious operations easier to mount, though without altering basic principles. The **Crimean** beach landings succeeded because they were virtually unopposed. Commander David Farragut's capture in 1862 of New Orleans during the **American Civil War** was unopposed after his naval bombardment had forced the Confederates to withdraw up the River Mississippi. But General Grant's subsequent waterborne attempts to capture **Vicksburg** were costly in lives because the obvious approaches were easily defended.

During **World War I** specialized **landing-craft** were virtually unheard of and techniques little changed since pre-1850 times. At **Gallipoli** rowing-boats and towed barges were the principal assault vessels, except for the steamer *River Clyde*, which had a **machine-gun** battery emplaced in its bows and holes cut in its sides with catwalks for the infantry to dash for the shore after the ship had beached. In the later raid on Zeebrugge, in 1918, men who were put ashore against fierce opposition came from an old cruiser placed alongside the mole.

Prior to **World War II** the Japanese and to a lesser extent the Americans, with their eyes fixed on **Pacific Ocean** domination based on islands, made such progress as there was with assault ships carrying mechanized landing-craft. It took the German conquest of Western Europe (**West European campaigns**) in 1940 (including the largely unopposed *coup de main* amphibious landings in **Norway**) to compel the British, soon assisted by the Americans, to project amphibious warfare beyond the techniques of 1918.

To launch **commando** raids in the aftermath of the evacuation from **Dunkirk** (a superb amphibious withdrawal) as preludes to the major invasions needed to defeat Germany and, in due course, Japan new doctrines, techniques and flotillas of hastily designed and built special assault ships and craft were required (*see* **Landing-craft and ships**).

Landing techniques were developed by trial, error and battle through minor hit-and-run raids from Britain and throughout the Mediterranean. Among the most celebrated were those at Saint-Nazaire and **Dieppe**, each of which suffered from immediate opposition. The American landing on Guadalcanal in August 1942 benefited greatly from troops getting ashore against minimal opposition, though what happened when the Japanese reacted was another matter, later resolved in the American favour as much by the solution of the **logistic** problems as by combat alone.

The pay-offs for all the preliminary operations, including the dangerous tasks of pre-assault, **hydrographic** and beach survey under the enemy nose, came during the sporadically opposed Allied landings in French North-West Africa in 1942: the more strongly disputed landings in Sicily, and the extremely tough fight at **Salerno** in 1943. Indeed, as the Allied ability to come ashore almost at will in every theatre of war became apparent to the Axis, they were forced to commit immense manpower and material resources to stop landings on the beaches in order to avoid being overrun in subsequent mobile battles inland. Yet, although the Allies got ashore almost trouble-free at Anzio in 1944, their failure to advance inland soon had them penned. But successive invasions of Pacific islands were usually met by fanatical resistance at the water's edge, since the Japanese could recognize, quite as well as the Americans, Australians, New Zealanders and British, which islands were strategically vital.

The invasions of **Normandy** and, on a lesser scale, of Walcheren in 1944 demonstrated not only how far techniques had developed but also how important was deception, to minimize opposition through surprise at the landing-place and fire-power to suppress it when unavoidable. The experience thus gained was used also against the Japanese in the Philippines and elsewhere in the East: and notably by United Nations forces during the **Korean War** at Inchon in 1950 as well as afterwards in the amphibious withdrawal, under strong enemy pressure, from Hungnam, which was a classic rivalling Dunkirk.

Since then the Suez landing, **Vietnam**, Grenada and the **Falklands** War have illustrated to what extent new technology, in particular the latest **communication** systems, the helicopter, VSTOL aircraft (*see* **Aircraft**), **rockets and guided missiles** have amplified, without by any means simplifying or making less hazardous, amphibious operations. That they are as important as ever is indisputable in the light of their recent repeated (indirect as well as direct) application in the exertion of political and military pressure.

Analysis, Operational. Operational analysis (more often referred to as Operational Research – OR) has been defined as 'the science of planning and executing an operation to make the most economical use of the resources available'. Before the 20th century new equipment and **tactics** were rarely, if ever, evaluated scientifically in order to reach properly thought-out decisions. Yet from the start these methods often were resisted by senior commanders.

One of the earliest examples of military OR occurred during **World War** I and was applied to the U-boat menace. Losses of merchantmen to **submarine** attack mounted frighteningly as the war progressed, since the ships sailed from port individually, and U-boat commanders had considerable freedom of action in picking off targets. Part of the problem lay in the British reluctance to allocate sufficient escort vessels to the guarding of merchant ships, plus the Admiralty's reluctance to adopt a **convoy** system. Admiral **Jellicoe**, First Sea Lord, believed that about 5,000 ships were leaving and entering British ports each week and he was convinced that an effective and suitably escorted convoy system would be impossible to organize on such a scale. There was no statistical branch in the Royal Navy at that time (and indeed no naval staff college prior to 1917) and hence no correlation of facts and figures to support Jellicoe's assumption.

It was left to a Commander Henderson to prove, by studying Ministry of Transport records, that weekly sailings amounted to only 120–140. Furthermore, he showed that those vessels which were escorted, such as troop transports, suffered few losses as the U-boats found it difficult to aim their torpedoes accurately when under threat. Nevertheless, it took an appeal from Henderson direct to the Prime Minister to convince Jellicoe that convoys should be tried, whereupon the accuracy of Henderson's calculations was dramatically confirmed. Within the year from May 1917, when convoys were started, monthly losses had dropped from 683,000 tonnes to about 300,000 tonnes; and this despite a significant increase in the number of U-boats at sea.

However, most of the ideas for military

OR emerged during **World War II**, particularly with respect to the operations mounted by the British and American forces, who found themselves committed to large-scale seaborne assaults many hundreds, even thousands of miles from their home bases. The resulting **logistic** problems required radical solutions which could only be found from preliminary mathematical studies. For example, the Allied invasion of Europe in the summer of 1944, with its subsequent build-up of men and stores to enable the break-out from the bridgeheads to take place, would never have succeeded without a detailed analysis of what the armies would need on each succeeding day of the operation, carried out months before the event.

Not so successful was the British analysis of the likely effect of strategic bombing. Thinking had been conditioned by the limited German bombing of London during World War I, which had resulted in some 4,400 casualties. Lord **Trenchard** became convinced that air power alone could destroy the **morale** of a civilian population and by 1938 the air staffs were estimating that London could expect to receive 3,500 tonnes of bombs in the first 24 hours of war and that after a month there would be over 25,000 dead; all this based on a yardstick of 50 casualties per tonne of bombs dropped. The eminent scientist Professor Haldane seriously estimated 20,000 dead from an average 500-aircraft raid. These figures were not supported by any sound statistical evidence and by 1941 the German bombing of London and elsewhere had shown that the earlier predictions were wrong. Despite this evidence, a large bomber force to shatter German morale was built. But the systematic bombing of German cities as the war progressed, with the loss of 59,223 aircrew, had little noticeable effect on Germany's morale or her ability to wage war. (*See* **Bombing, Strategic**.)

As World War II drew to a close much effort by the OR analysts went into detailed studies of particular aspects of the war. A typical example was the investigation, carried out in 1945, of a whole range of **tank** casualties resulting from engagements in North-West Europe. This provided detailed figures particularly on ranges of engagement for tank guns in close country, which has proved to be of value in more modern analytical work by **Nato**.

One outcome of the statistical work carried out during World War II was the setting up of establishments whose specific task was to provide the advice which defence chiefs were increasingly demanding on a wide variety of problems. In Britain, for example, the wartime Army Operational Research Group developed into the Defence Operational Analysis Establishment, to provide the service that was needed. While much of the work was statistical, on occasion complex field exercises, involving the deployment of large numbers of men and equipment, were carried out, carefully monitored by scientists from the establishment, to explore practically particular problems – to establish exposure times needed for a tank to fire its main armament effectively from behind cover, for example; or to assess the effectiveness of the armed **helicopter** as an **anti-tank weapons** system.

Operational research can have its macabre side too. It was in **Vietnam** that the United States Defence Department used the number of enemy corpses found after a battle – the 'body count' – as a chilling factor in the mathematical assessment of whether or not the war was being pursued in a cost-effective manner. Indeed it could be argued that Vietnam provided an example of how not to use OR techniques. It was under Robert McNamara, the then US Defense Secretary, that the Americans tried to conduct the war on a strictly financial 'profit and loss' basis. Having had it demonstrated statistically that the employment of the latest technology must inevitably give the US forces a crushing advantage over a technically unsophisticated opponent, the American defence chiefs seemed unable to grasp that **guerrilla** wars cannot be successfully fought that way. On the few occasions that the North Vietnamese army could be brought to battle in a major engagement the statistics could be thoroughly justified by the evident American superiority. But such occasions were rare indeed.

OR attempts to model the real world, so that the performance, either of an individual piece of equipment or of military forces carrying out a tactical or strategic task, can be optimized against agreed criteria. The model may be in the form of entirely theoretical computer studies or, more elaborately, by means of '**war-gaming**', where tactical situations are played out minute by minute using appropriately experienced military commanders to take the decisions at the level they would normally do, from their assessment of the situation fed to them by the exercise staff. The dispositions of troops which result would be plotted, fed into a **computer**, evaluated and a consequential new situation presented to the 'players'. Such war-gaming can be very time-consuming, with five minutes of play taking an hour or more to evaluate. Very effective assessments of new weapons systems and tactics can, however, be made by this method.

The techniques used in OR have been increasingly refined and it is usual to refer to that part of the real world in which the researcher has an interest as a 'system'. It is the art of mathematical modelling to identify the quantities associated with a system which, when linked mathematically with each other and to as few other variables as possible, will together form a model describing how the system behaves. By applying the factors in the system to specific situations a picture can emerge which can in turn lead to crucial and far-reaching military decisions. One area in which this can be especially valuable is the reliability of individual components in a complex piece of equipment such as an aircraft. Establishing the component failure rate can lead to the calculation of a mean time between failures, which can in turn pinpoint areas where remedial design work needs to be done or alternatively indicate the scale of spares that need to be carried in the logistic support system. And this in turn may effectively limit the tactical use of the weapon itself.

Typical of OR projects that might be set up would be one to examine the problems faced by the United States in reinforcing Europe in the event of war. The model could be required to determine deployment schedules, aircraft-lift capability and the system needed to meet the deployment requirement. Input to such a model would probably include such details as troop strengths, weights, dimensions, destinations; aircraft-lift characteristics, speed, load, load and unload times, capacity for different cargo types; and airfield restrictions, distances and attrition rates due to accidents and enemy action.

Studies such as those described have become an essential part of all major powers' defence planning.

Anti-aircraft weapons. In 1870 and 1871, during the Siege of **Paris**, the French used free-flight **balloons** to carry dispatches and prominent citizens across the Prussian lines to safety. To shoot them down the **Krupp** Company produced a 37mm *Ballonkanone*, pedestal-mounted in a horsedrawn cart – but without recorded hits. With the advent of **airships**, **aircraft**, **motor vehicles** and, above all, **World War I**, calls for static and mobile anti-aircraft artillery (AAA) became insistent. In 1906 a Rheinmetall Company 50mm *Flak* gun was turret-mounted on an Erhardt lorry. However, most AA guns used by both sides from 1914 onwards were adaptations of existing naval and field **artillery** mounted on trailers. Of these the Krupp 88mm naval gun of 1913, which in 1937 assumed a dual-purpose role as an **anti-tank** gun, would become most famous.

The effective vertical range of these guns was between 13,000 and 17,000ft and their rate of fire between six and fifteen rounds per minute (rpm). But since gunlayers could not compute the point of aim against a target moving at unknown speed at an unknown altitude, pre-planned barrages with shells set to explode in the target's assumed path, fired by groups of guns directed from central control posts, were the preferred solution – though even at low altitudes several thousand rounds per kill (rpk) were required.

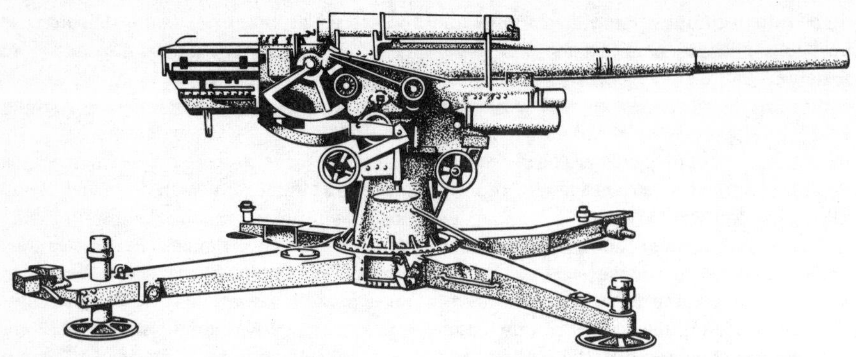

Dual-purpose anti-aircraft/anti-tank gun (German, 88mm)

Actually, only aircraft flying very low had much to fear from machine-guns, fired by good shots.

Prior to **World War II**, with guns in service of between 76 and 105mm and capable of firing vertically up 32,000ft, strenuous attempts were made with **sound locators** and **rangefinders** to fix the speed and altitude of fast-moving targets and feed that data, via a mechanical **computer**, to gunlayers. Yet, until the introduction in 1940 of ranging **radar** sets capable of an accuracy of 300ft at a distance of 42,000ft and giving 4,000rpk, the rpk was 20,000. A figure greatly improved upon by the introduction in 1944 of shells fitted with the **radio** proximity (Variable Time (VT)) fuse which proved extremely effective against the **V1** flying bomb at low altitudes, as well as many other kinds of target.

Similar rpk to the heavy guns were also achieved by light anti-aircraft guns at low altitudes, but in their case cost-effectiveness was better since they deterred and distracted low-flying bomber crews whose chances of finding and hitting pinpoint targets were the best. Belt-fed, multiple machine-guns of between 7·62 and 20mm were usually grouped with magazine-fed 37mm and 40mm guns to defend ships and vital ground targets. In Britain in 1940 they were supplemented by **balloon** barrages and by easily made *parachute-and-cable* devices – a linear arrangement of parachutes, to which light steel cables were attached, fired optimistically by rockets to a height of about 3,000ft in the path of aircraft. Later 3in anti-aircraft rockets fired barrages at greater altitudes with immense flash and noise, but scant material effect.

Nevertheless, as big guns became obsolete after World War II because they could no longer reach high-flying aircraft, their place was taken by *surface-to-air missiles* (SAM) (*see also* **Rockets and guided missiles**) which could 'home' or be guided on to their targets at low or extremely high altitudes. The German *radio-controlled beam-riding Wasserfall* single-stage missile, with a speed of 1,700 mph and a range of 16·5 miles, was the forerunner of four kinds of missile:

(1) The 'active' **radar** type, reacting to reflections from an 'illuminated' target, and the 'passive' type, reacting to reflections from the target.

(2) The **infrared** kind, responding to heat radiations (such as exhaust systems).

(3) The 'homer', which latched on to **radio** transmissions.

(4) The **television** type, with a camera in the nose and an operator monitoring the screen to steer the missile on to the target.

Small, short-range SAM can be shoulder-held; others statically emplaced or vehicle-mounted. All have the disadvantage of displaying clear firing signatures. Most to some extent are vulnerable to jamming, spoofing and evasion. Like all anti-aircraft weapons they are complementary to **fighter aircraft** and guns. During the **Arab–Israeli** and **Viet-**

nam wars, a combination of SAM with surveillance and tracking radars controlling light anti-aircraft weapons was often quite successful. However during the **Falklands War** the Royal Navy's lack of rapid-fire weapons and overdependence on SAM and fighters put warships at considerable risk.

Anti-ballistic missiles. It was in the 1960s that both the Soviet Union and the United States began to study ways of destroying the other's strategic ballistic missiles just after launch – the concept of the anti-ballistic missile or ABM. The idea was fraught with technical difficulties, involving a very short reaction time, highly complex **radars** and very-high-velocity missiles (*see* **Rockets and guided missiles**) capable of reaching the target before the latter could release its warheads.

The Soviets were first in the field with their Galosh system, deployed in 1968 for the air defence of Moscow. Each site had sixteen launchers and associated with them an acquisition **radar**, two battle management radars and four engagement radars. The missiles themselves were high-speed three-stage rockets, each with a nuclear warhead. On the US side the Nike–Hercules system, primarily developed for use against high-flying aircraft in 1961, had a limited capability against ABM but was dropped as ineffective. It was revived in the 1980s as the **Strategic Defence Initiative** (SDI) and also in the shape of the short-range Patriot (initially anti-aircraft) missile which, over Saudi Arabia and Israel in 1991, shot down many Iraqi long-range ballistic missiles.

Anti-submarine weapons. Since the Confederate **submarine** *Hunley* sank the Federal ship *Housatonic* in 1864, and was dragged to the bottom by her victim, anti-submarine measures have assumed growing importance in **naval warfare**. Moored **mines**, from the outset, have been and remain among the most potent killers, particularly in shallower waters. Nets have deterred, and ramming, to begin with the only way for surface craft to score a kill, has occasionally been effective. A surfaced submarine, of course, was vulnerable like any other vessel and, in **World War I**, **Q-ships** were developed to lure submarines to the surface.

The **depth charge**, first produced by the British in 1915 and consisting of a container with 120lb of **explosive** dropped into the water and set off by a hydrostatic pistol, was the most effective mobile weapon. However, to be effective it had to be within 21 feet of a boat, which raised the difficult problem of how to locate the submarine precisely. Seabed and ship-fitted **hydrophones** were tried before 1914, but results were usually discouraging. Until the invention of **sonar** in 1918, visual sighting by surface craft and **aircraft** was the best method, sometimes guided by knowledge of the victim's presence through direction-finding (DF) from intercepts of submarines' **radio** transmissions used during early attempts to coordinate operations.

Passive defensive methods were the most potent: zigzagging to defeat the submarine commander's predictions and aim; dazzle painting of ships to give a false impression of course and speed: and, most effective of all, the **convoy** system to reduce the number of targets and economically enhance defence by escorts. Surface escorts scored the most kills: **airships** and seaplanes made a contribution and there were examples of successful sub versus submarine encounters – the first on record being the torpedoing on the surface by the German U22 of her compatriot U7 in January 1915.

Between the World Wars remarkably little progress was made in submarine warfare. A consensus believed that the convoy system, allied to sonar's potential and to air escort, had overcome the threat. Only late in the 1930s were doubts expressed about sonar when it was discovered that Asdic was not nearly so good a detector as originally thought, owing to temperature variations of water layers at different depths. Nor had the potential of centralized direction by radio of submarine packs, as developed by Germany's Commodore Karl **Dönitz**, been visualized. Depth charges were still limited to a maximum 80 feet; escorts had been given a low priority by the British and the Americans – the latter even rejected the convoy system. A highly sophisticated organization to gather and synthesize **intelligence**, to control the evasive routeing of convoys and to initiate countermeasures was only in its infancy and made all the more inadequate in June 1940 when the Germans seized most of Europe's western seaboard and launched the **Battle of the Atlantic**.

That battle and the others which raged in the Mediterranean and, later, in the Pacific and Indian Oceans would be decided by the struggle between opposing control organizations to which radio intercept and direction finding, the deciphering of **codes** and **radar** were central in the detection process. New weapons also played their part from faster surface escorts and aircraft of longer range equipped with the latest detection devices. Depth charges were made to burst at greater depths and thrown, instead of released, at their targets by a multiple mortar called Hedgehog. Aircraft were given **rockets** to fire at surfaced boats. Primitive homing **torpedoes** began to make it possible even for one submerged submarine to sink another, with improved sonar making contact through the temperature layers at very long ranges and detector buoys, dropped from aircraft, pinning down the target at close range.

After **World War II**, and with redoubled urgency when nuclear-armed and -powered submarines up to 25,000 tonnes with extremely long endurance, deep diving capability below 4,000ft and underwater speeds of 45 knots came into service, both detectors and weapons had to be much improved. Sniffers to detect diesel-boat fumes on the surface; magnetic anomaly detectors (MAD) which record changes in the earth's magnetic field caused by a submerged submarine; complemented by improved radars and static, underwater hydrophone arrays laid in channels where boats often have to pass, all contribute to the game of hide-and-seek in which submarines themselves are probably the best anti-submarine weapon system in the ocean depths, in peace and war. **Helicopters** also are particularly useful with their ability, at the hover, to lower Sonar detectors below the temperature layers and then join in the attack with fast-running torpedoes or air-to-surface missiles, as happened during the **Falklands** War.

Yet these more sophisticated seek-and-hit methods are frequently defeated by quieter-running submarines equipped with detector and anti-detection devices and clad in sonar-reflecting materials. Even when a positive contact has been made, there is no guaranteed kill of a fast agile boat bent on evasion; or that, having struck, the weapon will be powerful enough fatally to rupture a single-, double- or (maybe with some Russian boats) triple-skin, **titanium**-alloy hull. As often as not the modern submarine is able to defeat the hunters and pursue its own objective. Thus thwarted pursuers may be compelled to resort to the ultimate weapon system – a nuclear missile or mine.

Anti-tank weapons. When **armoured fighting vehicles** (AFV) were first deployed in 1914, the principal threat to their 10mm **armour** at ranges of about 300 metres was from hardened **machine-gun** bullets or from high-explosive shells from field guns, the latter with a 30:1 chance of a hit above 400 metres. But by 1917 it had become clear that a specific anti-tank weapon was required to cope with thicker armoured and faster **tanks**. The Germans led with a clumsy 13mm rifle and, in the 1920s, were among several nations developing 20mm weapons on field mountings and in AFVs.

By the 1930s an accelerating gun-versus-armour race was in progress. Weapons of between 37 and 57mm firing 2 to 6lb shot at velocities of about 700m/sec came into service, capable of penetrating 40mm armour at less than 400 metres. Their chances of a hit/kill were, however, degraded by **gunnery** problems and defective **ammunition**. It was the Germans who drew ahead in the race when they decided to use their 88mm **anti-aircraft** gun in the same way as they had used 77mm AA guns against tanks at the Battle of **Cambrai** in 1917. Firing

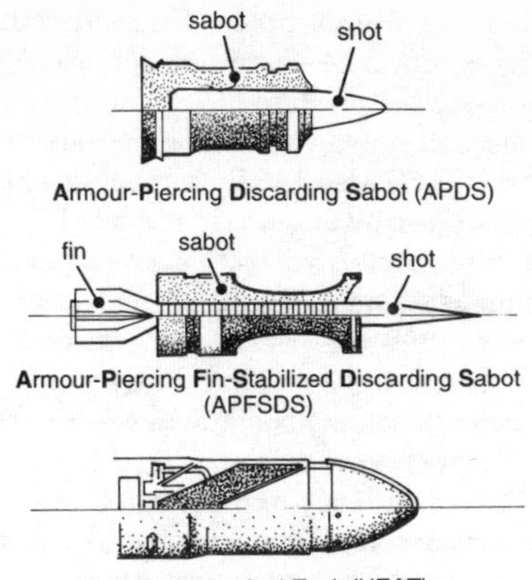

Anti-tank weapons: armour-piercing warheads

a 21lb shot at 810m/sec, the 88mm was capable of penetrating 60mm of armour at ranges greater than 1,200 metres. Like many new weapon systems, it was tried out successfully during the **Spanish Civil War** and, field and AFV-mounted, came to dominate the battlefields of **World War II**.

The Allies gradually caught up with the Germans by successively adopting guns of between 57 and 90mm to deal with ever-improving German AFVs. Most significant of all was the British 84mm gun, firing a 20lb shot, which came into service in 1949 and, in the 1950s, was vastly improved by an increase in calibre to produce the 105mm gun. This piece remains in service with a great many armies to this day, but is steadily being replaced by 120–125mm weapons as the gun–armour race continues.

Ammunition is, of course, crucial to performance. The solid shot fired by modern weapons has developed a long way from its crude beginnings to the highly successful use in the **UN–Iraq War** of fin-stabilized penetrators with a depleted-**uranium** core made to complex designs.

The development of the hollow-charge, chemical energy (CE) projectile (of which the **bazooka** was the first of such hand-held, short-range weapons) has provided a crucial factor in the development of the **rocket**-propelled anti-tank guided weapon (ATGW). The Germans had been first in this field during World War II with their planned wire-guided X-7 missile. However, the French and British took the design further, with the French producing the first practicable system in the 1950s with their SS10. Operating out to about 1,000m, it was the forerunner of a whole series of Soviet systems. While the UK produced the Vickers Vigilant and, later, the British Aerospace Swingfire.

It was during the Yom Kippur War in 1973 that the ATGW really showed its paces, the Egyptian infantry, equipped with Soviet 2,000m-range Saggers, being able to inflict heavy casualties on Israeli tanks, leading to some ill-informed commentators to suppose, mistakenly, that the day of the tank was done. ATGWs often have sophisticated guidance. Early systems such as those already mentioned used manual command to line of sight (MCLOS), which has the advantage of simplicity and accuracy with the operator tracking both missile and target; but it relies on considerable operator training. Later systems such as Milan and TOW use semi-active command to line of sight (SACLOS), technically more complex but simpler to operate. Maximum ranges in excess of 2,000m are usual with modern systems.

As protection increases with the use of special armours, the direct-fire weapon finds it increasingly difficult to penetrate the AFV's front and sides. So attention is now being paid to attacking top armour, inevitably thinner to meet overall weight restrictions. The bomblet, containing a small hollow-charge warhead and dropped from an **aircraft** in very large numbers to saturate an area containing AFVs, is one approach. Highly complex terminally guided munitions (TGM), fired from heavy, long-range artillery or MLRS, or delivered by aircraft (with far greater accuracy than the unstable free-flight rockets of previous generations), and which have proved their worth in Iraq, are another.

Mines too, which have always been a menace to AFVs by cutting tracks or removing wheels, now have a more lethal effect with elaborate fuses designed to penetrate the belly plate.

One thing is certain. The competition between AFV protection and the weapon to defeat it will continue.

Arab–Israeli wars. In the aftermath of the Arab revolt of **World War I** with its promise of Arab autonomy, the underhand Anglo-French grasping of mandates in the Middle East and approval for a Jewish national home in Palestine was a recipe for war. Incidents in the 1920s which swelled into **guerrilla warfare** in 1936, catching British troops in the middle, were first steps in the creation by the Jews of an underground army with the power to persuade the United Nations to cancel the British mandate. In May 1948 the State of Israel was declared, which acted as a signal for converging attacks upon it by Egypt, Lebanon, Syria and Jordan.

Defence of Israel, by poorly armed units, revolved around contested nodal localities. As the Arabs were worn down the fanatic Israelis acquired strength and sufficient weapons to assume local offensives. It was Israeli good fortune to be superior in mobility to the Egyptians in the south, to hold firm in the north and, after losing old Jerusalem to the Arab Legion in ten days of fierce fighting, nevertheless to retain modern Jerusalem. But most significant portents for the future were the events of October, when the Israeli army seized Beersheba and flung the Egyptians back into Sinai, and in March 1949 when they seized the Negev with its port of Eilat at the head of the Gulf of Akaba.

Truces, in the years to come, would be made and broken by both sides. Raids in the frontier regions, as all acquired arms from a variety of Eastern and Western bloc nations, became endemic. Tension came to a head on 29 October 1956 when, in collusion with the Anglo-French attack on Egypt at Suez, the Israelis struck deeply with mechanized and **airborne** troops, mainly equipped with obsolete weapons, to within 30 miles of the canal and also to capture the remainder of Sinai – only to withdraw under UN diplomatic pressure in March 1957.

The next bout – the so-called Six Days War of 5–10 June 1967 – was a much more sophisticated affair in that the Arabs had, for the most part, been rearmed with Russian equipment and the Israelis had acquired modern Western **aircraft** and **tanks**. Provoked by a strong buildup of Egyptian forces in Sinai and threatening Syrian and Jordanian moves in the north and near Jerusalem, the Israelis struck pre-emptively at the Egyptians, destroying their air force in the first few hours and, by overrunning Sinai, ruining their enemy in four days. At the same time, they defeated the Jordanians and captured Jerusalem, along with the entire Arab holding west of the River Jordan, also turning against the Syrians to eject them from the dominating Golan Heights.

The next six years are sometimes called the War of Attrition, during which the Arabs reconstructed their forces and planned retribution, and the Israelis fortified their new frontiers and strengthened themselves. Raid and counter-raid, by sea, air and land, erupted, to almost total Israeli surprise, into an all-out Arab offensive on 6 October 1973. The Yom Kippur War, so called because the Arabs made it coincide for their advantage with a Jewish religious feast, played havoc with Israeli dispositions. A massive Egyptian crossing of the Canal smashed the Israeli defences at the same time as Israeli air attacks suffered heavy loss from Russian-type **anti-aircraft** defences. The joint Syrian/Iraqi/Jordanian thrust in the north made headway but was blunted and then thrown back with crippling losses by a desperate defence and a brilliant counterstroke. Thereupon the Egyptians, against their better judgement, were persuaded to renew the offensive in Sinai, only to be defeated and themselves flung back to and across the Canal, bringing to an end a three weeks' battle which cost both sides dear.

Since then, despite a formal peace with Egypt, Israel has been engaged in sporadic conflict with her northern neighbours and

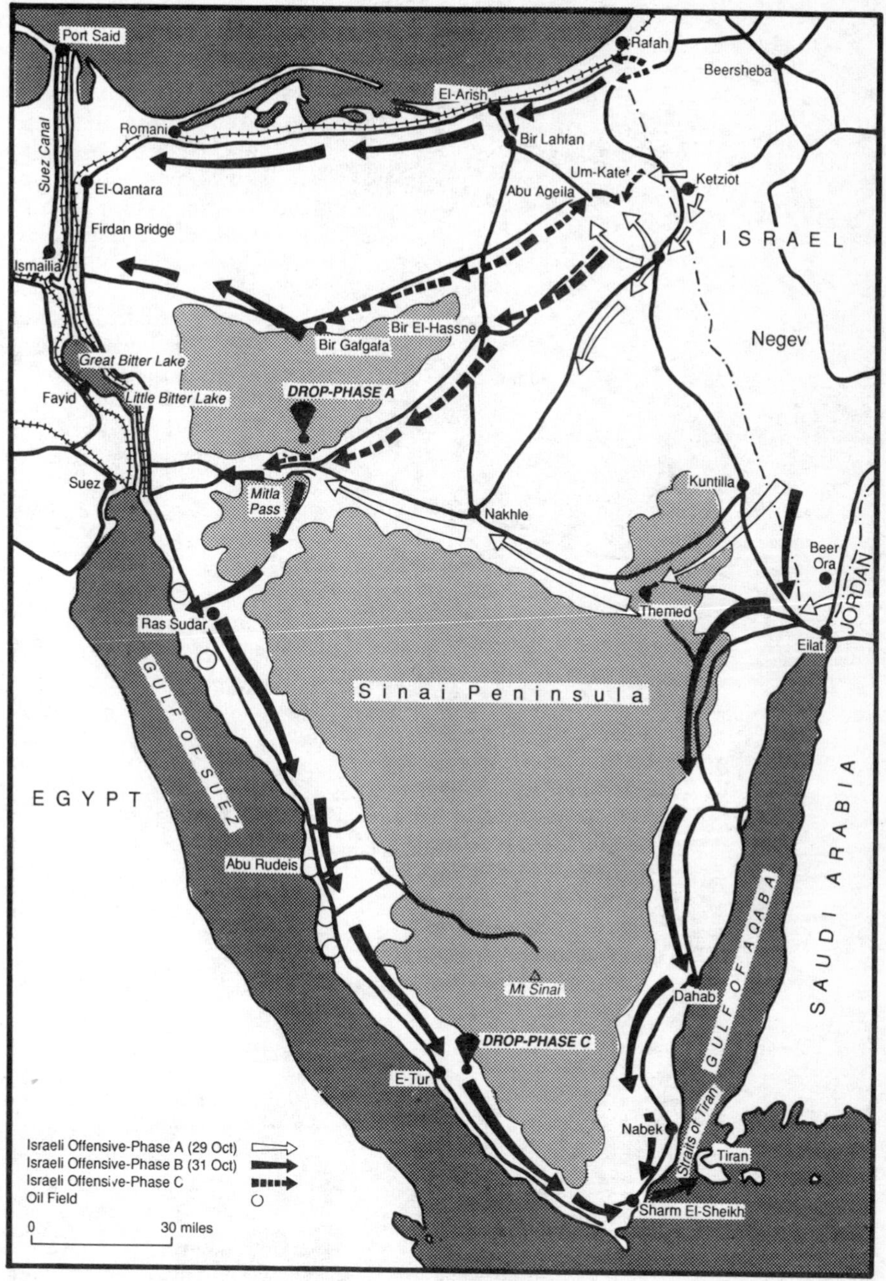

The Israeli offensive of 29 October to 5 November 1956

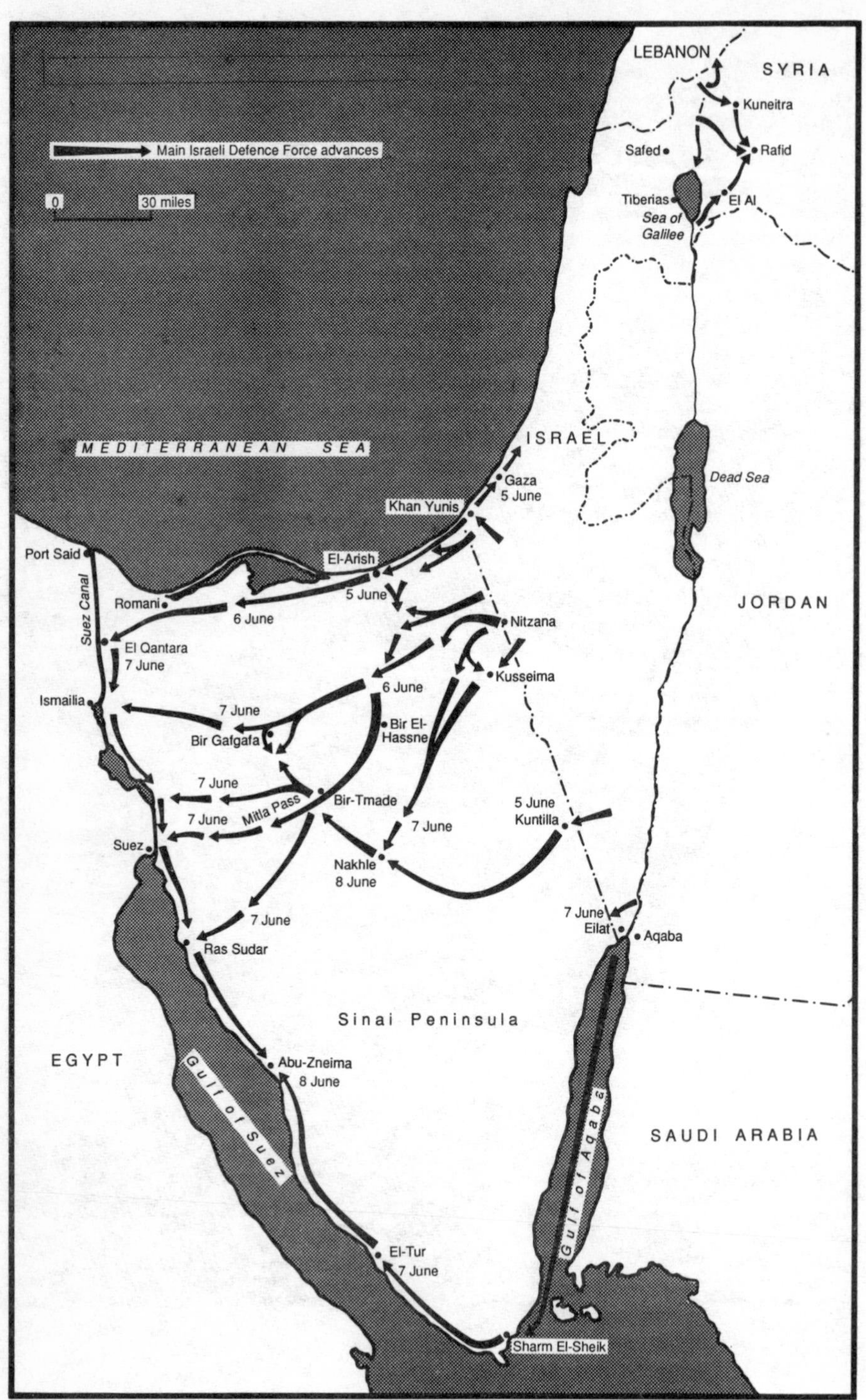

The Six-Day War, June 1967

the Palestinian Arabs. Confrontations emanating as often as not from Lebanon, where feuding Arab groups based themselves for raids into Israel, provoked Israel into large-scale action. Unfortunately their invasion of Lebanon on 6 June 1982, in an attempt to demilitarize the frontier regions, only made worse a complex imbroglio. Fighting against sophisticated Syrian forces, and others, was at times heavy. Finally touched by war weariness and a sense of frustration, the Israelis withdrew to the original frontier, which it continues to secure by a strategy of local defence and occasional punitive raids into Lebanon.

Arbela (or **Gaugamela**), **Battle of.** In 331 BC, on hearing that King Darius was assembling an army in Mesopotamia for a renewal of the **Greek–Persian War**, **Alexander III** sought battle. The armies met on 1 October on the Plain of Gaugamela. Darius, strong in **cavalry**, with 15 **elephants** and maybe 200 **chariots**, and, perhaps, 200,000 men of dubious training, stood on the defensive. Alexander, with fewer than 50,000 veterans approached the Persian left obliquely, to be attacked or threatened on both wings by the well-armoured Persian cavalry and, in the confusion of hand-to-hand combat, charged by the chariots. But the Greek **armoured infantry** phalanxes stood firm and repelled the chariots with arrows and javelins while the cavalry countered the Persian horse.

Seeing where Darius stood in a gap in the Persian line, Alexander led his élite Companion cavalry into it as vanguard for the phalanxes then deployed in line. Darius fled; the surviving Persian infantry ran or were slaughtered while their cavalry's last desperate attacks on the Greek flanks and rear were checked by Alexander's reserves. The Persian Empire collapsed, Darius was murdered and Alexander entered Babylon,

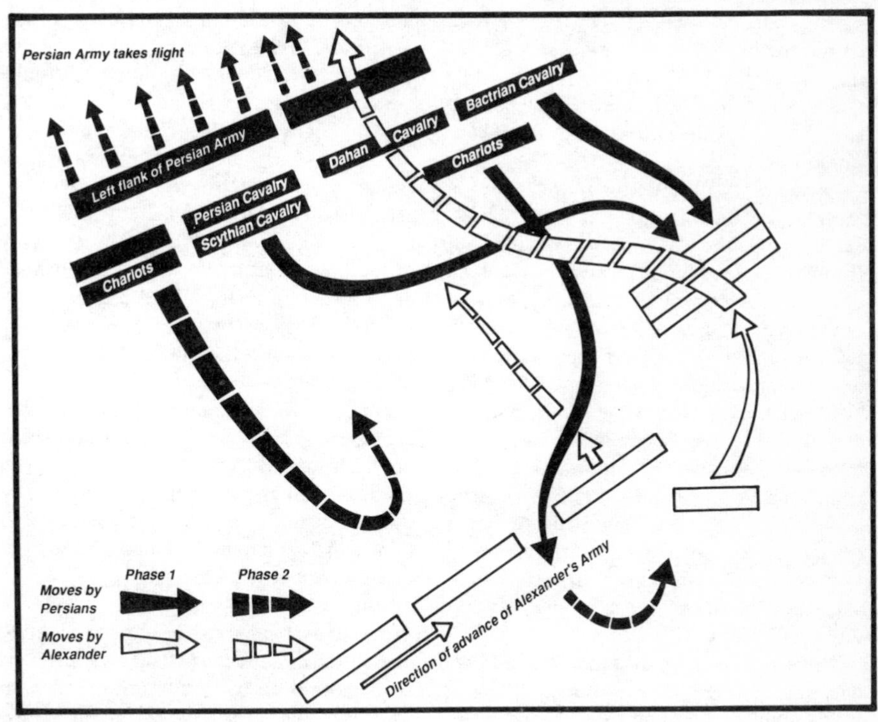

The Battle of Arbela, 331 BC

thrusting later into India in pursuit of surviving Persian forces.

Archimedes (*c.* 287–211 BC). A renowned mathematician and geometrician whose research significantly advanced science and technology. His demonstration of moving great weights by small forces and his invention of a helix to lift water from ships' holds were of great importance. His military fame rests on the engines devised to frustrate the Romans' siege of Syracuse (213–211 BC) – although he disclaimed it, perhaps because most of his devices were based on machines from a previous generation.

Arctic convoys. With the Allied supply route to Russia via Persia difficult and those through the Black Sea and the Baltic blocked, the best way was directly to Archangel and Murmansk after the German invasion of 1941. Nature and the Germans in Norway were hostile. Long summer daylight-hours aided **bombers**; long winter nights shielded surface and **submarine** attacks. The ice pack to northward and the Norwegian coast to the south canalized the approaches while the weather, rarely better than appalling, especially in winter, took its toll of both sides.

To begin with the Germans were slow to react to the delivery of aircraft by **aircraft-carrier** and the first eight convoys got through without loss. Only when the land campaign stalled and the long-term potential of the convoys became apparent did opposition increase. **Hitler** not only appreciated the **logistic** implications but worried perpetually over the convoys being a cover, in the aftermath of a few **Commando** raids, for an invasion of **Norway**. Admiral **Raeder** saw opportunities of sinking Allied warships and merchant vessels in advantageous circumstances. And while the British wanted to tackle enemy warships, they were handicapped by fuelling problems in voyages up to 2,000 miles long.

The eighth convoy in January 1942 was opposed, as were all subsequent attempts, with mounting fury and rising losses, many from air attack. The presence of the battleship *Tirpitz* injected such dread in the British Admiralty that, in July, a misconstrued report that she was at sea prompted Admiral Sir Dudley **Pound** (against advice) to order Convoy PQ17 to scatter – with the subsequent loss to aircraft and U-boats of 23 out of 33 merchant ships. Henceforward summer convoys were avoided.

The next winter convoy was threatened on 31 December by German heavy **cruisers**, but bold action by British cruisers and **destroyers** deterred them, making Hitler demand the scrapping of the big surface warships, to bring about the resignation of Raeder and the appointment of Admiral **Dönitz** in his place. Nevertheless, the surface vessels were reprieved and convoys continued to be run, until called off as the crisis of the Battle of the **Atlantic** was reached in March 1943 and the Russians obstructed British radio and air-support facilities.

Convoys were resumed in October with little opposition until, in December, Dönitz sent out the battleship *Scharnhorst* with five destroyers. The British were forewarned and at 0840 hours on the 26th their squadron of a battleship, cruisers and destroyers screening the convoy made contact by **radar**. German radar was either not working or damaged at once by fire from a cruiser. *Scharnhorst* manoeuvred and tried again to find the convoy until brought to action by the battleship *Duke of York*, whose 14in guns, at 1650 hours, scored hits with a first salvo. Her fate sealed, the battleship was eventually sunk by a combination of shell and torpedo hits.

Convoys would continue against fluctuating degrees of submarine and air opposition but with steadily decreasing losses. Between August 1944 and the end of the war more than 250 ships would carry over one million tonnes to Russia to make a vital contribution to the Soviet war effort.

Armaments industry. The earliest fighting men most certainly made their own weapons until, within each community or band of warriors, there emerged craftsmen whose products were so superior to others that they were best employed training and supervising design and manufacture. This industrial practice became all the more essential when the **wheel** and **chariots** were invented and as **wood** and **stone** gave way to **bronze** and **iron** weapons. Thus from as early as 3500 BC the semblance of armaments industries must have come into existence in various parts of the world. These assets became all the more influential as **armour**, more powerful **bows** and higher-grade metal weapons became crucial to success in battle. No doubt there were many instances, as there are to this day, of a craftsman-entrepreneur arms-maker persuading political and military leaders to back a new idea for offence or defence.

Certainly by *c.* 3500 BC there were Mediterranean **shipbuilders** whose wooden vessels were used for rather more than fishing and commerce – as the craft bearing the invading Sea People of *c.* 1200 BC undoubtedly were. Undeniably, between 900 and 600 BC, the highly organized societies of Babylonia and Assyria boasted a large metals industry to forge or cast large quantities of **edged weapons**, arrowheads, spearheads and armour of quite high quality for their persistently warring armies – in addition for use by hunters for food. By this time the Middle Eastern countries, along with some in the Far East, had evolved in sufficient breadth and coherence to communicate with or conquer neighbouring states and to hand to posterity almost indestructible records, many carved in stone or cast in metal, of their methods and skills.

Every martial state honoured and paid well the master craftsmen, whose names passed into legend. The names of those who invented mail armour are unrecorded, but the sword-makers who wrought the finest iron and **steel** blades in places such as Solingen in Germany and Toledo in Spain, were famous beyond their cities and commanded the highest prices. Among them were Clemens Horn, Johan Knobe, Paulus Willems, the Picino family and Thomas Tolido. Yet the survivors among these early industrialists were the few who adapted quickest to the invention of gunpowder and its impact on weapon design and construction. Where high-grade ferrous metals were already in production the gunsmiths would also thrive, although there would be casualties and many newcomers to the field as the industrial revolution of the 18th century gathered pace.

Coming as it did in the aftermath of the **Seven Years War** and on the eve of the **American War of Independence**, the publication in 1776 of Adam Smith's famous *Wealth of Nations* with its economic philosophy of *laissez-faire* and description of more efficient production through division of labour, could not have been better timed to stimulate the armaments industry. When, owing to improved **health** and **medical services**, populations were rising and **conscription** was mooted, demand for new ships and weapons grew strong. **Dockyards** flourished, the means of steam **propulsion** were boosted by James **Watt**'s inventions, iron foundries such as those established in Britain in the Severn Gorge grew and new kinds of factory were built. In 1796 the English firm of Boulton & Watt started the Soho engineering factory with a flow-line production system, and two years later Eli **Whitney** began mass production of muskets in the USA, at about the time Henry Maudsley, an ex-employee of the British government's Woolwich Arsenal, was inventing the first screw-cutting lathe, which allowed semi-skilled workers to achieve self-perpetuating production with a precision previously attainable only by craftsmen.

Henceforward industrial progress accelerated to keep pace with the huge demand for traditional weapons and development of the latest inventions such as **balloons** and **submarines**. Despite a reduction in hostilities and weapons production between 1815 and 1854, in that period the foundations for the next mighty phase of rearmament were laid. Iron warships and **railway** locomotives driven by steam power, breech-loading guns, the **telegraph**, **turbine** engines, the electric motor, the fuel cell, canned **food** and anaesthetics, in addition to many fundamental scientific and technological discoveries, were among those then invented.

Armaments industry

Beyond much doubt the second phase of the industrial revolution was spurred if not actually initiated by military demands. The **Crimean War**, preceded by the violent revolutions of the 1840s which destabilized European society, created a demand for the latest weapons to restock the obsolete armouries of the previous century, which were still in service. Henry **Bessemer**'s invention of a cheap steel-making process was a prime mover in the coming arms race, if only because his threatened competitors had to find ways of undercutting and outselling his products. The abrupt availability of cheap ferrous metals in large quantities made possible larger and more powerful armoured warships armed with much bigger **artillery** which required the latest **explosives** for propellants and projectiles. Increases in weapon power struck fear in peoples and politicians who, unknowingly for the most part, danced to the tune of armaments manufacturers, not a few of whom lacked scruples as to what goods they sold, or to whom.

Moreover, in the second half of the 19th century there began to emerge the great arms industrial conglomerates, of which the majority were British-controlled or influenced, as befitted the greatest industrial nation in the world which, by 1900 for example, built 80 per cent of the world's ships. European firms such as Armstrong, Whitworth, **Vickers**, Hadfields, Nobel, **Krupps**, **Siemens** and the **Le Creusot foundries** of **Schneider** sometimes shared knowledge and capital with each other and a few of the great American foundries, along with small-arms manufacturers like **Colt** and Gatling. Post-1869 the giants also helped emergent Japanese shipbuilding firms like Mitsu Bishi – a firm which, within sixty years as **Mitsubishi**, would comprise huge armaments manufacturers building ships, **aircraft** and a whole range of other weapons besides processing basic materials in their foundries.

Until 1900 the armament industries, including government-owned dockyards and factories, were associated with large-scale metal and explosives production and heavy engineering. But the next stage of the industrial revolution was looming ahead owing to the invention of **motor vehicles**, the **telephone** and **radio**; and on the eve of John **Fleming**'s invention of the vacuum diode valve and his coining of the word **electronics**; and the flight of the first heavier-than-air machine. Nearly all the companies running or supplying these new electrical or **aircraft industries** with new alloys, materials and services were newcomers. Rapidly **Daimler**, **Edison**, **Ford**, **Marconi** and **Wright** took on defence work to become major armaments producers. But of them all the most important were the aircraft firms such as **Zeppelin**, **de Havilland**, **Fokker**, **Sopwith** and Blériot who attracted much attention because of their novelty and the impact of **bombing** on the civil populaces. They expanded mightily although, in time, many would merge or would be absorbed by Vickers, Siemens, Krupp, Mitsubishi and other giants.

The arms and naval races are given as one cause of **World War I**, which eventually was extended into **World War II**. For this governments were responsible since, often in the interests of security or for predatory reasons, they could not resist sponsoring or ordering from their own industrial establishments the **research and development** (R&D) from which new weapon systems grew – especially as R&D became increasingly expensive and outstripped the financial means of even the gigantic firms. Throughout the World Wars the sheer size of orders and profit encouraged private ventures, but major developments of warships, aircraft, **armoured fighting vehicles**, electronics and **chemical** weapons, for example, demanded intensive government support.

The extreme example of state control was the Soviet Union once Josef **Stalin** made the armaments industry drive the nation's economy to help give full employment and boost Communist prestige. This policy created the massive but rather crude military forces that seemed to threaten Europe in the 1930s; saved only by the narrowest of margins the Soviet Union from the Germans in World War II; and contributed significantly to renewed fear of her aggressive intentions after 1945, leading to the **Cold War**.

The tremendous growth of the armaments industries in the Cold War was a phenomenon made all the more spectacular by the introduction of **nuclear weapons** and **space vehicles**, and by the US example of matching Russian brute strength with weapons of sophisticated technology.

When the USA entered World War I in 1917 the armament industries she had developed during the Civil War and the Spanish–American War of 1898 had, with the exception of shipbuilding and small arms, virtually disappeared. She depended almost entirely on Britain and France for heavy weapons and munitions; then, post-1918, largely closed the factories she had started to construct. Partly owing to the renewed threat of war in the 1930s, but also, like Germany and other nations, as a way to stimulate an economy in deep recession, America then began to build the vast armaments industries which, though to a far lesser extent than in Russia during and after World War II, drove the national economy.

The vital differences between the two giants were in fact those of scale, efficiency and value for money. To match the 18 per cent of gross national product (GNP) Russia devoted to defence, the much more efficient Americans required little more than 6 per cent GNP to sustain her allies, fight the **Korean War** and the **Vietnam War** and still enjoy the world's highest standard of living. They thus indirectly goaded into bankruptcy the inefficient Russians, with their low standard of living, until in the 1980s they could neither sustain the arms race nor dare risk a direct confrontation in war with the relatively inadequate weapon systems they had created. Thus one side's superior arms industry may be said to have defeated the other's without resorting to all-out war. It remains to be seen if the promised **disarmament** will be as effective as the preceding buildup.

Armour. From the earliest times ways of deflecting blows from missiles and **edged weapons** have been pre-eminent. To begin with, leather was the best armour, but with each advance in **metallurgy** the same material developed for weapons was at once utilized for armour. This competition between offensive and defensive measures has been crucial if only because the fighting man's **morale** is dependent upon its effects – be they related to protection of the person's body, his mount, vessel or the vehicle carrying him into battle.

Body armour is best designed to protect the most vulnerable parts and encumber as little as possible the wearer's mobility and hitting power. The shield carried on the disengaged arm was usually first choice, followed by helmet, upper torso, leg and lower torso armour, in that order. The material often depended upon a compromise between availability, price and effectiveness. **Bronze**-plate armour for Greek soldiers (the first well-armoured fighting men) was in service long into the **iron** age because wrought-iron plate gave only marginally better protection and was so difficult and costly to manufacture. Therefore the Romans preferred bronze, but carried more by adding scale armour, hung from the waist belt, even as iron became more common and very expensive **steel** daggers and **swords** appeared *c.* 1000 BC.

Inferior, though much cheaper, was cloth, padded armour (used also for horses and elephants). But the introduction of very expensive mail armour, *c.* 200, marked a turning-point because, although not as good as plate, when backed by cloth or leather to prevent the armour being driven into flesh, it gave reasonable protection against cuts and blows without seriously encumbering the wearer. Mail (derived from the Latin word *macula* for net) was made from small, wire split-rings 'knitted' into a close mesh and, therefore, capable of being worn like ordinary clothing. However, the improved performance of 14th-century **bows and arrows**, which easily penetrated mail, made plate fashionable again. Naturally there was a preference for the best-quality steel to defeat the high-velocity arrow and, in due course, the bullet. But high cost and far greater weight, without providing adequate protection, by the 16th century made full body armour obsolete. By the 20th century it was almost extinct.

In **World War** I, however, the steel helmet was reintroduced and, in the 1970s, sleeveless jackets made of specially woven, lightweight Kevlar capable of resisting small arms and splinters.

Ships' armour was virtually unthought of until the 1850s since the thick timbers of the larger vessels were often proof, coincidentally, against cannon shot. But threatened by **artillery** of increased calibre, in 1859 the French commissioned the first armoured warship, *La Gloire* (5,617 tonnes), a **frigate** with her timbers faced by 110mm iron plates, initiating a new gun/armour contest by provoking the British into commissioning HMS *Warrior* (9,137 tonnes), built entirely of 114·3mm wrought-iron plate facing 457mm teak – a combination which defeated 150-pounder guns' shot at 400 metres range. But after the 1860s the availability of large quantities of steel and the invention, in 1882, of armoured alloy steel led inexorably to the all-steel **battleship** with armour, in the case of the 1940s Japanese *Yamato*, up to 560mm, which made her impervious to 18in shells inside 34,000 metres, producing a revolution in **naval warfare** as well as **shipbuilding**.

Yet, with the supplanting of the gun by the aerial **bomb**, **rocket and guided missile**, and by the even more deadly **nuclear** threat and below-waterline attack from **mines** and **torpedoes**, large warships have become obsolete. In their place are **aircraft-carriers**, mainly with only their flight decks armoured, and smaller, fast vessels relying upon speed, agility and ECM (*see* **Electronics**) for their survival.

Land vehicles. Some **chariots** and the wagons of John **Zizka** were lightly armoured, as were a few **railway** trains during the **American Civil War** and numerous subsequent conflicts. With the internal combustion engine at last came the means once more to provide the soldier with armoured protection, through the introduction of **armoured cars** and **tanks**. Woefully thin as these first **armoured fighting vehicles** (AFV) were, they enabled men to cross fire-swept ground with a reasonable chance of survival. The chances improved because of the combat successes of AFVs during World War I and the subsequent development of properly engineered armour to help defeat the inevitable introduction of special **anti-tank weapons**.

The basis for vehicle protection became and still is rolled homogeneous armour, of which manganese steel (invented by the Briton Robert **Hadfield** in 1882) forms an essential part. At first the designers merely demanded thicker vertical slabs of it to counter the increasing threat. It was the American Walter **Christie** who realized that a sloped plate could provide much greater protection for a given thickness, as well as encouraging ricochets. The Russians showed what could be done with their T34 tank in 1941, combining as it did well-sloped armour plus an agility markedly superior to that of the current German tanks. The Germans, meanwhile, had developed face-hardened armour – steel heat-treated to give it a hard outer skin to shatter a projectile's piercing cap, with a softer thickness behind to absorb the projectile core as it attempted to penetrate. A form of this approach has been seen in modern Brazilian light armoured vehicles.

While steel armour has continued to find most favour, **aluminium** has been used successfully on light vehicles; and more recent years have seen, in the British development of Chobham armour, the introduction of composite armours using special materials to help disperse the effect of the molten jet produced by chemical-energy, hollow-charge warheads (*see* **Ammunition**).

An Israeli approach, soon copied by the USSR, the UK and the USA, has been explosive-reactive armour (ERA), consisting of boxes of **explosives**, hung on the outside of a vehicle, designed to detonate when struck by a hollow-charge projectile. These disrupt the latter's jet before it can attack the main armour behind.

Aircraft armour. In the air it has long been

recognized that the weight penalty would rule out complete armour protection for an aircraft. Vital components, including the pilot, can perhaps be shielded, but only by using extremely expensive lightweight materials such as **titanium**, used effectively on ground-attack aircraft such as the American A10 and in armed **helicopters** such as Apache, in which the pilot sits in an armoured tub.

Thus, in their different ways, naval and land forces have come full circle. Navies, with a few notable exceptions, have mostly abandoned armour, while armies, having done without it for 200–300 years, now try hard to send their soldiers into battle under armour, with dismounted infantry wearing partial body armour as of old. Air forces are left to depend mainly upon **stealth**, evasion and striking first.

Armoured cars. No sooner had **motor vehicles** demonstrated their practicality than arms manufacturers converted them into wheeled **armoured fighting vehicles** (AFV). In 1900 the **Vickers Company** invited Frederick Simms to build a 'war car' weighing 5·5 tonnes, powered by a 16hp Daimler internal-combustion engine, protected by 6mm armoured plate and armed with two Vickers Maxim **machine-guns** and a 1-pounder pom-pom gun. But although the Italians used two armoured cars in Libya during the Italo-Turkish War of 1912, armies showed little interest until the outbreak of **World War I** when, all of a sudden, a way to stop German **cavalry** from roaming at large in Belgium was urgently needed. Hastily the Belgians, British and French improvised armed motor cars shielded with **iron** and **steel** plates. In no time each dominated and came to be rated the equal of an **infantry** company.

Many varieties of armoured cars based on lorry and car designs followed. But wherever trench warfare set in, their vital role as provider of mobile, armoured fire support was forfeited, to be reassumed in 1916 by the fully tracked **tank**. Yet armoured cars flourished whenever it was possible for wheeled vehicles to move freely, mostly in the less trammelled spaces of the Eastern front or across desert. A great many were used by the Russians, and by the British in the Middle East. Strangely enough, however, the Germans built very few.

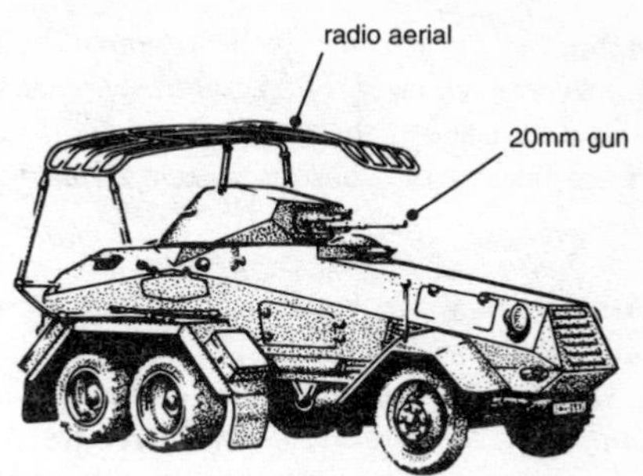

Armoured car, 1935

Between the World Wars armoured cars, because of their speed (up to 60mph), reliability and cost-effectiveness, were extensively developed for internal-security duties as well as for reconnaissance in general war. Armed with anything up to a 45mm gun, with their cross-country performance enhanced by six- or eight-wheel-driven running gear or, by French example, half-tracks, they helped keep the peace and played prominent roles in most campaigns.

During **World War II** most cars weighed about six tonnes, but as time went by less conspicuous and very quietly running, four-tonne armoured scout cars were found invaluable, at the same time as demands were made for 11-tonne, close-support cars with 75mm guns and 25mm armour to help fight for information. A trend projected into the 1950s when the British, for example, built their 11-tonne, six-wheel Saladin with 76mm gun, 32mm armour and a road speed of 45mph; and in the 1970s the Germans built Luchs, a monstrous 20-tonne, eight-wheel amphibious vehicle with a 20mm gun.

Although a tendency developed in the 1960s towards tracked vehicles for reconnaissance, most nations continued to use armoured cars, especially for internal-security duties when it came to be realized that tracked vehicles were often regarded as politically provocative, the antithesis of desirable minimum force; whereas wheeled vehicles were not.

Armoured fighting vehicles (AFV). **Chariots** were the first fighting vehicles but mostly their crews seemed to have depended on body **armour**. On the other hand John **Zizka**'s war **wagons** of the 15th century with their **artillery** were genuine AFVs, with the disadvantage, like chariots, of being powered only by unprotected animals. Not until the steam tractor evolved in the mid 19th century was a self-propelled, cross-country AFV feasible – as James Cowen proposed in 1855 with the concept of a Boydell tractor, with footed wheels, armed with cannon and encased in **iron** plate.

It was the first practical internal-combustion engine in 1883 which made possible a functional AFV, but it was not until 1900 that one appeared in the form of the Simms **armoured car**. And it was 1912 before an (Italian) armoured car went into action, in Libya during the Italo-Turkish War. Yet the success of Belgian, British and French armoured cars in the opening weeks of **World War I**, and subsequently on most fronts with reasonably good going, indisputably demonstrated the motor-propelled AFV's combat potential. This was a vital step towards the invention in 1916 of a petrol-engine-driven, tracked, armed and armoured box, called **tank**, tasked to cross disrupted, fire-swept ground in order to crush **barbed wire** entanglements and overrun entrenched machine-guns and **artillery**.

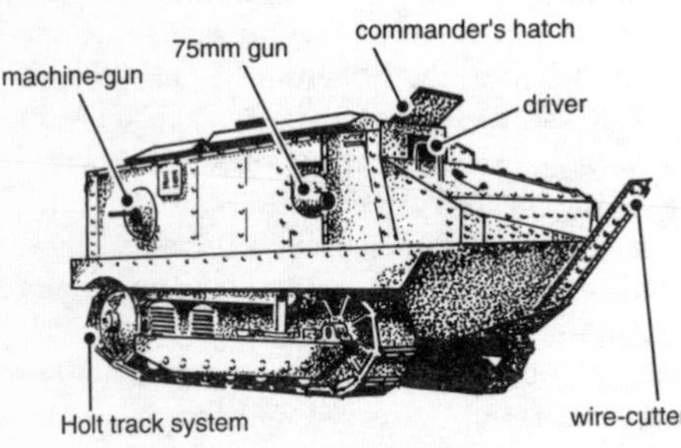

Schneider M16 tank (French)

Influenced by numerous people, two different kinds of AFV running on endless tracks were developed, respectively, by the Hornsby and Holt companies. In Britain a team led by Colonel Ernest Swinton and with the inventive genius of Walter **Wilson** designed an entirely original **steel**-plate track link for a 28-tonne rhomboidal-shaped machine called Mother, crewed by eight men, armed with two 57mm cannon and/or machine-guns, which first saw action on 15 September 1916. The French, inspired by an artillerist, Colonel Jean-Baptiste Estienne, favoured armoured boxes on Holt tracks armed with a single forward-pointing 75mm cannon – in effect what came to be known as an assault gun – which first saw action on 16 April 1917. Neither vehicle moved faster than 4mph and both were mechanically very unreliable and vulnerable to artillery and armour-piercing bullets (see **anti-tank weapons**). Yet their débuts were sufficiently successful to encourage further development, including thicker armour, more powerful engines and improved cross-country performance. Each, however, would be complemented and, in due course, superseded by smaller and faster tanks of which the two-man French 6·5-tonne Renault FT, with its single machine-gun in a rotating turret, was a trend setter.

At the battles of **Cambrai** and **Amiens** the future of armoured forces and warfare was settled. After 1918, however, prolonged debate, experiments and trials (especially during the **Spanish Civil War**), in a climate of financial stringency, conditioned vehicle design. Two schools of thought predominated. One preferred heavily armoured, slow-moving tanks for tactical support of **infantry**; the other envisaged, as did adherents of Colonel J. F. C. **Fuller**, fast-moving, far-ranging armoured forces dedicated to strategically decisive operations. As a general rule, however, there was agreement on the need for armoured cars and light (6-tonne) tanks for internal security operations and

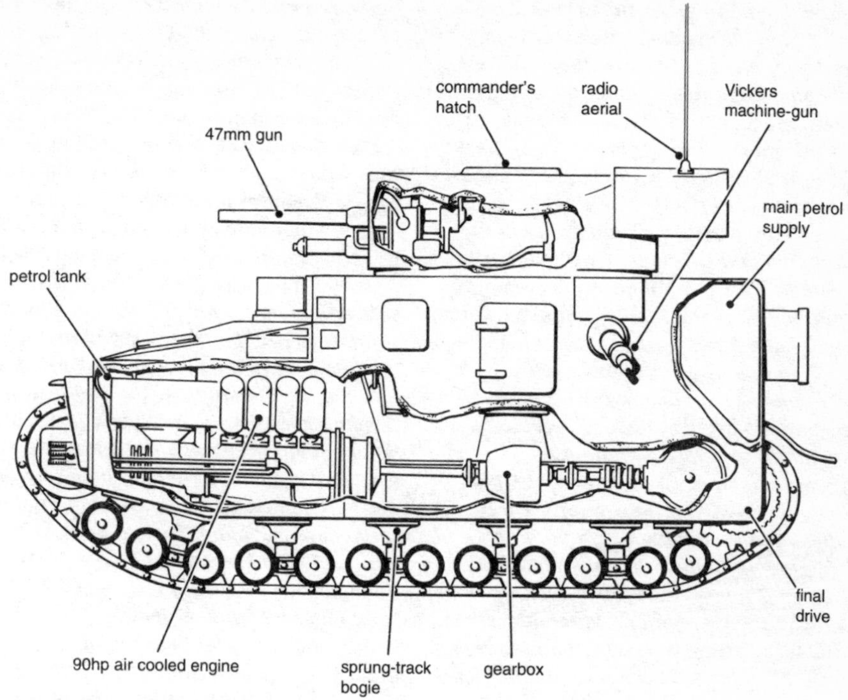

Vickers 13-tonne so-called Medium tank (c. 1923) – the model for future turreted AFVs

reconnaissance; and medium (20-tonne) and heavy (up to 70-tonne) main battle tanks (MBT). The light, because of cheapness, were favourites, and the heavy, except in Russia where every sort were built in large numbers, outsiders because of expense.

At the start of **World War II** Germany and Britain had numerous light tanks, a number of useful mediums and, in the case of the British, a single good heavy type, the 26·5-tonne Matilda II with a 40mm gun and 78mm armour. The French, on the other hand, had concentrated mainly on infantry support machines of which the 32-tonne Char B with a 75mm gun and 60mm armour was formidable. Few were faster than 20mph, most had crews of between three and five men and all tended to unreliability. The Americans, meanwhile, possessed only a handful of useless light tanks. The Russians, on the other hand, had copied in large numbers the **Vickers Company**'s 6-tonne tank, called T26; plus J. Walter **Christie**'s fast-medium, the BT tanks; as well as several heavy tanks, mostly based on Vickers' ideas. By 1941, however, the Russians had developed the BTs into the fine, three-man, 26-tonne T34 with a 76mm gun, capable of firing both high explosive and solid shot, and with 45mm sloped armour. And they had also produced the formidable 43·5-tonne KV1 tank with 90mm armour which was proof even against the German 88mm dual-purpose gun.

World War II has been described as a 'tank war', and certainly this is born out by AFV dominance of the initial campaigns in Poland (1939), Western Europe and **North Africa** (1940), and the Balkans and Russia (1941 and 1942), when German armoured forces, with air power, enjoyed runaway victories against outclassed opponents. Yet in

1942 the tide was on the turn in company with the traditional contest between gun and armour, led by the German reaction when outmatched by the Russian T34s and KV1s.

One German response was to build easily manufactured assault guns – tank chassis fitted with the most powerful guns available, but forward-pointing with limited traverse. Another was to up-gun and up-armour their existing PzKw III and IV MBT and also build the 56-tonne Tiger I, with an 88mm gun, and the 43-tonne Panther with its long 75mm gun. Against this the Russians gave the T34 an 85mm gun and produced the 46-tonne JS2 tanks with a 122mm gun. These developments left the British and Americans far behind until the end of the war, when the former produced their reliable, 32·5-tonne Comet, with a 77mm gun; and the Americans the 41-tonne Pershing with a 90mm gun.

After 1945, shadowing the **Cold War**, the **Korean War** and the **Arab–Israeli wars**, among others, the tank-design contest continued apace. Humiliated by the inadequacy of their own tanks and determined never again to be outgunned, the British built the good 49-tonne Centurion, armed with an 84mm gun and, in the 1950s, up-gunned it to 105mm. In 1954 they introduced the 65-tonne Conqueror with a 120mm gun and, in 1966, the 54-tonne Chieftain, also with a 120mm gun. They were followed in the 1980s by the 60-tonne Challenger with its 120mm gun and special Chobham armour.

The Americans, meanwhile, developed the M47 (Patton), derived from Pershing, and M48, both with a 90mm gun, the latter frequently modified until it became known as M60. At the same time their heavy tank projects fell by the wayside along with some impractical weapon systems, compelling them to adopt the British 105mm gun for successive versions of M60 and introduce the Shillelagh gun/missile system for the A2 model (*see* **Rockets and guided missiles**). Not until the late 1970s did they at last catch up, with the good 55-tonne Abrams with Chobham armour, but only with a 105mm gun until fitting a German smooth-bore 120mm in the 1980s.

The Russians, though lagging behind technically, continued to up-armour and up-gun. JS3 appeared with ingeniously sloped armour and 122mm gun in 1945 and T10 was an improvement on that. Mainly they concentrated on developing medium tanks, still with Christie suspensions but with greatly improved protection, a 100mm gun in T54 and T55 and a 115mm in T62, followed by the radically redesigned T64, T72 and T80 with thicker armour, lower silhouette, 125mm gun and automatic loader.

Germany, on the other hand, drew rather dubiously on experience against the Russian T34s. In the 1950s they opted for their 39-tonne Leopard I with speed of 45mph, maximum 60mm armour and the British 105mm gun. But, in the 1970s, they changed policy and gave the rebuilt Leopard II much thicker special armour and a smooth-bore 120mm gun, a philosophy copied to some extent by the French with their 34-tonne AMX30 of the 1950s and in subsequent projects.

Concomitant with vehicle development arrived far more sophisticated accessories. **Rangefinders**, **surveillance** and **night-vision devices**, **computer**-assisted fire-control systems and automatic fire-prevention equipment were widely adopted in the interests of striking power and survivability. Propulsion units increasingly went over to diesel as power output rose to 1,500hp in the 1970s and the Americans fitted a fuel-thirsty gas-turbine engine to the Abrams.

A demand has also arisen for cheaper AFVs in lieu of costly MBT, with the result that several kinds of limited-traverse assault gun have appeared as well as lightly armoured cars with high-power guns in traversing turrets.

Indeed, there thrives a demand for fast, light reconnaissance AFVs. Britain continued to develop the armoured car, though realizing that its cross-country performance would always be inferior to that of a tracked vehicle. So she also developed light tanks such as the 8-tonne, 60mph Scorpion series. These vehicles could be fitted with a 76mm or 30mm gun or guided missiles, and had a nominal ground pressure less than that of

Abrams tank (USA)

a man's foot and in boggy going in the **Falkland** Islands went many places where infantry could not go. Often these light vehicles have **aluminium** armour and an amphibious capability. They are only intended to search, not fight, for information and require excellent surveillance and communications equipment. The French six-wheeled AMX10RC with skid steering and 105mm gun is an example of a cross between a reconnaissance and combat AFV.

Another important category of AFV is the armoured personnel carrier, invented by Britain in 1917 and further developed by most nations during and after World War II. At first little more than an infantry armoured taxi with a machine-gun for local protection, it has gradually grown into a combat vehicle from which the infantry can fire their personal weapons. Since the 1950s models with turrets, a variety of guns and guided missiles and an inherent swimming capability have appeared. Typical of what are often called infantry fighting vehicles (IFV) are the tracked, 13·5-tonne Russian BMP1 with Sagger guided missile and 73mm gun and carrying eleven men; and the tracked, 21·3-tonne American Bradley with TOW missiles, 25mm chain gun and nine-man crew.

MBT have played important roles in recent wars, notably in the Middle East and the **UN–Iraqi War**. But they are very expensive and impose a considerable **logistic** load. The political tendency in peacetime is likely to favour purchase of cheaper, lighter and less combatworthy AFVs – regardless of the risks. Yet whatever happens in the future, a role for AFVs must continue as heavy-weapon carriers and if only to enable men to move across the battlefield.

Armstrong, Sir William (later **Baron**) (1810–1900). An English lawyer and engineer who experimented with a hydroelectric machine to generate electricity, gave up the law to build high-pressure hydraulic machinery in 1847 and then, prompted by the **Crimean War**, turned to the development and construction of **artillery** with rifled barrels greatly strengthened by shrinking metal bands round an inner steel tube. By 1878 his firm was a world leader in building 100-tonne breech-loading guns of 450mm calibre, controlled by hydraulic recoil mechanisms, for the latest **battleships**.

Army organizations. An army's organization is shaped by the need to control, train, supply and deploy in peace and war large numbers of properly motivated soldiers with their equipment. A judicious proportion of skilled and determined officers to men at the various levels of command has always been crucial to optimum achievement of those functions and in the building of **morale**. Likewise, armies require an optimum mix of the combat arms at their disposal – of **cavalry**, **infantry**, **artillery**, **engineers**, **fighting vehicles**, **communications** – and of **logistic** services.

To begin with military forces were based on strongholds commanded by warlords. Later, as in Assyria, they controlled the state, usually under a monarch or dictator. Sometimes, as in Greece, they came under committees. But whether or not remaining defensively within **fortifications** or breaking outwards, they relied above all upon an adequate logistic system in order to sustain operations; and upon clear signal **communications** to make best **strategic** and **tactical** use of their resources. Very often a regular élite, such as the Persian and Greek Companions, was formed to guard the leader and stiffen and train the less motivated masses – a system which survives as Guards, Household troops or SS (*Schutzstaffel*).

The army of Philip of Macedon, later used so effectively by **Alexander the Great**, was revolutionary because, unlike its predecessors, it attempted to combine the operations of its combat arms. See how, at **Arbela**, Darius III employed his **chariots**, cavalry and infantry separately, to be defeated by the cooperation of Greek cavalry and infantry phalanxes all operating under Alexander's direct command. And note how the armoured-infantry phalanx, standing or moving eight deep in close formation, used javelins and long pikes before closing with **swords** for the mêlée. These organizations and methods were copied and refined later by the **Romans** as they were compelled to use new methods and technology to cope with warfare in many different circumstances.

The dissolution of the Roman Empire, which partly came about from the exhaustion and collapse of its rigid military organization, presented highly mobile and flexibly organized forces with greater freedom to raid and invade. **Attila**'s Huns, the **Vikings**, the **Muslim** Arabs and the **Mongols** provided prime examples of military organizations inspired by the expansionist urge, greed or evangelical enthusiasm, motivated in much the same way as the feudal **Crusaders** of **chivalry** were when they reacted to the Islamic threat. But the feudal system, with its reliance on armoured cavalry, was little more than an élite under self-interested warlords, and was so dependent on the technology of the **edged weapon** and armour that it fell into decline during the **Hundred Years War** once the longbow, artillery and firearms supervened.

The **wars of the 16th century** mark the commencement of what may be called the modern era, when army organizations resembling those of today evolved. For although, initially, the limitations of firearms dictated that troops should fight in close formation to develop sufficient firepower and shock action, a gradual thinning out and expansion of frontages occurred as weapons improved, static warfare loomed ahead and the size and complexity of armies multiplied. This process was accelerated by the invention of rapid-fire artillery and **machine-guns**, which induced widespread dispersion (*see* **Empty-battlefield theory**) for survival, and the demise of horsed cavalry, simultaneously with the adoption of self-propelled **armoured fighting vehicles** and **aircraft** to regenerate mobility, surprise and offensive action.

In the 20th century the lowest sub-unit of an arm might be a troop or platoon, of which two or more might be under the command of a squadron or company (if, respectively, cavalry or infantry) or a battery if artillery. Squadrons, companies and batteries came under the command for operational, training and administrative convenience of either a regiment (cavalry, artillery and engineers) or a battalion (infantry) or their own arm, and termed a 'unit'. Two units or more, grouped, became 'formations', called, in ascending order, a brigade (sometimes known as a regiment and usually single-arm composition); a division (consisting usually of all arms); a corps; and an army (all, except the last two, being of established composition of all arms). Elements of logistic support would be found at all levels, including the individual soldier with his ammunition and emergency ration. Command, control and **communications** (in modern parlance, C^3) of units and formations would be exercised by the Commander, through his **staff**. Armies would be instructed and administered by War Departments of State which, with the introduction of the **telegraph** and **telephone** began to exert direct control.

By 1918 the basic organizations had survived, but many specialized arms and services had been created to handle new technology. **Railways** needed operators and engineers; the latest signals **communications** demanded corps as did **aircraft**, **anti-aircraft** artillery, **tanks** and **machine-guns**. The latter mainly were founded by infantry, the rest by engineers, artillery or newly recruited civilian technicians. **Logistics** also underwent a revolution.

Although at the beginning of **World War I** each arm tended to fight independently of the other, this practice soon gave way to the necessity for all arms cooperation to be controlled by doctrine worked out by the General Staffs, the military colleges and arms' schools (*see* **Education, Military** and **Training**). In 1917 the Germans took a lead with the institution of ad hoc battle groups assembled as required from units or sub-units of all arms to cope with local situations. Cavalry, infantry and artillery had to cooperate among themselves and with tanks and aircraft – and so on.

Between the two World Wars the new organizations were consolidated and rationalized at the same time as army air corps began to convert to independent air forces and, again by German example, joint-service High Command systems were created to make combined use of available resources. Command and control systems proliferated as more special formations, such as armoured, anti-aircraft and **airborne** forces appeared. Voracious demands were made for complex signals facilities, extensive **intelligence** backing and enormous quantities of machinery, to make workable close collaboration within the army and with the other services. The requirements were easy to formulate but often very difficult to arrange without friction when change was rapid and tensions high.

Since 1945, under the pressure of the **nuclear** threat, vast progress with technology and numerous wars, the trend towards established, mixed battle and brigade groups of all arms has been consolidated, despite numerous lapses when, at their peril, arms and services have gone their own way. Joint-service defence departments and headquarters have eliminated some army establishments. But so far only the Canadian Army has found itself wearing almost the same uniform as the Navy and Air Force – an experience from which it was saved after ten long years by the strength of an irrepressible **regimental system** it never abandoned.

Arnhem, Battle of. As the pursuit from **Normandy** gathered pace in August 1944, the use of First Allied Airborne Army to maintain momentum was frequently foiled by circumstances. Early in September Field Marshal **Montgomery** obtained General Eisenhower's consent to drop the entire Army ahead of Second British Army to form a corridor for a strong advance into Holland and northern Germany with a view to cutting off the Ruhr industrial complex and ending the war at a stroke. Planning for Operation Market Garden was rushed: there were too few **aircraft** to lift the whole force at once: **intelligence** about the German defences and, in particular, those at the vital bridges crossing the Rivers Maas at Grave, the Waal at Nijmegen, and the Neder Rijn at Arnhem was indifferent; and **logistic** support tenuous.

On 17 September the bridge at Grave was seized intact at once but failure to follow up initial successes and swift German reaction thwarted the design. The bridge at Nijmegen was not secured until late on the 20th (and could have been blown if the Germans had not been so intent upon preserving it for their own use). The bridge at Arnhem remained in German hands while they wore down the 1st British Airborne Division whose task was to capture it. By then adverse weather had delayed the fly-in of reinforcements and supplies to Arnhem and the Germans had won time to block the Anglo-American forces trying to get through from Nijmegen, and also to launch damaging strokes against the narrow corridor from the south. Inadequate **radio** communications with the British at Arnhem sealed their fate. Those few who survived were evacuated on the night of 25/26 September, leaving behind more than 7,000 of their number. Hope of ending the war had then to be postponed until 1945.

Arras, Battles of. Sited at a crossing of the River Scarpe, the city of Arras fell to Louis XIII of France in 1640 during the **Thirty Years War**; and in 1654, during the continuing war between France and Spain, was the scene of a victory by the French, under Marshal Henri Turenne, over the Spanish under General Prince Louis II of Bourbon and Condé. Both actions were typical of 17th-century wars (*see* **Wars of the 16th, 17th and 18th centuries**) and preceded the construction of strong **fortifications** by Marshal Sébastien **Vauban** later that century.

During the War of the Spanish Succession, Arras was a key position in the French *ne plus ultra* lines of defence but was skilfully bypassed by the Duke of **Marlborough** in 1710. It was also spared enemy occupation during the **Franco-Prussian War** of 1870. In 1914, prior to the Battle of the **Marne**, it was occupied only briefly, by **cavalry**, until evacuated on 9 September. In October, however, the Germans attempted to encircle it with a strong thrust from **Vimy Ridge** which nearly succeeded. For nearly four years the city would be pounded by **artillery** while beleaguered.

Twice in 1915 the French tried, with enormous losses, to drive the Germans back, but made only marginal gains without relieving the city from direct observation by the enemy. In April 1917, however, a large-scale offensive by the British and Canadians, respectively, threw the Germans back six miles to the east and captured Vimy Ridge to the north. This operation, one of the most successful founded upon artillery tactics, was also assisted by use of the extensive ancient tunnels under the city which were extended to the front and also to the crest of the Ridge. Once more, during the **Hindenburg** offensive in 1918, the Germans tried to seize the city but were repulsed within hours. But when the British and Canadians began probing eastwards on 19 August, it was to find an enemy willing to give ground prior to the withdrawal of his entire front in the aftermath of defeat at **Amiens** to the south.

In 1939 GHQ of the British Expeditionary Force (BEF) was located, along with base organizations, in and around Arras. When in May 1940 the German thrust to the Channel threatened the city, in desperation an improvised garrison was assembled in time to repulse General Rommel's 7th Panzer Division on the 20th. At the same time the British 50th Division, with two tank battalions, was ordered, together with French armoured forces, to attack southwards next day against the German flank. Time was short and the attack was launched in some confusion, but it made progress when it cut into Rommel's troops as they advanced round Arras to the west. The Germans suffered heavy casualties and abandoned their attack in order to counter-attack. The British attack was stopped but their garrison held out until withdrawn under pressure on the night of the 23rd/24th. In this battle the Germans suffered badly because their 37mm anti-tank guns could not penetrate the thick British and French armour; it was their dual-purpose 88mm guns which decided the tank battle. Of profounder importance were the psychological shock waves from the rebuff. These contributed to the worries of General von Rundstedt and **Hitler** in persuading them to call the halt which made evacuation from **Dunkirk** possible.

Artillery is an omnibus word, often employed to include almost any projectile-throwing system other than personal arms. This entry will concentrate mainly on cannon or guns, but reference should also be made to **anti-aircraft** and **anti-tank weapons**, **bows and arrows**, **grenades**, **mortars**, **rockets and**

Artillery piece, 14th century

guided missiles and **siege engines**, as well as in conjunction with **ammunition**, **explosives** and **gunnery**.

Close combat, though sometimes extolled as the acme of the martial spirit, has regularly been prudently avoided in favour of hitting the enemy from beyond the range of his weapons. Thus, artillery development has been governed by the search for longer ranges of engagement besides increases in rate of fire and the lethality of projectiles.

Tube-launched projectiles awaited the invention of gunpowder in the 12th century, although, controversially, it seems to have been *c.* 1325 before a practical cannon entered service, maybe in 1340 at the sea Battle of **Sluys** and certainly on land in 1345 at **Crécy**. The early pieces seem to have been small-calibre muzzle-loaders, made of **bronze**, brass or **iron** and firing metal arrows or **stone** balls. Their range of 200 metres was about that of the longbow, but their accuracy and rate of fire much less owing to the difficulty of estimating the charge required, the poor ballistic properties of handmade projectiles, and low velocity induced by loss of impetus from windage – the escape of expanding gases through the essential gap between projectile and barrel walls. For if an oversize projectile was not rammed hard against the charge, an uncontrolled explosion was liable to burst the barrel and injure nearby gunners – which only the enemy wanted.

Requirements for the engagement of men in the open, as well as smashing ships' timbers and battering stone walls, led to the construction of rapid-fire artillery – such as multi-barrel systems of rows of small-calibre guns firing in the lower register (less than 45 degrees above horizontal), in contrast to *Dulle Griete*, a 13-tonne bombard from Ghent of 25in calibre which lobbed a 700lb stone in the upper register. Although breech-loading was available at a very early stage, muzzle-loading was the norm. By trial and error, master gunners learnt how to mix, improve, handle and measure gunpowder while the smiths experimented with **metallurgy** and designed to order larger and stronger pieces. It was their welding together of wrought-iron bars in a circle 'like the staves of a barrel' which formed the tube that became known as the gun barrel, a defective method which gave way to stronger, cast-iron barrels when the technique was mastered in the 15th century.

Until the 17th century navies and armies tended to prefer bows and small arms to cannon – except in **siege warfare** in the 15th century when big guns with ranges out to 600 metres already were superior weapons. On warships the encumbrance and inefficiency of artillery militated against its popularity because it hampered the ship's working and threatened fires and explosions. On land the cannon's immobility and inflexibility were handicaps until the appearance in the mid 15th century of *field artillery*: **horse**-drawn, wheeled carriages (in place of oxen-drawn sledges). The provision of trunnions (lugs attached to the gun barrel and mounted in simple bearings) enabled quicker and more precise elevating and depressing; the shock of firing still had to be absorbed through the mounting, but the more mobile carriage enabled the weapon to be run back more easily for the next shot.

Normally individual gun-founders in different nations followed their own whims and produced an infinite variety of weapons. Standardization was unheard of – with consequential inertia in the evolution of **gunnery** and **tactical** doctrine. The demand by the French in the 16th century for the new field artillery with a range of 500 metres not only became universal but evolved in parallel with official disciplines on the burgeoning ordnance industry in deciding terminology,

Field artillery piece: 18th-century muzzle-loader

nomenclature (usually by weight of shot), the science of ballistics, techniques and **tactics**, leading to the establishment of schools of instruction for artificers and gunners.

Likewise, navies had to take heed of the destructive power of swivel guns mounted on the upper deck and wheeled carriages on the gun deck below to fire broadsides through ports cut in the ship's sides. The revolutionary improvement was demonstrated by Vasco da Gama with his Portuguese caravels against Arab dhows in 1502 off the Malabar coast; and as was confirmed beyond doubt in 1588 when English ships fired stand-off, massed broadsides against the Spanish Armada and outmatched its **galleons** without resort to traditional boarding. Until the mid 19th century warship design would be dictated by the increasing power of the muzzle-loader, firing broadsides of iron shot or shell weighing as much as 42lb at ranges increasing to 2,000 metres.

The **Thirty Years War** and the **Spanish** and **French wars** of the 17th and 18th centuries concentrated the minds of gunners on specialization as well as improved performance. Although long-barrelled guns firing in the lower register became increasingly popular, the need for the **mortar**, firing in the upper register to attack targets located behind cover, was always recognized – even though observation and correction of fall of shot was often difficult because of **surveillance** and **communication** limitations. Yet efforts to reduce the weight of guns (as with the Swedish mobile, two-man-crewed, copper-tubed, leather-covered gun invented by Colonel Wurmbrandt, weighing only 90lb) were not matched by elimination of basic design faults, such as the fundamental frailty of trunnions mounted below the bore, instead of on a level with its axis.

The sciences of ballistics and **gunnery** were significantly advanced in the 1740s by the mathematician Benjamin Robin's explanations (using French data) of how his *ballistic pendulum* could calculate the exact amount of gunpowder needed to produce, for ranging purposes, the exact vital velocity; also, how a spinning projectile was ballistically superior to an unspun one; and the principles governing penetration of wood or iron by shot. As an immediate product of Robin's work there appeared the Scottish Carron Company's carronade, a short-barrelled gun firing a very heavy ball. First used in action by the Royal Navy in 1779, its advantages as a ship 'smasher' were sometimes forfeited because of its short range – since ships with carronades alone suffered badly when outranged by conventional guns.

After the Battle of the Saintes in 1782, the Royal Navy's artillery assumed a considerable superiority over the French. This was partly due to intense **training** but also to the adoption of the flintlock ignition system, already in use with hand guns, and using block and tackle quickly to run out guns after firing and loading, instead of by slow handspike leverage. For the first time, too, William **Congreve**'s **rockets** were available for service with ranges out to 2,000 metres, but also with dubious accuracy.

Meanwhile, on land the French Army, driven by Jean de **Gribeauval**'s reforms, reduced to three the classes by calibre of its guns and insisted upon the production of truly shaped balls for better performance. De Gribeauval also improved transport and, 140 years after **Gustavus Adolphus**'s rejection of loose powder, copied his system of issuing it in precisely weighed bags tied to the projectile. Then in 1784 Henry **Shrapnel**'s air-bursting, high-explosive ammunition appeared on the scene as a deadly anti-personnel projectile.

Yet, even by the mid 19th century, with its numerous technical and organizational innovations and the clear distinctions drawn between field, siege and coastal artillery (*see* **Coastal defence**), tactics still remained much the same as during the previous 200 years. Prevented by technical deficiencies from indirectly engaging targets out of sight and also from generating rapid fire, guns still had to be sited on crest lines or forward slopes, secured by infantry or obstacles against enemy **cavalry** and **infantry**.

Technical progress accelerated as a result of Daniel Treadwell's attempts in the 1840s to build up iron or **steel** barrels from hoops shrunk round a central cylinder; and the experiments of Captain Thomas Rodman, US Army, to cast hollow barrels round a water-cooled core. Partly inspired by the latest breech-loading **rifles** and improved ammunition, and also influenced by artillerists like Rodman, industrialists like Alfred Krupp and Henry **Bessemer** risked investment on new projects; the German in drilling out an expensive, single-piece, cast homogeneous **steel** gun in 1851; the Englishman seeking a better material than cast-iron for spun shells for the **Crimean War**, discovering how to make **steel** by decarburizing molten pig-iron with air blasting to produce a metal which was both malleable and cheaply produced.

Chiefly, however, it was the onset of re-armament, spurred by the rising nationalistic struggles of the 1830s and 1840s, which encouraged entrepreneurs and technologists. Then the Crimean War inspired William **Armstrong** and Joseph **Whitworth**, respectively, to begin the manufacture, in Britain, of rifled barrels with spiral grooves or a hexagonal bore. In 1858 an 18-pounder Armstrong gun shooting at 1,000 metres was 7 times more accurate than a normal smooth-bore muzzle-loader, and 57 times better at 3,000. Moreover its screw breech gave such excellent sealing and high rate of fire that this type of gun made all existing pieces obsolete – as shown whenever employed in the **American Civil War**.

A further twist to the artillery revolution resulted from the introduction of **armoured** warships and the ***Merrimack* versus *Monitor*** encounter in 1862 when the former's pair of smooth-bore, 11in breech-loaders failed to penetrate the latter's hull armour; and the latter's rifled guns did no better against the former's armoured, rotating turret. But with the turret's feasibility proven, the day of sailing wooden walls was over. Within three decades there were steam-powered, armoured **battleships** and **cruisers**, mounting in their turrets much-larger-calibre guns with stronger, wire-bound barrels (in which the windings enclose the inner, rifled tube) to withstand higher firing pressures. Inevitably the naval gunnery art was changed beyond recognition.

The same trend was only slightly less evident with armies. The new breech-loading field artillery was much lighter and handier, and more quick firing, than its predecessors. Yet heavier pieces from within **fortifications** dominated through their greater range and hitting power, and the mobility of heavy siege guns was considerably enhanced by **railways**. These trends were clearly revealed during the American Civil War and the **Franco-Prussian War** and became the subject of trials during the 1880s in efforts to discover if forts could be made shell-proof – with conflicting results followed by decisions to build both stronger forts as well as guns and howitzers weighing anything up to 75 tonnes and firing a 1-tonne shell 16,000 metres.

Greater attention was focused on field guns, especially after the revelation of the **Schneider** Company's 'secret' screw-breech-loading 75mm gun of 1897. Weighing 2,700lb, using fixed ammunition with smokeless powder, and firing a 16lb shell 7,500 metres, its vital component was the barrel's semi-automatic return to battery after recoil; an action facilitated by the hydraulic recoil system to give a rate of fire of six aimed rounds per minute. Other nations were quick to learn and develop their own versions of the 75. The basic idea was soon transferred to larger guns, and adapted not only for warships but also as **anti-aircraft weapons** over land and sea as soon as **aircraft** became a threat.

Yet the artillery arm entered **World War I**

as a subsidiary to infantry, in the doctrinal belief that men imbued with the offensive spirit would overcome material. But come the end of 1915, when this false notion stood exposed and a shell shortage affecting every contestant had been overcome, artillery became the dominant partner in trench stalemate akin to **siege warfare**.

From 1916 onwards the use of massed artillery to destroy the enemy's guns, **barbed wire** obstacles and entrenchments became a false article of faith since the task of breaching enemy lines *and* maintaining the momentum of the advance was beyond artillery alone and forfeited surprise. But the introduction of the **tank** to crush wire and, with its gun under armour, tasked to fire direct at enemy targets, not only reverted the artillery role to that of a complementary vital weapon system, but also compelled it to work as an **anti-tank weapon**. Nevertheless, under the influence of artillerists such as the British General Noel Birch and the German Colonel Georg Brüchmuller, scientific methods of arranging *indirect* fire by massed artillery *without* the warning given by previous registration by observation of shots, restored the possibility of achieving surprise. At the same time **smoke** and poison-gas rounds also extended the versatility of the guns' effect.

At sea, paradoxically, long guns of between 11 and 16in calibre, when laid and fired by the latest gunnery methods, proved a disappointment – partly because of natural effects and partly through technical problems such as ammunition defects and barrel bend. The naval lessons of the **Russo-Japanese War** in 1904 and 1905 had seemed to indicate the supremacy of the big gun against armour at ranges of about 7,000 metres (even though the percentage of hits to rounds fired was remarkably low). In World War I, with ranges in the region of 20,000 metres during successive naval engagements, battleships withstood the relatively few hits they received even at the shorter ranges. And at the Battle of **Jutland**, all three British **battle-cruisers** lost blew up as a result of magazine explosions. Defects with British armour-piercing ammunition and their own safer magazines were two reasons why the Germans escaped destruction. But it was **torpedoes** and **mines**, not big guns, which sank most warships in that war – although many sailors chose to ignore it and pinned their hopes on big guns and unsinkable battleships in **World War II**.

Armies also tended to increase the calibre and range of field artillery, but put less store by heavy pieces, including coastal guns. In 1939 the calibre of light field guns was between 75 and 105mm firing 35lb shells to ranges of between 11,000 and 18,000 metres. Loading and rates of fire were also greatly improved by the introduction of the sliding, semi-automatic breech. But when positional gave way to mobile warfare, the need for artillery in self-propelled armoured vehicles, engaging targets direct as well as indirect, became important.

Post-1945, with the overshadowing of the big-gun ship at sea by **bomber aircraft** and the advent of **nuclear** projectiles, armies have clung to towed artillery, even though mobile doctrines call for more of the expensive self-propelled kind. Indeed, the arrival of **rockets and guided missiles** has meant that rapid-fire naval guns now have mainly an air-defence role, with dependence on **radar**-assisted detection and laying. By contrast armies, which also demand more rapid fire assisted by semi- or fully automatic loading, have adopted a higher proportion of guns with calibres of 150mm or more in order more easily to destroy hard targets.

But, with the successful introduction of MLRS (Multiple Launched Rocket System) in the **UN–Iraq War** and of TGM (Terminally Guided Munitions), the big gun firing conventional munitions may be approaching eclipse.

Asdic. *See* **Sonar**.

Atlantic, Battle of the. With fewer than 40 operational U-boats and 15 major surface warships available in 1939, Germany could not mount decisive attacks in the Atlantic against the Anglo-French navies which controlled the Western approaches. But when

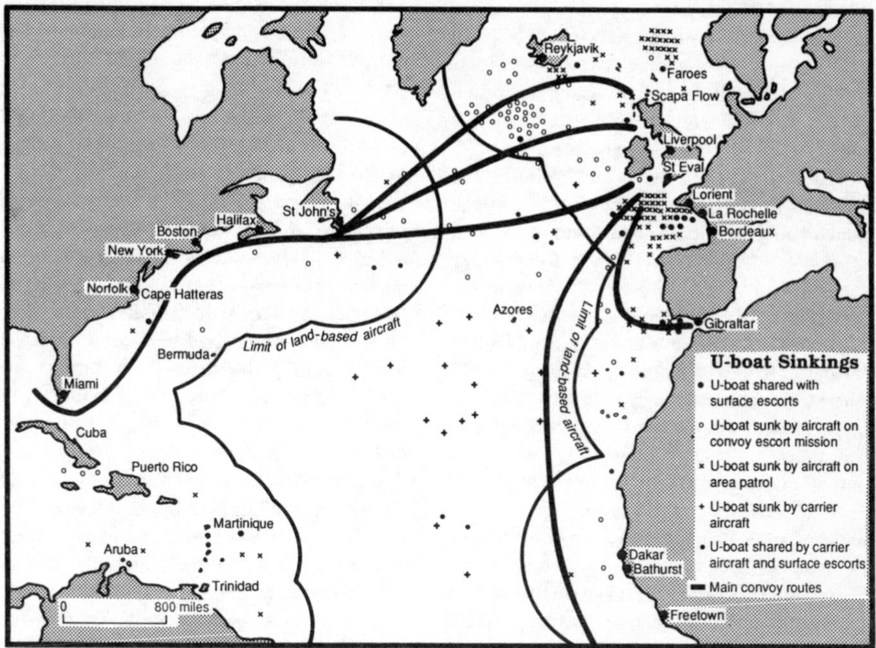

The Battle of the Atlantic, 1939–45

France collapsed and almost the entire Western European coastline fell into German possession in June 1940, the strategic balance changed. Henceforward British **convoys** had to run the gauntlet of enemy **aircraft**, **submarines** and surface raiders which enjoyed almost unlimited access to the Atlantic. The Germans could also lay **mines** and bomb ports with comparative ease, while the British found their **anti-submarine** resources stretched beyond the limit.

Moreover, aircraft which at first enjoyed some successes were soon checked, though continuing to provide valuable **reconnaissance**. Yet although the German battleships, cruisers and other raiders did get loose and sink several ships in 1940 and 1941, their depredations were cut short when the battleships *Scharnhorst* and *Gneisenau* were damaged and shut up in French ports, the *Bismarck* sunk in May 1941, and their supply ships rounded up. Admiral **Dönitz** believed that his system of U-boats hunting in 'wolf packs', which were centrally controlled by encoded **radio** signals, would provide the answer: and that the convoy escorts could be defeated by night attacks on the surface. Pack attacks began in September 1940 and in 1941 sank the great majority of 1,299 ships put down by various means for the loss of 51 submarines.

But in 1942, although the U-boats continued on top (despite losses of 128 with many experienced commanders), countermeasures were taking their toll. Direction-finding of radio signals was rendering them vulnerable to surface and air attacks as the increasing number of sea and air escorts received improved **radar** and more effective **depth charges**. Both sides benefited sporadically from reading their opponents' **codes** and each enjoyed successes due to the other's stupidity or won by their own guile. It was the obdurate unwillingness of Admiral Ernest **King**, US Chief of Naval Staff, to implement the convoy system in 1941 which gave the U-boats their happiest of 'happy times' against unescorted shipping off the American coast in 1942. And it was a well-timed change of compromised codes which enabled Allied convoys to conceal their movements in the approach to the invasion of French North Africa in November 1942 at a time when the U-boats still were on top.

The turning-point arrived between February and May 1943. Dönitz, with 240 U-boats, could keep 50 at sea at once supplied by so-called 'milch cow' U-boats. In February they sank 48 ships and lost 22 U-boats, in March it was 105 (72 of them in convoys) against 16 – a seemingly disastrous Allied set-back which suggested convoys might have to be abandoned, thus almost conceding defeat in the one battle which could lose the Allies the war. Suddenly in April the pendulum swung the other way; that month only 25 ships went down and 16 U-boats were missing, figures which worsened from Dönitz's point of view in May when 46 U-boats failed to return and he was forced to withdraw his packs from mid-Atlantic and tacitly admit a defeat. Worst of all, Dönitz was unaware of the reason: a combination of closer-knit Allied defences based upon the use of small escort **aircraft-carriers** to cover the so-called mid-Atlantic gap; the latest 10cm radar fitted to long-range Liberator **bombers** which carried **searchlights** to illuminate and attack surfaced U-boats; and improved **Sonar** and voice radio to assist the more skilled commanders of the hunting groups. Try as the Germans would to overcome their difficulties, their losses remained high while Allied shipping increased in tonnage as the massive US building programme got into its stride. Far too late, Dönitz attempted to hasten the development and production of the schnorkel, homing torpedoes and the faster U-boats which could outmanoeuvre escorts. Only slowly did he appreciate the 10cm radar menace and never the penetration of radio security which divulged so much that was vital to the enemy.

In the last few months of the war, amid defeat off the **Normandy** coast and after many Channel ports had been captured by the Allies, the new U-boats did prove their worth. But the convoys kept sailing and U-boats were still sunk in large numbers, bringing the total from all causes in all theatres of war to 785 out of 1,162 built.

Atlantic Wall. A German linear **fortifications** system covering the West coast from northern Norway to the Spanish frontier. Started in 1940, work was intensified as a result of the successful **commando** raid against Saint-Nazaire in March 1942. Based on thick **concrete** bunkers covering sea and land, minefields, anti-tank ditches and barricades, it was strongest around ports and in those sectors deemed the most likely targets for Allied raiding and invasion. Gradually the coasts between the ports were filled by lesser emplacements which were deepened in 1944 when an Allied invasion seemed imminent. Spurred on by Field Marshal Erwin Rommel, anti-**glider** poles were erected and reinforced **armoured forces** stationed inland for the counter-penetration role.

On 6 June (*see* **Normandy, Battles of**) Rommel's plans were incomplete because of shortage of materials, weapons and forces. Within a few hours wide and deep penetrations had been made. Nevertheless, as the Allies advanced they were embarrassed by the resistance of port garrisons which might have caused fatal **logistic** consequences if they had not provided the **Mulberry harbour**; and very costly to assault without a powerful **siege** train, comprising specialized **armoured fighting vehicles**, under General Sir Percy **Hobart**.

Atom bomb. It is one of the great ironies that those intimately involved with the discoveries leading to the manufacture of the atom bomb were concerned purely with the healing and peaceful possibilities of their knowledge: the discovery of radioactivity by Antoine Becquerel in the 1890s; of radium by the Curies; of X-rays by Wilhelm Röntgen; and work on alpha and beta rays by Ernest **Rutherford** in 1900. It was Lise Meitner, Otto Hahn and Fritz Strassman in Germany in 1938 who showed that **uranium** atoms could be split. From this developed the realization that a chain reaction would follow from the released neutrons, in turn splitting other atoms, thus releasing even more neutrons plus immense quantities of energy.

By 1939 war with Germany was imminent. Though the theory was known in Britain, scientific opinion was somewhat sceptical of the practicality of applying it to a bomb to release all this energy. However, two physicists in Birmingham – Otto Frisch and Rudolph Peierls – were able to support the theory with figures showing that a 'super bomb' was feasible and official interest was aroused. Similar work had been going on in America and in 1943 **Robert Oppenheimer** led a team of Europeans and Americans to produce a practical weapon. By the summer of 1945 a bomb weighing about 4,000kg, in which a few grams of uranium could, by shooting one subcritical mass into another, initiate fission, was ready. On 6 August this device (code-named 'Little Boy') was dropped on Hiroshima and three days later 'Fat Boy' followed it on to Nagasaki. Explosions resulted equivalent to igniting 20,000 tonnes of TNT, coupled with a literally blinding flash and a devastating shock wave, with radiation effects from the deadly gamma rays still being suffered by the survivors half a century later (as indeed Frisch and Peierls had predicted in their original paper). The world had entered the age of nuclear war. *See also* **Hydrogen bomb**.

Attila (d. 453). Like the nomadic Hun tribe of superbly trained mounted archers he came to lead between *c.* 435 and 453, details of Attila's background and life are somewhat obscure. Related to the Hsiung-nu tribe which irrupted out of Central Asia into China *c.* 135 BC, the so-called Black Huns moved westwards to attack eastern European tribes in AD 370 before turning southwards under Attila to invade the East Roman Empire. Attila's **logistic** system of living off the country while keeping on the move, and his **tactics** of charging under cover of a shower of well-aimed arrows and slipping away when progress was checked, won him many victories in the field. But, lacking **siege engines**, he was baffled by **fortifications**. Even so he crushed the Roman army at Gallipoli in 443 to impose a large annual tribute on the Empire. And in 447 he attacked and defeated the Romans on the Utus before invading Greece, until checked at Thermopylae. A skilled and bullying negotiator, he nevertheless managed to impose crushing terms, forcing the Romans to evacuate vast territories south of the Danube. In 451 he tackled the European tribes and their Roman allies in Gaul (France) and came close to complete success until foiled in an assault on Aurelianum (Orleans); and then was heavily defeated, for the only time, in the Battle of the Catalaunian Plains, location unknown. But he had ruined the West Roman Empire.

The following year he overran northern Italy until forced to withdraw owing to famine and an outbreak of disease. But in 453, the night after marrying a young girl, he died from causes unknown. His empire was divided between numerous sons, who destroyed it by fratricidal wars which led to the absorption of the Huns by the other European tribes and also gave the East Roman Empire a new lease of life.

AWACS. The acronym for an Airborne Warning and Control Station, an American requirement during the **Vietnam War** to provide **surveillance**, warning, fighter and missile control over extensive areas. AWACS aircraft flying at 15,000ft have been equipped with **radar** developed to give, for example, a horizon range of 300 miles, making it possible to identify enemy aircraft taking off and, with computer assistance, quickly arrange interception by fighters or missiles. AWACS also can direct strike aircraft to their targets, deep in hostile terrain, simultaneously bypassing detected defences – as was performed for the Indian Air Force by a Russian Moss during the Indo-Pakistan War in December 1981.

During the **UN–Iraq War** of 1991 AWACS flew patrols designed, successfully, to deceive the Iraqis, prior to launching the initial air attack which was largely controlled by them, despite attempted jamming of their electronic systems.

B

Balaklava, Battle of. An attempt by the Russian forces to capture the British base at Balaklava, and thus break the Allied siege of **Sevastopol** and, perhaps, decide the **Crimean War**, brought on this battle for control of the vital Vorontsov ridge, commanding the Balaklava–Sevastopol road, on 25 October 1854. Russian attempts with 3,000 cavalry to exploit their initial, almost unopposed advances to the Fedyukhin heights were checked by the 93rd Highlanders, whose position (to quote a witness) was secured at a range of about 800 yards by 'a ravine deep enough to swallow up the whole lot of them had they charged'. But the Russians were thrown back by a most determined charge by 900 men of the Heavy Cavalry Brigade. It is, however, the subsequent charge of the 670-strong Light Cavalry Brigade, along with the legend of the 'thin red line' (created by the journalist William Russell for the glory of its commander, his kinsman Sir Colin Campbell), for which the battle is renowned, a futile charge which might have lost even more than the 40 per cent of those engaged had it not been for a supporting charge against the Fedyukhin heights by the French Chasseurs d'Afrique. As it was, the Vorontsov Ridge remained in Russian hands.

This battle foreshadowed the demise of horsed **cavalry** at the hands of **artillery** and **infantry**, the latter armed with the latest rifled musket firing the Minié ball projectile (*see* **Rifle**).

Balloon. The feasibility of ascent by an impermeable 'envelope' filled by hot air was first demonstrated with a model by Father Bartolemeu de Gusmão in 1709. In France on 4 June 1783 a 33ft-diameter bag was filled with smoke from a straw-and-wool fire by the **Montgolfier brothers** and rose to a height of 1,500ft – a phenomenon ascribed to the heat, not the gas produced. Two months later a balloon filled with inflammable hydrogen (molecularly the lightest of all gases which had been isolated by Henry Cavendish in 1766) went aloft. And in November a hot air balloon took up François de Rozier.

The **air warfare** potential of balloons was immediately perceived. In 1794 a tethered one was used for observation at the Battle of Fleurus. But the problem of steerable flight was not solved until 1852 when Henri Giffard flew 17 miles beneath an envelope powered by a 3hp steam engine. In the meantime, during the siege of Venice by the Austrians in 1849, a score of paper balloons, intended to drift over the city and release 30lb **bombs** in response to a time fuse, were launched. No lives were lost, little damage was done and initial alarm soon subsided.

During the **American Civil War**, under command of Thaddeus Lowe and Count Ferdinand **Zeppelin**, tethered balloons (sometimes towed from barges) were used for directing artillery fire. And in 1870–71, during the **Franco-Prussian War**, 66 free-flight types were launched from Paris during the siege carrying microfilm messages, homing pigeons (for replies) and very important people. A great majority reached safety; one drifted 1,400 miles to Narvik!

With the advent of the **airship**, balloons lost their military attractiveness, although the tethered kind were extensively used for artillery observation in **World War I** and in **World War II** flown over vital ground targets and **convoys** to deter attack by low-flying aircraft. In both roles they were usefully effective, although very vulnerable to **fighter** attack, their crews being among the first to be provided with **parachutes** for escape. Carrying instruments aloft to very high altitudes they continue to be used for the gathering of meteorological information.

Barbed wire. Wire manufacture is among the earliest of **metallurgic** manual crafts. It assumed great military importance when utilized for mail **armour** but was not machine-made (water-powered) until the 14th century in Germany. Not until the 1840s were large quantities required for wire rope and **telegraph** lines. The cheaper, mass-produced ferrous metals of the 1850s prompted the invention in 1874 by the American Joseph Glidden of a machine to make barbed wire for farm fencing, helping to start prairie range wars. Basically, two-point barbs were wrapped around twisted galvanized wire and coiled on a reel for handling.

The first extensive use of barbed wire for fortifications was by the Russians during the **Russo-Japanese War**. Because it was so difficult to clear away, even by **artillery** fire, it vastly strengthened defences and was a principal reason for **tactical** stalemate during **World War I** and the invention of the **tank** to crush it. It is still used along with even more resilient steel sawtooth varieties which are very hard to clear.

Batteries, Electric. Modern electric storage batteries consist of voltaic cells which convert chemical energy into electric energy. There is a suggestion that the Parthians knew about batteries *c.* 250 BC. But it was Alessandro Volta in 1792 who first noted the effects of an electrolyte on metal and in 1800 described his first primary cell, consisting of a silver cathode, a zinc anode and a salt-solution electrolyte which produced an electric current of significant voltage.

Volta's batteries were vital to future discoveries by Humphry Davy and Michael **Faraday**. The early cells, which were essential to the **telegraph** and could not frequently and easily be recharged and discharged, were made obsolete with Gaston Planté's discovery in 1859 of the lead–acid system which, among other uses, powered the ignition system of internal-combustion engines (*see* **Propulsion, Means of**), **telephones**, **submarines** and **radio**. Heavy and liable to spilling as Planté's wet cell was, its use for **motor vehicles** was crucial and it continues in service to this day. In **aircraft** a wet battery was obviously unsuitable. However, the aqueous dry cell, as described in 1868 by Georges Leclanché (and which was already in use for **railways** and the telegraph), overcame the disadvantages of size, weight and safety to make them suitable not only for aircraft but a host of applications in industry and the home. Ninety per cent of today's batteries are the dry-cell type with solid electrolyte.

The discharge and charging rates of batteries cannot safely and efficiently be raised above certain limits. **Research and development**, driven by the aircraft and **space** industries, has therefore tended to concentrate on new materials, weight-saving and improved charge retention; and by the utilization of, for example, solar, light and nuclear sources for charging.

Battle-cruiser (originally **Cruiser-battleship**). The concept of a warship that would combine the hitting power of a **battleship** with the speed and light armour of a **cruiser** seems to have originated in the mind of Émile Bertin of France in 1896. But the first example of such a vessel was not laid down until 1906, by the British, under the influence of Admiral Sir James **Fisher**, who believed speed gave protection. HMS *Inflexible*, with its 41,000hp steam-turbine engines, had a speed of 25 knots and the capability, with eight 12in guns, of engaging battleships. But

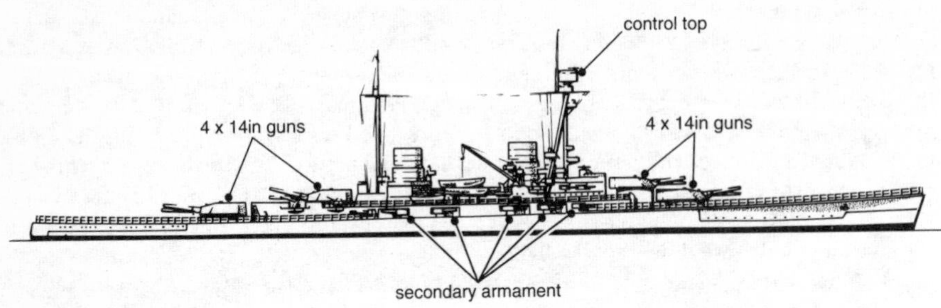

The battle-cruiser Derflinger (*German, speed 26½ knots*)

a mere six inches of side armour amidships, tapering to four inches at bow and stern, left it vulnerable even to an armoured-cruiser's 8in guns. By contrast Germany's battle-cruisers, of which *Von der Tann*, laid down in 1907, was the first, had ten inches of armour amidships and 11in guns which were as effective as the British 12in.

Although two British battle-cruisers proved their worth, in December 1914 at the Battle of the **Falklands**, by sinking two German armoured cruisers, the vulnerability of the class was exposed at the Battle of **Jutland** when three were destroyed by magazine detonation from penetration by gunfire; as compared with only one German battle-cruiser lost from numerous hits, any one of which might have been fatal had not the German magazines been safer and British shells inferior.

Battle-cruisers went out of fashion once the necessity for thick armoured protection of high-value battleships was proven. Three were retained by the British, of which HMS *Hood* blew up after hits from the German battleship *Bismarck*, in May 1941, and HMS *Repulse* went down under Japanese air attack later that year off Malaya.

Battleship. In 1863 the term battle ship (*sic*) was coined officially by the Royal Navy to describe a ship of the line of 74 guns or more. Soon, however, it would mean the heaviest armoured warships afloat, of which a few already were in existence. For in 1859 the French had launched the steam-powered *La Gloire*, an **iron**-clad, timber-built warship of 5,617 tonnes, designed by Stanislas Dupuy du Lôme, and armed with 36 163mm guns. With a speed of 13 knots, she could outmanoeuvre and outshoot all wooden sailing vessels and was likened to 'a lion among a flock of sheep'. Within two years the British replied with the 9,210-tonne, 14½-knot *Warrior* with 40 heavy guns and 4½in-iron **armour** backed by 18in of teak. These were revolutionary capital ships to set in motion a new and very expensive naval technological race.

Although the desire to increase speed and improve mechanical reliability was strong, endeavours to augment armament to overcome the latest **armour**, and increase protection to counter better **artillery**, were prime. In 1862, John **Ericsson**'s *Monitor* demonstrated the advantages of a rotating turret, although it had been preceded in action by Captain Cowper Coles's turreted raft, *Lady Nancy*, during the **Crimean War**. Turrets, however, were incompatible with masts and it was almost as much due to them as to improvements to engines that, in 1871, the 9,330-tonne HMS *Devastation* was launched – the first battleship without sails. with four guns in two turrets.

There were diversions from the mainstream of development, such as the fitting of **rams** in the aftermath of the sinking by ramming of the Italian flagship at the Battle of **Lissa** in 1866. Mainly there was seen a rapid increase in size and weight of guns,

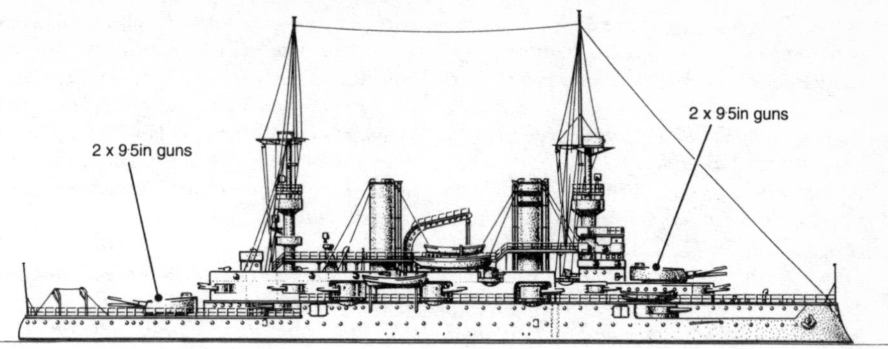

German battleship, 1900 (11,150 tonnes)

up to 17·75in calibre. mounted in turrets of improved design in **steel** frames, and armoured ships fitted with **torpedoes**. At the end of the century battleships such as the Americans employed in the Spanish–American War, and those used in the **Russo-Japanese** War of 1904, weighed anything from 8,000 to 15,000 tonnes, with 12in guns and speeds of up to 18 knots. But combat demonstrated how vulnerable they were below the water-line from **mines** and torpedoes; quite as much as above when guns, after enormous ammunition expenditures even at ranges down to about 6,000 metres, managed only occasionally to hit their targets.

In 1906 came a fresh revolution, the launching by the British, under the genius of Admiral Fisher, of HMS ***Dreadnought***. This 21,250-tonne steam-**turbine** ship with a speed of 21 knots, thick armour and 10 12in guns in five turrets, made all existing battleships obsolete. Moreover, the system of main armament control from a central director, fed with data (from an optical **range-finder**) transmitted to the guns, promised significant improvements in the accuracy of **gunnery**. Copied by all other nations, the dreadnought-type battleship, along with the thinner-armoured **battle-cruisers**, was steadily improved and built in large numbers, dominating naval warfare until 1940. But no amount of armour and internal watertight compartments could secure them from the underwater attacks which caused the majority of their losses during the two **World Wars**. By 1918, within two years of the battleship-dominated Battle of **Jutland**, fears of mine and torpedo, along with the threat of air attack, were having a crucial impact on design.

To thickened top armour, against **bombs**, had to be added batteries of **anti-aircraft weapons**, despite the reluctance of many sailors to recognize the air menace. When the great expense of a battleship-building race, instigated by the Japanese, led to the Washington Naval Treaty of 1922, with its limitation of 5:5:3, respectively, in battleship tonnages, to the USA, Britain and Japan, with 1·67 each to France and Italy, it merely promoted extensive scrapping of obsolete vessels and the construction of fewer modern warships, together with **aircraft-carriers**. Moreover, despite a restriction of 35,000 tonnes for capital ships, this limit was flagrantly breached at the start of the 1930s arms race when, as articles of faith in the battleship, the Italians laid down two 41,000-tonners, the Germans two 40,000-tonners and the Japanese the mighty 65,000-tonne *Yamato* and *Musashi* with their 18·1in guns.

Nevertheless, although Britain, France and Germany had between them in 1914 57 dreadnoughts (with 22 much improved types building, and 91 pre-dreadnoughts), in 1939 (including Italy, the USA and Japan) the main combatants' totals were 56 obsolescent and modern battleships (with 24 building), which in many respects was an expression of the escalating cost of warships whose size and complexity, in order to improve their chances of survival and striking power, had risen beyond sensible bounds. For not only were such high-value targets expensive to man, service and protect whenever they put to sea, their combat-effectiveness also was suspect. True the initial effect of their existence as part of fleets 'in being' was considerable. But soon the air threat made its presence felt, notably over the North Sea and in November 1940 at **Taranto** when carrier-borne British torpedo aircraft sank or crippled three Italian battleships in port. And although Germany's three so-called pocket-battleships of 12,000 tonnes and her two 32,000-tonne battleships scored limited successes against commerce and managed to pin down disproportionately large elements of the Royal Navy, the dispatching of the *Bismarck*, after she sank the battle-cruiser *Hood* (as the first of only two examples of battleship gunfire being exclusively responsible for sinking one of its kind in this war), was a body blow to the theory of the 'unsinkable capital ship'.

Pearl Harbor, the sinking of the *Duke of York* and *Repulse* off Malaya, the battles of the **Coral Sea** and **Midway** simply restated the battleship's decline. Even when, in the first night battle conducted with **radar** on 12 November 1942, the US battleship *South Dakota* crippled by gunfire the Japanese battleship *Kirishima*, the *coup de grâce* was scuttling by the latter's crew. So it was perhaps symbolic that the last battleship versus battleship action of all should have ended in a blaze of fire on 25 October 1944, during the climactic Battle of **Leyte Gulf**, when the guns of six US battleships pulverized and capsized the Japanese *Yamashiro* in 30 minutes at 22,000 yards. For the rest of the Japanese battleship fleet it was to be destruction by other means.

Scrapping began again after 1945, although Britain, France, Russia and the USA clung to a few battleships, the latter employing them for offshore bombardment during the **Korean War**. By the 1980s, however, the USA alone possessed them, and went so far as to take some out of mothballs, fit them with the latest missiles, besides their 16in guns, and deploy their impressive presence adjacent to such trouble spots as the Lebanon, where they fired occasionally, with dubious political impact, on land targets. And during the **UN–Iraq War** the USA used them again to launch **cruise missiles** against land targets.

Bayonet. A short, thrusting **edged weapon** fitted to a musket (*see* **Rifle**) and said to originate in Bayonne *c.* 1640. Intended as a substitute for the **pike**, the original plug bayonet was inserted into the muzzle, thus preventing firing. This was soon followed by the ring bayonet that fitted round the muzzle and superseded in 1688 by Marshal **Vauban**'s socket bayonet which slotted into a stud on the barrel. Various kinds of blade were developed, sometimes as dual-purpose entrenching tools or can-openers. In some armies the bayonet became an article of faith in offensive action through close combat. They are still produced but since **World War I** have been used infrequently for their original purpose.

Bazooka. The first of a long line of infantry hand-held anti-tank weapons; so named from its likeness to a home-made trombone used by radio comedian Bob Burns, the bazooka was the brain-child of a Colonel Skinner, US Army. With no funds and a staff of one, he designed and made a tube-launched rocket using the M10 hollow-charge grenade as a warhead. Demonstrating his device at Aberdeen Proving Ground in 1942, the bazooka was so effective that it was instantly accepted for service and first used in the **North African** campaign of World War II. The bazooka continued in service until the **Korean War** of 1950, when its small warhead proved inadequate against the North Koreans' T34s. It was superseded soon afterwards by the 3·5in rocket-launcher.

Beams. *See* **Radio** *and* **Radar**.

Belisarius (*c.* 505–65). An outstanding tactician and **cavalry** leader who served Justinian I during Byzantium's most testing years. Worshipped by his troops, who usually were mercenaries and whom he led from the front, he defeated the Persians at Daras in 530, ruthlessly suppressed a revolt in Constantinople in 532 and at Tricameron, in 533 with 15,000 troops, routed 50,000 Vandals. In 535 Justinian, forever jealous of Belisarius, ordered him, with a mere 8,000 superbly trained troops, to drive the Ostrogoths out of Sicily and Italy. After several battles and sieges he captured Rome in 536 and for the ensuing five years, still denied adequate resources, outfought the Goths through operational mobility and sound **logistics**. Thereupon Justinian dispatched him to Persia, but returned him to Italy in 544 to deal with a resurgence of Gothic power. Finally recalled to Constantinople in 549, he spent the next decade in retirement until called up to eject a Bulgarian army which had penetrated the city walls. As a final humiliation in 562, Justinian imprisoned Belisarius and seized his fortune – happily to reverse the decision in 563 in order to permit a most loyal general to end his days in repose.

Berlin Airlift. As **Cold War** political tension between the Western allies and Russia increased in June 1947, Josef **Stalin** imposed a **blockade** on the allied garrison in Berlin. Caught unprepared, the Americans and British, with limited French support, began supplying the city by air through three corridors connecting with West Germany. Fortunately, a sophisticated **radar** and **radio** traffic-control system was in place. At remarkable speed, led by the Americans, a great fleet of two- and four-engine (10-tonne lift) **transport aircraft** was assembled to meet a target of 12,000 tonnes per day. Deliveries rose as economy measures took effect in the city and the aircraft flew through bad winter weather and 'buzzing' by Russian aircraft. Stocks of heating fuel ran low. But morale rose at the same time as the weak East German economy began to collapse under collateral strain. In January 1949, when 172,000 tonnes were delivered, the Russian resolve and **propaganda** faltered. In April, with cargoes coming in at 13,000 tonnes per day, they gave in. The blockade was lifted on 5 May. Superior organization and technology had lost the Russians an important battle in the Cold War.

Bessemer, Sir Henry (1813–98). A self-educated engineer who made his fortune with a 'gold' powder made from brass and bronze. During the **Crimean War** his attempt to make a spun **artillery** shell was frustrated because **iron** cannon were too weak and **steel** scarce and expensive. By 1856 he had invented the revolutionary converter which mass-produced steel by blowing air through molten pig-iron. The process took eight years more to perfect but, in modified forms, serves to this day. In later life he built a stabilized steamship, a big telescope and a solar furnace, none of which was wholly successful.

Bibliography of war. The number of books and documents generated by war is so immense as to give mental indigestion when listing them, which is mainly why, in addition to the tyranny of available space, this Encyclopedia's bibliography (see page 373) has been made select and concise – little more than a bibliography of bibliographies, in fact.

Modern fashion provides extensive bibliographies and notes to many learned works on war: Professor Paul Kennedy's *The Rise and Fall of the Great Powers* has, for example, 40 pages of one and 80 of the other. Few bibliographies are analytical or critical: rarely are there guarantees of integrity or perfection. Far too many military histories are uncritical regurgitations, tending merely to perpetuate legends and myths.

Since research into documents as often as not begins among private collections or in libraries or the great national archives, the task is both laborious and expensive. Normally, even when an index of documents is provided, their contents remain obscure, with the result that much time can be lost, with consequent frustration of all but the most dedicated researchers working in ideal conditions. Ignorance of languages also bedevils. Glance at the majority of military histories in English, for example, and it will be seen that information from works in foreign languages is mainly from all-too-familiar ones picked from notably few popular translated publications. At least one honest historian of repute has admitted that his extensive bibliography is restricted to books in English.

There are people who assume that official histories, based predominantly upon public records, are repositories of truth. But official historians, like, for example, regimental historians and biographers, are subject to vested interests, national security and the laws of libel – hazards which are harder to overcome the more contemporary the events happen to be. Some overlook the fallibilities of war diaries, which are sometimes written in the aftermath of battle or under stress; under political and emotional pressures; or suffer misguided 'weeding' of 'irrelevant' material; and from genuine desires to avoid giving pain to individuals. It is unchallenged, for example, that parts of the British official history of **World War I** were deliberately equivocal for political reasons, and that nearly all Allied histories of **World War II** avoided divulgence of the breaking of the principal enemy **codes and ciphers**. And it is noteworthy that, while American official histories of that war are profusely authenticated with footnotes, the British, apart from a few unofficial, hand-annoted copies, are not – with the notable exception of the recent, extremely revealing five-volume *British Intelligence in the Second World War*. Easier freedom of access to official records will illuminate as well as complicate a labyrinth.

Biological warfare. There are records from earliest times of attempts to induce disease in the ranks of the enemy – by throwing dead animals into wells to poison the water and catapulting rotten carcasses into besieged cities in the hope that the starving inhabitants would eat them and succumb. As **medical** knowledge improved and the real causes of disease were identified and viruses isolated, the possibility of projecting these viruses into enemy lines to do him damage was often considered. As recently as the **Korean War** in the 1950s there was a propaganda ploy from the Communist side that the Americans were employing 'germ warfare' by dropping **bombs** containing a deadly virus on the North Koreans.

All this was entirely without foundation, not least because at that time no one had found satisfactory solutions to the three fundamental problems facing anyone planning to use a projectile to carry a suitable organism: (1) how to protect one's own side from contagion when filling the bomb or shell; (2) how to keep the organism alive after filling until required for use; (3) how to protect the organism from the shock of impact (plus, in the case of an artillery shell, the shock of discharge), so that it could do its deadly work. Quite apart from any moral questions which might prevent a government from pursuing the development of biological weapons, for many years these practical considerations proved to be too daunting.

Typically, biological agents are bacteria, rickettsia and viruses and they could be disseminated by spray, shell or rocket, in the form of a liquid suspension or as a powder. The effectiveness depends on the incubation period of the organism, which could be a number of days. The respirator may provide a defence if it is closely fitted, but vaccination, if practical, would be the best way. However, the real problem for the defender lies in detecting the biological cloud in sufficient time to take evasive action; since the human senses could not detect such an attack, some form of air sampling would be necessary.

By 1969 improvements in microbiology had allowed the USA to develop biological weapons (later unilaterally renounced and destroyed) and the USSR is known to have carried out military and civil defence training on this subject. Furthermore, such agents could now be manufactured with comparative ease and without sophisticated technology, as evidence from Iraq proves. The threat therefore remains, though the USA, Britain and the **Warsaw** Pact have all signed the UN convention banning production and use of biological weapons.

Birch, General Sir Noel (1865–1939). A brilliant British gunner who rose from chief artillery officer in the Cavalry Division during the retreat from Mons in 1914 to become artillery adviser, in May 1916, to GHQ British Expeditionary Force. Under his leadership in a war dominated by **artillery**, revolutionary organizational, **gunnery** and **logistic** problems were solved. From 1923 to 1927, as Master General of the Ordnance, he tackled the problems of **mechanization** and the replacement of **horses** with **armoured cars**, **tanks** and **motor vehicles**. He then became a director of the **Vickers Company** with responsibility, among other projects, for the first self-propelled artillery and new tanks.

Bismarck, Prince Otto von, (1815–98). As prime minister of Prussia from 1862, Bismarck's way of unifying the German nation and pursuing an adventurous foreign policy, contrary to the instincts of his king, was founded on 'blood and iron'. In nine years, with the encouragement of the Army and its Chief of Staff, Helmuth von **Moltke**, he had, by aggressive diplomacy and war, seized Schleswig-Holstein from Denmark, won the Austro-Prussian War of 1866 and the **Franco-Prussian War** of 1870. By 1871 he had acquired Alsace-Lorraine from France (thus unnecessarily exacerbating ill will between the two nations), and engineered the creation of the German Empire, under the Prussian Kaiser Wilhelm I, with himself as Chancellor. Until dismissed by Wilhelm II in 1890, Bismarck dominated Germany, developed her overseas Empire and played a leading role in European politics with the series of interlocking, and often secret, treaties which became major factors in the start and course of **World War I**. Ever flexible in his policies, trickery lay at the heart of his tactics whenever necessary for the current aim.

Blake, Admiral Robert (1599–1657). A British Member of Parliament who as a general during the Civil War defended Bristol and Lyme against the Royalists and, in 1644, seized Taunton and held it for two years. In 1649 he became a 'General-at-Sea' with command of the Fleet. In that role he pursued the Royalist fleet from Kinsale to Lisbon and finally to Cartagena where, in 1650, he destroyed his quarry – having in the meantime captured the entire Portuguese treasure fleet from Brazil, as **blockade** runners, a feat he repeated in 1657 against the Spaniards.

In 1652 he commanded the Fleet in the first Dutch War in a series of battles he usually won. His victories were very much the product of the tactical *Fighting Instructions* he wrote to innovate the line-ahead formation (as a rather rigid aid to manoeuvre to bring **gunnery** broadsides to bear on the enemy), and as standing orders for captains to obey on receipt of a gun and flag signal. The orders were enforced by *The Articles of War*, which he also wrote to give the Admiral, for the first time, judiciary powers to command the Fleet absolutely. No longer would captains fight as they chose!

Blitzkrieg. This German word for 'lightning war' was probably coined by Adolf **Hitler**, prior to 1939, as **propaganda** in support of a political bluff to conceal Germany's, as yet, military unreadiness for total war. As originally propounded by J. F. C. **Fuller** and then practised in Poland and subsequent campaigns, it stood for the rapid disruption of enemy defences by a combined air and mechanized land offensive. Strokes against key targets and communication centres, deep in the enemy rear, were executed at high speed with the intention of stunning a surprised nation into submission. Linked to deception, to propaganda and to **fifth column** activity, the sheer pace and violence of coordinated attacks by all means available was a formula commanders had long desired, but been denied by inadequate signal **communications** and **logistic** services.

Later the British shortened the word to 'Blitz', as the name for the various day and night air attacks upon their country.

Blockade. Prior to the 17th century there were many instances of the **siege** of ports (*see* **Dockyards and naval bases**) being supplemented by naval operations. Not until the construction of good seagoing vessels in the 16th century, however, was it feasible to sustain the blockade of a whole coastline. The French word *blocus* dates from then.

The act of 'surrounding or blocking of a place by enemy to prevent entry and exit' has, since the 17th century, been governed by civil laws as well as military rules. The 'right of blockade' was established by the British, in 1650, to implement the interception of neutral shipping and seizure of cargoes – acts which have usually been disputed in prize and international courts.

Blockade was frequently resorted to during the Dutch Wars, the **French wars, 1635–1783**, the **American War of Independence** and the **French revolutionary wars**. Yet it never was more than an embarrassment because there were never enough ships; sailing vessels were always at the mercy of weather; coordination of operations was extremely haphazard owing to meagre signal communications (*see* **Communications, Signal**); and **intelligence** was usually sparse and inaccurate. The advent of the **telegraph** and seaworthy steamships changed all that – without making blockade foolproof.

During the **Crimean War**, **Sevastopol** was both blockaded and besieged although, as usual, supplies and men slipped through.

The blockade of the Confederacy during the **American Civil War**, though decisive in the long run, by denying essential materials, could never be complete. To begin with the 90 obsolete warships at Federal disposal could not stop determined blockade runners reaching ports along such a vast coastline. But while access from the hinterland and Mexico, particularly while the River Mississippi was in Confederate hands, was easy, their initial underestimate of the blockade's potential for denying cotton to European manufacturers, in order to apply political economic pressure, was a mistake. Eventually the blockade inflicted fundamental damage on Confederate **logistics**. Yet the hazards of a 'state of blockade' such as the Federals had to declare in order to comply with the law were made plain by delicate negotiations with objecting foreign statesmen and traders, while wrestling with the military problem of a dispersed blockade net being contradictory to the principle of concentration.

Few wars of the 19th and 20th centuries have been without a blockade if one side or the other possessed a navy. Once the Spanish squadrons had been destroyed in the Philippines and Cuba at the start of the Spanish–American War, the Spaniards had to recognize the implications without an official 'state' being declared. Throughout the South African War (1899–1902), even European nations with Boer sympathies were reluctant or unable to run the British blockade – although **medical** supplies were allowed through. And while any Russian hope of blockading Japan evaporated at the start of the **Russo-Japanese War**, the ability of the Japanese to impose a blockade as well as a siege of **Port Arthur** only shortened its garrison's existence without embarrassing the relieving forces, who were supplied by railway from the west.

Allied attempts to blockade the central powers during **World War I** raised extremely complex problems exacerbated by the 1909 Declaration of London. This treaty included a list of strictly military contraband which could be seized from neutral ships, and a humanitarian 'conditional list' of supplies, for the civil populace, which were excluded if not consigned to the enemy's armed forces, or a department of state or through an enemy port. Thus Germany could legally import strategic goods through neutral countries, such as Holland and Denmark, and put Britain and France in the wrong when they interfered. But when the British, in 1914, declared the entire North Sea a 'military area', where **mines** would be laid, and the Germans said in 1915 they would destroy every merchant vessel found in British waters, the way was prepared, by stages, for unrestricted **submarine** warfare in 1917.

It was impossible for either side to impose an absolute blockade, although the plight of the central powers progressively worsened as Italy, Greece, Romania and the USA joined the Allies, and as the Allies worked out indirect blockade measures, such as placing embargoes on facilities for neutral shipping and on the movement of strategic cargoes, and by blacklisting those who broke the blockade. At the same time, vital materials, as well as food, were rationed by all the belligerents to help defeat the blockades – of which the system finally evolved by the Allies was a major factor in the defeat of the central powers.

Within the different circumstances of **World War II** both sides tried blockade by very similar methods to those of 1918, with the notable addition of **aircraft** to help both make and break sieges. Yet, although Britain and the Axis powers were embarrassed or brought close to collapse by blockades, their respective countermeasures in breaking the enemy cordon, by rationing and control of distribution, and by use of synthetic materials, including **oil** fuel, kept Germany going until she was all but overrun and Japan in the fight until the **atom bomb** made her prospects hopeless.

Blockades of varying intensity and effect continue to be imposed, sometimes by direct action to prevent supply to belligerents, such as during the **Korean War**, the **Vietnam War** or the **Falklands War** and the **UN–Iraq War**, often by quicker-acting threats or embargoes on weapons, fuel or financial subsidy, such as by the USA against Britain and France during the Suez War of 1956; and by the superpowers in relation to the Indo-Pakistani wars and various outbreaks in Africa, the Middle East and South-East Asia. But ways round are often discovered to frustrate a way of war which is slow-acting, rarely decisive on its own and usually inflicts considerable misery.

Boeing Company. In 1916 William Boeing and Conrad Westerman formed the Boeing company to build their B&W float plane. When war came in 1917 they built trainers and flying-boats, and in the 1920s and 1930s, expanded their products to include **fighters**, **torpedo** and dive **bombers**, heavy bombers and **transport aircraft**. During **World War II** 12,723 B17 (Flying Fortress) and 3,480 B29 bombers were built. After it the company expanded to build the eight-engine B52 (Stratofortress) as well as the famous 707, 727, 737, 747, 757 and 767 airliners and the 717, which became a flight-refuelling tanker KC135, a long-range troop transport and, under the title E3A, an **AWACS**. Now the world's largest aircraft-builder, it purchased the Vertol **helicopter** company in 1960, and participated in the design and construction of **rockets and guided missiles**, including the Apollo/Saturn moon rockets and Minuteman intercontinental ballistic missile (IBM).

Bomber aircraft. Almost any **aircraft**, including **balloons** and **airships**, are capable of dropping **bombs**. Free-flight balloons in 1849 scattered a few on Venice and some were dropped from primitive aeroplanes during the Italo-Turkish War in 1911. Airships were used as bombers by the Germans at the outset of **World War I** and employed for strategic bombing (*see* **Bombing, Strategic**) until outclassed by heavier-than-air machines.

The sturdy French single-engine Voisin I of 1914, with a bomb-load of 132lb, was among the first to be worthy of the title, as most certainly was Igor Sikorsky's mighty Russian four-engine tractor biplane or the later twin-engine German Gothas and British Handley Page 0/100s. But with speeds under 100mph, ranges hardly more than 300 miles with loads of about 1,000lb, the majority even of the purpose-built machines were rarely of more than nuisance value, especially since, with primitive bomb-sights from altitudes above 12,000ft, their oxygen-starved crews were unlikely to find, let alone hit, the easiest of targets. At the same time the smaller, single-engine machines which attacked with greater accuracy at the lower altitudes were more vulnerable to ground fire and **fighter aircraft**. Indeed, the faster fighters (a few armoured against small-arms fire in 1918) carrying a small bomb-load were increasingly employed for daylight bombing at short range.

By 1919, bombers (including seaplanes) had sunk or damaged few ships by bomb and torpedo. While over land, industrial and residential targets had been harassed but without immense loss of production or life. But in the battle zone targets of all kinds had been frequently hit with impressive results. Yet, although by 1918 the proponents of strategic bombing managed to encourage the development of big bombers with ranges close to 1,000 miles and bomb-loads of 6,500lb, it was cheaper medium and light machines which, post-war, were required for policing restive colonies. And most of these, therefore, were slow, single-engine biplanes of wood and fabric construction with speeds below 200mph.

The rearmament programmes of the 1930s, however, promoted a fundamental change to multi-engine, streamlined, all-metal monoplanes with internal bomb stowage. By 1939 medium bombers capable of 250mph with ranges up to 1,000 miles and bomb-loads of 4,000lb were in service and threatening to outpace existing fighters. Some **transport aircraft**, such as the German Junkers Ju52, were adapted from civil airliners and served as a bomber during the **Spanish Civil War** – thus encouraging a false impression that airliners were suitable for this role.

But in addition to the mediums, specialized dive-bombers, led by the Curtis OC2 fighter-bomber pioneered by the American **Marine** Corps, were introduced with the ability to reduce errors in aiming by using a modified gun-sight to point the aircraft (and thus the bomb) straight at the target. Copied by the Germans with their Junkers Ju87, for close support of land forces, and by the Japanese with their Aichi D3A for anti-shipping strikes, machines of this kind proved effective day bombers, although they were very vulnerable to fighters unless escorted. As also were the fast medium bombers, even if flying in tight formations and fitted with power-operated multi-machine-gun turrets.

And as for the four-engine heavies which were designed for strategic bombing with speeds of about 300mph, ranges in the neighbourhood of 1,500 miles and bomb-loads, in the case of the American B17F, of 8,000lb, and the British Lancaster B1 of 18,000lb, they too were far from ideal. The British, unable to operate without escort by day in face of enemy fighters, were compelled mostly to use their heavies by night; with the result that, even with **electronic navigation** and bomb-aiming equipment, they could only make area attacks – and at increasing risk against radar-assisted night fighters. Similarly, the American Fortresses, even when escorted, also suffered dreadful losses in attempts to fight their way through. And the Norden **gyro**-stabilized, optical bomb-sight, which was unusable only at low level, did not always hit the target from high level.

The invention of the **jet engine** revolutionized aircraft. Speed, payload and ceiling were increased so dramatically that it was possible after **World War II** gradually to dispense with specialized heavy bombers and rely more on the fighter-bombers which, since 1940, had gradually evolved into multi-role combat aircraft. Subsonic, jet-propelled heavies capable of carrying **nuclear** bombs, such as the eight-engine American Boeing B52 with its payload of 60,000lb, and the British four-engine Vulcan, payload 21,000lb, gradually gave place to smaller, more agile supersonic multi-role machines with far higher chances of survival against modern defences. With smaller crews and faster mission turn-round time, sophisticated machines such as the variable-geometry American F111 and British Panavia Tornado were superior in every respect to their predecessors – and proved it by results with low casualties over Iraq in 1991. As did the unique US Fairchild Thunderbolt, a twin-jet-engine, single-seater, heavily armoured ground-attack aircraft with external weapons stowage.

Regardless of type, the employment of highly sophisticated electronics and power-assisted controls (such as fly-by-wire) to ease, without necessarily simplifying, aircrews'

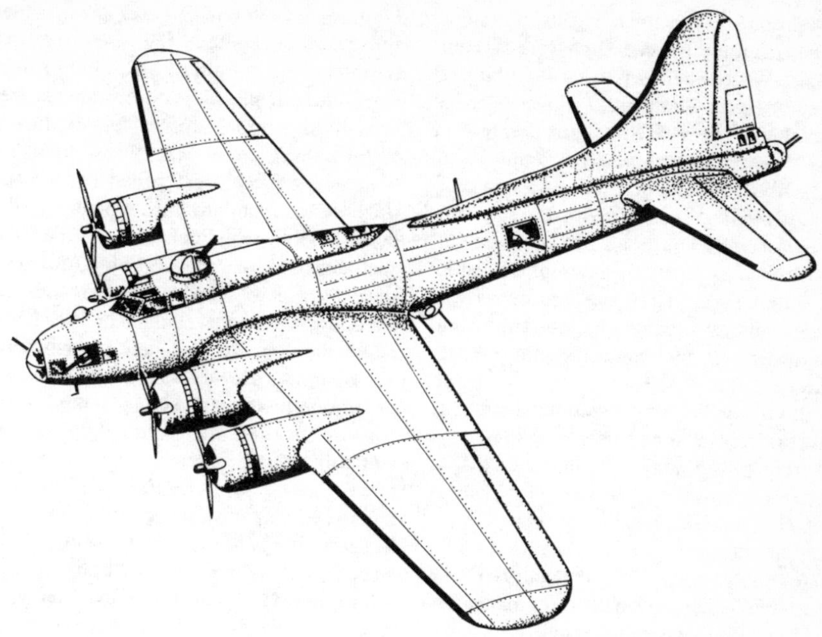

Bomber aircraft: Boeing B17F (US)

tasks, predominated in bombers. Carrying smart munitions, air-to-air missiles, **electronic** countermeasures (ECM) and **surveillance equipment**, these formidable machines could as often as not steer clear of trouble, hit their targets from long range and take care of themselves in combat if necessary.

Bombing, Strategic. The concept of strategic bombing, mistily foreseen from the moment man first went aloft in **balloons**, did not emerge as doctrine until after **World War I**. Even then it was more like a theory than strict military **strategy**. The closest thing to a definition is to be found in General Giulio **Douhet**'s essay of 1921, *The Command of the Air*, in which he postulated the winning of air superiority by attacks on enemy air forces in the air or on the ground, followed by massive 'terror' bombing with high-explosive, gas and incendiary bombs on the population, with a view to shattering their **morale**. He argued that if this worked navies and armies would have no need for air support.

In fact the experience of **air warfare** in World War I already had exposed the flaws in Douhet's theory. Although the first scattered attacks by **airships** and **bombers** had caused immense alarm and despondency and diverted considerable defensive resources from the battle front, complete air superiority had been well beyond all air forces, the volume of explosives dropped small and scattered, and civilian morale only ruffled. Britain's newly formed Royal Air Force with its **Independent Air Force**, under General Sir Hugh **Trenchard**, only had in mind reprisals for German raids on Britain in 1918 with the bombing of strategic targets, and scored few successes. But in the war's aftermath Trenchard, fighting for the RAF's continued existence, made strategic bombing a main plank for his case – and therefore welcomed Douhet's ideas in support.

By the mid-1930s, Britain was alone in almost wholly embracing Douhet's theories and, with Italy and France, was in a minority of nations to adopt an air force separated from Navy and Army control. The

Americans lagged behind in strategic developments and the Russian airmen were 'discouraged' by Josef **Stalin**'s purges. But the Germans, realizing Adolf **Hitler** was bent on an early war, created an Army-support air force which was capable of 'terror' attacks to weaken enemy national determination. In fact, General Albert **Kesselring**, who cancelled German plans for a multi-engine strategic bomber force, was heavily influenced by the realization that existing technology was unequal to the task without a design and production effort beyond the nation's resources.

Britain took the lead in creating a strategic bomber force of four-engine bombers as a deterrent to Germany and, if that failed, in the hope of substituting bombing for involvement in another blood-letting in Europe. Without resort to scientific operational analysis (*see* **Analysis, Operational**) dedicated RAF officers and certain scientists persuaded already frightened government ministers to believe that 'bombers always would get through' and would wreak catastrophic damage and shatter the morale and resistance of the people. Untested assumptions and guesses about the destructive power of **explosives** were accepted without challenge, and repeated luridly by the press. Enormous outlay was committed to the building of a strategic bomber force, to strike back, and a **civil defence** organization as a shield in readiness for **World War II**.

In the event the early bomber attacks were not nearly as concentrated or crushing as expected, because of the limitations of bombers and **bombs** and the effectiveness of defensive measures. For nearly every weapon has its counter and there is a limit to the losses any air force can withstand. Moreover, civil morale, though often shaken, was never broken – not even in Germany where, in the final months of the war, the combined British and American strategic bomber forces laid waste to the country. Nor in Japan when her cities were incinerated and two **atom bombs** wiped out Hiroshima and Nagasaki in 1945.

In fact, in numerous wars since, strategic bombing by aircraft and **rocket and guided missiles** (but without the use of nuclear weapons) has not forced a nation to surrender, although sometimes it has seemed very close to the mark. This was notably the case in North Vietnam in December 1972 when, after a relatively short period of intensive, unrestricted bombing (in which 17 B52 bombers were shot down by missiles), the Vietnamese were 'bombed to the conference table' – but later allowed to recover because the country was not occupied by land forces. It also occurred in Iraq in 1991 when her air force was neutralized by bombing and her infrastructure wrecked – though without collapse of morale as required by Douhet, and without subsequent occupation by land forces.

The bombing of strategic targets nevertheless has always been a potent factor in war, even though it has failed to bring about a lasting conclusion to a war by itself. As a deterrent, however, be it with **nuclear**, **chemical** or smart munitions, it is extremely effective, especially since, by area or precision strikes, it now threatens the top warmongers in their deepest shelters and maybe terrifies them sufficiently to make them stand back from armed aggression.

Bombs, Aerial. In March 1849 the Austrians bombarded Venice with 30lb bombs released by a time fuse from free-flight **balloons**. The experiment was not a success. But in 1911, also for experimental purposes, bombs were dropped in a trial by an American aircraft and that same year the Italians unloaded a few bombs and **grenades** on the Turks during the Italo-Turkish War, and that was significant.

Sometimes there was a tendency to adapt **artillery ammunition**, but almost at once, to satisfy the laws of ballistics and deal with an enormous variety of targets – naval, battlefield, residential and industrial – specifically designed missiles were deemed essential. Usually, aerial bombs consist of a finned, streamlined casing (thin-walled for blast effect, thicker for armour-piercing or fragmentation) which contains the fuse and detonator

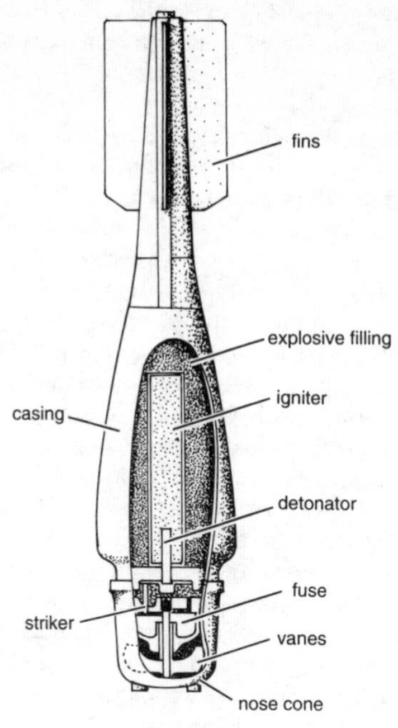

World War I aerial bomb

to activate the main **explosive** charge or incendiary material, as well as a safety mechanism to prevent premature detonation. When released, bombs are affected by yaw from the aircraft's forward motion and by drift from crosswinds. From high levels, free-fall bombs are therefore almost impossible to aim precisely, although when launched in a dive at higher velocity accuracy improves. At low levels, with more chance of a hit (along with the possibility of the safety mechanism failing to disengage), they are almost as menacing to the bomber as the enemy.

At the start of **World War I** requirements emerged for darts and high-explosive bombs for the attack on military and land communications targets; plus incendiary bombs against **airships** (in the air or on the ground) which already were dropping bombs on cities. As the war progressed, considerable improvements as well as increases in weight were introduced. In 1914, 10lb bombs (4·5lb explosive) were sometimes dropped by hand of observer over the side of his cockpit. But this crude method was replaced by release mechanisms for bombs stowed internally or externally on racks. On 15 September 1915 a German zeppelin dropped a 660lb bomb on London and by 1918 the British had a 1,650lb bomb in service. Moreover there had been a universal improvement in the reliability and destructive effect of bombs – though no great advance in accuracy of delivery.

Prior to **World War II** bomb development, constrained by parsimony, was sporadic. Indeed, British bombs in 1939 were largely inferior to those of 1918. Much attention was focused on armour-piercing, **anti-submarine** in **naval warfare**, **gas** and incendiary bombs – these for attacks on cities and as target markers, the latter scattered from containers known as 'Molotov bread baskets'. The combustible filling of the last was usually metallic (using thermite and magnesium, for example), jellied petroleum (*see* **Flame warfare**) or chemical (using elements like carbide and phosphorus).

During the war more powerful explosives and sophisticated fuses were introduced and greater attention given to blast-bombs for use against strategic targets. For example, the Germans, without accuracy, used sea-mines (*see* **Mines, Naval**) suspended from **parachutes**. The British produced so-called high-capacity (HC) bombs weighing 12,000lb (5,200lb explosive) and a mighty 22,000lb medium-capacity (MC) bomb (9,135lb explosive) with deep-penetration fuses to destroy hardened targets such as **battleships** and concrete submarine shelters. There were, indeed, many kinds of specialized bombs, such as the spectacular 9,150lb 'bouncing bomb' (designed by Dr Barnes Wallis to destroy dams); and rocket-assisted bombs to improve accuracy and penetration.

Yet precision on target remained elusive, compelling the delivery of excessive quantities of explosive by so-called area bombing to achieve the desired results. And calling also for the development of elaborate pyrotechnic, sky or ground target indicator (TI) bombs for aiming-marks. The introduction of **atom** and **hydrogen** bombs simply carried area bombing to excess, albeit with persuasive political **deterrent** effect and as a spur to invention of precision bombs.

Not until the **Vietnam War**, with the invention of **rockets and guided missiles** along with smart munitions, was precision of delivery assured, by accurately navigated aircraft brought within reach of the target, leading to the further improvements so devastatingly and economically demonstrated during the **UN–Iraq War**. Nevertheless, a requirement for area weapons remains, such as cluster, anti-personnel and **anti-tank** bomb dispensers; and the British JP233 airfield-attack weapon system which, when delivered from as low as 200ft, scatters a mixture of 30 parachute-retarded, concrete-piercing submunitions, to disrupt runways, and 215 delayed-action mines to hamper repairs.

Bows and arrows. Flint arrowheads have been discovered from the palaeolithic period, though nothing is known about the bows. Stone carvings show the Egyptian armies of the pharaohs armed with a short bow of unknown material with a string which was only drawn to the chest; therefore probably ineffective over 150 metres. The Assyrians made much use of bows, sometimes fired from **chariots**. Greek archers provided covering fire for **infantry**; but the Romans neglected the art, perhaps because they were unwilling to spare time for the necessary intensive training.

Skill was a major factor in the contest between the *crossbow* (a short wood or **steel** bow mounted on a stock, effectively firing metal-tipped arrows 300 metres (probably invented in Italy *c.* 1100 AD) and the 5ft 6in Welsh wooden *longbow*, with 250 metres effective range, of *c.* 1160. The crossbow was less demanding of physical strength, easier to release by a trigger from the prone position and from behind cover. The more accurate longbow had four times the rate of fire and, as used by expert massed British archers, often was likened to **artillery**; although arguably a shower of arrows is better likened to **machine-gun** fire (*see* **Hundred Years War** and **Crécy, Battle of**). By the 17th century both weapons were made obsolete by firearms.

Braun, Werner von (1912–77). A German physicist who specialized in rocket motors and in 1934 became technical director of the Army rocket establishment at Peenemünde. Here, against his desire to develop rockets for peaceful purposes, he was ordered to create the series of missiles that included such surface-to-surface missiles as the **V1** and **V2**, and numerous anti-shipping and **anti-aircraft** weapons.

In 1945 he was taken to America where he continued to work on rockets for both military and civil use. When the Russians launched their first satellite in 1957, von Braun was told in November to match it – which he managed to do within two months. In 1960 he came under the National Aeronautics and Space Administrations (NASA), with the task of developing the big rockets that not only had important military roles in placing guided missiles, **surveillance** and **communication** satellites in **space**, but also, in 1969, carried the first men to the moon as well as mapping the surface of Mars. Thus he began to satisfy his original ambition of exploring space for peaceful purposes.

Breitenfeld, Battle of. During the **Thirty Years War**, in an attempt to coerce Saxony, Marshal Johan Tilly of the Catholic League invaded with an army of 21,000 **infantry**, 15,000 **cavalry** and 27 guns. He was opposed by 16,000 Swedish infantry, 10,000 cavalry and 50 guns, plus 20,000 low-grade Saxons with 20 guns, all under command of **Gustavus Adolphus**. On 17 September 1631 Tilly deployed conventionally with infantry squares in line and cavalry on the wings. Employing his new, more flexible mix of units of greater mobility supported by light and field **artillery**, Gustavus placed infantry squares in depth with cavalry on the wings and in reserve.

Due to poor **C^3**, Tilly attacked piecemeal. His left-flank cavalry charged and was repulsed by artillery and cavalry. Then the

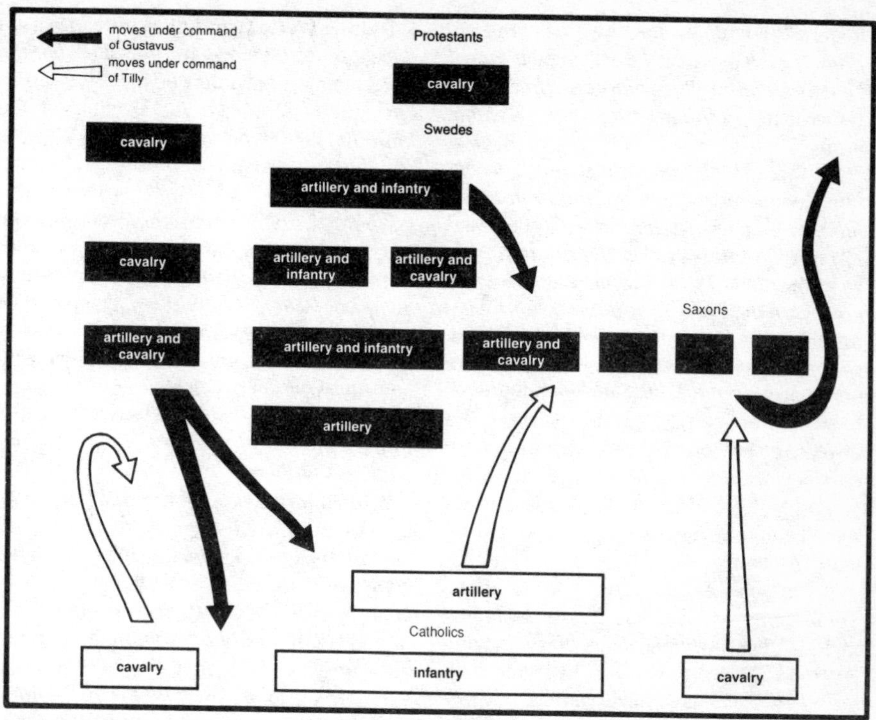

The Battle of Breitenfeld, 1631

right-flank cavalry hit the Saxons and drove them off; persuading Tilly to incline his right-flank infantry against the now exposed Swedish left. But they were met by artillery, which generated three times his guns' volume of fire and threw them into confusion. Whereupon, as his artillery now pounded Tilly's centre, Gustavus led his massed cavalry in a well-coordinated, overwhelming charge, followed by a pursuit which captured all the Catholic guns and **logistic** train and killed or wounded 13,000 survivors.

Bridging, Military (*see also* **Engineering, Military**). Almost inevitably during military operations the **tactical** and **logistic** need to bridge gaps arises. So long as loads are light and the gap narrow, a bridge of beams and planks with a trackway is generally sufficient. But in early times **siege engines** and, later, **artillery**, **railways**, **motor vehicles** and **tanks** called for much heavier materials and more elaborate equipment, especially if wide gaps had to be tackled and rafting or ferrying was inadequate. Furthermore, if bridging is needed in an assault, speed of construction is crucial. Indeed, there is almost always pressure for haste and improvisation to satisfy operational requirements.

The simple log bridge or stone piers of the Stone Age, cut or quarried from nearby sites, Darius the Great's floating bridge across the Bosporus and the Emperor Xerxes' very elaborate double bridge across the Hellespont in 480 BC, made up of 774 anchored boats, with bows facing the current, are examples of what could be done in ancient times, and which remain basic methods to the present day. Bridging poses immense logistic and **training** demands, which call for economy measures. **Alexander the Great** tried to economize by employing specialists to make best use of the inflatable skins carried plus materials found on site. Also, in order to simplify bridging, heavy siege engines would be either broken down into components for transporting or built entirely on site from local materials.

By the end of the 17th century, wood-frame pontoons covered by leather or metal and horse-drawn on carts were to be found in most European armies' logistic tail. Basically they look much the same in the 20th century, except that canvas or rubber substitutes for leather and metal for wood and, sometimes, watertight compartments are included. Sometimes, also, the pontoons have been open boats, hand-propelled by paddles or oars, or by outboard motors. These were relatively quickly assembled – unlike trestle or pile bridges which in the 19th century sometimes took days or weeks to build by manual labour.

The much heavier railway loads of the 19th century and road loads, including **tanks**, of the 20th created a military bridging revolution. Allied to the weight factor, calling for heavier and bulkier bridge components, was the demand, along with labour saving, for high-speed construction to maintain the momentum of operations made possible by mechanized mobility. To span short gaps tank or tractor-borne quick-laying box-girder or tubular-steel bridges were developed, though rarely used, during **World War I** and these were developed most successfully to carry 40-tonne loads for assault crossings during **World War II**.

But for gaps in excess of 10 metres and/or loads of more than 40 tonnes, the British panel bridge was developed. The invention of Donald Bailey in 1940, it was a so-called 'through' structure with a single-track road or railway between two main side girders built up of easily assembled 10ft x 5ft steel panels that could be manhandled. All manner of panel permutations could construct all kinds of bridge, including pontoon and suspension, carrying loads from 9 to 120 tonnes, depending on length of span. And they could do so very quickly indeed.

The need for concealment or deception purposes to conceal bridges from observation led during World War II to water-gap bridges which could be sunk and refloated (as the Russians did sometimes) or rapidly taken to bits and reassembled. An example of the latter is the Frenchman, Colonel Gillois's amphibious bridge comprising individual, self-propelled, wheeled units driven, independently, straight into the water at the site for use as a raft or, linked to other units, a continuous bridge which can swiftly be uncoupled and driven away. The vast majority of rivers and canals, however, are less than 20 metres wide, enabling mechanically operated, folding or expanding vehicle-borne bridges to be easily laid and recovered.

Britain, Battle of. Delayed until 12 July 1940, as was **Hitler**'s directive to invade Britain, and disputed as is its starting date, the battle actually began on 5 June with scattered Luftwaffe night sorties. Both sides realized invasion was impracticable without German air superiority and that the key was the **fighter** struggle. But Reichsmarschall **Göring** thought the battle might be won by **strategic bombing** alone. With 2,790 aircraft available, including 760 excellent single-engined fighters, and adequate reserves, he took a chance against Air Marshal Sir Hugh **Dowding's** RAF Fighter Command, with 900 single-engined fighters and meagre reserves, particularly of pilots.

Attacks against English Channel ports and shipping began in earnest on 10 July. RAF defence of the convoys was costly against a slightly superior enemy and called into question the wisdom of fighting at a disadvantage in an attritional struggle. As it was, the Royal Navy was compelled, by air attacks on 29 July, to withdraw its anti-invasion destroyers from Dover. Meanwhile, Göring settled upon 10 August for Eagle Day, when the onslaught upon Fighter Command would start but, owing to bad weather, had to postpone it until the 13th. By then the attrition rate in fighters was in German favour and bombing of six **radar** stations on the 12th had knocked out one for 11 days. But, mistakenly thinking such targets were too hard to destroy, the Germans mostly left them alone thereafter.

In heavy air combat between 13 and 23 August, in which they penetrated close to London and lost aircraft disproportionately to results achieved against Fighter Command, the Germans forfeited some advantages by ignorance of where the enemy fighters were based, and how they were controlled and commanded by radio from sector airfields. As a result, their bombers often wasted effort on secondary targets, and their losses rose further when the fighters ran into trouble at their extreme range of endurance.

After 24 August the Germans placed their daylight offensive under Field Marshal **Kesselring**'s 2nd Air Fleet, and concentrated its attacks, with bombers more closely escorted by fighters, against fighter airfields. Fighter Command deteriorated as damaged airfields and weary pilots, of whom there was an acute shortage, impaired its combatworthiness. Shortage of **anti-aircraft** guns often compelled fighters to concentrate on defence of their own bases and factories. By 6 September the fighter contest was well in the German favour and the vital sector stations in a parlous state – though the Germans neither knew of nor suspected their significance.

Next day everything changed. Stung by an air raid against Berlin, and choosing to believe Fighter Command was beaten, Göring, at Hitler's command, switched the attack to London. The pressure was taken off the airfields and transferred to the German fighters, which lost their tactical advantage at extreme range and were unable adequately to protect the bombers. Though causing heavy damage to a few strategic targets, the Germans were incapable of a decisive strategic bombing effort. Losses of 60 aircraft during a maximum effort on 15 September, when Fighter Command intercepted in apparently massive strength (and lost only 26 fighters), convinced Hitler, as the weather got worse, that invasion was impossible. He postponed it on 17 September and cancelled it on 12 October.

Until May 1941 the night bombing of Britain, which had been stepped up on 24 August, would take over (as part of the diversion for the planned invasion of Russia) – the inconclusive so-called Blitz of London and other major cities. Between 10 July and 31 October, Fighter Command had 449 aircrew, including Canadians, New Zealanders, Belgians, Czechs and Poles, killed. Up to 30 September it had lost 678 fighters against German losses of 1,099 machines of all types. It had won a victory which changed the course of the war.

Bronze. As a rare alloy of copper and tin in variable proportions, bronze is said to have been discovered *c.* 6,500 BC, but it only came into more general and military use *c.* 4,000 BC. In all likelihood it was first used for **edged weapons** and arrowheads, and only later as cast **armour**, when its hardness and resistance to blows from both bronze and **iron** weapons was realized.

To make armour-plate, tin was added to copper in the proportion of about eight to one, whereas in the softer alloys sometimes used for the barrels of early **artillery** the proportion could be as little as one to 16. But long before bronze was tried for gun-making it had been supplanted, as armour, by **iron** and then **steel**.

In due course, however, when alloyed with metals such as steel, phosphorus, antimony, manganese and **aluminium**, it has acquired numerous important functions in machines for bearings, bushes, valves and pump plungers, which require strong metal and can be relatively easily cast and worked. Also, alloyed with silicon greatly to increase tensile strength, it has been used for **telegraph** wires.

Brooke, Field Marshal Alan (Lord Alanbrooke) (1883–1963). A talented gunner who, in the 1930s, played a leading part in the development of **artillery**, **anti-aircraft** fire and **training** of the British Army. His leadership of II Corps in the retreat to **Dunkirk** and withdrawal from France in June 1940 was brilliant, but were his last actions in the field. Appointed to command Home Forces in July 1940, with the tasks of defence against imminent invasion and of reforming and training a defeated Army, he had made impressive progress before Winston **Churchill** made him Chief of the Imperial General Staff in December 1941. Thereafter, as Chairman of the Chiefs of Staff Committee and a powerful member of the Allied Combined Chiefs of Staff, he would play a dominating role in shaping Allied strategy. By reaching a working agreement with the American Chief of Army Staff, General **Marshall**, he was able to persuade the Americans to invade North Africa, instead of a very risky invasion of Europe in 1942; and to curb Churchill's often dangerous impetuosities while he skilfully steered a course towards a victory of which he was very much a master architect.

Buller, General Sir Redvers, VC (1839–1908). An officer of wide experience and outstanding bravery in colonial wars who, as Quartermaster-General and Adjutant-General successively, implemented many much-needed British Army reforms in the 1890s, including the **logistics** system. Too late in life he was made C-in-C in South Africa in 1899 against the Boers. Lacking sufficient troops and adequate logistic support, his dispersed forces suffered a quick succession of humiliating defeats. Replaced by Field Marshal Lord Roberts, he remained in field command and, after much travail, conquered the east Transvaal. Upon return to a command in England, he was forced to retire by strength of public indignation at his re-employment.

Bureau, Lord Jean (d. 1463). A highly innovative engineer and artillerist whose handling of the French **artillery** was decisive in the closing stages of the **Hundred Years War**. Between 1439 and 1450 at Meaux, Creil and Pontoise he cultivated accuracy. At **Caen** in 1450 the English surrendered to gunfire alone. And that same year at Cherbourg he emplaced 'waterproofed' guns below high water, reopening fire when the tide went out. Finally, at Castillon in 1453, he won the war's last battle by repulsing the English relief force with intense fire from his **siege** guns, thus demonstrating artillery's decisive power and versatility.

Burma Road. When the Japanese cut off China from the sea in 1939, the Chinese began to use a narrow, 681-mile winding road from Kunming to Lashio. But in July 1940 the Japanese politically coerced the British into closing it. It was a matter of priority for the Allies in 1942, after Burma had been seized by the Japanese, to reopen the road. But not until January 1945, after hard campaigns, was this accomplished. In the meantime work on a 478-mile spur from Ledo was put in hand, but of far greater immediate use was the air lift over the 21,000ft so-called Hump. General Stilwell initiated it in June 1942 to provide fuel for American aircraft operating in China. Mostly civil aircraft were used. Sometimes over 6,000 tonnes a month were carried in all sorts of weather.

Burma wars. The steep-sloped jungles of Burma, intersected by rivers and paddy-fields, have frequently been the scene of internal struggles and wars with China, Laos and Manipur. Often these wars were (and are) **guerrilla** in nature. Increasingly after the *First War* with Britain (1823–6), when Burma invaded India, the rivers were vital to mobility. In 1824 Britain captured Rangoon and in subsequent **amphibious** operations advanced up the Irrawaddy River to Prome. In the *Second War* (1852–3), after further difficulties with the aggressive Burmese, the British repeated their intervention up the Irrawaddy and Salween rivers. Finally, after the *Third War* in 1885, they annexed the country once an amphibious force in 55 steamships had captured the capital, Ava.

World War II brought a return of major conflict in 1942 when the Japanese invaded, outfought the British and occupied the country. The Japanese held on to their possessions when British/Chinese/American **jungle warfare** offensives in 1942, 1943 and early 1944 attempted to reopen the supply route to China via the **Burma Road. Logistics** ruled.

One aspect of the 1944 offensive was General Orde **Wingate**'s **glider**borne interdiction of Japanese lines of communication in north Burma, which was aborted when the resupply aircraft were needed to help stem the Japanese invasion of India, in March, via Kohima and Imphal. For General Sir William **Slim** relied upon **transport aircraft** to supply the British garrisons cut off by the invaders. He succeeded after a two-month siege, compelling the starved and disease-ravaged Japanese to retire when their supplies ran low and because, unlike the Allies, they did not possess anti-malaria medicine.

Slim's pursuit further caught the Japanese by technological surprise when he continued operations during the monsoon. Supply aircraft somehow got through and prodigies of **engineering** improvisation opened routes through the jungle and along the rivers to move men, vehicles and supplies to Mandalay and Meiktila, which fell in March 1945. Thereupon a pincer movement to recapture Rangoon was won by its amphibious claw, sent by sea from India, and the surviving Japanese were marooned in the jungle.

Bushnell, David (*c.* 1742–1824). A Harvard graduate, Bushnell in 1775 demonstrated that dynamite would explode under water. The inventor of the first operational **submarine**: in 1776 during the **American War of Independence**, he built *Turtle*, a one-man, hand-cranked, submersible, spherical boat equipped with an **explosive** charge with screw-attachment device. The crewman, Sergeant Ezra Lee, used a glass window in the small conning tower and a compass for **navigation**. Twice by night, after getting beneath the British flagship, Lee found that the screw would not penetrate the ship's copper-sheathed bottom. The attempt was not repeated after *Turtle* was pursued on the surface. But the British felt constrained to withdraw to seaward, thus weakening their **blockade** of the Hudson River.

C

C^3. Modern jargon abbreviation standing for the joint functions of Command, Control and **Communications**, and therefore a subject of increasing and vital importance as the sheer volume of information generated and fed into military communication networks by surveillance and other sources of information threatens to overwhelm commanders and staffs. Considerable attention is being paid to rationalizing the presentation of essential information, through ADP (Automatic Data Processing) and visual displays, and how to separate essentials from non-essentials. Necessarily, **electronic** and mechanical aids are required. But since these often need a stable environment in which to operate, their use in forward combat zones, where machinery must be extremely simple and rugged, has to be limited. Over-dependence upon mechanical means in unstable situations, when human judgement is so often at a premium, can be very risky for the commander, who also has to take into account such variables as human performance, mechanical reliability and unperceived enemy intentions and moves. (*See also* **C^3I**.)

C^3I is simply the inclusion of the vital **intelligence** factor as a very important adjunct of C^3. It is easy to forget the number of times in which the lack of close and discerning integration of Intelligence with Command and Control has led to military disasters, and how often it has been failures in human as well as signal **communications** which have caused this. It must not be overlooked that, like C^3, C^3I in one form or another has been a vital element in military operations since the beginning of time.

Caen, Battles of. In 1346, during the **Hundred Years War**, the fortified bridges over the River Orne at Caen blocked **Edward III**'s advance to **Crécy** but fell after a sharp **infantry** assault. In 1417, however, Henry V was compelled to mount a three-week **siege** with **artillery** to breach the walls, the converging assault on the city culminating in great slaughter of the French and the surrender of the castle without further resistance. In 1450, however, it surrendered to the French after three weeks' bombardment by Jean **Bureau**.

On 6 June 1944, the communication centre of Caen was General **Montgomery**'s main objective on the first day of the Battle of **Normandy**, an ambitious target for I British Corps. Against a surprised enemy, whose reactions to the landings were by no means positive enough for the occasion, its attainment was, however, an extremely close-run thing. Troops who had suffered from a rough landing in surf and had been delayed breaking free of the beaches, came within two miles of the city at last light. By then, however, a German armoured counter-attack had forced the leading British and Canadian elements to pause, giving time for a reinforced perimeter defence.

Hitler and Field Marshal Rommel were as anxious to hold Caen as was Montgomery to capture it, each realizing that possession was the key to Allied exploitation of the more open terrain to the southwards and deep into France. But although I Corps tried next day to reach the city, it was also distracted by fierce German attacks and the need to help link up with XXX Corps on its right. While directing that Caen must be taken, Montgomery regarded the city not only as a vital pivot but also, as planned from the outset, a strategic bait. A threat there would distract the Germans from First US Army as it consolidated on the right, tackled Cherbourg and prepared for the decisive break-out. Every effort by I Corps to work around Caen's flanks, with thrusts towards Troarn and against Carpiquet airfield, lured the German armoured forces into an attritional struggle which was very expensive for both sides.

Between 25 and 29 June, VIII British Corps advanced to Hill 112, simultaneously beating off Rommel's major counterstroke. The Germans now knew Caen was untenable, but Hitler would not permit withdrawal. On 4 July the Canadians stormed Carpiquet airfield in a desperate fight, but still without persuading the Germans to evacuate the city. So, on the evening of 7 July, as a prelude to assault next day, heavy **bombers** dropped 2,560 tonnes of bombs which hit the city much harder than its defended perimeter and largely blocked access. When three divisions advanced next day it was against stiff resistance from fanatical Germans in the centre, but yielding troops on the flanks. Mainly it was craters and ruins blocking the roads which made it easier for the Germans to slip away on the 8th/9th, blowing the river bridges as they departed.

Caesar, Gaius Julius (100–44 BC). A member of the Roman nobility, Julius Caesar was an able lawyer of immense political ambition whose first military operation was at the head of a private naval force he raised to execute pirates who had captured him. He developed into a military genius, the most distinguished of whose numerous campaigns on his way to becoming Dictator of the Roman Empire were the suppression of uprisings in Spain (61–60) and the defence of Northern Italy against raiding Helvetii, Germans and Gauls. The conquest of Gaul (58–51) was far more important than the incidental though related **amphibious** raids on Britain in 55 and 54 BC.

More by improvement than innovation, Caesar trained his legions to a high pitch of mobile efficiency, based on an excellent **logistic** system linked to extensive road-building between the numerous **fortifications** he established to control his conquests. In **siege warfare** he had few equals. His seven-volume *Gallic Wars* is a masterpiece of **propaganda**. His tragedy was the fact that his numerous victories in the Roman Civil War (50–44) ended in his assassination because he permitted his enemies too much latitude.

Cambrai, Battles of. The city developed as an important route centre from Roman times and was fortified by Marshal **Vauban** in the 17th century. Not until **World War I** were large battles fought in its immediate vicinity.

The important route centre of Cambrai fell to German cavalry in August 1914 and, as trench warfare began, became a main base for the German armies in Northern France. When construction of the **Siegfried Line** began in 1916, one of its strongest sectors shielded Cambrai, a quiet front to which German divisions were sent to recuperate, earning it the name 'Sanatorium from Flanders'.

In June 1917 Lieutenant-Colonel J. F. C. **Fuller**, senior staff officer of the British Tank Corps, submitted a plan for **tank** raids on favourable ground, such as at Cambrai. Then, and again in August, it was turned down by the British Expeditionary Force's C-in-C General **Haig**. But in October, when the third Battle of Ypres was plainly a disastrous failure, Haig, yearning for a whiff of success, revived the idea but augmented it to a full-scale offensive at Cambrai of two days' duration. Third Army was to attack at dawn on 20 November with six divisions in the assault, supported by the entire Tank Corps with 476 tanks, 1,003 artillery pieces,

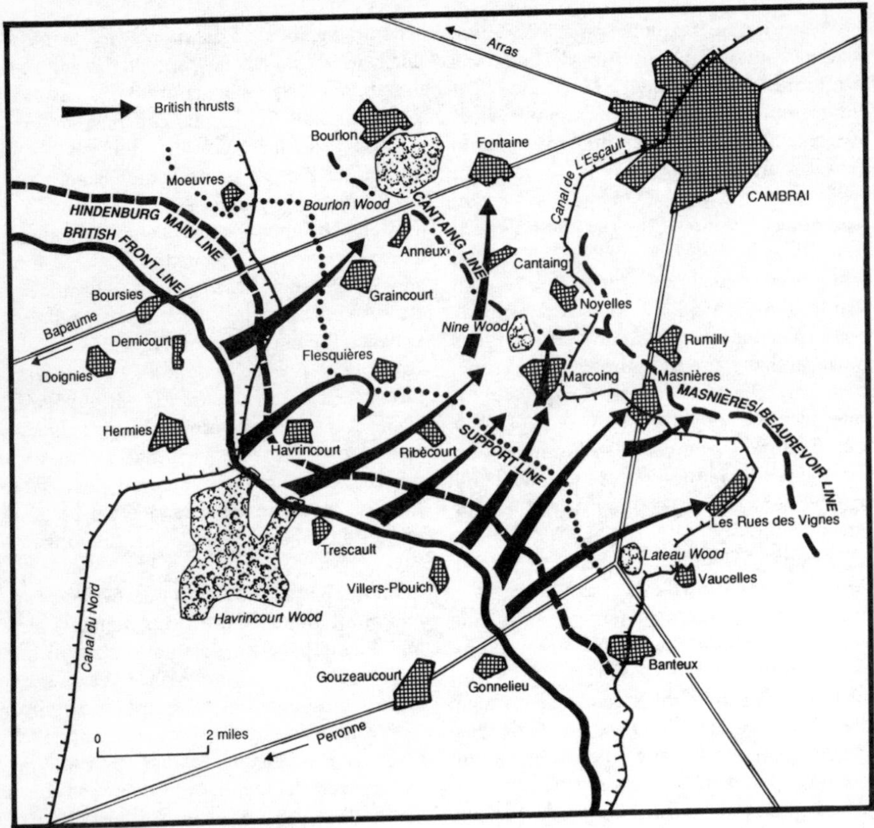

The Battle of Cambrai, 20 November 1917

massed machine-guns and air bombing. When the Siegfried Line had been breached, the Cavalry Corps was to pass through and seize the Bourlon Ridge and Cambrai, thus isolating the Germans' Arras front. The greatest secrecy was observed in assembling the troops by night; and, to obtain surprise, the **artillery**, for the first time, adopted fire by silent registration, off the map, in order not to disclose the intention.

Fuller dictated the **tactics**. A wave of tanks headed the attack, subduing enemy machine-guns, crushing the deep wire entanglements and filling the deepest trench with the wood fascines carried on top of each tank. A second wave, with infantry in close attendance, then occupied the trenches, whereupon tanks drove to the final objective.

Almost total surprise was achieved. The British tactics worked well. Nearly all the day's objectives were reached. British losses were remarkably few in men and tanks while those of the enemy were over 10,000 and included 123 guns. But the Germans, by staunch defence of Flesquières, managed to prevent the British reaching the Bourlon feature; and although the **cavalry** was poised to overrun a broken enemy, it failed to seize the opportunity. In consequence, the Germans, although on the point of a wholesale withdrawal from the Arras front, held on and, in the ensuing seven days, checked desperate British attempts to seize the Bourlon Ridge.

On 3 December the Germans counter-attacked. Using their latest infiltration tactics behind a surprise bombardment, they drove the British back to Flesquières and seized ground on the southern flank which had been in British possession before 20 November. On balance of losses and prestige, honours were even. But tank dominance and surprise tactics marked a turning-point in the history of war.

Cambrai would be the scene of renewed heavy fighting in March 1918 at the commencement of the **Hindenburg** offensive when the British were pushed back towards Bapaume. And yet again when, on 8 October, a British attack on an 18-mile front, which included 82 tanks (all that could be made fit after an exhausting pursuit), broke through, brushed with German tanks, and entered the badly damaged city on the 9th.

Battle came to the city once more on 8 May 1940 when a German Panzer division, taking advantage of the total French collapse, stormed in with scarcely any resistance. Four days later, however, French tanks broke back into the city as spearheads of a small French counter-attack which savaged a German infantry division. The Germans countered with bombing, strafing and **anti-tank fire** from 150 yards which finally drove the French off.

Cameras (*see also* **Photography**). The camera is an optical instrument for taking photographs, film or **television** pictures. Developed from the *c.* 16th-century camera obscura (which projected an image through a hole or lens on to the wall of a darkened room), the modern camera may be said to date from 1822, when Joseph and Claude Niepce managed to copy a permanent photograph on chemically treated, light-sensitive glass. Initially envisaged as an art form, it was not until 1858, when Gaston Tournachon, using a wooden box camera, took aerial pictures from a **balloon**, and 1859 during the Italian War of Independence, that a military, plus a **propaganda** role, for the camera had been established.

By then the need for lenses along with demands for easier operating, greater robustness, reduction in size and weight (with help from collapsing bellows or telescopic boxes), and speed of exposure and developing of film were having their effect. The invention by Leon Warnerke in 1875 of a spool of film giving 100 exposures was a major step towards the Eastman Company's simple, mass-produced Kodak No. 1 camera in 1888. This took circular, 2½in pictures on paper film, which was developed after camera and film were returned to the manufacturer for processing. Miniature cameras, much in demand by spies, would soon follow in company with focusing lenses and mechanical shutters to regulate aperture size and exposure time.

Meanwhile the invention of transparent film and the urge to produce 'moving pictures' led (under the sponsorship of Thomas **Edison**) to W. K. L. Dickson's invention in 1889 of the kinematograph camera, the forerunner, along with 35mm film, of cinema photography and a vast new technological industry for scientific as well as entertainment purposes. For, as time went by, it became apparent that high-speed, high-definition cine-cameras, sometimes enhanced by flash lighting, had a role in the monitoring of **research and development** projects.

Television, image intensifiers, infrared, X-ray and all manner of electronic cameras in the 20th century simply augmented the ways of employing (and misemploying) the camera which, as soon was demonstrated, could be used for faking and all manner of deceptions in the presentation of **intelligence**. But for **surveillance** from space as well as on earth, the camera is a most potent, though not infallible, instrument.

Cannae, Battle of. In the Second **Punic War**, after the Carthaginians had crossed the Alps and beaten the Romans at Lake Trasimene in 217 BC, **Hannibal**, with 40,000 **infantry** and 10,000 **cavalry**, baited a trap at Cannae. He chose a day when it was the hot-headed Consul Terentius Varro's turn (instead of the more cautious Aemilius Paulus) to command the Roman army of approximately 70,000 infantry and 7,000 cavalry. Deploying his army in the bend of the River Aufidas, Hannibal not only heavily reinforced both flanks

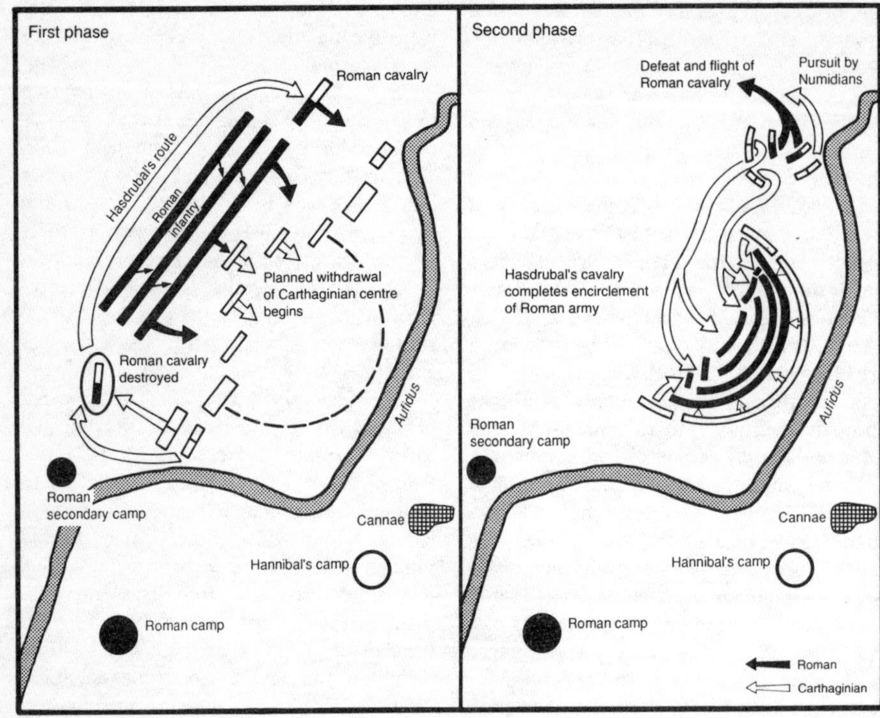

The Battle of Cannae, 216 BC

with infantry and his entire cavalry, but led a weakened centre against Varro, whose cavalry was also on the wings, guarding the flanks of infantry concentrated in three lines.

Falling to the temptation to overwhelm Hannibal's centre, Varro advanced his infantry as Hannibal's cavalry charged the Roman cavalry and drove it from the field. Then, as the Carthaginian centre withdrew, their overlapping infantry mass on the wings, joined by the cavalry returned from pursuit, swung inwards. The result was a classic double envelopment and destruction of the encircled Roman army, less than 10,000 of whom fought clear. This battle is regarded as the acme of total victory.

Caporetto, Battle of. After the eleventh Battle of the Isonzo, in September 1917, both the Italian and Austrian armies in Italy were war-weary, the latter near collapse. General **Hindenburg** decided to back a sound Austrian plan for a counter-offensive on the Isonzo with 14 divisions in a newly formed Fourteenth Army under General Otto von Below. It began in vile weather on 24 October on a frontage of 25 miles employing General Oscar von Hütier's new infiltration **tactics**. In difficult terrain behind a surprise bombardment, General Luigi Cadorna's armies were routed and rolled back to the River Tagliamento. The insubordination and indolence of Italian commanders, with a breakdown of **communications**, compounded the disaster. The Austro-Germans, to their surprise, advanced more than 25 miles in a week, eventually capturing 275,000 prisoners, 2,500 guns and masses of equipment.

Responding to desperate pleas by the Italians, who showed signs of total disintegration, the British and French sent 11 divisions to the River Piave front where Cadorna made a stand. But although their enemies pursued with vigour, **logistic** problems soon inter-

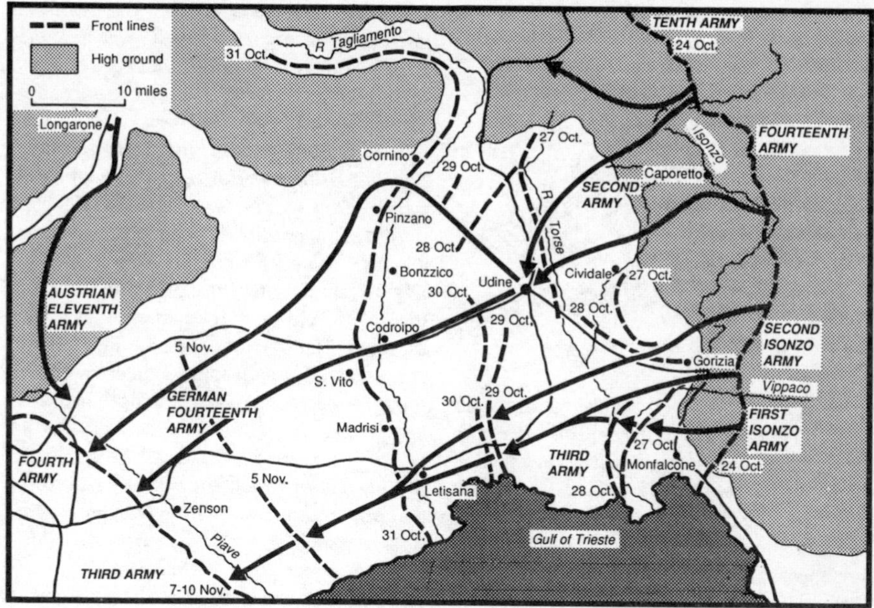

The Battle of Caporetto, October 1917

vened and the pace slackened. Cadorna was replaced on 8 November by General Armando Diaz, who confirmed existing arrangements and garnered the credit for Cadorna's leadership in the final days of retreat. For on the Piave the Italians, without direct Anglo-French support, held off an enemy who had outstripped his supplies. A renewal of the offensive by von Below on the 23rd was soon brought to a halt; but already the Germans were withdrawing their best troops to Germany.

Meanwhile, as a direct result of the Caporetto experience, a Supreme War Council was formed to improve the Allies' coordination.

Carnot, General Lazare (1753–1823). An engineer officer and expert in **fortifications**, Carnot was a Republican who became a deputy in the French legislative assembly in 1791. As a leader of its Military Committee at the start of the **French revolutionary wars**, he also commanded none too skilfully in the field. But as a man of principle, steeped in political intrigue, his work in the Ministry of War to restore the Revolutionary Army to a sound organization and administration under professional officers he released from prison, gave France, and **Napoleon**, vital strength. In 1794 he formed divisions of all arms (as prescribed by Duc de Broglie in 1759) in which the **infantry** element operated in flexible battalion columns (*see* **Army organizations**).

Carrack. A three- or four-masted sailing ship of the 16th century weighing up to 1,200 tonnes which was a development of the *c.* 50–150-tonne caravel promoted for voyages of exploration by **Henry the Navigator** of Portugal in the late 15th century. Unlike the caravel with its simple curved stem and plain transom stern, the square-rigged carrack was often provided with high fore- and after-castles where, usually, **artillery** was mounted. With its speed and capacity to sail closer to the wind than clumsy Spanish **galleases** and **galleons**, the English carrack had a useful tactical advantage against the Spanish Armada in 1588. Frequently modified to the

Carrack, 16th century

point of obsolescence (notably through Sir John Hawkins's elimination in the 17th century of the forecastle, significantly to improve the English galleon's seaworthiness), the carrack nevertheless counts as the forerunner of all major commercial or warships until the appearance of steamships in the mid 19th century.

Cavalry. As the principal arm of land-warfare mobility and, occasionally, shock action, the cavalry had to rely, until the 20th century, on the very vulnerable and physically limited **horse**. Throughout its existence, horsed cavalry's combatworthiness depended upon **sword**, lance, **bow**, **pistol** and **rifle** for armament; upon compromises between speed and **armour** for protection; and upon cosseting and extensive **logistic** services to sustain its operations. Before efficient bows and arrows and good armour were available, lightly equipped mounted men could operate fairly effectively against infantry in the open. In close or mountainous terrain or against infantry in position, however, it was rarely decisive, and against its own kind prey to the normal fortunes of generalship and combat chances.

Inevitably, therefore, cavalry did best when operating under **Alexander the Great** in open country and was of less use to the Romans when entangled in European mountains or forest. Similarly, the successive mounted hordes, such as Huns and **Mongols**, which swept in from the east, flourished on the steppes but failed if they did not adapt to European conditions. In fact there was no such thing as a golden cavalry age as sometimes is claimed. Mounted men operated best in special roles, notably for reconnaissance, or in suitable terrain; and desirably in collaboration with **infantry** and archers or **artillery**. It could be a powerful complementary arm which rarely was decisive on its own.

All the more absurd, therefore, have been cavalrymen's efforts irrationally and emotionally to perpetuate their existence. At the door of **chivalry** and aristocracy must be laid the blame for trying to perpetuate, beyond the point of practicability, the role of the horse in battle against the long bow, artillery, small arms and gas. The manner in which the French failed initially to bow to the lessons of **Crécy** is a prime example of doctrinal obstinacy. And the way great commanders such as **Gustavus Adolphus**, **Marlborough**, **Frederick the Great** and **Napoleon** used cavalry as a complementary combat arm to extend its viability well into the 19th century taught lessons in strategic and tactical wisdom.

Owing to significant improvements in firepower by the 1850s, however, horsemen were finding it increasingly difficult to carry out their traditional mobile roles of reconnaissance and shock action. It was often preferable to dismount for scouting and the discharge of weapons, and usually suicidal to charge with lance or sword. The **Crimean War** is better known for cavalry disasters than successes. The last successful massed, close-order charges took place, with 45 per cent losses, at Vionville in 1870, during the **Franco-Prussian War**, when General von Bredow's brigade carried out its celebrated 'Death Ride' against emplaced artillery, only to be run down by French cavalry.

An interim way ahead appeared in the **American Civil War** when cavalry found room to manoeuvre in vast undefended spaces where they could live fairly easily off the country. Nevertheless, great cavalry commanders, such as Generals J. E. B. Stuart, Philip Sheridan and Nathan Forrest, found no real profit in the charge. They preferred reconnaissance, the screening of infantry formations, deceptive manoeuvres and swift raids against enemy lines of communication (mainly **railways**). Carbine and pistol superseded lance and sword as the armament of what became mounted riflemen. And though these lessons were largely ignored by the apostles of shock action in Europe, British troops in colonial wars and, notably in the South African wars, used the same tactics in the veldt's vastness when hunting Boer **guerrillas** whose raiding depended upon the horse.

The application of the internal-combustion engine to land warfare sounded the knell of horse cavalry and transport. By 1914 it was apparent to the enlightened how much more reliable and easy to fuel, handle and maintain was the motor vehicle than the temperamental horse, which consumed so much bulky fodder and whose endurance and range of action was extremely limited by comparison. Throughout the opening, mobile phases of **World War I**, when 10 German cavalry divisions faced 10 French and one British; and when the Russians had 24 cavalry divisions to 10 Austrian, horsemen achieved but little. No sooner did artillery or machine-guns open fire than they checked. The only occasion a German cavalry formation had an opportunity to ride through a gap during the advance to the **Marne**, the horses needed their shoes replacing. French massed cavalry achieved nothing and the British division did its best work with dismounted rifle fire. On the Eastern front, at **Tannenberg**, cavalry's best service was by a single German division holding off one Russian army while the other was defeated.

Yet cavalry divisions, despite repeated demonstrations of their inability to survive under fire and the appalling wastage of horses' lives in the hardships of a war environment, remained in being – the Russians raising 54 in all. But not once on the Western Front did cavalry manage to find the G in Gap (as was the term) after infantry, artillery and tanks had made progress – not even at **Arras** and **Cambrai** in 1917, nor **Amiens** in 1918 when wide gaps were opened. Even in the Middle East, where cavalry found freedom for manoeuvre in the desert, they relied largely upon **artillery**, aircraft and **armoured fighting vehicles** (AFVs) to prepare the way.

Between the two **World Wars**, despite illogical opposition by cavalrymen but hastened by significant reductions in the horse population following mechanization of civil transport, most cavalry units were either disbanded or mechanized. By 1942 only a few horsed divisions were to be found on the Russian front, where space was available. In 1939 Britain's cavalry units became part of a Royal Armoured Corps along with the Royal Tank Regiment; in 1940 the US Army's Armored Branch superseded the Cavalry Branch. Traditional horsed operational functions, of necessity, were taken over by AFVs in all armies. Only the old traditions remained to cavalry regiments which survived in name.

Indeed, the concept of Air Cavalry, now sponsored so strongly within the US Army and demonstrated in **Vietnam** and elsewhere, is possibly closer in affinity to latter-day horsed than present-day armoured cavalry doctrine, whereby lightly protected, moderately hard-hitting **helicopters** and light aircraft, with their inherent speed and mobility, can carry out the reconnaissance, screening, raiding and mounted-weapon roles, but become involved at their peril in a face-to-face fire-fight in the manner of tanks.

Chaffee, General Adna (1884–1941). A US cavalryman who served with distinction in France in 1918 and in the 1930s became the leading American advocate of **armoured** forces. Though opposed by vested interests in the Cavalry Branch in particular, he managed to generate such strong interest in experiments with mechanized cavalry that in May 1940 he persuaded the US Army to create an entirely new Armored Branch, the foundation of a mechanized force which expanded to 16 armoured divisions and 139 armoured battalions during **World War II**.

Chariots probably date from the invention of the **wheel**, in the 4th millennium BC, possibly in Mesopotamia. They could have run on two or four wheels, sometimes were drawn

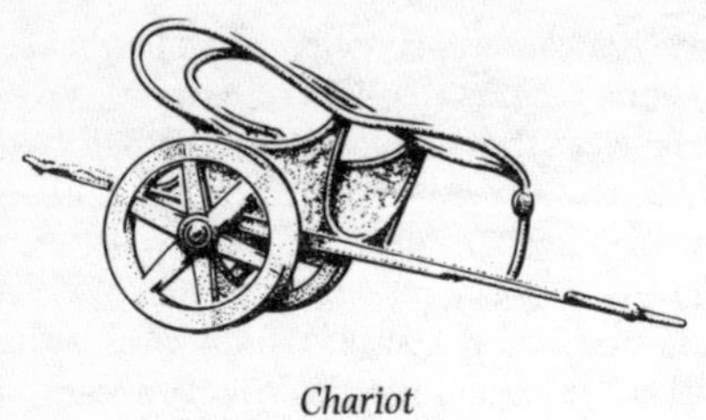

Chariot

by onagers but usually by one or more **horses** and, as a result, suffered from the same vulnerability and physical limitations as horsed **cavalry**. As a mobile weapon platform and a command-and-control vehicle (sometimes partially, lightly armoured) they were superior to the horse and best used to outmanoeuvre and harass the enemy from long range. Found in China and northern Europe, they were virtually extinct by AD 400, although wagon **laagers** and motorized **armoured fighting vehicles** became their true successors in providing a combination of mobility, fire-power and protection to man.

Charts and maps. There is evidence that even primitive people drew charts and maps of a kind, otherwise they could not have navigated as they did. Not until *c.* 600 BC, however, were rudimentary examples published in Greece, probably for trade and military purposes. But henceforward the **Greek–Persian wars** and, most important of all, **Alexander the Great**'s journey to the Caspian Sea, northern India and the Indian Ocean lent impetus to scientific mapping by geographers as well as by traders and invaders. By AD 150 Ptolemy's *Geography* had guessed, very inaccurately, at the shape of the known world. Until the 14th century, despite information from the **Crusades**, the results of travel remained rough and ready.

The urgent necessity for safe **navigation** at sea naturally spurred pilots to prepare ever more accurate, annotated charts based on observations and compass bearings and dead reckoning as their voyages, promoted by **Henry the Navigator**, extended beyond the Mediterranean and Europe in the 15th century. But although most charts and maps prior to the 16th century were highly unreliable, it is noteworthy that Matthew Paris's maps of Britain in the 13th century were quite accurate; and that in mid 14th century **Edward III** found his way to **Crécy** with a directness suggesting that he must have been in possession of a reasonably good map of France.

Gradual rationalization of presentation took a long step forward with the publication of Mercator's chart of the world in 1569. By then, indeed, cartography had become an extensive, international industry serving exploration, trade and war – particularly the latter as it gradually was realized that sea power was fundamental to national survival and that **artillery** required reliable maps for its employment. Thereupon navies and armies contributed strongly to **hydrography**, survey and the production of better charts and maps. The **French wars** of the 18th century and **French revolutionary wars** had enormous impact on all forms of cartography.

By the mid 19th century, under the impetus originally of years of European war, the Ordnance Survey in Britain was well established under firm military control; similarly the Institut Géographique National of France (**Napoleon**'s 1:100.000 survey of Europe was well under way at the time of his downfall) and the Landestopographie of Switzerland are other examples. Gradually the civilian element took over in these organizations (the British Ministry of Defence no longer has control of the Ordnance Survey) and in the USA, where military considerations were less paramount, the US Geological Survey had the responsibility. Only during **World War II** did the US military become closely involved in mapping with the Oceanographic Office (navy), the Aeronautical Chart Service (air force) and the US Army Topographical Command.

World War I, and more especially World War II, brought great progress in mapping, with vast areas of the world, previously unmapped, being covered, much of it by aerial survey. Post-1945, continuing tension between East and West has maintained an interest in and continuous improvement of maps and charts. Though mostly generated for military purposes, much of the work has played its part in opening up more remote areas of the world, most of which is now covered by terrain data at least: but many areas remain sketchy and Antarctica, for instance, will not be completely mapped for some years to come.

While the more detailed maps are used by land forces, equally important are the charts used by navies. Surprisingly perhaps, nautical chart coverage of the world as a whole leaves much to be desired, though the areas bordering continents and islands are reasonably well covered. For general navigation, ships are increasingly reliant on satellite fixes and on **radar** coverage, but the chart on which positions are plotted remains an essential ingredient.

Aeronautical charts are in effect small-scale topographic maps on which navigation aids have been superimposed. Principal features of the land that would be visible in flight are shown, to the exclusion of much other detail. Worldwide coverage is now extensive.

For military purposes specialized maps are often needed and one example developed during World War II was the normal topographic map with, printed on the reverse, a photomosaic of the same area showing much more detail than could be included on the standard map. Similarly, three-dimensional or relief maps, often crudely made on the ground (the 'sand-table' model) have long formed a useful aid to military tactical planning. An accurate relief map was costly and time-consuming to build: nowadays such maps can be constructed much more easily by the pantograph-router, which cuts a model, typically from plaster, as the contours are followed by the operator on a topographical map.

While computerization and **electronic** measuring have revolutionized the accuracy and speed of survey, the problem for an army in the field remains its ability to produce the enormous number of map sheets needed to meet, at short notice, the requirement for a particular operation. Modern printing and photocopying techniques have speeded up the process, but nevertheless the map reproduction element of any major headquarters is a crucial part of its successful prosecution of its tasks.

Chemical warfare. Many times prior to the 20th century **flame** and **smoke** generated by chemicals have been used in war. Inevitably, post 18th century, the vast amount of **research and development** devoted to the growth of the chemical industry, particularly in America and Germany, led to suggestions for the employment of noxious or poisonous gases. Indeed, British proposals to release burning sulphur against the Russians at **Sevastopol** in 1855, and a Federal idea to fire poisonous chlorine gas in shells during the **American Civil War** were put forward and rejected.

There was considerable repugnance at the idea of chemical warfare, exemplified by the 1899 Hague Convention renouncing 'the . . . diffusion of asphyxiating or harmful gases'. Britain and the USA were not originally signatories of this Convention, though the former did sign in 1907.

Despite the Convention, experiments were made before 1914 by the Germans, French and British and the stalemate on the Western Front led both sides to look again at the possibilities. The German General Staff, presented with a working system, were reluctant to embark upon the use of gas; they were persuaded that the Allies had no means of retaliation and so gave it a try. They first used an irritant tear gas against the French at Neuve-Chapelle in October 1914, then against the Russians at Bolimov in very cold weather in February 1915; and then in support of a limited attack in the Ypres salient on 22 April 1915 with 150 tonnes of chlorine. Two French divisions panicked but there was no follow-up and remarkably few casualties among totally unprotected Allied troops.

The gas was carried to the battle zone by rail in liquid form, filled into cylinders at the railhead and then discharged as a cloud at the front line. The inconclusive results of the first attack led the Germans to redeploy the special unit formed to operate the cylinders to the Eastern Front, where more success was achieved during the attack on Warsaw.

The German assessment of Allied capability proved to be very wrong and by July 1915 the British had formed several special companies for gas operations and the French had three. It was at first thought that a cloud expelled from cylinders would be the best method of dissemination, but the British found that gas-filled shell and special projectors firing gas bombs were more effective. Success of the operation depended entirely on the wind direction being both steady and favourable: there were a number of disasters on both sides resulting from the gas cloud being blown back on to friendly forces.

Both sides at first used chlorine but new developments were quickly in hand. The British favoured phosgene, which could be inhaled without immediate effect and was thus less easily detected; the Germans used this gas also. As the Allies developed more effective respirators, so the Germans sought for more effective agents. Two scientists – Lommel and Steinkopf – suggested the use of dichlordiethyl sulphide, given the code-name LOST from their two names, and this was developed into an effective gas known as 'mustard' to the Allies from its smell and the yellow cross used to identify its canister. Difficult to detect, it had the additional and unpleasant characteristic of causing blistering on exposed skin and with its low vapour pressure and solubility in water could lie for days on the surface of the ground. Mustard gas was used by the Germans up to the end of **World War I**, while the Allies had supplies of it available from early 1918.

The Italians had some successes with poisonous gases, including mustard, against unprotected people during the Italo-Ethiopian War in 1935–6. This episode between the World Wars, contributed to a terror of gas in the minds of the populace and politicians, with an impact on strategic considerations and **deterrence**. Undoubtedly it was fear of retaliation which deterred the use of gas in **World War II**, despite the availability in Germany of deadly new 'nerve' gases – Tabun, Sarin and Soman, which are lethal within fifteen minutes of skin contact or inhalation.

The first recorded use of nerve gases on a large scale was in the Gulf War (**Iran–Iraq War**) in 1984 when Iraq, exhausted and bankrupt, appeared to be losing the struggle against fanatical but unprotected hordes of Iranians. Along with mustard gas, these agents caused heavy casualties to soldiers and civilians, bringing several offensives to a halt and, no doubt, influencing a cessation of hostilities in 1988 – by which time improved protective clothing was mitigating the effect of the gases.

The creation by Iraq, with foreign help, of an extensive and sophisticated chemical-warfare industry, and her oft-boasted, but unfulfilled, intention to dispense its products by aircraft and rockets, needless to say was a major politico-strategic terror factor during the **UN–Iraq War**.

Chivalry. A code of conduct introduced by the Christian Church *c.* 700 in an endeavour to curb the savagery of the warlords who dwelt within **fortifications** and behaved virtually as they pleased after the collapse of the Western Roman Empire. In practice the rules, myths and rituals which it cynically evolved served to legalize malpractice under a cloak of courtesy. The feudal, armoured knights of various orders of chivalry invested in **cavalry** for financial profit by raising capital to equip themselves for warlike ventures in the hope of winning a dividend from pillage or ransom from affluent captives. Prisoners lacking affluence as likely as not were killed to minimize overhead costs.

Chivalry and the **armaments industry** flourished during the **Crusades** until the longbow (despite a typically ineffectual **disarmament** move to outlaw it) and **artillery** portended the decline of the knights and their castles. Constantinople's fall in 1453 and the final collapse of the Eastern Roman Empire marked its end as a military system.

Christie, J. Walter (1865–1944). An American inventor of front-wheel drive

motor vehicles who turned to self-propelled **artillery** in 1916 and in 1919 produced a $13\frac{1}{2}$-tonne **tank**. After 1921 he invented **amphibious** vehicles, and in 1928 built a tracked **armoured fighting vehicle** characterized by its shock-absorbed big-wheel suspension and capability to run on tracks or wheels. This was the prototype of the revolutionary $10\frac{1}{2}$-tonne T3 tank of 1931 with sloped **armour**, a turreted 37mm gun and a speed on tracks of 46·8mph. His tanks always required extensive redesign. T3 was rejected by the Americans but bought by the Russians as a model for their BT tanks and subsequent main battle tanks from T34 to T62; as well as by the British for their cruiser tanks Mark III to VII.

Churchill, Winston Spencer (1874–1965). Commissioned into the Hussars in 1895, Churchill's apprenticeship in war began as a correspondent during the Cuban War of 1895; as a soldier in the Malakand in India, 1897; as a correspondent in the Sudan in 1898 (when he charged with the cavalry at Omdurman); and in the Boer War as a correspondent, when he was captured but escaped.

It was when he became First Lord of the Admiralty in 1911 that his military talent was given rein: when **World War I** broke out he could claim credit for having the Royal Navy at a high pitch of readiness. And during the initial battles he played a personal part in the defence of Antwerp and the founding of Britain's air defences, while also setting in motion the development of **armoured fighting vehicles**. Churchill's sponsorship of the disastrous **Gallipoli** venture in 1915 put a temporary end to his higher direction of the war. For a few months in 1916 he served in the trenches and in 1917 became Minister for Munitions, in which capacity he had a voice in strategy before becoming Minister for War in 1919, charged with demobilization of the Army, British involvement in the Russian Revolution, and military assistance in the policing of the Empire, notably in India.

After 1921 Churchill was either in civil ministries or out of office. But he kept in touch and, with the emergence of aggressive Japanese, Italian and German designs in the 1930s, became a lonely and largely ignored voice of warning. Once more appointed First Lord of the Admiralty at the outbreak of **World War II**, he immediately took the lead in proposing aggressive strategies and, through radio broadcasts, reacquired a strong following inside Britain and abroad. When the Chamberlain government fell in May 1940, on the eve of Allied defeat in the West by the Axis, and Churchill became Prime Minister and Minister of Defence, it was his determination and inspiration which kept Britain in the war.

Lacking a classical education, his military training and experience inculcated in him a strong innovatory technical awareness and sense of inquiry. This often influenced his dominating, all-embracing and often controversial conduct of the war, its operations and technological developments. A pronounced tendency to interfere in detail and to propose unsound schemes frequently dismayed military leaders. But for the most part his conduct of grand strategy was sound and vital to victory, notably by his shrewd wooing of President **Roosevelt** to bring the USA into the war; his immediate acceptance, against the political grain, of Josef **Stalin**'s USSR as an ally; and his adherence, once convinced after prolonged debate, to the strategies and plans of his professional advisers. Above all, his projection of character kept the people steadfast in the darkest hours and held the Alliance together when the stresses were overbearing, through a personality which he impressed upon all by indefatigably travelling the world and the battle fronts to see things for himself and keep in touch with the leaders and the fighting men.

It was a sad irony that having seen Britain to victory in Europe in May 1945, the electorate denied him the opportunity of remaining Prime Minister when Japan sued for peace, and to help lay the foundations of the post-war world.

CIA (Central Intelligence Agency). The USA's principal intelligence organization, under the

National Security Council (NSC), formed by the National Security Act of Congress in 1947. As a distant successor to the Office of Strategic Services (OSS), with many members of that wartime organization involved, its tasks include: security; coordination of intelligence departments; gathering, evaluation and distribution of intelligence; and performance of special tasks as directed by the NSC.

In practice the CIA delves in every form of covert state activity, sometimes stormily crossing the path of the Federal Bureau of Investigation (FBI). It wages **guerrilla** and anti-guerrilla warfare, and not only against the Communists, who were its principal targets. Little escapes its agents' attention. Many are the crises it has either caused or stepped into, including undermining the Iranian government in 1954 and overthrowing the Chilean one in 1973, quite apart from its interminable contest with Soviet and Chinese Communist agents throughout the world and in such hot spots as Cuba, **Afghanistan**, Central and South America. Assassination and abduction are among its trades, along with dissemination of **propaganda** of all shades.

Frequently portrayed by its enemies as a thoroughly corrupt and destabilizing influence, one capable of acting independently without presidential approval, it has only occasionally been challenged by Congress during such dramas as the Watergate affair in the 1970s and the 'Iran–Contra' scandal of 1987. In a traditionally dirty trade in which all secret services play shady roles, the CIA has a vital role in defence of the West.

Civil defence. Civil populations have invariably undergone fluctuating fortunes in time of war. Within primitive tribal groups they as often as not took their chance with the warriors, the feeble, the women and children falling prey to the victors as the latter chose. Such protection as was provided would usually have been used to conserve the most precious productive captives, such as weapon-makers and other artisans vital to the community – a tendency which grew in parallel with civilization and humanitarianism, even during and after **sieges**. Civil defence in its modern form, however, dates from the response to the dropping of aerial bombs by German aircraft in 1914 and, in particular, from the bombing of London and Paris by Zeppelins and long-range **bombers** throughout **World War I**. To begin with, people were warned of approaching danger and the need to take cover in cellars and beneath stairs by the sounding of sirens. These manifestations of governmental care usually averted panic and sustained **morale** against light and sporadic raids.

The threat prior to **World War II** of intensified **air warfare** as postulated by the apostles of strategic bombing (*see* **Bombing, Strategic**) of cities, such as Giulio **Douhet** and others, compelled governments to give high priority to civil defence, or 'Air Raid Precautions' as, for example, they were first called in Britain. Badly scared by the protagonists of bombing, who produced spurious evidence (much culled from the **Spanish Civil War** and the Chinese–Japanese wars) of the devastating effects of high explosives, **chemical weapons** and incendiary **bombs**, authorities spent vast sums on building public shelters, issuing respirators and forming organizations to evacuate city populations to safer areas. Also they formed rescue, fire-fighting and **medical** services to deal with damage and casualties on a huge scale to maintain the nation, industry and law and order.

Although the effects of short, sharp attacks on Warsaw in 1939 and Rotterdam in May 1940 tended to endorse the Cassandras, the prolonged air attacks on Britain – the so-called Blitz between August 1940 and May 1941 – brought matters into perspective. Gas was not used and damage from high explosives and fire, though often extensive as the result of concentrated attacks, was neither so crippling as forecast, nor so expensive in lives or erosive of morale. As the civil-defence services improved with practice (in particular in post-raid welfare, repair and rehousing) and as people learned to take shelter and 'live with the Blitz', the dangers of a collapse in morale and serious loss of production receded. The experience of Germany and Japan, as their cities were razed under the most intensive bombing of the war, was on a far worse scale. Provision of very strong shelters helped in Germany and Draconian measures by the police and military stamped out such signs of relapses of steadfastness as became evident among pre-eminently apathetic people.

The dropping of the **atom bomb** on Hiroshima and Nagasaki in 1945 and the subsequent development of the **hydrogen bomb** and extremely deadly chemical weapons stimulated greater interest than ever in civil defence, even though the prospects of survival against weapons of mass destruction seemed problematical, to say the least. The reinforced organization and measures of nations who saw themselves as possible targets were mainly scaling up of what previously existed, with greater emphasis on post-attack survival than anything else. While some nations, such as Britain in the 1960s, virtually abandoned efforts to prepare for the effects of nuclear attack by disbanding their major civil-defence services, the Soviet authorities seemed to reinforce theirs. It was suggested that, in the event of a nuclear holocaust, surviving Russians, sustained by efficient civil defence, would have a decisive advantage over unprotected opponents, a contention which declined in credibility in the aftermath of the Armenian earthquake of 1988 when Soviet rescue and welfare services failed badly. In effect, civil-defence plans provide little more than guidelines for action in a major catastrophe.

Clausewitz, General Karl von (1780–1831). A Prussian officer of middle class who took part in the Jena campaign of 1806 and, as a staff officer, was prominent in reforming the Prussian Army. He then entered Russian service and distinguished himself in that country's defence against **Napoleon**'s invasion in 1812, the campaigns of 1813 and 1814, and as chief of staff to a Prussian corps during the Waterloo campaign.

As chief administrator of the Prussian War Academy from 1818 until his death from cholera, he wrote histories of the **French revolutionary wars** as the basis of his famous theoretical study, *On War*, which analyses the psychology of operations and **logistics**. Stating that war is a continuation of politics by different means, he argued that the defence was militarily and politically the stronger; that every effort should be directed against the enemy's forces, resources and will to fight; and that direction should be in the hands of the civil power. He emphasized the effects of random 'frictions of war', 90 per cent of which were caused by **logistic** problems, **technology** and chance.

Over the years *On War* became much misinterpreted, partly because it was rather long-winded but also because the Prussian general staff deleted the vital principle of political control from the 1853 edition. But its influence on the wars of the 19th and 20th centuries was immense.

Clocks and watches. Prior to *c.* AD 100 in China the only timepieces in use were the sundials and water clocks of the ancient Egyptians, the Greeks and the Romans, all of which were far too impractical and inaccurate for military, **navigation** and scientific purposes. Indeed, not until the late 13th or early 14th century is there evidence of large clocks, powered by weights and regulated by **iron**, mechanical escapements. And not until the late 14th century before smaller, portable clocks appeared. Accurate timekeeping was still at a premium, however, which the efforts of Galileo Galilei and Christiaan Huygens in the 16th century to invent the pendulum only partially improved because, like all uncompensated metallic regulators, its effective length altered with changes in temperature. A problem solved in 1726 by an English carpenter-cum-horologist, John Harrison, with the gridiron pendulum.

Yet although since 1530, when the astronomer Gemma Frisius showed that longitude could be determined at sea by a timepiece, the need for a very accurate time-measuring instrument for navigation was apparent and many inventors had tried to make one, it was not until 1761 that Harrison demonstrated the first of four practical chronometers, as such machines came to be named after their escapement. This expensive machine had a steel-and-brass balance mechanism (to compensate for changes in temperature) which was kept horizontal by suspension in gimbals; it was to revolutionize navigation from the time of its use by Captain James Cook in his voyages of exploration of the 1760s and 1770s.

The need for accurate portable timepieces for land warfare was not apparent until the mid 19th century, even though the first pocket watches appeared shortly after Peter Henlein invented the compact mainspring *c.* 1500. The French showed the way when they precisely coordinated their **artillery** and **infantry** attacks at **Sevastopol** in 1855 with synchronized watches.

The operations of **World War I**, however, introduced a new dimension with its immense fire programmes from hundreds of guns, all out of sight from the infantry they were supporting, all of which had to be carefully coordinated with the forward movement of attacking troops. Accurate timing became vital and a good watch became an essential part of an officer's equipment.

Even so, up until 1925 the *pendulum clock* remained the most exact timepiece available, with rates constant to about 0·001 sec per day. No longer used for precise timekeeping the *pendulum clock*, and indeed the ordinary *mechanical watch*, have been supplanted by the *quartz-crystal clock* and *watch*, operating either a conventional hands-and-dial counter, or with a digital read-out. A difference in electrical potential applied across the face of a quartz crystal will deform the crystal – the piezoelectric effect. This enables the crystal to control the frequency of an electrical circuit; since frequency is measured in cycles per second (or hertz) the property can be related to a visible time-scale in a timepiece. The best such devices have very low accelerations of the order of one part in 10^{11} per day.

An even later development has been the *atomic clock*, which uses energy changes within atoms to produce coherent, regular, electromagnetic radiation waves. These can be counted to give a measure of time. Small versions weighing about 30kg have been made, accurate to about two parts in 10^{12}.

With military operations in the 1990s ranging far and wide over land, sea and in the air, and with missiles aimed at targets on the other side of the world, such accuracy of timing becomes an essential feature of military planning and of the execution of those plans.

Clothing, Military. Perhaps some élite Greek soldiers of **Alexander the Great** and certainly those of the **Roman** Empire must have been the first members of a uniformly dressed army bearing indications of rank and status. The **armour** they wore or bore had something to do with it, of course, because clothing manufacture was to remain a simple craft until the 15th century and always lagged behind hand-spinning and cloth-weaving techniques. Clearly the Roman leaders understood the importance of ensuring their men could easily recognize each other in battle, and also that **morale** was enhanced by smart and practical dress.

Following the Roman decline uniformed armies virtually disappeared, except when the men wore fairly standard mail armour, as did some Normans who invaded England in 1066 and those of their successors who took part in the **Crusades**. Indeed, many feudal retinues were provided with colourful cloth uniforms for prestige and identification purposes, a tendency which extended into the 17th century when standardized uniforms, which reflected the sartorial taste (or lack of it) of leaders but not necessarily their suitability for campaigning, spread throughout Europe.

The appearance of **iron** needles in the Middle Ages and, particularly, the invention in 1830 of a sewing-machine with a barbed needle by Barthélemey Thimonnier were revolutionary. Furthermore, despite a mob of tailors, in fear of redundancy, wrecking Thimonnier's machines when he contracted to make uniforms for the French Army in 1841, the appearance of several improved machines ensured that by 1860 tens of thousands were at work throughout Britain, the USA and Europe.

The **American Civil War** saw the Federal forces clad in sober blue, while the Confederates wore a generally grey uniform, which often degenerated into a locally dyed 'butternut' colour as the war went against them. The extent to which Southern grey was a stylistic choice rather than a conscious effort at camouflage is not clear, but gradually it was beginning to dawn on soldiers that smart parade uniforms were incompatible with the battlefield.

Nevertheless, the British Army clung to their eye-catching red coats to the end of the 19th century, suffering accordingly at the hands of the Boers, an enemy clothed in a way which allowed for easy concealment. But change was on the way and it stemmed from the Indian Army, where the Corps of Guides had been raised in 1846. Their first commanding officer, Lumsden, decided that his Corps should be clothed in brown – to some extent, it seems, to be different from Coke's Rifles, who wore green. He ordered the cloth expensively from Britain, but before it could arrive the Guides were in action dressed in locally dyed material, which subsequently they continued to use. Sir Charles Napier, C-in-C India, commented after the Kohaut Pass campaign: 'The Guides are the only properly dressed Riflemen in India.'

Gradually, khaki was adopted in the Indian Army and, during the Boer War, by the British Army also. An Army Order of 1902 laid down regulations for the new form of dress, described as 'service' dress, for use in the field, the old forms being retained for use in peacetime conditions.

Khaki service dress clothed the British Expeditionary Force that went to France in 1914 and photographs of that period show how almost equally unsuitable it was for fighting in (apart from the colour) as previous styles had been: tight-fitting; choker collar (for non-commissioned ranks); cumbersome puttees; peaked caps. The Germans, for their part, had gone into a field grey uniform, broadly similar in style to the British; only the French were clinging to their unchanged blue coat with red facings. **World War I** produced only one major change in military clothing, the introduction of the steel helmet: the German 'coal scuttle' (probably the most effective); the British 'tin hat' (clearly derived from the minimal armour worn by British infantry in the Middle Ages), and the distinctive French helmet.

The next major change came about through mechanization in the British Army, when it was realized that service dress was ill-suited to men fighting in armoured vehicles. The Royal Tank Corps, the Royal Artillery and other mechanized units thereupon put their men into one-piece 'combination overalls', shapeless garments such as garage mechanics wore. From this evolved the battledress adopted throughout the British Army in 1938 – the first really practical combat clothing to be designed (though the doctors were concerned about the lack of climatic protection to the lower abdomen given by the blouse top). The Germans put their tank crews into black overalls similar to those of the Tank Corps, but they, and most other armies, retained for most of their troops the service-dress-type jacket which by this time was in almost universal use.

In the latter half of the 20th century it has been belatedly realized that, if the soldier is to maintain his efficiency on the battlefield, he must have properly designed clothing for the purpose. Formal wear remains broadly unchanged, but almost all armies now have special tough, lightweight, water-repellent, windproof, camouflaged combat clothing, worn in action with a helmet and, often, some form of body **armour**. Especially difficult has been the search for a satisfactory boot, one that is simultaneously hard-wearing, waterproof, comfortable and easy to put on and remove. The Soviet soldier retained his high boots – an unsatisfactory design for long distance marching – while in the West calf-length lace-up boots have found most favour. Headgear, apart from the combat helmet and full-dress caps in various forms, is almost universally the French beret, in the style first adopted by the British Tank

Corps and worn now even by the Royal Air Force (whose uniform has generally followed Army patterns) and by the Royal Navy.

Underlying all this is the need to distinguish the soldier from the civilian during active operations. It has been tacitly recognized since the 18th century, when warfare became more formalized in Europe, that the uniformed soldier was a legitimate target while the civilian should be spared, even when found on the battlefield. At the same time it also became usual for the soldier fighting in uniform, and subsequently captured, to be treated humanely as a prisoner of war; but if fighting when not in uniform, be he soldier or civilian, he became liable to execution as a **guerrilla** or spy. In the 20th century, and especially with the development of the Resistance movement in occupied Europe during **World War II**, the definition of what constituted a uniform, or indeed a soldier, became less clear; an armband, such as worn by the British Local Defence Volunteers (later the Home Guard) was generally considered an acceptable minimum. The distinction between soldier and civilian was formally defined for the first time by the Geneva Convention of 1949, the soldier being required to wear clothing indicating he was 'of the military', to be under formal command and to carry his arms openly.

As for naval uniform, sailors have traditionally worn more casual and practical dress than soldiers and it was not until the 19th century that members of the lower deck had a formal uniform of any kind. Blue has not surprisingly been the predominant colour and while full dress for officers has, like that of the Army, generally reflected national styles, embellished with their own particular trimmings, working dress has always been and remains much less formal. In the modern Royal Navy sailors carrying out operational tasks ashore have sensibly adopted the soldiers' standard combat clothing.

Coal is a fossilized plant material which has been used as a fuel for over 2,000 years. It came into essential general use in the 18th century to power the Industrial Revolution and continued in that role until cleaner and more easily handled **oil** became more and cheaply available early in the 20th century. Usually difficult and often expensive to mine, coal is found at various depths and in different parts of the world. Many of these, like the naval coal dumps established worldwide, became strategically important as sources of power and heat as well as a feed stock for synthetic materials.

Coastal defence. Any country with a coastline knows that it is impossible to guard every conceivable landing-place. Therefore, although efforts have usually been made to secure vital points, above all **dockyards and naval bases**, the most cost-effective method has always been the employment of mobile naval and, since the introduction of **aircraft**, air forces. It is noteworthy that the great Mediterranean naval battles – **Salamis** (480 BC), Ecnomus (256 BC), Actium (31 BC), Sena Gallica (AD 546), **Lepanto** (1571), Nile (1798) and **Lissa** (1866) – were strictly speaking coastal-defence actions. While **Marathon** (490 BC) was a quite rare example of a major battle fought on the shoreline.

King Alfred's defence of England's coastline against the **Viking** invaders (871–99) was a classic example of naval action based on coastal **fortifications**. Never sure of preventing the enemy landing and penetrating inland, he frequently defeated or deterred them at sea and thus weakened those forces which did get ashore but were unable to seize vital shelter from the weather. This strategy was repeated by **Edward III** during the **Hundred Years War** when he reinforced the fleet to defeat the French at **Sluys** in 1339 and effectively deter cross-Channel raiding. And employed again against the **Spanish Armada** in 1588 to prevent an invasion of England which otherwise could not have been long resisted on land. And repeated once more when **Napoleon** planned invasion with his nigh-invincible army during the **French revolutionary wars**. Not once in the periods of crisis was any serious attempt made to defend the beaches. Only the ports would be denied to the enemy, to give time for the main army to arrive and deal with the enemy already ashore.

What could happen if both fleet and port defences were neglected was painfully demonstrated in 1667 when the Dutch destroyed the unmanned British fleet in its River Medway anchorage against little opposition from the shore, the sort of triumph the British were denied three times in 1673 when the Dutch Fleet stopped them invading the Netherlands.

Needless to say, **artillery** had a significant impact on coastal defences. Shore batteries firing at established ranges had a considerable advantage over sailing ships on the move and could also call on heated shot to pose a dire threat to tinder-dry vessels, such as they rarely faced against other ships. But it was the increased range of guns and the introduction of **mines** in the 1860s which led to a radical reassessment of coastal fortifications.

The realization that French ships might land and establish shore batteries to dominate dockyards (such as did occur in 1904 when the Japanese besieged **Port Arthur**) persuaded Lord Palmerston to have built a series of very expensive forts (known as 'Palmerston's follies'), manned by volunteers, to keep the invaders at a distance of 6,000 yards at Portsmouth and Plymouth while the army mobilized to counter the menace. In due course **mines** would be laid in the seaward approaches. There was also a proposal to launch a wire-guided **torpedo** from slipways in the River Thames estuary.

Vital-point defences were the normal systems in use by most combatants in **World War I** and seriously tested only at **Gallipoli** (1915) and the Zeebrugge and Ostend **amphibious** raids of 1918. Throughout **World War II**, however, Britain, Germany and Japan tended to build linear fortifications in order to stop invasion on the beaches – even though air power and mechanized land forces promised significantly to reinforce traditional naval action. Under test the German **Atlantic Wall** proved a vastly expensive and ineffective barrier, while fortified Japanese islands in the **Pacific Ocean War**, although extremely costly to overcome by amphibious assault, invariably fell for lack of a mobile reserve.

As the **Falklands** War in 1982 showed only too clearly, if an invading force cannot be intercepted offshore it is still extremely difficult to prevent a landing and therefore defence must be based on direct defence of key points (that includes anti-sabotage patrols) with a mobile force for reinforcement and counter-attack.

Codes and ciphers. Until *c.* 3100 BC, when the first word-syllabic systems led to fully developed writing by the Chinese and Sumerians, pictures were used for communication. And to begin with, because only a handful of people were literate, what was written remained fairly secure. The spread of literacy, however, created demands for secrecy and brevity in communication for diplomatic, military and, in due course, commercial use and fighting crime.

Secret messages sent by hand or pigeon were always liable to interception and therefore had to be encoded. Even so, the most ingenious codes were always vulnerable to decoding. One famous example among many was the interception and decoding of messages sent by Mary Queen of Scots in 1586 to those plotting the assassination of Queen Elizabeth I by the agents of Sir Francis Walsingham – with fatal consequences for Mary. Great ingenuity has been applied to interception and decoding, especially since the introduction of 'open' **electronic** communications by **telegraph**, **telephone** and **radio**.

Although ways of concealing writing (*steganography*), such as use of invisible ink, have been used with some success, the most secure, rapid and efficient methods have been by *transposition* of letters by jumbling them or by *substitution* by other letters, numbers or symbols. Samuel Morse invented his code of dashes and dots in the 1830s for standard non-secret simplicity in transmitting telegraphic pulses. At once hundreds of different ways of encoding messages in Morse code were invented; at the same time, people skilled in decoding concentrated upon breaking each system. To begin with code systems either were written, recorded in documents and books, or were mechanical, such as the compact-wheel cipher invented by Thomas Jefferson at the end of the 18th century. Usually these systems were bulky, complex and vulnerable to decoding if they or their key fell into enemy hands, or if they could be rapidly broken. A code's value is in proportion to its degree of resistance to attack.

Codes were extensively used during the **American Civil War**. The Stager transposition cipher employed by the Federals could be carried on a man's back and was both easy to use and reliable. The very secure Vigenère cipher in Confederate use was complicated, prone to corruption and easy to break once the keys were detected by Federal decoders. Most codes were vulnerable but interception of messages was dependent on wire-tapping (mainly in cavalry raids) and capturing messengers – methods which produced only sporadic results. Indeed, not until radio came into adequate operational use during the **Russo-Japanese War** was the prospect of continuous intercept possible.

Codes were employed from the start to disguise radio messages, although to begin with most systems and many operators were slipshod. The Russian Fleet's signals en route to **Tsushima** were monitored and understood by the Japanese. At the beginning of **World War I** messages were as often decoded as not. The Russian defeat at **Tannenberg** was largely due to extremely lax security discipline which precisely disclosed to the Germans their deployment because messages were faultily encoded or even sent 'in clear'. On the Western Front the French, who had established an elaborate monitoring system, obtained remarkably correct intelligence of German organization, deployment and intentions as they advanced to defeat at the **Marne**. While at sea the British, having obtained a copy of German naval codes from a wrecked cruiser (and, later, other vessels), read much German naval traffic for the rest of the war, to enormous advantage in the Battles of **Dogger Bank** and **Jutland** as well as in **Blockade** operations.

Between the **World Wars** a number of sophisticated electromechanical encoding machines were developed, of which the German *Enigma* was the most famous. Assumed to be unbreakable by the Germans, the combined work of the Poles and the British had, by 1940, produced an electromechanical **computer** (the Bombe) which could rapidly break the majority of German Enigma keys as well as those used by the Italians and Japanese on similar machines. As a result the Germans, who made only limited penetrations of enemy high-grade codes, were at a war-losing disadvantage to the Allies, who for most of the war read a vast quantity of high-grade, top-level messages by opponents who used radio with abandon. At the same time simplified codes were in widespread use for operations at lower levels and by agents and partisans operating behind enemy lines. Of these the only ones which were both compact, reliable and unbreakable were the one-time-use kind – sometimes, in simplest form, known as OTP (One-Time Pad).

In the **Cold War**, with espionage and outbreaks of violence, the struggle between code-makers and code-breakers intensified. To baffle more sophisticated computers and methods, immensely complex and more secure encoding systems were developed, such as manpack radio sets with an inbuilt computer capable of informing a receiver of its automatically selected random key at start of transmission. Such systems are difficult to break. They place enormous demands on technology and code-breakers, besides employing enormous numbers of people and machines to handle the ever-growing volume of traffic sent by hand and electronic means. Codes will be broken but by saturation a vast amount of intelligence will get through unread or unreadable.

With so vast a subject only an outline can be given here. Since David Kahn's *The Codebreakers* was published in 1966 a subject previously little known has become popular, yet remains deep in the secrecy upon which it thrives.

Cold War. A term describing the state of mistrust and hostility which existed between Russia and her principal Western **World War II** allies. Sometimes considered to have started at the time of friction during the Warsaw Uprising in 1944, its name was coined by a journalist, Herbert Swope, and first used politically in 1947. Since then it has collectively described frigid diplomacy, political manoeuvring, economic pressures, subversion and almost any conflict, usually fuelled by **propaganda**, which falls short of outright war. Although during brief periods of *détente* in the mid-1960s and the late 1980s there were periods of thaw, the state of Cold War remained.

The Cold War was created by mutual suspicion between Britain and America, on the one hand, and the USSR on the other. To Joseph **Stalin**, America's possession of the **atom bomb** posed a threat to security. To President Harry **Truman**, Soviet occupation of Eastern Europe by large military forces was no less menacing. A host of unresolved difficulties which could not be settled by negotiations (which, as often as not, were themselves integral with the Cold War) contributed to a series of major confrontations fomented by escalating disenchantment.

Some battlegrounds were located at United Nations meetings and peace-treaty conferences. Elsewhere overt and covert campaigns were conducted on the slightest pretext with the Russians encouraging Communist activists to stir up dissent and strikes wherever blossoming nationalism or anti-colonialism were taking root. For example, as part of the Greek Civil War, the siege of **Berlin**, and in the subversion of Czechoslovakia in 1948. Or in Korea (**Korean wars**), where Russian and Chinese support of the North Korean ambitions in South Korea got out of control and led to **limited war** in 1950. On the other side, American and colonial powers' initiatives to defeat or pre-empt further Russian and Chinese-sponsored take-overs were quoted as blatant examples of acquisitive Western motives and exploited to aggravate the situation (notably in the case of ineffectual American attempts to frustrate Communist domination of Cuba in the 1960s).

An early major turning-point was the immense success of the American Marshall Plan, with its revitalization by 1952 of the European economy, and the simultaneous setting up of **Nato** and Seato as military bulwarks of joint defence by threatened nations. Thereafter the Communists lost the initiative and, in turn, were brought under pressure by Western, often **CIA**-inspired, up-risings such as in Berlin in 1952, Hungary in 1956 and Czechoslovakia in 1968. At the same time the possession of the **hydrogen bomb** by both sides in the mid-1950s made it far less likely that global war would occur, and thus more possible for limited and cold wars to be preferred as ways of continuing diplomacy by other means.

Cold War strains can be detected in virtually all the major as well as minor confrontations of the post-World War II period, penetrating almost every aspect of evolving high technology in the struggle for weapons superiority as part of the balancing act of **deterrence**, infected by espionage and corruption on a vast scale, and often becoming an integral way of life in international relationships when attempts by one party or another to lessen tension tend to be interpreted as just another round in the continuing contest. Thus, for example, **disarmament** and arms-limitation treaties were frequently regarded as means to an end in the struggle for military advantage.

Since the political revolutions of 1990 the Cold War has receded, but tensions continue and threats of war remain.

Colt, Samuel (1814–62). The inventor in 1835 of a **pistol** with a simple, multi-chamber breech which revolved, was locked and unlocked by the cocking action. His own firm to manufacture the revolver failed in 1842 but in 1846, on the outbreak of the Mexican War, he received an order from the US Army for 1,000, which Eli **Whitney** Co. undertook. But in 1847 Colt started his own mass-production factory which grew into the largest privately owned one of its kind in the world. He also invented electrical firing mechanisms for sea-**mines** and ran a **telegraph** business with the world's first underwater cable.

Combat Fatigue. A psychiatric disorder which, under the title 'shellshock', came to notice in **World War I** as the result of men enduring prolonged periods of stress in warships, the trenches and **aircraft**. Its manifestations include 'jumpiness' from hypersensitivity to noise; loss of appetite and sleep; irritability; and ailments such as ulcers and headaches in conjunction with nervous collapse, all of which can be accumulative and reduce efficiency. Little sympathy was shown to numerous sufferers, who usually were regarded as malingerers or cowards, and quite frequently court-martialled and shot.

As a result of publicity between the **World Wars**, public concern and a better-enlightened understanding of nervous disorders induced a more sympathetic and constructive approach to the recognition and treatment of what, in **World War II**, came to be known as 'battle exhaustion'. On medical advice, attempts were made to limit men's exposure to stress, chiefly by reducing time spent in combat (for example, by restricting the number of operations undertaken by air crew) and providing periods of rest and recuperation in comfortable accommodation at a distance from the front. Treatment of bad cases included prolonged periods under sedation and electric-shock therapy to both cure and return the patient to duty, a practice which, in its modern form, has been improved upon in some forces by the establishment of stress-management centres close to the front line where minor cases are dealt with at the earliest moment, often returning a high proportion of cases to the front within 72 hours.

Command, Control and Communications. *See* **C³**.

Command, Control, Communications and Intelligence. *See* **C³I**.

Commando. A Portuguese word adapted by the South African Boers as the title of militia units, based on electoral districts. In the Boer War they won a great reputation, despite a certain indiscipline, for their outstanding marksmanship, horsemanship and endurance in prolonged combat and, subsequently, **guerrilla warfare** against the British.

In 1940 the British used the word to describe élite Army **infantry** units earmarked for **amphibious** raiding against Axis forces in Europe. All-volunteer and of high standard, they were well disciplined and specially trained in rigorous conditions (often under live fire and to the point of exhaustion) to use their initiative. Their 'battle drills' became models for all **infantry**. But valid objections were raised by those who feared that ordinary combat units would suffer from the loss of so many potential leaders. Had it not been for the efforts of Winston **Churchill** and Admiral Sir Roger **Keyes** the Commandos would have been disbanded.

As it was, they produced a cadre for many specialized British 'private armies', such as the **airborne forces**, the Special Boat Squadron (SBS) and Special Air Service (SAS) and were copied by other nations, notably the USA, which raised Marine Corps 'Raiders' and Army 'Rangers'. Also, in Britain, the Royal **Marines** who, by tradition, were the exponents of amphibious warfare, began forming their own Commandos in 1942. After the war, when the Army Commandos were disbanded, they continued to perform this role.

Commandos, most inadequately equipped and trained, and carried in inferior **landing-craft**, started raiding the French coast on 23 June 1940. Thereafter the number and magnitude of their forays only slowly increased as they also were sent to work in the Mediterranean. Indeed, although in 1941 only a few small raids were launched from Britain, it was in the Mediterranean and Middle East where, sometimes as ordinary infantry, they were most active – with patchy results at a high price. As techniques and training improved and more special craft and equipment became available in 1942, the scale of raiding increased, rising to a peak with the

Dieppe Raid in August and culminating in the invasion of North-West Africa in November when SBS and Commandos spearheaded the seaborne assaults.

In all the forthcoming major landings in Madagascar, Sicily, Italy (Italian Campaign), France and Holland, Commandos were in the lead and usually fought on as ordinary infantry until exhausted or required for another task. Likewise in the **Pacific Ocean War**, Marine Raiders, who raided Makin, led the way in 1942 at Tulagi and on Guadalcanal, while Army Rangers were to the forefront in the Philippines in 1944. Simultaneously, however, small-scale raiding was restricted, the majority of Commando operations being concentrated upon pre-invasion beach reconnaissance and the pilotage of main assault flotillas to landing-places. There was a tug of war between those who wanted to raid and those who feared that larger plans would be disclosed or that 'stirring up' coastlines would endanger the movement of intelligence agents and guerrillas into and out of occupied territory. By the war's end in Europe and **Burma**, Commandos were chiefly employed as shock troops in conventional land operations.

Since 1945 amphibious warfare has, in the main, been performed by Marines. British Royal Marines have been in the thick of the Malayan War of Independence, Korea (**Korean wars**), Suez, Irish Reunification and the **Falklands** War of 1982. The US Marines played a leading part in Korea and in **Vietnam**, and were involved in lesser operations in the **Lebanon**, Grenada and Somalia. They now are lifted in more sophisticated craft than originally and, since Suez, have adopted the **helicopter** as a prime means of mobility and logistic support between ship and shore.

Communications, Signal. Before the arrival of a practical electrical **telegraph** in the 1840s, **strategic** and **tactical** military communications depended upon personal contact, couriers and liaison officers, drums, trumpets, homing pigeons and such visible signalling instruments as semaphore, flag or hand, light and smoke. Most were hampered by inherent limitations of speed and capacity in transmission. Time of day and weather, quite apart from obscuration in battle from dust and smoke, also degraded performance.

In primitive warfare, as conducted by native American tribes in the 19th century for example, personal contact, drum beats, beacon fires and smoke signals were evidently pre-eminent. **Alexander the Great** simply improved on these means by forming an élite corps of liaison officers and messengers to carry both strategic and tactical messages – the former no doubt briefed to explain what their general had in mind. The system was improved upon by the **Romans**, whose network of trunk roads eased the passage of horsemen carrying strategic messages from stage to stage across the Empire.

The breakdown of the Roman system coincided with the deterioration of communications in the wars of the **Middle Ages**. The encroaching Barbarians who swarmed to where food and loot were available operated with a minimum of intercommunication. Raiding **Vikings** came and went by sea with little contact once they set forth from Scandinavia. That the **Crusades** were rarely models of military coordination can be laid as much as anything else at the door of poor signalling – making it all the more remarkable that in the same period, **Edward III** managed to coordinate intricate operations on exterior lines during the first phase of the **Hundred Years War**. The feat may be ascribed to clear and foresighted instructions (often carried by sea) and his commanders' insight rather than to explicit signal messages.

Exaggerated claims are sometimes made concerning the battlefield communications of the **Mongols** as they migrated from Central Asia and entered Eastern Europe in the 13th and 14th centuries. In fact these relatively small groups of invaders had to depend upon shouts and drum beats, supplemented by the waving of black and white flags, to implement their ingrained operational procedures and combat drills. For they were completely illiterate!

Indeed, not until 1653, when Admiral **Blake**'s *Fighting Instructions* were issued, along with rudimentary flag-signal procedures, was progress made. Already, as their exploratory ocean voyages extended far beyond the shores of Europe, admiralties and ships' captains were reduced to intuitive control of operations because strategic messages could only be carried by ships, which often took weeks to reach their destinations. And tactical communication by hail, gun and flag signals was very inflexible until the Vicomte Sébastien Morogues, founder of the Académie Marine at Brest, and his collaborator, Monsieur de la Bourdonnais, published in 1763 a code book of flags, numbered 0 to 9. The British Admirals Richard Kempenfelt and Richard Howe improved upon this unofficially between 1776 and 1790 until, at last in 1799, the Admiralty adopted their system. In the years to come this was developed into a system using 25 flags with a dictionary of 1,000 words, a method still in use during **World War I** and employed at the Battle of **Jutland**.

In 1792 the French built chains of semaphore stations, designed by Claude Chappé, across the length and breadth of their mainland. These were signalling places which could at a speed of 150mph relay messages in 15 minutes (provided visibility was not impeded) to provide central government and the army with unprecedented information throughout the **French revolutionary wars** and beyond. Semaphore stations were also adopted by the British on land and at sea, although the system suffered from the disadvantage of low capacity with only one message in transit at a time.

Not until the electric **telegraph** came into operation in the 1840s, to be served by Morse code post-1850, and in due course given multi-channel capability, were the semaphore's main disadvantages – on land only – overcome.

The primitive telegraph system first used in war by the British and French during the **Crimean War** provided a poor service, but it did, nevertheless, introduce close government interference with operations. Really it was the fast steamship service across the English Channel, linking with the **railway** to Marseilles and the longer steamship voyage to Scutari and the Crimea, which carried the bulk of messages at revolutionary speed. Courier dispatch and visual signals would still be needed, as they are today, but less so as land and submarine cable networks carrying multiplex Morse code messages and, in the 1880s, speech by **telephone** had come into general use. These changes were revolutionized by **electronics** with the introduction of **radio** transmissions early in the 20th century.

Navies and armies were quick to exploit the importance of these inventions, taking their cues from the alacrity of merchant-shipping firms and the railways, for example, to enhance their control of business and traffic. Operational uses, including the security problem posed by the adoption of suitable **codes and ciphers**, were slower to develop. It was 1885 before the Germans used the telephone for the direction by forward observers of **artillery** fire. And, despite unsuccessful British attempts to use radio during the Boer War, not until the end of the **Russo-Japanese War** was it seen how effective in naval war this method was. Thus it was all the more remarkable that, although the Germans pinned great hopes on control by radio of the armies advancing into France at the start of **World War I**, they failed utterly to devise workable operating procedures, a lapse which certainly contributed to their defeat at the **Marne**.

No doubt it was the ramification of rapid signals growth which baffled military leaders, few of whom were technically educated. Since there were many officers who delegated speaking on the telephone because they thought it beneath their dignity, it was hardly surprising that so many problems remained unnoticed prior to 1914. Take, for example, the unwillingness of the German General Staff to discuss their plans with the Signals Corps, thus denying forward planning of cable routes to replace those destroyed by the retreating enemy; and the tardiness of the Royal Navy to take full advantage of their insight into German codes, a reluctance which probably cost them victory at **Jutland**.

During World War I there was such a vast expansion of signals services that those armies like the British which had not already (as had the Germans and Americans) formed a Signals Corps felt compelled to do so afterwards. The corps could not only improve the handling of signals traffic but also give advice on and lead development of the flood of new technology. As knowledge of the ionosphere increased, worldwide communication networks improved, leading to 'beam' transmissions. In the 1920s the next generation of smaller, more robust, quenched radio sets, some with crystal tuning, working at ever higher frequencies of 40 megahertz (MHz), enormously improved reception. In conjunction, voice radio sets became commonplace for **aircraft** and land vehicles, and were complemented by high-speed **teleprinter** transmissions carried on cable which could be laid mechanically at 100 miles per day.

Military signalling in **World War II** benefited immeasurably from pre-war integration with civil systems. In Germany the armed forces dominated. But so keen was awareness of signals potential that problems of generating sufficient capacity often arose at the same time as vulnerability to intercept increased. For example, the dependence of German U-boat operations upon radio was a weakness not only in providing a plethora of *Enigma*-encoded messages for the Allies to attack, but also frequently disclosing the location of boats from direction finding (DF). When secure cable routes were either destroyed or overloaded, additional use was made of less secure multi-channel radio. Net radio, however, was vital to the tactical control of ships, aircraft and of armies (particularly fast mechanized forces and artillery), in addition to being crucial to the control and logistic support of clandestine and **guerrilla** forces behind the enemy lines. Only strictly enforced operating procedures and traffic limitation could mitigate the risks.

Since 1945 utilization of immensely improved and sophisticated signals systems has gone on apace, a revolution to which **space** satellites have contributed more than anything else. Vastly quicker methods of transmission with higher capacity working are allied to the sending of documents by facsimile and the capability of reducing the threat of DF by high-intensity radio 'burst'

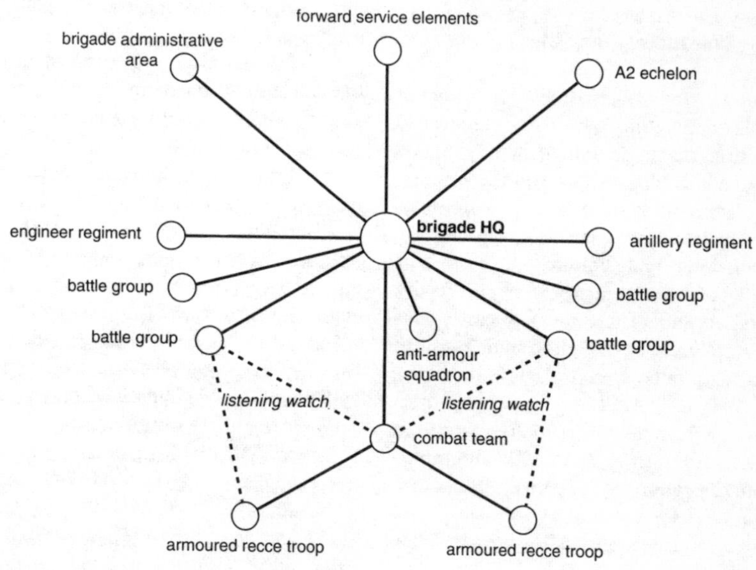

Specimen Army brigade radio net, c. 1980

transmissions. When integrated with **television** displays and **computers**, modern signal communications provide commanders and administrators with a tool of immense power. Not only is it possible to solve and disseminate the most complex operational and **logistic** problems at immense speed, but greater opportunities to assert minute-by-minute combat control from the highest level are provided, a practice frequently indulged in by **Hitler** (often with disastrous results) and during the **Falklands** battles when, had she wished, the British Prime Minister could have spoken direct to commanders at sea.

Nevertheless, electronic signalling has its vulnerabilities still. Sunspot activity can blot out or interfere with transmissions, as can electronic countermeasures (ECM), by jamming and **nuclear** explosions in the upper atmosphere. Security, despite stringent precautions, will always be threatened – not simply from the danger of a few messages out of thousands being decoded, but also from DF. For it is the detection of grouped transmitters, even more than individual stations, which can provide priceless **intelligence**, since a pattern of transmissions from related sources is often a clear indication of the identification, location, layout and intention of whole units or formations.

Computers and calculating machines. A clear distinction must be made between a calculating machine, which simply makes mathematical calculations, and a computer, which stores and analyses data, presents visual displays and controls machines, besides making calculations. The simple, manual abacus dating from before the 6th century BC and of unknown ancestry was the first calculating machine and is still in use. It was overtaken in 1642 by Blaise Pascal's digital machine, which counted integers, and thereafter by many types from different inventors to cope with logarithms and multiplication, leading to 20th-century manual/electric calculators which could quickly and accurately add, subtract, divide and multiply.

It was Charles Babbage who in the early 19th century designed a mechanical tabulating machine which worked to eight places of decimals for the purpose of removing inaccuracies from logarithm tables. From this start he proceeded to build the first mechanical computer (a 'difference engine'), programmed by punch cards which Joseph Jacquard had invented for the automation of weaving looms, and what Babbage rated as a brain.

Babbage's machine failed because of inherent engineering deficiencies and lack of financial support, and it was not until Hermann Hollerith produced his punch-card, electro-mechanical machines to handle the 1890 US census counts that possible commercial, accounting and military applications for computers emerged. For not only could the relatively slow Hollerith machines analyse information and store personnel records, they were adapted in the 1930s by the Poles in their Bombe computer to analyse and break the German *Enigma* machine **codes**. This process was speeded up in the following decade by the British who, during **World War II**, were driven to seek even faster ways of cracking ever more complicated *Enigma* machines and the very complicated *Geheimschreiber*, which could encipher teleprinter messages.

The outcome was three successive types of semi-electronic machines designed by T. H. Flowers and Alan **Turing**, the third of which, Colossus II, had 2,400 diode valves, binary adders and decade counters, and a limited memory on punched tape. This, the first programmable electronic digital computer, was ready for use on 1 June 1944, just in time to decrypt the vital information passing along the *Geheimschreiber* link from Paris to Berlin. At the same time the Americans had been working on ENIAC, a huge computer with 18,000 valves plus 20 adding machines designed rapidly to integrate ballistic equations for **artillery** firing tables. But it was not ready until 1946 and was far from reliable.

The next breakthrough came with the transistor, which began to replace the valve in the 1950s. This development heralded the rapidly increasing miniaturization and reliability of **electronic** components, making the general application of electronics to military uses even more attractive. The initial impetus for miniaturization had been the **space** programme, but the incorporation of on-board computers as part of the guidance system for long-range missiles quickly followed. While inertial **navigation** equipment had been available in 1954, the new developments enabled much more accurate and less bulky systems to be developed for both **submarines** and surface vessels.

For land forces developments took rather longer to reach fruition but obvious requirements were soon identified: **logistics** systems; control of predicted fire from field **artillery**; fire control for **tank** guns; improved **communications.** It was for control of resources that, in the British Army at least, computers were first introduced and it is hard to imagine, for example, how the British in 1982 could have assembled and launched, at under a week's notice, the task force to recover the **Falkland Islands** without the aid of a computerized logistics system to control the correct provision and loading of all the men and stores needed. Or the UN have been ready in 1990 in Iraq.

The arrival of the microprocessor integrated circuit, invented by Robert Royce – a cheap complete computer on a single tiny silicon chip – has done much to enhance weapon performance. FACE (Field Artillery Computing Equipment) has been in service with the Royal Artillery for over 20 years and provides an effective analysis of such aspects as target acquisition, ballistic data, fire planning and logistics to enable the battery to make the best use of its limited resources. In tank **gunnery** too the computer is now a well-established component, analysing such variables as ambient temperature, **ammunition** characteristics, wind speed and range: all in an effort not only to reduce human error but also, so vital in a tank engagement, to reduce the interval between seeing the target and obtaining a hit.

It is, perhaps, in the field of communications that the computer has allowed the most significant improvements to be made. Military communications need secure means of transmission from many mobile stations. The staff need to process the information flowing along the communication links; **reconnaissance** data from all **surveillance** sources, to be compared with the operational plan; decisions to be made from that comparison; consequential orders to be passed to the forward troops for action. Improvements in ADP (automatic data processing) mean not only that much more data can be handled, but also that it can be handled much more swiftly. In the British Army the WAVELL system has been developed for this purpose, the world's first automated battlefield command and control system to enter service at corps level and below, providing secure data links between formation headquarters.

Such developments point the way for the future, when a commander can almost take for granted that he will be able to pass information and orders, by voice or other means, directly and securely to his subordinates, with reports and returns continuously updated and available to all staff levels. Similarly it is certain that weapon performance will continue to be enhanced by electronic means.

Condor Legion. This air formation was formed in November 1936 to command the units of the German Air Force already involved in the **Spanish Civil War** in support of Fascist Nationalist forces. Led initially by General Hugo Sperrle, it comprised Ju52 **transport aircraft** (also used as **bombers**), **fighters**, seaplanes, communications and ground-support units, including **anti-aircraft** artillery, the 88mm version of which was used also for **anti-tank** work. Employed in support of sea and land operations, its anti-shipping role was in a minor key and **strategic bombing** tasks very few, even when the latest medium bombers were supplied and tried out. In the main the Legion was utilized as a proving unit for the latest machines and techniques, besides giving invaluable training to the future leaders and instructors of the Air Force. It played a vital part in establishing the role of the dive-bomber as heavy artillery in support of ground forces – thus tilting the Air Force's strategy in the direction of tactical as opposed to strategic policy. It was withdrawn in 1939.

Congreve, Colonel Sir William (1772–1828). A British artillery officer, who improved on the **rockets** which Hyder Ali used so effectively against the British in India at the end of the 18th century. By 1813 he had developed missiles of up to 42lb, with explosive and incendiary warheads and ranges out to 2,000 yards. They were frequently fired in salvoes from boats into towns (for example Boulogne in 1806 and Copenhagen in 1807, when immense damage was done); and on land, notably at **Leipzig** in 1813 and Bladensburg in 1814, when they helped the British to capture Washington from the Americans. In 1805 Congreve seems also to have been the first to propose **armour** for floating **mortar** batteries, although this was not acted upon. In 1814 he became comptroller of the Royal Laboratory, Woolwich, and Superintendent of Machines.

Coningham, Air Marshal Sir Arthur (1895–1948). A British pilot with a fine record in **World War I** and between the wars who, in 1941, took command of air forces in support of the British Army in **North Africa**. Progressively he developed the organization and techniques of tactical air forces in support of the land battle, successively commanding them in the Mediterranean theatre of war, the **Normandy** battles and the closing battles of **Germany**.

Conscription. In connection with the enlistment of soldiers, the word seems to have come into use in the 16th century, although compulsory service, linked to slavery/serfdom, had been imposed by élitist overlords since at least the second millennium BC. Assyria, Greece and Rome made use of it, although occasionally recognizing that volunteers fought better when motivated by self-preservation, patriotism, religious fervour, glory, plunder and other inducements. Nevertheless, the feudal systems of Europe until the 16th century and in certain Asiatic countries, such as China and Japan, into the 20th century, not only forced citizens into the armed forces but required them to give their labour to the overlord.

The mass of followers to the **Crusades** simply abided by their feudal obligations. Few among the well-trained British archers who fought in the **Hundred Years War** were volunteers. Nor were the Swiss **infantry**, who fought so well throughout the 15th century, volunteers, but conscripted mercenaries, paid by client nations, and who departed when the money ran out.

The Renaissance and radical change in social attitudes impelled the replacement of conscripted feudal forces with national standing armies tasked to disband mercenary bands and the remaining feudal armies. From the 15th until the late 18th century these expensive uniformed volunteer forces were, of necessity, better trained than their predecessors because of the need to cope with the technology of **artillery** and firearms. Henceforward, in the **Wars of the 16th, 17th and 18th centuries**, they were fairly carefully conserved by their commanders. Costly battles were avoided because manpower was in short supply; better **logistic** arrangements were made to save lives and maintain **morale**. **Limited war** was the norm until 1772 when the Comte de **Guibert**'s concept of the conscripted nation in arms was received with approval in France.

In fact the Royal Navy had for long impressed sailors in wartime since the extreme hardships of the service deterred volunteers in sufficient numbers. But when in 1793 the French government enacted the *levée en masse* to protect the Revolution and, under **Napoleon**, begin their wars of aggression, other nations had to follow suit with their own systems of compulsory service. As a result vast pools of manpower were tapped for the squandering in unlimited wars. The French system of registration of men by age groups which were called up for training and service for specified periods of time and then transferred to the reserve, with the obligation of recall in emergency or for retraining, was copied in peace and war by the majority of military nations.

It was conscripted armies which, for the most part, faced each other in the major European wars after 1850. Each nation, of course, had its own special variations. The French balloted for those called up and placed themselves at a disadvantage in the **Franco-Prussian War** of 1870 against an enemy who conscripted in mass. The British eschewed conscription in peacetime until 1939, although they had been forced to introduce it during **World War I**. But in **World War II** they widened its scope to include women recruited for the armed forces, agriculture and industry.

The USA also managed to resist what was called 'the draft' until World War I; but having revived it during World War II, with a ballot provision, kept it in force until the early 1970s, with serious political repercussions during the **Vietnam War**. Russia tended, even after freeing the serfs in 1861, to practise a sort of random impressment, with men snatched off the streets for lifetime service. Gradually, after the Revolution, this practice was modified but conscription remains universal to this day, as it does in China, which also maintains mass forces.

Russia, China and their political allies shrink from exemptions on conscientious as well as medical grounds. More liberal nations, such as the USA, Britain, France and Germany, have usually been far more sympathetic with proven conscientious objectors and flexible on medical grounds.

The ethics and sense of conscription in the age of discernment and high technology are open to question. Undoubtedly, large numbers of trained reserves will be needed in time of exacting emergency, and judicious military **training** benefits the individual's education. Yet states which insist on periods of less than two years' service by unwilling people all too easily produce inadequately trained personnel to operate very complex weapon systems. Arguably a well-equipped, volunteer force with high morale will defeat a much larger conscript one. But it must be backed by adequate reserves.

Convoys. The convoying of groups of animals, vehicles and ships is among the most ancient operations of war, dating from the first tribal migrations. Then as now the escorting of weak groups in motion by mobile combat units was not only **tactical** but, quite often, a challenge to battle with **strategic** intent.

On land. The passage of convoys before **artillery** and small arms could be used usually depended upon close escort and the prior securing of likely ambush places. In theory those manning convoys should contribute to their own defence, but frequently they are helpless, and all the more so under the threat of long-range artillery and small-arms fire. It then became vital for escorts to secure both the route ahead and the ground dominating it. In mountainous country the picketing of heights became standard practice, while in open terrain to interpose troops between the convoy and the enemy was also essential.

Convoys, in close order or well spaced, need to maintain formation to facilitate command and control. When road-bound they are compelled to move in line ahead, but to traverse desert terrain and the like they usually form into several columns with a commander centrally placed.

There was a fine example of a convoy action of strategic and tactical importance in 1708 when the Duke of **Marlborough** resupplied Prince **Eugene**'s forces besieging Lille by passing a convoy from Ostend via Wynendael. When 40 French **infantry** battalions threatened it, Marlborough doubled to 24 battalions the escort, which routed the advancing French in a defile before Marlborough's main force arrived. The convoy got through unscathed; Lille's eventual fall was assured.

In the 20th century **mechanization** did not alter the principles of convoy movement. Trucks moved faster and, given good cross-country performance, could sometimes avoid the roads. Air power, however, called for greater dispersion and cover from detection, thus complicating control and making night driving more necessary, in addition to requiring the emplacement of **anti-aircraft** weapons to guard defiles and other vital points.

At sea principles very similar to those on land apply – and were established in 256 BC off Mount Ecnomus when the Carthaginians tried to intercept a Roman troop convoy from Sicily to North Africa. Each of the three Carthaginian columns was tackled by Roman ships which, by boarding, sank or captured 90 ships for the loss of 24 ships of their own, and without harm to the convoy.

The development of the caravel and **carrack** as multi-purpose cargo and fighting ships, along with the expansion of world trade, made the protection of commerce of increasing strategic importance. Ships tended to sail individually until threatened by pirates or, as in the case of Spanish treasure ships, by the British. In the late 16th century it became necessary for the treasure ships and others to group for mutual protection, a trend further developed in the First Dutch War by Admiral Cornelius van Tromp when, in February 1653, his fleet tried disastrously, against Admiral Robert **Blake**, to escort a Dutch merchant fleet through the English Channel. Thereafter convoying became vital in time of war. As for example at the battles of **Lissa**, between Austria and Italy in 1866, and the Yalu River, between China and Japan in 1894. Needless to say, the convoying by warships of merchant ships, particularly neutral vessels, has given rise to legal wrangles over 'search' and so-called 'rights of convoy' under conditions of **blockade**.

It was thus all the more surprising that when, in **World War I**, the Germans posed a major threat to the Allies (Britain in particular) by unrestricted **submarine** warfare, the British Admiralty was most reluctant to institute convoys. In part this was because **destroyers**, the most effective hunter and escort vessels, were literally tied up in case of need to support the Grand Fleet. But also because Admiral Sir John **Jellicoe**, Chief of Naval Staff, reasoned that not only was the method dubious in the circumstances but technically impossible. Only after statistical **analysis** of operations and a trial had proved him wrong in all respects was the system fully adopted.

Not only did the presence of warships deter attack by surface as well as submarine vessels, the act of concentrating hundreds of merchantmen, which normally sailed individually as they chose, into a few compact groups also made the enemy's task of detection far harder and prevented him attacking with unopposed gunfire when surfaced, thus forcing him to attack with expensive **torpedoes** when submerged. Prior to effective use of convoys in June 1917, monthly shipping losses averaged 750,000 tonnes. Subsequently they declined to below 300,000 tonnes per month at the same time as U-boat losses rose from about three a month, mid-1917, to double figures in May 1918.

Between the **World Wars** improvements in anti-submarine techniques and technology (meaning detection by **sonar** and attack by **depth charge**) persuaded some sailors that the submarine threat was neutralized. Therefore interest in convoys declined with only low priority given to the construction of escort vessels. Britain, for example, did not hold a single convoy-protection exercise in this period. Nevertheless at the start of **World War II**, convoys were formed and proved reasonably protective against enemy surface raiders and submarines (which mostly hunted singly) and aircraft, whose crews' techniques were raw. Not until the Germans built more U-boats, seized control of the European seaboard and began to control packs by **radio**, in conjunction with **aircraft**, did the convoys come under intense pressure. But by then techniques had been rationalized, security raised and the system of fast and slow convoys introduced to improve average transit times and thus reduce waste of shipping.

Come April 1943, the Battle of the **Atlantic** was the one campaign which could have lost the Allies the war. As it was, the pendulum of success swung their way through a combination of technical and operational innovations which prevented the fatal abandonment of the convoy system which, because losses seemed out of control, was mooted in March 1943. The keys to victory were more escort vessels with improved sonar and weapons working in concert; voice radio to facilitate hunting in groups; direction finding (DF) of submarine radio signals to provide information enabling diversion of convoys round lurking packs; breaking of **codes** for insight into enemy intentions; introduction of longer-range aircraft and light **aircraft-carriers** to provide cover in mid-ocean; development of centimetric **radar** to detect surfaced submarines (and even their periscopes), thus enabling illumination and attack by night.

Even though the crucial battles ended in victory for the convoy system in the Atlantic and in the Mediterranean (where the principal menace to both sides was aircraft based at relatively short ranges), it had become apparent by 1945 that faster German submarines, fitted with improved sonar and weapons, posed a serious threat to convoys whose speed, after all, was restricted by that of the slowest vessel. In the Pacific, moreover, US Navy submarines did much as they chose in a slaughter of Japanese convoys whose weapons, detectors and techniques were poor. With the arrival since 1945 of vastly improved nuclear submarines and torpedoes, the future safety of convoys hangs in the balance. For their position is easily established by reconnaissance satellites (*see* **Space vehicles**); their merchant ships are not that much faster than their predecessors; and their escorts are stretched to the limit to find, track and kill hunters who move fast, dive very deep and can detect foes at extreme range.

Coral Sea, Battle of the. The Japanese intention to consolidate their conquests in the South-West Pacific prompted the planned capture of Port Moresby in May 1942. Their invasion convoy was covered by three **aircraft-carriers** in two task forces which entered the Coral Sea

The Battle of the Coral Sea, 1942

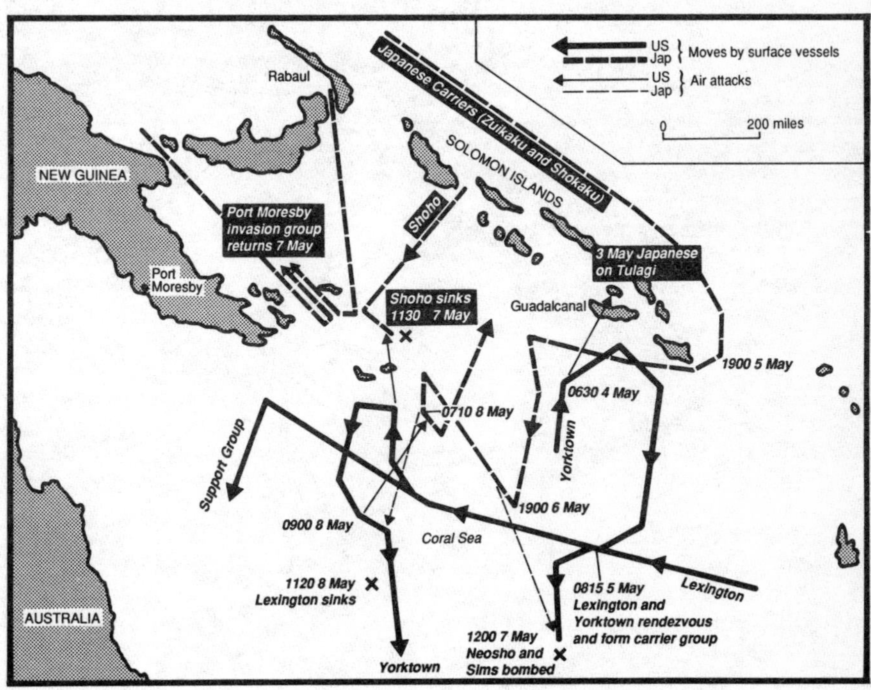

on 5 May. Awaiting them were two American carriers which were well forewarned by **radio** intercept of their enemy's intentions. In a battle during which surface vessels never came to blows, **aircraft** did all the damage. After successful air attacks on a small Japanese invasion **convoy** at Tulagi on 4 May, the American carriers headed west, hunting the Port Moresby **amphibious** force. On 7 May, when aircraft from the *Yorktown* sank the *Shoho*, the Japanese crippled an American tanker in the belief it was a carrier. Next day, after the Japanese accurately located the American carriers, there were concentrated exchanges of attacks from 69 American and 51 Japanese dive-and torpedo-bombers. They caused the loss of the USS *Lexington* and damage to *Yorktown*, and severe damage to *Shokaku*. But in what may be judged a drawn tactical battle, the Japanese had to concede strategic defeat by calling off the invasion of Port Moresby and domination of the Coral Sea.

Corvette. An 18th-century French, three-masted, single-deck ship with about 20 guns. Based on the **galleon**, the type was adopted by the British, particularly for service in hot climates where their good ventilation was beneficial. Rated lower than **frigate**, the last went out of service in 1903.

The name was revived in 1940 for an easily constructed ocean escort vessel. Of a whale-catcher design with excellent, if lively, sea-keeping qualities, the first of the *Flower* class were of 925 tonnes, lightly armed, equipped with **depth charges** and **sonar**, with a speed of 16 knots and a crew of 85. As the war progressed, 150 were built and steadily improved upon by the addition, among other things, of **radar**. In effect they were far more

The Battle of Crécy, 25 August 1346

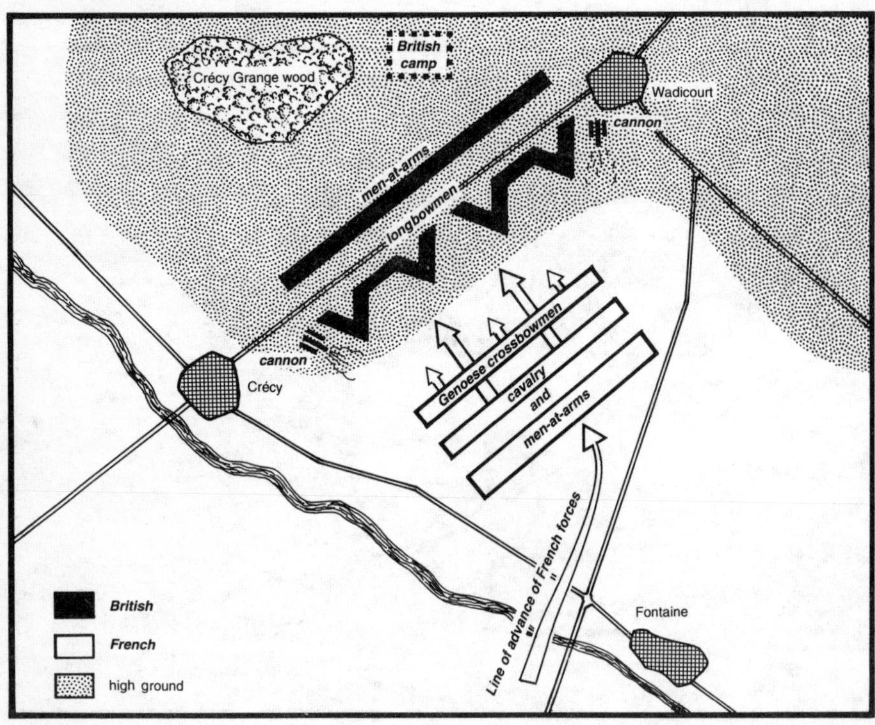

economical, efficient and deadly **submarine** hunters than **destroyers** – especially when later increased in size and renamed frigates.

Crécy, Battle of. In 1346 the British under **Edward III** landed with 9,000 archers, 4,000 **infantry** and 2,000 heavy **cavalry** in the Cotentin peninsula and marched via Abbeville towards Calais; pursued by a French army, under King Philip VI, of 6,000 mercenary Genoese crossbowmen and about 29,000 cavalry plus an unknown number of unreliable conscripts.

On 25 August Edward occupied a front of 2,000 yards between Crécy and Wadicourt overlooking a valley at 300 yards range. Keeping only a small mounted counter-penetration reserve in rear, he placed the remainder of his force, including a few pieces of **artillery** making their début, half-way down the slope. The French, led by the Genoese and hustled by some impulsive knights, advanced across rain-soaked ground to within 150 yards of their enemy when the British archers and artillery opened fire. The Genoese broke, fled and were ridden over by waves of French knights who also were slaughtered by a hail of arrows as, time after time, they closed for hand-to-hand combat. The attacks went on by moonlight but all failed with losses amounting to some 10,000 against 200.

Crete, Battle of. The German invasion of Crete on 20 May 1941 was conceived as an exploitation of the conquest of Greece to

The Battle of Crete, 1941

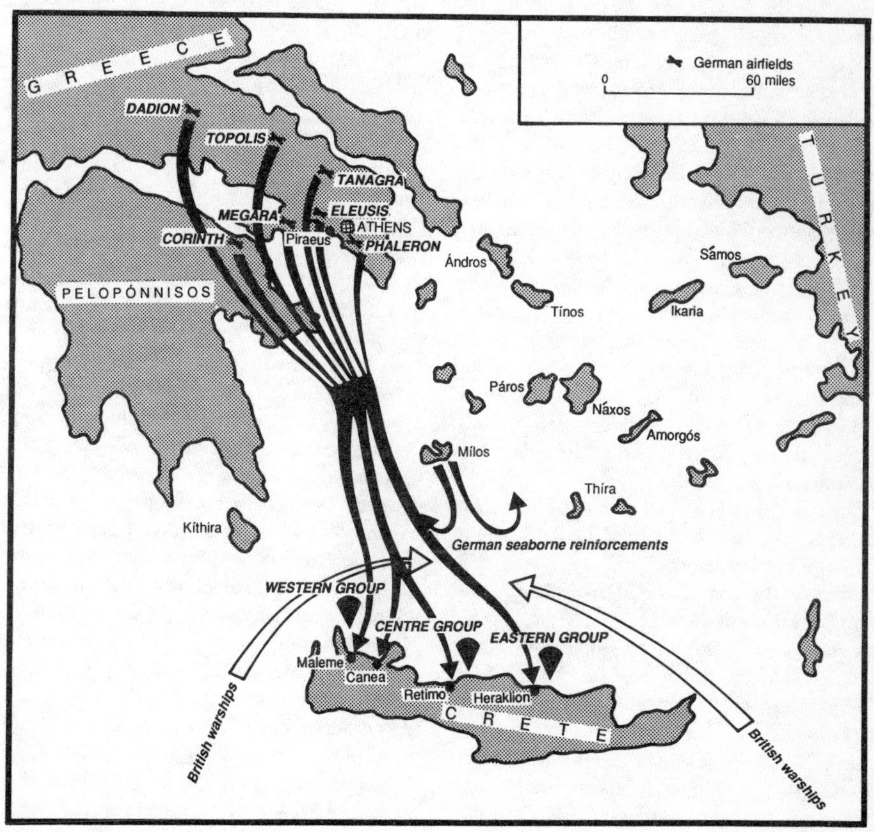

acquire dominance of the eastern Mediterranean and the Middle East, and also to enhance the prestige of Hermann **Göring**'s German Air Force. The Air Force, under General Alexander Löhr, had available 430 **bombers**, 180 **fighters**, 500 Ju52 **transports** and 100 **gliders**. They supported and lifted XI Airborne Corps of one parachute and one air-landing division, under General Kurt **Student** – its heavy equipment to be conveyed by sea.

The island's garrison, under Major-General Bernard Freyberg, consisted of 27,500 inadequately equipped troops (of whom 15,500 had recently been evacuated from Greece), 14,000 Greek soldiers, a small **fighter** force (which was withdrawn on the 19th under overwhelming German air attack) and strong naval units without air cover. Well supplied with **intelligence** from **radio** intercept, Freyberg was able to concentrate his troops at known enemy landing-places.

As a result, although the Germans had complete air dominance to pound the defences as they chose, and to drop parachutists at small cost in Ju52s, they had it far from their own way on the ground against alert garrisons. By nightfall landing grounds had not been secured. In consequence, when Student persisted in dropping more parachutists and flying in the air-landing division next day, their losses in men and machines were extremely heavy. Moreover, the sea convoy was intercepted on the night of the 21st/22nd and turned back with loss. Indeed, there was a period when it only required a determined (and feasible) counter-attack by the British to remove a German toe-hold at Maleme, thus making further reinforcement impossible and sealing the fate of the invaders.

But Freyberg vacillated, despite clear intelligence of the situation, and the German crisis passed when large air reinforcements were landed, forcing the garrisons back. Evacuation to Egypt began on the 28th under deadly air attacks. By attempting to operate in broad daylight, the Royal Navy exposed its ships to constant bombing which eventually sank four **cruisers** and nine **destroyers**, besides damaging many more, including two **battleships**. The evacuation ended on 31 May. British losses were over 18,000; those of the Germans numbered about 13,000. So heavy, indeed, was the loss of parachutists that **Hitler** ruled out all such future use of them.

Crimean War. This limited war, fought chiefly in the Black Sea and on the Crimean peninsula, came about in October 1853 when Turkey declared war on Russia, whose political threats she could no longer tolerate. On 30 November a Russian naval squadron of nine vessels, armed with the latest shell-firing cannon, overwhelmed in Sinope harbour, in six hours, a Turkish squadron of 10 wooden-hulled warships. As an expression of concern at this destabilizing event, a combined Anglo-French fleet entered the Black Sea in January 1854, an act which led to a formal declaration of war on 28 March, seven days after a Russian army invaded Bulgaria to attack the Turks.

The Allies aimed to support Turkey and, on 29 April, having already severely damaged Odessa by naval bombardment, established a base at Varna to prevent the Russians advancing on Constantinople. Coming under pressure also from Austria, Russia withdrew from Bulgaria and sought a diplomatic solution. But France and Britain now decided to punish Russia by destroying her power in the Black Sea through destruction of the naval base at **Sevastopol**. On 13 September their armies, already suffering from cholera and with the British component in **logistic** chaos, landed unopposed in the Crimea.

There now ensued a sequence of costly battles, fought by weakened troops at the end of tenuous lines of communication, for possession of Sevastopol. The Russian army, under the command of Prince Alexander Menshikov and by no means strong, first tried to halt the Allies, under General Lord Raglan and Marshal Armand Saint-Armand, at the Alma River on 20 September. They were pushed back with heavy losses. The Siege of Sevastopol began on 8 October. But it was rarely complete; a naval bombardment by the Allies on 17 October was a costly fiasco; while attempts by the Russians to drive the British back to their base were only just thwarted at the desperate Battle of **Balaklava** on the 20th, and at Inkerman on 5 November.

What followed that winter represented a nadir for the British Army. Its ranks were catastrophically depleted by cholera and other diseases, its wounded were scarcely tended and its logistic organization virtually collapsed when supply ships in open waters were wrecked by a storm. In due course this disaster was to lead to a radical overhaul of its supply services and, under Florence **Nightingale**'s urging, the administrative reform of the Medical Branch. The scandal, well publicized in *The Times*, brought down the government of the day. Indeed, it was a Turkish force which defeated the next attempt by the Russians (now under Prince Gorchakov) to attack Balaklava, and the better-serviced French, who carried the main burden of the bombardment of Sevastopol's defences, which might have fallen in April had not governments in Paris and London set a precedent by interfering by **telegraph**.

As it was, the capture of Kerch on 24 May at last cut Sevastopol's communications, although it was not until 9 September, after the exhausting Battle of the Traktir on 16 August, that the city fell. Thereafter fighting continued in a desultory fashion in the Baltic and the Caucasus, as well as off the Black Sea coast (when ironclad French warships were used for the first time to bombard forts at Kinburn), until peace was restored in February 1856. By this time estimated Russian losses amounted to 256,000 and those of the Allies to 253,000 – of whom a high proportion died from disease and deprivation.

The Crimean War marked a turning-point in the technology and conduct of war. Not only were governments taught that the development of **communications** meant they deprived their armed forces of reasonable logistic support at their peril, and that the semi-moratorium on modernization since 1815 was at an end. Henceforward they and the admirals, generals and captains of industry, would be swept along by a well-nigh irresistible torrent of inventions and **research and development**.

The Crimean War, 1854–6

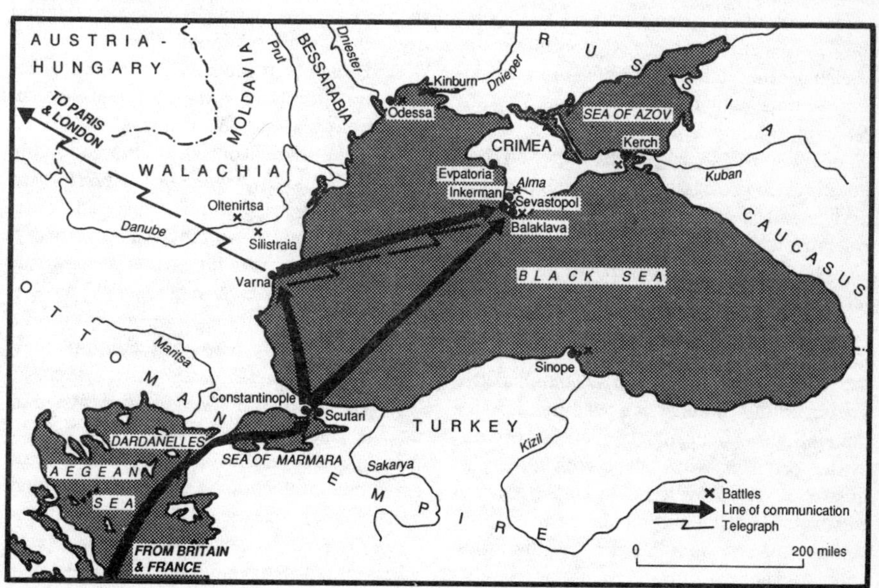

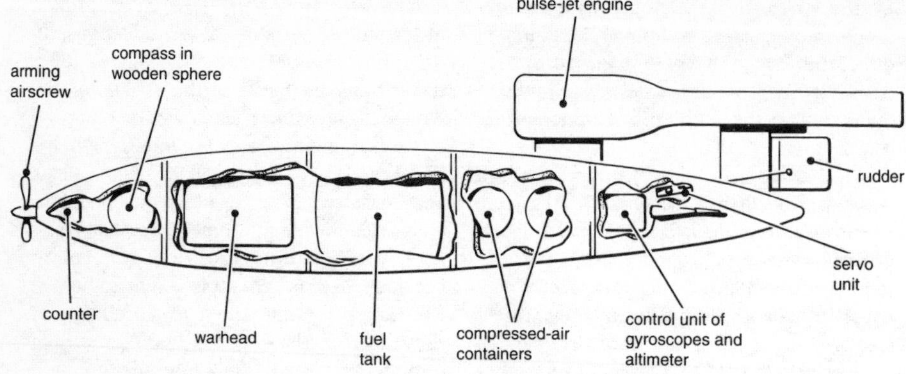

V1 cruise missile, 1944 (*German*)

Cruise missiles. The precursor of the modern cruise missile was the German low-flying, comparatively slow **V1** 'flying bomb' developed during the later stages of **World War II**. Powered by a ram-jet engine, V1 was a crude weapon of limited accuracy and with no on-board guidance to correct any in-flight drift from its pre-set course. However, V1 was capable of hitting an area target such as a city and its large high-explosive warhead could inflict considerable, if indiscriminate, damage.

The low-altitude flight path of the cruise missile means that it has a good chance of escaping observation by the enemy's air defence **radars**, which are usually designed to detect high-flying aircraft or incoming ballistic missiles with their high trajectory. When this characteristic is coupled with comparatively small size and a sophisticated guidance system such weapons, air, ground or sea-launched, can pose a real threat in both the tactical and strategic spheres. Post-1945 the Soviet Union were first in the field, with several examples such as their air-launched Kitchen with a range of 720km. Better known is the more advanced United States Tomahawk with a range of about 2,400km.

The problem with the low-flying cruise missile is to provide a guidance system capable of steering the missile clear of intervening obstacles to allow it to hit the target with sufficient accuracy to allow its warhead (which may be **nuclear**) to do the required damage. The development of terrain-following radar (originally a requirement for low-flying **fighter** aircraft) pointed the way. Also incorporated in the US Tomahawk is an inertial guidance system (TERCOM) linked to a pre-recorded radar map to enable the missile to find its own way to the target. For such a device to work properly an accurately surveyed start point is essential, not so easy to achieve for air and sea launches. Furthermore, the radar map is less effective if much of the flight takes place over the open sea.

Development problems have been immense, particularly with the naval version. But in the **UN–Iraq War** of 1991 US Tomahawks, launched from **battleships** and **submarines**, found their way over featureless sea and desert to strike pinpoint targets in cities at long range with remarkable accuracy, thus adding a new dimension to strategic bombing (**Bombing, Strategic**).

Cruiser. Probably the most amorphously named class of warship, the cruiser, as classified by the British in 1887, was designed as a general-purpose warship for policing the seas in peace and scouting for a fleet and commerce protection in wartime. They have ranged from **frigates** in the days of sail, to 21,000-tonne heavy cruisers armed with guns, **torpedoes** and missiles in the mid 20th century. By definition fast with medium armament, cruisers were (as first designed by the British – HMS *Shah* for example – in 1868, with steam power and sail and a speed of 16 knots) built of **iron** or mild **steel**, with a displacement of about 6,000 tonnes. The inevitable demand for better protection, however, soon produced vessels with armoured decks and armoured belts intended to shield magazines and machinery, raising displacements considerably. Likewise, armament could be anything between the two 10in guns and torpedo tubes in HMS *Shah* and the 14,200-tonne HMS *Powerful* of 1895 with its two 9·2in, 16 6in and 16 3in guns, its 6in armoured deck and speed of 22 knots.

Protected cruisers were followed by 'armoured' cruisers, such as the German *Scharnhorst* and *Gneisenau* of 11,240 tonnes which won at Coronel but were sunk by **battle-cruisers** at the **Falkland** Islands in 1914. With eight 8·2in and six 5·9in guns they were best suited for commerce raiding, but with a speed of only 21 knots useless for scouting. Indeed it was in the scouting role, as well as commerce raiding and protection, that 'light' cruisers – the 3,000–4,000 tonne, thinly protected, very fast kind with speeds of about 30 knots, armed with six or more 6in guns, plus torpedoes – which would prove most effective in **World War I**. In major battles such as Coronel and the Falkland Islands, they located the enemy and hunted his light cruisers; at **Jutland** they were the eyes of the battleship squadrons and potent in the anti-torpedo-boat role. In numerous actions, when battleship squadrons were absent, they were often decisive by their presence.

In 1922 the Washington Naval Treaties stipulated that no British, American, French, Japanese or Italian warship (other than capital ships) should displace more than 10,000 tonnes or mount a gun with a calibre greater than 8·1in. In fact, many cruisers of only 6,000–7,000 tonnes with guns of less than 8in calibre were built between the wars and, in **World War II**, were successful, with their speeds in excess of 30 knots, in the commerce protection and **anti-aircraft** roles. Nevertheless, Washington Treaty violations were common enough, notably by the Japanese. The Germans, too, managed to conceal their breach of a 10,000 tonnes upper limit on warship construction, imposed by the **Versailles Treaty**, to build three fast, welded-hull 'armoured' cruisers (nicknamed 'pocket battleships') which, with six 11in, eight 5·9in and six 4·1in guns, plus eight torpedo-tubes, eventually displaced 12,000 tonnes.

Cruisers in the nuclear and missile age seem in decline, displaced, for the most part, by more ubiquitous, almost equally deadly and considerably less expensive **destroyers**. Nevertheless, both the US and the Russian navies keep numbers in service, their role as headquarters and missile ships of undeniable service in support of **amphibious** forces engaged in such operations as fire-support off **Vietnam** and **convoy** escort in 1990 during the **UN–Iraq War**.

Crusades. It was the victory by the Muslim Seljuk Turks over the Byzantine army at the Battle of Manzikert in 1071, with the subsequent loss of Anatolia, which brought calls for a Crusade from Europe to restore access by Christian pilgrims to Palestine (the Holy Land) and Jerusalem. But at root lay the bellicosity inspired by Christian **chivalry** and the knights of the feudal system and the menace of **Muslim expansion** westwards.

The First Crusade (1096–9) was about 50,000-men strong under several strong leaders, who regularly fell out among themselves. They were opposed by more numerous, lightly **armoured** Turks whose mounted archers proved formidable. But the well-armoured Crusaders usually prevailed and after two prolonged sieges and four major battles entered Jerusalem in July 1099. By then a mere 10,000 strong they still were able to rout 50,000 Turks at Askelon in August.

The Second Crusade (1147–9) was necessary because the Muslims took Edessa in 1144. Despite several defeats in battle it reinforced the garrison of Jerusalem before foundering from internal dissent.

The Third Crusade (1189–92) was the outcome of the overrunning of Palestine and capture of Jerusalem by the redoubtable Kurd, Saladin, in 1187. An army of 30,000 under the Emperor Frederick I of Swabia, including a force of crossbowmen, fought its way to Acre where it was joined in 1191 by a seaborne force under King Richard I. A master of tactics and **logistics**, Richard took Acre, won the Battle of Arsouf and, with an outnumbered force, marched to Jerusalem in 1192 – where his resources were unequal to a long siege. Nevertheless he concluded a treaty with Saladin giving rights to pilgrims.

The Fourth Crusade (1202–4) captured and sacked Constantinople after a fierce siege, but got no farther. While the badly led *Fifth* (1218–21), which went by sea to Acre and via Egypt, was a disaster. But the *Sixth* (1228–9), without combat, regained possession of Jerusalem as the result of shrewd diplomacy by its excommunicated leader, the Emperor Frederick II.

The Seventh Crusade (1248–54) was another attempt to recapture Jerusalem (which had fallen to Turks in 1244) via Egypt. Undermined by numerous blunders, it collapsed after the Egyptians annihilated the Crusader army at Fariskur on 6 April 1250. And the *Eighth* in 1270 withered away from disease shortly after its start, whereupon, by 1303, the Muslims cleared Palestine of Christians and resumed their expansion into Europe. Nevertheless, various minor Crusading ventures were launched as late as the 16th century, even as the **Ottoman Empire** expanded after the collapse of Byzantium in 1453.

Militarily and technically there was give and take between the contestants. The Muslims learnt the value of **armour** and the importance of cooperation between the **cavalry** and **infantry** arms. The Crusaders came to appreciate the quality of the scimitar made of Damascus **steel**, the subtle effect of manoeuvre by light cavalry and the need to counter mounted archers with the superior crossbow and, in due course, longbow. They also incorporated in their European **fortifications** the multiple, vertical stone walls and high towers favoured by Byzantium for the security of cities.

Cunningham, Admiral Sir Andrew (later **Lord**) (1883–1963). Andrew Cunningham joined the Royal Navy in 1898 and served with distinction in **World War I**. Between the wars he helped develop naval gunnery and night-fighting capability. As C-in-C Mediterranean Fleet from 1939 until 1942, he at once won moral superiority over the Italians at such battles as Calabria and **Taranto** in 1940 and **Matapan** in 1941. But the heavy losses off **Crete** and towards the end of 1941 in the Eastern Mediterranean were the nadir of his fortunes. They were compensated for when, as Allied Naval Commander in November 1942, he led the invasion of French North-West Africa and, in 1943, Sicily and Italy, prior to becoming First Sea Lord and Chief of Naval Staff in 1943 until 1946, thus steering the Royal Navy through its final struggle against Germany and Japan and into its peacetime reorganization.

Currie, Lieutenant-General Sir Arthur (1875–1933). A Canadian militia artillery officer who, at the beginning of **World War I**, came to Europe as a battalion commander and was appointed to command a brigade in 1915. Made Commander 1st Canadian Division in 1916, he was soon given command of the Canadian Corps, an appointment he held until the end of the war when he became Inspector Canadian Militia. Currie not only commanded Canadians with great skill in all their major actions on the Western Front, he also managed significantly to stimulate Canadian nationalism. The brilliantly executed capture of **Vimy Ridge** in April 1917, his insistence thereafter that the Canadian Corps must be used only as an entity, and the part the Corps played at the Battle of **Amiens** and in the final advance to victory in 1918, all strengthened Canada's post-war claims to nationhood.

Cyrus II (the Great) (530 BC). The ruler of the Persian tribes from 559 BC and acknowledged as a fine statesman and general who created an empire from Egypt to the Indus. His campaigns against Medea and other nations were invariably victorious owing to his skill in coordinating the overhead fire of archers with the manoeuvres of **chariots**, **cavalry** and drilled **infantry** in close formation. He thus formalized existing loose practices and laid the foundations of organized land warfare, but made no significant technical innovations.

D

Daimler, Gottlieb (1834–1900). A German engineer who, as director of the Gasmotorenfabrik Deutz, developed existing internal-combustion engines (*see* **Propulsion, Means of**) in collaboration with Karl Benz and others. In 1883 at his own workshop in Cannstatt he began work on the successful high-revolution, four-stroke engine which established his fame. In 1883 it powered a bicycle, in 1886 a four-wheeled carriage, in 1887 a boat and in 1888 an **airship**. In 1890 he founded the Daimler Motor Company which, with Mercedes Benz, became a leader in motor technology.

Darius I (**the Great**) (d. 486 BC) inherited **Cyrus the Great**'s empire and army when he came to the Persian throne in 522 BC, as well as his expansionist aims once he had subdued dissidents within the kingdom. Mainly he sent forth invading armies under other generals. However, he led in person the well-planned and technically innovative invasion of Europe in 511 BC, which was distinguished by the **bridging** of both the Bosporus and the River Danube. But his northward thrust was defeated when the Scythians withdrew, laid waste the country and ruined his **logistic** plan. Likewise his attempts to subdue Greece were frustrated by defeat at **Marathon** in 490 BC (*see* **Greek–Persian wars**).

Darlan, Admiral Jean-Louis (1881–1942). As Chief of Staff, 1936–9, and then C-in-C, he made a great contribution to the efficiency of the French Navy prior to **World War II**. In the aftermath of the fall of France in 1940, he had the difficult duty of keeping his ships out of German and British hands, a political task he fulfilled with what was left of the Fleet under Vichy government control between July 1940 and November 1942. As a minister in the Vichy government, his activities earned him much abuse, even after he became commander of all French forces in 1942. Undeniably his presence in Algiers, when the Allies landed in November 1942, was crucial in preventing prolonged French resistance and in ensuring the French ships at Toulon were scuttled before the Germans arrived. Yet, politically, his assassination on Christmas Eve was a relief for the Allies.

Davis, Sir Robert (1870–1965). The inventor of diving suits and underwater escape apparatus from **submarines** and amphibious **tanks**. He was chairman of Siebe, Gorman & Company, which concentrated (prior to **World War II**) on the closed-circuit system supplying the diver with pure oxygen, the carbon dioxide being exhaled through a soda–lime filter. His apparatus was adopted by many navies, even though it was restricted to a depth of 100 feet. For use at greater depths, Davis developed mini-submarines with a chamber for exit and entry by divers and **frogmen** wearing compressed-air suits. He also was involved with respiratory equipment for high-altitude working.

Defence ministries and departments. Some sort of defence department, even if comprised of only a handful of leaders and administrators, has always been required. Naturally, the need increased with the growth of states, the complexity of their forces and of technology. *C.* 495 BC **Darius the Great**, a better administrator than general, decentralized control of his forces when maritime operations became essential to satisfy his expansive aims. **Shipbuilding** and the crewing and maintenance of warships called for care that was both special and expensive. Coordination of sea and land operations was tentative whenever the major arm (until the 15th century usually the Army) claimed precedence.

The advent of the **carrack** and the consequential rising importance of trade and sea-power in the 16th and 17th centuries made essential in Portugal, Spain, Britain and Holland the creation of navy departments of state – admiralty as it was called in Britain. Indeed, throughout the **Wars of the 16th, 17th and 18th centuries**, when technical effect (*see* **Technology, Effects of**) was steadily imposing its revolution, the creation of military bureaucracies in which a political and military head worked through a council served by permanent staffs, who gathered **intelligence**, planned, arranged finance, manning and procurement, and transmitted requirements to subordinate formations, became irresistible – a system which burgeoned at the end of the 18th century when France adopted Claude Chappé's semaphore **communications** system and enacted mass **conscription**.

Rarely until the **telegraph** did the departments have a direct operational function. But with the arrival of **telephone** and **radio** the opportunity to interfere was often irresistible to ministers and heads of state, who also were commanders-in-chief.

When a requirement arose for joint or combined operations involving both services, it was customary, whether or not the head of state was also C-in-C, to form committees which would settle who should make the plan, appoint the commander, issue the orders and allocate resources. In 1902 the British formed a Committee of Imperial Defence which evolved gradually into a sophisticated joint staffs committee system. Nations with limited or no access to the sea at all had fewer problems: the army dominated. But for the Americans, for example, in the **American Civil War** and the Spanish–American War, intricate problems of allocation and command and control had to be solved. These were matters which readily produced clashes and rivalries and which could only by resolved at the pinnacle of power, often by a person in ignorance of the subject who lacked unbiased professional advice.

As **aircraft** were added to navies and armies, the traditional systems bent under a weight of complexity beyond their organization and flexibility to sustain. The **Gallipoli** battles in **World War I** were chaotic examples of the existing departmental and command machinery's inability to mount operations in which naval and land forces, with new technology available, had complementary parts to play. As air power not only demonstrated its impact upon sea and land warfare, but also (and of fundamental political importance) attacked homelands, the need for separate departments of aviation controlling a third, independent service became evident. The British creation in 1918 of an Air Ministry running the Royal Air Force (RAF), which combined the Royal Navy's Air Service (RNAS) and the Army's flying corps (RFC), inaugurated a revolution – but without making any easier combined or joint operations between all three Services. Indeed it complicated matters and exacerbated rivalries.

Between the World Wars attempts to create ministries of defence to coordinate the three services were usually either blocked or accepted, in diluted form, as secretariats without power. Only in Germany was the subject grasped with any determination, when the Wehrmacht (Armed Forces) was created in the mid-1930s under OKW (Armed Forces Command). But with **Hitler** in charge as head of state, supreme commander, minister of war and, after 1941, Army C-in-C, it caused as many problems as it solved. For one thing the Army, Navy and Air Force retained their autonomy in almost every respect, including the vital area of weapon procurement, ensuring that interdepartmental rivalries and waste from duplication flourished as always. Only later, during **World War II**, was a central Ministry of Armaments given the authority to rationalize the allocation of materials and production facilities. Operational command to begin with was orthodox, although the Air Force intruded wherever it could (as in **Crete**). But soon the practice of creating so-called OKW Theatres of War, such as **Norway**, reflected Hitler's internal, divisive, political whims and determination to subjugate the Army.

The principal Allied nations, with the exception of Russia where the Red Army was predominant under Josef **Stalin**, moved cautiously and empirically towards centralized systems based on small, coordinating ministries of defence and joint committees. In Britain joint staffs were highly developed, but the divorce of design and supply of weapons from the defence departments proved as much a disaster as did the pre-war unwillingness of the RAF to provide the Royal Navy with suitable **aircraft**. In the USA, on the other hand, where the Navy, the Marines and the Army retained their own air forces and right of direct access to the President, waste in the acquisition of supply continued.

Since World War II all the major nations have created separate air forces and merged them within departments of ministries of defence which have control over design, development and procurement of equipment. The USA took the plunge by Act of Congress in 1947, when they also formed a separate air force. The British moved reluctantly to a fully merged Ministry of Defence in 1964. But although each service, at first, largely retained its old structure and influence within the new agglomerations, remorseless rationalization, notably in **logistics** and procurement of weapons, whittled away ancient privileges and traditions. As an example of extreme rationalization, the Canadian decision in the 1960s to unify their three services, dressing them in one uniform, and centralizing finance, weapons procurement and administration in a National Defence HQ, was not a success. The imposition of an all-embracing, amorphous bureaucracy eroded the basic loyalties which are so vital to combat efficiency, making reconsideration essential in the 1980s. Yet there is much to be said for standard logistic services, wherever possible, for sea, ground and air forces – provided a doctrinal approach does not stifle initiative in the extraordinary and exacting circumstances of war.

De Gaulle, General Charles (1890–1970). An infantryman who was taken prisoner at **Verdun** in 1916. In the 1930s he emerged as a leader of French military thought with advocacy of mobile defence by armoured forces in his book *Vers l'armée de métier*. Taken note of by the Germans, he was given command of the incomplete 4th Armoured Division in 1940 in time to lead it against the Germans in May. In battle it made some impression, but on 6 June he was made Under-Secretary of War as France entered her death throes. In Britain on 18 June he raised the Free French movement of his countrymen who wished to continue the fight.

Four years later, after struggling to build up his forces and acquire political status, he returned to France determined to revitalize his country. Rejected in 1946, he returned to power in June 1958 when France stood on the verge of civil war. After winning adoption of a new Constitution, he became President in December, a post he held until December 1969, having extracted France from the Algerian War of Independence and improved her self-confidence significantly.

De Havilland, Sir Geoffrey (1882–1965). An engineer who began work on **motor vehicles** but in 1908 designed an **aircraft** engine. Before **World War I** he joined the British Army Balloon Factory and designed tractor biplanes, transferring later to the Aircraft

Manufacturing Company to design very successful **fighters** and **bombers**. Between the World Wars his company produced light biplanes, airliners and a fast, long-range racing monoplane – the Comet, which was forerunner of the famous wood-built Mosquito fighter-bomber. During **World War II** he began work on jet fighters and **jet** engines and afterwards built the revolutionary though ill-fated Comet jet airliner.

Depth charges were 120lb containers of **explosive** invented by the British in 1915 to counter the submerged **submarine**. Dropped overboard, they could be detonated by a hydrostatic pistol at 40ft or 80ft; to be effective had to burst within 21ft of the submarine; and claimed only 22 out of 59 German U-boats sunk in **World War I**, at a ratio of two per 2,000. A marginally better ratio was achieved in **World War II** as the result of increases in depth, the 30 per cent more effective Torpex explosive in place of Amatol, and firing them broadside and ahead in a pattern.

Between the world wars ineffective and inefficient air-dropped depth charges up to 450lb were developed by Britain. The 250lb type with Torpex, dropped from low altitude at great risk to the aircraft, proved best after 1942. Today the depth charge is largely outmoded (*see also* **Anti-submarine weapons**).

Desert warfare. Campaigning in country which aptly has been described as the tactician's paradise and the quartermaster's hell is among the most exacting of military arts, in which **logistics** and, above all, the supply of water are always vital. Prior to **mechanization**, therefore, large armies shied away from a military theatre of such arid terrain, depending on small-scale forces, as often as not supplied by or mounted on camels. For example, the desert campaigns of **Cyrus the Great**, **Darius the Great**, **Alexander the Great** and the **Crusades** never strayed far from rivers or the coast; and 'oasis hopping' was rare indeed. But when General Sir Horatio **Kitchener** reconquered the Sudan in 1898 with over 25,000 men, his lines of communication depended on river steamers and the **railway** he built as he advanced.

World War I was distinguished by the pioneering of large-scale desert operations in the Middle East employing **motor vehicles** and **aircraft**. The Turkish attempt in 1915 to cut the Suez Canal would have benefited from greater mechanization as it crossed the Sinai desert. The British invasion of **Mesopotamia** would have been impossible without the great rivers and the building of a railway, while the use there in 1917 of **armoured cars** and light lorries conferred hitherto unattainable tactical mobility on long-range forces. But already in 1916, armoured cars had soon put paid to a Turkish-inspired insurrection in Cyrenaica. Supported by aircraft, they would greatly enhance the striking power of the camel forces of the Arab Revolt and the British advance into Palestine across Sinai, which also benefited hugely from the rapid laying of a pumped-water-supply pipeline.

Experiments between the World Wars in the desert with rough-terrain transport, principally by the British, French and Italians, demonstrated not only the importance of track-laying vehicles (especially French half-tracks), but such things as improved air filters to minimize engine wear, sealed cooling systems to save water and the sun compass for accurate navigation in the wastes at a distance from the coast where most traffic moved. Mastery of these techniques, many of which were associated with unusual **medical** and psychological problems, made survival possible in North Africa during **World War II** for the men of entirely mechanized forces when operating, in strength, deep in the interior and under extremely testing conditions.

Tactically, too, mechanized vehicles, above all the **armoured fighting vehicles**, were the key to success in terrain with few 'tank proof' places for **infantry** to shelter, and in which major natural obstacles were few and far between. Mobility was all-important to bypass and encircle strong points, which fell as soon as their water ran out. Aircraft played a crucial role because they could so

The destroyer Hibiki *(Japanese, 1,700 tonnes)*

easily detect and attack targets in wide open terrain. **Photography** was all-revealing. The side without air cover was only safe by night. Concentrations and convoys of supply vehicles were especially vulnerable unless widely dispersed, placing demands on highly disciplined combat **drills** and **C³**.

It goes without saying that the lessons learnt prior to 1946 were fundamental to the subsequent **Arab–Israeli** and **UN–Iraq** wars.

Destroyers. The invention of the **torpedo**, and the call for special warships to carry it into action, led in 1878 to the first fast **torpedo-boat** (HMS *Lightning*, speed 19 knots) and, later, to stealthy **submarines**. Countermeasures were at once imperative and included increases in secondary armament on **battleships** and **cruisers** and in 1893 the first unarmoured torpedo-boat destroyer (HMS *Havoc*, speed 27 knots, armament four quick-firing guns), soon to be known as 'destroyers' and made even faster with the installation of steam-**turbine** engines.

In the normal course of development, to satisfy demands for more seaworthy boats with improved armament and protection, displacement weights increased from the initial 300 tonnes. By 1912, for example, the French had built a 700-tonne ship with a speed of 34 knots and two 3·9in and four 9-pounder guns, plus four torpedo-tubes. After the **Russo-Japanese War**, ships such as these were built in large numbers to escort the battle fleets. No sooner had the submarine threat been recognized at the start of **World War I** than the requirement became insatiable for destroyers for use as escorts for commerce and, in due course, **convoys**. Come 1918 destroyers had sunk battleships and cruisers, fought one against the other, helped curb, none too efficiently with **ram** and **depth charge** the submarine menace, and shot down **aircraft**. As maids of all work they had acquired light **armour** and come close to the equivalent of pre-war light cruisers with displacements of 1,500 tonnes, four 4·7in guns and secondary weapons, plus six torpedo-tubes and depth charges.

This enlargement, at ever-increased expense, continued in the inter-war period. The need for **anti-aircraft** escort ships called for high-angle guns and multiple cannon and **machine-guns**, all of which, along with ammunition, required extra space. Speeds remained around the 30-knot-plus mark but displacements rose to over 2,000 tonnes. Indeed, so complex were these ships that no longer were they best suited as convoy escorts. In **World War II** sloops, **corvettes** and **frigates** assumed the principal anti-submarine role.

The so-called destroyers of 3,700 tonnes or more built in the 1950s, for example, by Britain and the USA had, with their extensive **radar**, numerous guns and **rocket missiles**, sophisticated **sonar** and anti-submarine weapons, grown into light cruisers. Indeed, the name 'destroyer' lost its original meaning as battleships and cruisers became outmoded, and as the submarine began to assume an important anti-submarine and **deterrence** role. So as economics, as well as the tactical requirements of the **nuclear** age, indicated that smaller and yet more effective surface vessels, backed up by aircraft, were necessary, the nomenclature 'frigate' and 'corvette' came back into vogue.

Deterrence is a word coined in 1861 in connection with the war against crime; it has only since the 1930s assumed a politico-military designation as prevention by fear. But, needless to say, any adoption of an aggressive demeanour has always tended to provoke a deterrent reaction.

For example, at the start of the **Greek–Persian Wars** the Greek defensive measures were intended, without much hope against so powerful an enemy, to deter **Darius the Great**. But by the time **Alexander the Great** invaded Persian territory the balance of power had swung the other way. Another sort of deterrent appeared during the European **Wars of the 16th, 17th and 18th centuries**: the convention of **limited war** through abstention from battle and restraint in **sieges** in order to conserve manpower and minimize destruction. The rise of humanitarianism in the 19th century imposed another limiting (though by no means total) effect on waste of life among massed **conscripted** armies in the aftermath of the well-publicized horrors of the **Crimean War** and the Battle of Solferino in 1859. These examples, among many, simply show that the factor of fear was too weak decisively to influence the public and their rulers.

Until 1859 the only likely effective deterrent was the size and supposed efficiency of units and forces. But that year the appearance of the French ironclad *La Gloire*, with its potential to outmatch all existing warships, compelled navies instantly to react to an irresistible weapon system. The experience set in motion the technological race at sea, on land and in the air which has escalated ever since because, even if one nation gained a seemingly decisive military (and therefore political) advantage, the others needed to demonstrate equal retaliatory capability to maintain the status quo.

The pre-**World War I** naval races between Germany and Britain; among the major powers before **World War II**; and since the 1950s between Russia and Nato, all contained strong overtones of national deterrence. The threat to homelands by **bomber aircraft**, which some assumed in the 1930s to be unstoppable, prompted Britain, in particular, to build better bombers as a deterrent – even though the bomber's effectiveness was overrated and the **fighter**'s underrated. Likewise, controversy over the offensive and defensive merits of the **tank** enabled the Germans to create armoured forces against which, in 1939, there was no credible deterrent.

It was not, however, until the arrival of **nuclear** weapons (*see* **Atom bomb**) that the concept of deterrence assumed more than a passing effectiveness. Demonstration of the **hydrogen bomb**'s total destructiveness made governments far more careful to control such weapons and to defuse potentially menacing situations which, in years gone by, would have led to major clashes of global dimensions. They concentrated more on achieving their aims by diplomacy, by **cold** and **limited wars** – though at the same time the Americans, British, Russians, Chinese and French, who possessed nuclear weapons, paraded their willingness to use them.

The intellectual and strategic debates at the heart of an enormous accumulation of nuclear weapons by the major powers have tended to focus excessive attention on their means of delivery and the defence against them. Jargon and clichés litter this battleground in which the contenders, on the whole, attempt to maintain stability through mutual deterrence to prevent inflicting unacceptable damage on each other. Fresh concepts or emerging technology, such as the theory of **nuclear winter** or the anti-missile **Strategic Defence Initiative**, can either destabilize or enhance deterrence. But integral with the quite ludicrously excessive accumulation of nuclear weapons, intended to ensure no power can out-obliterate the other, is the equally dangerous buildup of forces whose highly accurate, non-nuclear weapons pose a far more realistic threat of destabilization if one rival or another feels able to win a war by conventional means alone. Effective deterrence always demands unambiguous willingness to meet force with force.

Dieppe Raid. Despite instructions to execute large-scale hit-and-run **amphibious** raids across the English Channel in 1942, to take the strain off Russia and bring the German Air Force to battle, the British had failed to do so by mid-July. At that time Admiral Lord **Mountbatten** won acceptance for revival of a very recent, last-minute cancellation of a raid against strong defences at Dieppe. The plan called for three **Commandos**, the better part of the 2nd Canadian Division (with **tanks**), heavily supported by naval and air bombardment, to capture the port by a landing at dawn on 19 August, and withdraw that day.

Surprise was achieved, but by a series of accidents the enemy was partially prepared. **Intelligence** about the beaches and defences was faulty. Only one Commando reached its objective, the rest of the landing force being slaughtered either at sea or on the beaches. In addition to sailors from 38 sunken vessels, 3,403 Canadians, 247 Commando soldiers, 13 US Rangers and 153 airmen were lost, along with their equipment. In aircraft the score was 48 to 88 in the German favour. But the lessons learnt by the Allies of the need for avoidance of very strong defences, complete beach intelligence, heavy, carefully planned naval and air bombardment, greatly improved signal **communications** between all three services and each echelon of attack, and exact **navigation**, along with beach-traffic control, handsomely justified an important aim of an operation which was regarded as an essential experiment. The lessons were to be applied in numerous future landings.

Director Fire Control System (*see also* **Gunnery**). By 1890 the Royal Navy had concluded that the heavy **artillery** they were installing in their major warships had little chance of scoring hits beyond a tenth of their maximum range. A **rangefinder** with a 'plotter' designed to plot the course of the target in relation to the course of the ship made scant improvement in 1892. In 1905 Arthur Pollen, managing director of the Linotype Company, demonstrated a plotting table which made use of bearings to the target. It failed mainly because it was found impossible to take sufficiently accurate bearings when the ship was pitching and rolling. So Pollen designed a **gyroscope**-stabilized table with a rotating arm, holding a pencil, linked to the range finder; plotting was made fully automatic except for the need to feed in the ship's speed.

In competition with Pollen, Lieutenant Frederic Dreyer RN developed an even more complicated plotting table controlled by a mechanical calculator, but lacking the vital gyroscope. This was adopted in 1913 despite its many shortcomings – as revealed in **World War I**.

Disarmament. Before the 19th century disarmament as deliberate political business between nations was virtually unheard of. Until the 1850s, when nearly all weapons suddenly became obsolete, nations had retained equipment that was more than 50 years old, safe in the knowledge that it would be effective. But when, of a sudden, rearmament of great magnitude set in there was no checking it (*see* **Armaments industry**) and a new approach was needed.

In the past whenever arms limitation or disarmament had occurred, it was as often as not *unilateral* and either came as an imposed by-product of defeat by the victors, or out of self-interest. For example, when the knights of **chivalry** outlawed the longbow the motive was not altruism but the fact that their costly **armour** no longer offered adequate protection, thus seriously threatening their dominant role. Needless to say, nobody took any notice of these rules for restrictive practice.

Usually, of course, political economy dictated. No sooner was peace declared than there was a tendency to dispose quickly of the most expensive weapon systems. This, until **aircraft** arrived, often meant the laying up of warships and demobilization of their crews. The aftermath of nearly every major war since the **Crimean** has witnessed this process; indeed it is simpler to quote occasions when it has not happened – such as after the **Russo-Japanese War** of 1904–5, when neither side had won a clear-cut victory; and **World War II**, when Josef **Stalin**'s paranoia, allied to traditional suspicion, secretiveness and fear of the **atom bomb** in American possession, persuaded the Russians to raise rather than greatly reduce their forces in the manner of their erstwhile allies.

The urge to disarm, by people sickened of excessive violence, is of ancient standing. Yet not until the Hague Conferences of 1899 and 1907 were international attempts made to limit armaments – and with scant hope of success at a time when popular glorification of war, along with a burgeoning arms race in Europe, was in progress. Called by Tsar Nicholas II, who was aware of Russia's military inferiority, they were hardly likely to appeal to nations, particularly Germany, France and Britain with forces that were growing increasingly strong. Hague declarations in 1899 to prohibit poison gas (*see* **Chemical warfare**) and expanding bullets were not renewed in 1907. A ban on the dropping of missiles from **balloons** looked absurd in the age of aeroplanes.

After **World War I** the 1922 Washington Naval Treaty achieved a measure of *multilateral* limitation – after the League of Nations had begun in 1920 to seek large-scale disarmament. But by 1932 when an international Disarmament Conference at last was arranged, rearmament in Japan and Europe was under way, the Washington Naval Treaty had already been broken and an atmosphere of heightening tension precluded agreement between the 59 nations – who tended to propose reductions of their rivals' most cherished weapons while resisting attempts to dispose of their own. Germany's flouting of the **Versailles Peace Treaty** which, among other penalties, limited her Army to 100,000 men and denied her heavy **artillery**, **tanks**, gas, an **air force**, **submarines** and war-ships in excess of 10,000 tonnes, put the Disarmament Conference in suspense.

At the core of resistance to disarmament were nationalist fears for security, mutual distrust and inability effectively to monitor agreements. After **World War II** the United Nations became the forum for international efforts to limit and reduce armaments. **Deterrence**, the advent of **nuclear** and **rocket** weapons and the outward movement into **space** were productive of inevitable deadlock among frightened politicians and peoples. The testing of atomic devices continues under tenuous control by agreement. Explosions in space are outlawed, largely because of disruption of vital signals **communications**. **Cold** and **limited wars** bred weapons of unheard-of power, making armament control all the more difficult. Not only did the Great Powers, Russia to the fore, vastly increase their armouries, they armed the smaller nations whose nationalist aspirations drew them into a continuing worldwide struggle, despite three major Limitation Treaties signed since 1969.

Of the two **Strategic Arms Limitation Treaties** (SALT 1 and 2) negotiated between 1969 and 1979, SALT 2 has not been ratified by the USA. Meanwhile the 1971 Seabed Treaty, which made illegal the emplacement of nuclear weapons on the seabed, outside the 12-mile limit, had only limited application and no guaranteed method of enforcement.

Nevertheless, the Strategic Arms Reduction Talks (START), which gathered momentum in the 1980s under the impulse of Russian willingness to disarm unilaterally, were maintained in the 1990s as the Soviet state began to fragment. For example, the American announcement in 1991 of an intention to dispose of tactical-nuclear-weapon systems prompted a positive Russian response.

But the key to what little, though much publicized, disarmament has taken place would appear to be Russia's short-term need to put her overstretched economy to rights, rather than any long-term change of heart. The major portion of military formations and weapons designated for abolition are obsolete; only a few of the modern kind have been disposed of. Meanwhile the admission to international inspection of military installations, although restricted, improves safeguards against subterfuge and surprise attacks – with the caveat, based on the experience of UN teams trying to disarm Iraq in the aftermath of the **UN–Iraq War**, that inspectors can be frustrated and misled. Disarmament is desirable – but not to the detriment of security and **deterrence**.

Dockyards and naval bases. No doubt the first boats were hollowed out from tree trunks and built on river banks within close reach of forests. But as time went by and the embryo **shipbuilding** industries had to satisfy the need for larger vessels, shipyards became more complex and had to be located alongside wide waterways or within sheltered estuaries and bays, which in turn needed protection from predators as well as the elements. For virtually always, fighting and merchant vessels were built in the same yards, and as often as not naval bases were established in the same places, where skilled labour and the special facilities could be found and shared.

At a very early stage, therefore, **fortifications** were required to defend dockyards, especially the coastal ones. Prior to **artillery** in the 14th century, a boom across a harbour entrance and wood or stone walls circling the vital points usually sufficed, their strength and complexity commensurate with strategic importance. The strong defences of Constantinople, for example, reflected the vital multi-purpose importance of this political, strategic, shipbuilding, commercial and cultural centre. Inevitably the growth and spread of maritime trade under the burgeoning influence of caravels and **carracks** led to the development of numerous fortified dockyards, especially by Mediterranean countries and in France, Britain and Holland.

The arrival of **artillery** naturally demanded expansion of existing fortifications as well as the creation of arsenals within naval bases. To keep warships out of bombardment range, coastal artillery was required at harbour entrances and on headlands to dominate the approaches (*see* **Coastal defence**). Adjacent to the shipyards, with their slipways and dry docks, specially protected magazines and armouries had to be developed to make and service munitions and armament, often on a grand scale. Starting in the 16th century great naval bases such as Cadiz, Plymouth, Portsmouth, Brest, Rotterdam and St Petersburg flourished in support of wooden sailing ships and reached a zenith during the **French revolutionary wars**.

In the mid 19th century, however, these vital shore establishments were pitched into revolution as steam-powered, ironclad warships began to supplant the old 'wooden walls'. Not only had wharves, slipways and dry docks to be greatly enlarged to accept bigger ships, and the engineering facilities completely changed to deal with machinery and iron instead of sailcloth and wood, the labour force had to be retrained in totally different skills and the warehouses stocked with an immense variety of stores and spare parts.

The **artillery** revolution also had a radical effect, one which went beyond the need to cope with the more complex rifled, breech-loading pieces which replaced the old muzzle-loaders, but to withstand what the British envisaged as a new threat to security. No longer was a harbour kept safe by batteries guarding its entrance. Henceforward it might be possible for an enemy (the French) to put ashore shell-firing guns and deny main base facilities to the Fleet, which then would be unable to repel an invasion. At Portsmouth, and elsewhere, rings of expensive **fortifications**, nicknamed 'Palmerston's follies', were constructed to keep enemy forces at a distance from the docks.

The technological revolution and added complexity of ships enforced increased reliance on sophisticated dockyards where ships tended to spend longer times for maintenance. Since steamships could no longer remain at sea for long periods, more dockyards with coal (and later oil) bunkering facilities were required at strategic locations to satisfy need. Bases, along with suitable commercial ports, became targets for **blockade** and attack by the enemy from sea, land and, in due course, the air – calling for **anti-aircraft** defences to supplement the forts.

Just how vulnerable navies were becoming from threats to their dockyards was demonstrated in the **Crimean War**, the **American Civil War**, the Spanish–American War and the **Russo-Japanese War**, when ships were penned in and destroyed at anchor, occasions when rings of forts merely delayed the inevitable crushing of squadrons which would have done better manoeuvring at sea than sitting as targets in harbour. Nevertheless, it was appreciated prior to **World War I** that ports themselves were becoming additionally vulnerable to moored **mines** and **torpedoes** – the former laid secretly by sea (in due course by air) in the approaches, the latter fired by **submarines** infiltrating the netted and mined seaward defences. Henceforward defence had to be all-round and umbrella-like. Ships had to be at short notice to escape to sea.

In **World War I** it had been realized that flotilla craft, such as **torpedo-boats**, **submarines** and **landing-craft**, required base and repair ships to provide mobile accommodation and technical facilities. These auxiliary support vessels were the forerunners of the huge fleet trains, including floating docks, which accompanied fleets and **amphibious** forces during **World War II** when they operated for weeks on end at long distances from shore bases. By their mobility the trains provided a security not always absolute in heavily defended ports, where major warships were several times sunk – by **aircraft** at **Taranto** and **Pearl Harbor**, submarines at Scapa Flow, Alexandria, Diego Suarez and Singapore, and raiders at Saint-Nazaire.

The need for well-defended shore bases as well as fleet trains is greater than ever to serve warships whose requirements are increasingly complicated. How important the latter are has been demonstrated on innumerable occasions since 1945, notably, for example, off **Vietnam**, in the Persian Gulf and off the **Falkland** Islands.

Dogger Bank, Battle of. Faced by the immense strength of the British Grand Fleet in **World War I**, the German High Seas Fleet's only hope of scoring successes was by hit-and-run attacks designed to lure enemy ships to destruction on minefields (*see* **Mines, Sea**). On 3 November 1914, under the command of Rear-Admiral Franz Hipper, four German **battle-cruisers** and four light **cruisers**, covering minelayers, ineffectually bombarded Great Yarmouth. Mines sank one British **submarine** and one German cruiser. Despite intelligence of German intentions by decryption of intercepted **radio** messages, the British failed to bring Hipper to action; and bungled again on 16 December when he successfully bombarded Hartlepool (a defended port), Whitby and Scarborough.

On 24 January, supported by the High Seas Fleet, Hipper, with three battle-cruisers and the armoured cruiser *Blücher* went hunting British ships and was intercepted near the Dogger Bank by five British battle-cruisers and a light-cruiser squadron under Admiral Sir David Beatty. As Hipper ran for home, *Blücher* was fatally hit. But a few minutes later Beatty's flagship also was slowed by three hits, whereupon his orders to the remaining four battle-cruisers, to 'attack the enemy rear' were misinterpreted by his second-in-command, who allowed Hipper's battle-cruisers to escape while concentrating on *Blücher*.

Dönitz, Admiral Karl (1891–1990). A U-boat officer in **World War I** who specialized in the technical and tactical development of German **submarines** and, in September 1935, was given command of the first U-boat flotilla. Dönitz was convinced that Germany's main weapon against merchant shipping was U-boats which, controlled by **radio**, should operate in so-called 'wolf packs' and concentrate on hunting by night.

Throughout **World War II** he developed these tactics, first as Commander, Submarines; later, after January 1943, as C-in-C of the Navy. From beginning to end he was deprived of sufficient resources to win the Battle of the **Atlantic**. But he came close to doing so in 1943 – and might indeed have succeeded if he had sooner appreciated and taken measures against the deadly enemy **radar**, and had given much higher priority to the development of the schnorkel and much faster U-boats, all of which could have been in service in 1943 instead of, respectively, 1944 and 1945. As one of the few military leaders who could handle **Hitler**, he became Head of State after his Führer committed suicide in Berlin.

Douhet, General Giulio (1860–1930). An artillerist who commanded Italy's first aviation unit from 1912 to 1915, but whose criticism of his country's conduct of **World War I** brought about a court martial and imprisonment in 1917. Exonerated by the débâcle at **Caporetto**, he was brought back to command the Italian Air Force. In 1921 he published *The Command of the Air*, which postulated a doctrine (already strongly voiced in Britain) of strategic **bombing** against enemy cities by independent air forces. The book won international fame but did more to terrorize politicians and populaces than influence **air force** leaders.

Dowding, Air Chief Marshal Sir Hugh (later **Lord**) (1882–1970). An artillerist who, serving with the Royal Flying Corps in France throughout **World War I** and, between the wars, in senior staff appointments, was to have a profound influence on the development of RAF doctrine, training and, as director of **research and development** from 1930 to 1936, equipment policy. In that post and the next as Air Officer Commanding Fighter Command, he was responsible, among many vital innovations, for adopting the eight-gun Hurricane and Spitfire **fighters**, **radar** and the air defence system he was to command during the Battle of **Britain**. It was he who dissuaded Winston **Churchill** from dissipating RAF fighters during the Battle of France; who was quick to jam enemy **radio** beams; and who, although misled by faulty **intelligence** concerning German air strength, adhered, under intense pressure, to the strategy and tactics which saved Britain.

Dreadnought, HMS. The launching in 1906 of the eighth Royal Navy ship of this name was the result of British reaction to the threat of German naval expansion. Designed by William Gard and William Watts to the specification of Admiral Sir John **Fisher**, she was the first **battleship** to mount a main armament of guns of the same size – ten 12in pieces – plus a secondary armament of 27 12-pounders (later increased to 4in) and five **torpedo**-tubes. With **armour** up to 12 inches thick and a displacement of 18,100 tonnes, she was powered by 23,000hp Parsons, **coal**-fuelled, steam-**turbine** engines and had a top speed of 21 knots. She started trials a year after laying of the keel.

Dreadnought established the shape and characteristics of all future battleships and **battle-cruisers**, and made obsolete nearly every warship afloat. Major navies felt compelled to copy her and subsequently scrap many existing ships, the Royal Navy alone disposing of 154, 17 of which were battleships. In **World War I** she rammed and sank one German submarine but took no part in the Battle of **Jutland** and was scrapped in 1920.

Drill, Military. To make the best use of weapons and technology, as well as to facilitate the execution of tactical manoeuvres, the ingraining of established combat drills is essential – if only to help commanders and men under stress to overcome fear and concentrate on their tasks. Beyond doubt warriors in ancient times spent much time on repetitive weapon practice and rudimentary combat teams and groups were exercised in attack and defence. The first-recorded drilled, close formations are those of Persia and Greece (*c.* 500 BC). The well-disciplined hoplites of the Greek phalanx were well drilled to fight in close formation with **spear** and **sword**, and later were excelled by the Roman legionnaires. Later, the unsurpassed British longbowmen in the 14th and 15th centuries owed their superiority to regular and repeated practice of efficient techniques – the sort of drills which became even more important, on sea and land with the first complicated **artillery** and small arms, in order to maintain high rates of accurate fire.

The survival of infantry with early firearms

depended upon the alacrity with which, while packed together, they could manoeuvre as well as perform numerous operations in sequence to load, aim and fire matchlock and flintlock muskets – combat drills which continue to be displayed on ceremonial occasions. Likewise, every piece of equipment demanded foolproof drills for effective operation – be it a **motor vehicle**, **torpedo**, **fighter aircraft**, **radio** or **computer**. And every tactical manoeuvre has a better chance of success if executed within the framework of a well-known procedure.

The introduction of rapid-fire weapons in the 1850s began to eliminate the need for close formations. But, until late in the 20th century, there were many who argued that private soldiers would yield under fire if not packed crowded together under their leader's personal supervision. With greater prescience, they believed close contact between officers and men in combat was synonymous with drilling, a foundation of **morale**. It took the experiences of the Boer War, the **Russo-Japanese War** and **World War I** to show not only that close-order tactics were exorbitant in lives, as well as tactically outmoded, but also that educated soldiers, properly trained, motivated and led, would fight well without close supervision. It was hard for many traditionalists to abandon the rigidity of tactical formations, although in essence nearly all successful operations in the later stages of World War I depended, in the final analysis, on personal initiatives by lowly individuals operating in loosely knit groups.

Enlightened leaders who survived World War I and recognized the value of flexibility in the control of mechanized operations as well as tactics, devised procedures which applied efficient drills to almost every function at sea, on land, in the air, in headquarters, on the lines of communications and airfields and at bases. Repeatedly it was shown how intelligently directed junior commanders in command of individual ships, small army detachments and single aircraft could be persuaded to fight without coercion. Nevertheless, it is undeniable that usually a commander has to be heard clearly even if not seen. Voice **radio** made this possible, enabling the leaders of, for example, submarine-hunting groups, armoured units and aircraft formations, to command and inspire their subordinates as if by remote control. Hence the subtler disciplined drills of officers using signals **communications** were often substituted for raised parade-ground-type voices. This system makes it increasingly feasible, as radios become smaller and more reliable, for leaders to command any subordinate at the touch of a switch.

Dung, General Van Tien (b. 1917). An enthusiastic member of the Vietminh **guerrillas** who, from 1936, unrelentingly fought the French and Japanese in French Indo-China (Indo-Chinese War of Independence). As Chief of Staff of the North Vietnamese Army in 1953, he played a prominent role in the French defeat at Dien Bien Phu and later in preparations for the campaign of the 1960s to conquer South Vietnam (**Vietnam War**). North Vietnam's costly failures led in 1974 to Dung replacing General Vo **Giap** as C-in-C whereupon he began limited attacks in December which evolved into the offensive which overran the South by April 1975. In 1980 he took over as Minister of Defence and became a member of the Politburo.

Dunkirk, Battles of. This Flanders port, sheltering behind a maze of sandbanks, first assumed military importance when, in the 12th century, it became an anti-Norman-pirate base. During the **Spanish Wars**, it was seized and burnt by a French raiding force in 1558; taken again by the French in 1582; and retaken by the Spaniards a year later. During the **Thirty Years War** the French, in 1646, once more seized it; lost it again in 1652; but recaptured it with British assistance in 1658 after the Battle of the Dunes. Then it was strongly fortified by **Vauban**, but in 1706 fell to **Marlborough** during the War of the Spanish Succession; the port subsequently being demolished by the British under the terms of the Treaty of Utrecht to prevent further piracy by Jean Bart. Not until after 1983 was the port rebuilt.

During the whole of **World War I** it was within artillery range of the Germans, garrisoned by the French and British. The latter used it as a forward base for light coastal forces and aircraft engaged mainly in defence of Britain and the Straits of Dover and in attacks on the ports of Ostend and Zeebrugge.

In **World War II**, when the Germans broke through in Holland and the Ardennes in May 1940 and headed for the Channel ports, it became plain that the Allied forces in Belgium would soon be cut off from their bases in France. The British nominated Boulogne, Calais, Dunkirk and Ostend as alternative, temporary supply ports at the same time as plans were made to evacuate the armies through Dunkirk – if possible. In the event, Ostend was never used, and Boulogne and Calais were surrounded on the 23rd and captured by the 26th. On the 24th, with the Allied armies still well to the east in Belgium and only a small, ill-organized garrison holding Dunkirk, General von Kleist's overwhelmingly strong Panzer Group had only to advance a few miles to cut off the Allies from the sea.

He was prevented from doing so by General Gerd von Rundstedt, who called a halt because he feared for his left flank. He was supported by **Hitler**, who agreed with Marshal Hermann **Göring** that the German Air Force should complete the job for the Wehrmacht. But the task proved beyond the Air Force, which was over-extended and incapable of operating at night when the bulk of Allied troops were being evacuated by sea. On the 26th, when Hitler rescinded the halt order, it was too late. By then the Allies had rushed strong forces back to Dunkirk and had let in the sea, making the task of immediately seizing the port impossible for the German Army, thus giving time for the British to retreat within the perimeter and carry out a famous naval operation which, by 4 June, had evacuated 338,000 Allied soldiers – with hardly any equipment.

For the rest of the war, Dunkirk was regarded by the Germans as a base for naval operations against Britain, a staging post for coastal **convoys** and a bastion of their **Atlantic Wall**. In that capacity it was frequently bombed by the British. When the Allied armies advanced from Normandy in August 1944, the task of clearing the Channel ports fell to First Canadian Army. But the exacting business of capturing Le Havre, Boulogne and Calais, and clearing the approaches to Antwerp, induced the decision to leave the port encircled for the rest of the war. On 10 May 1945 it was the last French town to be liberated.

Dynamo. When Michael **Faraday** demonstrated in 1831 that a so-called direct electric current (DC) could be generated by rotating a copper disc between the poles of a permanent magnet, he stimulated widespread technological research, as well as examination of the **electromagnetic effect** and strong demand for electric power. A year later a generator with a wired armature moving in a permanent magnetic field was invented by Hippolyte Pixii; vastly to be improved on in 1845 by Charles Wheatstone using an electromagnet.

One improvement followed another, until in 1877 the Hopkinson brothers devised a practical dynamo, which Thomas **Edison** took in 1878 as the basis for the central generating system required for commercial application of his electric-light bulb. Other major steps were taken in 1877 when dynamos producing alternating current (AC) were invented; and in 1888 with Nikola Tesla's invention of the induction motor. With immaculate timing generators were ready to make feasible the internal-combustion engine and all the other new means of **propulsion** powering the devices, machines and the **electronics** of the 20th century.

E

Edged weapons. No doubt a **stone** dagger was the first crude edged personal weapon in use at the same time as arrowheads were being shaped from flint (*see* **Bows and arrows**). Inevitably, of course, these primitive arms became obsolete with the appearance of **metallurgy**'s first products, *c.* 3500 BC, with the making of **bronze** and **iron**, followed *c.* 1200 BC by **steel**. From then on, in close consultations between warriors, armourers and craftsmen, the efficiency of arrowheads, daggers, spears, **swords**, **pikes**, battleaxes and lances was improved as each sought ways of enhancing lethality as well as overcoming the increasingly upgraded **armour** protection being worn or carried by adversaries.

Interest and experiment usually centred upon the conflicting merits of thrusting and cutting weapons. A well-aimed thrust could kill instantly, but the blade was often difficult to withdraw from flesh and bone, or might be deflected. A slash from a cutting weapon, on the other hand, might only gash or lop without immediately causing incapacitation. Thrust or cut, high-grade metals, which did not easily shatter in the manner of cast bronze and, to a lesser extent, iron, and retained their cutting edge were in demand for all weapons, including the battleaxe, which relied more upon brute force to achieve maximum effect, but was least popular among the cutting weapons, though crippling in the hands of strong men. None, however, was effective unless the shaft was strong and well designed. And there was always an appreciation that stand-off weapons – the arrow, spear, pike and lance – had distinct advantages over those dedicated to hand-to-hand combat.

The introduction of hand-held firearms and **artillery** in no way made edged weapons obsolete, since the chances of hand-to-hand combat remained as high as ever until rates of accurate fire and **gunnery** techniques were improved. Before the invention of the **bayonet**, towards the end of the 17th century, pikemen continued to be mixed with musketeers in **infantry** formations. Swords went on being used by officers in the **Crimean War** and sometimes were carried into action during **World War I**. And both sabre and lance were sacrificially employed by Polish and Russian **cavalry** at the start of **World War II**.

For the infantryman the bayonet remains in service to the present day. But the length of blade has varied considerably: for example, the long British bayonet of 1914 was superseded in World War II by an eight-inch spike; and in recent years by a broad-blade-knife type which can be used as a dagger (the traditional hand weapon of **special forces**), a tin-opener and, with its scabbard, a wire-cutter.

Edison, Thomas (1847–1931). An entrepreneurial American inventor, prolific in ideas, whose contribution to military technology was mainly indirect. Largely self-educated, he learnt about electric **communications** devices when a **telegraph** operator. Beginning in 1868, he successively invented automatic telegraphic ticker-tape machines, an office duplicating machine, improved **telephones**, a wax-cylinder recording machine and a dictating machine. After costly **research and development** by his ever-growing laboratory and workshop, he made in 1879 a carbon-filament, vacuum electric-light bulb and at once concentrated upon the development of **dynamos** to power public and industrial electricity systems. In 1885 his interest focused on an induction telegraph which could transmit signals from a moving railway train, or between ships at sea – a step towards experiments with rudimentary **radio** valves, intermingled with the patenting in 1891 of a motion-picture camera (*see* **Photography**).

In all he took out 1,033 patents, the majority of which had military applications, including **batteries** and chemical substances, such as phenol.

Education, Military. It was the appearance of standing armies in the 17th century which vitally stimulated formal military education. Until then written works on **strategy**, **tactics**, **drills** and **training**, such as those compiled by **Sun Tzu** *c.* 500 BC and **Vegetius** in AD 390, were known to only a very few experts and mainly disseminated by example and word of mouth. Not even the invention of the **printing** machine *c.* 1450 and the publication of many more books were able to produce a dramatic improvement in the spread of knowledge from practical philosophers such as Niccolò **Machiavelli** in the 16th century. The latter nevertheless undoubtedly spurred the modernization and **technological effect** which inspired the military institutions during the **wars of the 17th century**, and such manuals as the British and French naval *Fighting Instructions*.

The founding in 1660 of the Royal Society from an existing nucleus of 'divers worthy persons' of scientific bent, proved a powerful stimulus to European education and investigation in a period of revolution and war, albeit when Peter I of Russia established in 1698 the first military academy it was still only for the privileged nobility. Wider dissemination came from General Maurice de **Saxe**'s *Reveries on the Art of War* in the mid 18th century and **Frederick the Great**'s *Instructions to the Generals*. These were authoritative textbooks, which reflected the latest advances in technology, proposals for **conscription** of steadily growing populations, and better control methods and **logistics** for enlarged forces in the **wars of the 18th century**.

The employment by **Napoleon Bonaparte** of a Chief of Staff to regulate his armies, and the establishment of the École Spéciale Militaire in 1802 (the same year in which West Point Military Academy was formed in the USA) to train officers, pointed the way to the creation of professional General Staff Corps – like that of Prussia, formed in 1801, which required specialized Staff Colleges to train an élite. Throughout the 19th century, naval and army academies and colleges for commissioned officers proliferated, yet hardly at all for other ranks, who continued to learn their trades within their own units.

Scholarly analysis and the publication of such books as Karl von **Clausewitz**'s *On War* in 1832 and Antoine **Jomini**'s *Précis de l'art de la guerre* in 1836, significantly advanced the study of military philosophy, strategy, tactics and organization. Simultaneously, science, new technology, weapons and equipment imposed considerable demands upon technicians – of whom navies and armies possessed few. In an age when classically educated people predominated and gentlemen who were officers rarely were involved with industry or commerce, it was mainly civilians and non-commissioned military men who grappled with engines, mechanized vehicles, the latest weapons, signalling systems and suchlike.

Paradoxically the Royal Navy, which in 1917 was almost the last to form a Staff College, was among the first, in 1830, to form a Gunnery School – a mere 500 years after the first use of cannon! Its task was not only to teach the handling of the latest complicated pieces and **gunnery** techniques, but also to establish uniform doctrine and training in a vital subject which, until then, was taught at the whim of individual captains.

In armies it was the principal technical branches, the Artillery, Engineers and Transportation Corps, which were ordered to study, operate and teach about the majority of new weapons and devices. Since the 14th century, gunners and sappers had been treated as specialists and paid extra. In the 17th century the French military engineer, Sébastien **de Vauban**, educated his officers in **fortification**, though without greatly raising the status of technically qualified people or overcoming a snobbery which exists to the present day in many armed forces.

Nevertheless, schools and specialized organizations for each new weapon system had to be formed. Normally it was artillerists who dealt with **rockets**, missiles, some **mortars** and **anti-aircraft weapons**, and engineers who took on signalling, **chemical** warfare and, in conjunction with transportation experts, helped develop **armoured fighting vehicles**, as well as, through work on **balloons**, providing the nucleus of **air forces**, leaving **machine-guns** and also some mortars to the **infantry**.

Self-evidence of the need for improved military educational standards and instructional methods created pressures for well-educated recruits. To begin with, illiteracy could no longer be tolerated. Where parents or civil schools had failed, the military had to put matters right before progressively educating its members in all subjects. Throughout the 20th century, the extent to which any nation's armed forces have raised their standards of education has become a benchmark of progress. By sponsoring students at universities and technical colleges, for example, the armed forces have done more than stimulate their own **recruiting** and complement the work of their own colleges and schools. They have reinforced national standards of education. Indeed, in many ways they have pioneered personnel selection, enlightened discipline, syllabus planning and instructional methods which have been copied in civil schools and industry.

It is no exaggeration to say that the considerable advances in military education have fundamentally altered the relationship between officers and lower ranks. The old-fashioned 'do as I say, not as I do' discipline is outmoded in well-educated forces. Authority becomes based on knowledge and reasoning as much as on rank. The educated officer who realizes he cannot 'know it all', listens carefully to the advice of subordinate experts – who, when ordered, are expected to carry out tasks in a skilful way. In this environment criticism and insights which at one time would have been rated as insubordination simply become part of the decision-making process.

Edward III (1312–77). As King of England from 1327, he ranks among the great generals for his **strategic** and **tactical** conduct of the first phase of the **Hundred Years War**. Having decisively defeated the Scots at Halidon Hill in 1332, he drifted into a war with France which, until the naval Battle of **Sluys** in 1340, was mainly a conflict of minor **amphibious** and land raiding. Edward's tactics with a balanced, disciplined and well-trained force of **infantry**, longbowmen (*see* **Bows and arrows**) and armoured **cavalry**, who often fought dismounted as so-called men-at-arms, were vastly superior to the French mounted-cavalry tactics.

Edward's military fame rests chiefly on the decisive victories at **Crécy** in 1346 (when he introduced **artillery** to European land warfare) and Poitiers in 1356. As a result the brilliance of his strategy on external lines, conducted through shrewd operational instructions to great commanders such as the Black Prince, the Duke of Lancaster and Sir John Chandos, has frequently been undervalued. His pioneer use of artillery in **siege warfare** has also been overlooked; and his tactical employment of the bow as a rapid-fire, infantry weapon system misunderstood.

Electromagnetic effect (*see also* **Electronics**). Electromagnetic radiation is the propagation of energy through space by means of electrical and magnetic fields that vary in time. An electromagnetic wave travels through space or the atmosphere at 300,000 metres/second and the number of such waves passing a given point per second gives a measure of the frequency in hertz (cycles per second).

The discovery and evaluation of these effects by successive scientists, of whom Humphry Davy, Michael **Faraday**, Karl Gauss, James Maxwell and Heinrich **Hertz** are among the most important, has had the profoundest influence on the development of warfare, especially in the field of **communications**. Because electromagnetic waves travel in straight lines, Guglielmo **Marconi**'s pioneering of wireless **radio** telegraphy was derided by some of his contemporaries, until his success proved them wrong, because, in fact, radio ground waves are normally limited to a range of about 160km by the earth's curvature. However, the discovery by Oliver Heaviside and Arthur Kennelly of the ionosphere (*c.* 1901) and its ability to refract electromagnetic waves and return them to earth (sky waves), indicated that radio signals have an almost unlimited range.

The electromagnetic spectrum runs from one hertz, covering at the lower end radio, then radar and infrared, through ultraviolet, X-rays and to gamma rays at about 10^{11} gigahertz (1 gigahertz = 10^9 hertz). Radio waves, occupying a fairly wide part of the lower end of the spectrum, illustrate well the problem the electronic engineer has in harnessing the electromagnetic effect to military needs. Radio bands range from low frequency at 30–300kHz up to extra high frequency at 30–300gHz. The various bandwidths between these two extremes are becoming increasingly crowded, with demands for space for different military requirements; typical channel bandwidths are: teleprinters, 150Hz; facsimile, 1.5kHz; **computer** data, not less than 2.4kHz; high-frequency radio, 3–6kHz; very-high to extra-high-frequency radio, 12.5–25Hz; **television**, 5.5 MHz. Thus a television channel, which needs a wide bandwidth, would take up the space required for about 250 voice channels – which is why television is rarely used for military communications purposes.

While the electromagnetic effect has encouraged the development of sophisticated **radio**, **radar** and **computer** equipment for military use, there are dangers from it too. Since electromagnetic radiation depends, broadly, on the excitation of electrons, it follows that there can be significant effects from a **nuclear** explosion (*see* **Atom bomb**). Though often ignored by the layman, to whom the blast and radioactive effects on humans and objects gainsay all else, in the context of military operations the electromagnetic effect can be of dramatic importance. An airburst at over 100 kilometres above the ground can produce an electromagnetic pulse covering an enormous area, perhaps a whole continent. As well as damaging equipment the pulse can make changes in the electrical conductivity of the ionosphere. This could make high-frequency communications completely unworkable and would reduce radar effectiveness; very-high and extra-high-frequency links should not be affected. The damage to electronic components following a **nuclear** explosion, from electromagnetic radiation, could be equally catastrophic since it would affect computers as well as radios and radars. The phenomenon is known as Transient Radiation Effects on Electronic Equipments (TREE). Typically, computer magnetic stores would be corrupted, relays tripped and there could be insulation failures and burn-outs in signals equipment.

The overall effect of a nuclear explosion could be to disrupt communications in the widest sense over a whole theatre of operations, if not worldwide. So dependent on sophisticated electronic equipment is modern warfare that it is hard to see how coherent operations could be continued under such circumstances, perhaps one of the most cogent military reasons for not crossing the nuclear threshold.

Electronics. The electronics age may be said to have been ushered in with the invention of the vacuum diode valve in 1902 by the Briton John **Fleming** (himself coining the word 'electronics'), the immediate application being in the field of **radio**. It was the invention of the transistor in 1948, however, which saw the emergence of electronics as a particular branch of electrical engineering, now showing every indication of becoming the biggest single industry in the world, permeating as it does every aspect of daily life. The effect of electronic developments on the art of war have been similarly dramatic. Not only has miniaturization reduced the sheer bulk of radio and other equipment, but the introduction of the transducer as a means of changing energy from one form to another and the development of solid-state circuitry has meant a dramatic improvement in reliability – always a high priority for military equipment.

Perhaps one of the most significant electronic developments from the military point of view was the invention of the cathode-ray tube, patented in 1923 by the Russian-American Vladimir Zworykin as the iconoscope cathode-ray camera. This made feasible the development of **television**; for the military it was **radar** that beckoned.

The cathode-ray tube fires electrons on to a photoelectric luminescent screen to produce an image, not necessarily a picture but perhaps a displayed radio pulse or other electrical phenomenon. The impact of this essentially simple device on the location of objects, rangefinding (*see* **Rangefinders**) and **navigation**, for example, was immense.

Electronic engineering lies at the heart of the weapons system that is the modern naval vessel, with its automatically controlled rapid-firing guns and its various missile systems (*see* **Rockets and guided missiles**) and electronic **anti-submarine** equipment. The modern naval battle could hardly be fought without such aids, which form a major part of the equipment on both surface vessels and **submarines**. Military **aircraft** too would fare ill without an impressive **array** of electronics: the speed of modern **fighter** aircraft is such that a pilot who could not rely on, for example, onboard **computers** to determine the best moment for an attack, coupled with a head-up (i.e. at eye level) display to present the information he needs without the distraction of studying his instrument panel, would be at a serious disadvantage.

Such dependence on electronics means that the enemy will always be looking for countermeasures to enable him to dodge the (electronically) guided missile that seeks to destroy him or to disrupt his opponent's radars, **communications** and **computers** so as to render him ineffective. The *electronic battlefield* is now a reality and it means that a whole new aspect of war has emerged – *electronic warfare*, a powerful weapon indeed. All electromagnetic emissions may be intercepted and can be of value to an enemy; formations may be identified, locations revealed and intentions discovered. Certain specialist electronic equipment, such as air defence radars, if detected, could identify particular units and help to disclose the order of battle.

The importance of electronic warfare began to emerge during **World War II** with the rapid development of radar, with improvements in radio communications, and with the Allied successes in intercepting and decoding German encrypted transmissions. Post-1945 electronic warfare has become an accepted part of Nato's naval and air operations, not only for intercept but also as a means of jamming guidance systems and similar electronic emissions: the ground forces have tended to lag behind in this field, but are fast developing the necessary techniques. The Warsaw Pact armies (*see* **Warsaw Treaty Organization**), on the other hand, developed electronic warfare into an important **intelligence**-gathering operation, so as to improve their offensive capability.

Electronic warfare may be divided into three main elements:

– *Electronic support measures* are mainly concerned with producing operational intelligence and with providing advice for countermeasures, warning of the need for higher-level surveillance and the deployment of target acquisition systems when it seems necessary.

– *Electronic countermeasures* (ECM), as the name implies, aim to disrupt the enemy's transmissions by jamming of radios, radars or missile-guidance systems, or to mislead him by deception.

– *Electronic counter countermeasures* (ECCM) are designed to foil the enemy's countermeasures deployed against one's own forces. Encryption of transmissions is one obvious, though expensive, means. Frequency-hopping techniques where transmission frequencies are changed automatically and frequently, help to confuse the enemy by forcing him to search a wide spectrum. A similar system for radar allows transmissions over a sweep of frequencies. Free-channel search involves the selection of one unoccupied frequency from a set of frequencies and using it for one transmission only: this helps to disguise the structure of a radio net since adjacent nets may share the same frequency allocation.

Computers come into a category of their own, though they are equally vulnerable to electronic attack if countermeasures are not taken. An unprotected computer terminal is vulnerable to interception at quite long range, while material is being displayed and worked on by the user, unless the terminal is kept inside a specially screened area. The cost and complexity of such screening means that only terminals and computers handling sensitive material are likely to enjoy such protection. For most defence computers security is maintained by forbidding the input of material above the appropriate classification. Nevertheless, the problem is severe, since developments in electronics technology in the second half of the 20th century mean that most staff officers in a military headquarters will have a computer terminal on their desks and can expect to be using it constantly.

Elephant. As the largest land mammal, the elephant, both in its European, Asiatic and Indian species, has frequently been given a military role. When used as combat vehicles their size often terrified the enemy; although their own natural fear of fire and loud noises usually revealed their vulnerability and unreliability. Since the advent of firearms they have perforce been relegated to **logistic** tasks and in that role were used in the **Vietnam War.**

Empty-battlefield theory. The widespread dispersal of men in combat, brought about by increased fire-power (*see* **Drill, Military**) is said by proponents of this theory to have produced 'a void on the battlefield'. They can point to the Battle of Agincourt in 1415, when British archers, who suffered negligible casualties in an hour's combat, stood seven to the yard against an enemy who was packed so tightly he could hardly wield his weapons; and show that in the **American Civil War** the ratio of men to space was 1:257 square metres and at the end of **World War II** 1:27,500; while in three days at Gettysburg in 1863 Federal casualties were 17 per cent; and in eleven days at **Kursk** in 1943 3 per cent per day. It is argued that the increased lethality of modern weapons compels soldiers not only to disperse but also to take cover behind **armour** or below ground as they cling to the desire to survive. Hence the vast majority of fire is aimed indirectly at the void, with minimal chances of killing an invisible enemy who also rarely aims directly to kill.

Engels, Friedrich (1820–95). The collaborator with Karl **Marx** in the formulation of the Communist philosophy with its emphasis on class struggle and total war by insurrection of the masses. As an industrialist and one-time Prussian soldier, he argued on historical grounds that conventional warfare by traditional armies would be outmoded by the social unification he assumed would be brought about by industrialization. Nevertheless he envisaged a kind of total warfare which struck at nations' brains (instead of their limbs) and as an extension of the revolutionary movements of 1848. Towards the end of the 20th century the truths as well as the flaws in his assumptions are all too apparent.

Engineering, Military. Throughout history, élite military engineers have worked at the heart of communities, planning, improvising and building. They were responsible for the **fortifications** which protected the early settlements they had designed and fabricated.

When warriors from those settlements took the offensive, those same engineers would be involved with **logistic** support and **bridging** for obstacle crossing, supervision of field defences and assembly and operation of **siege engines** should the need to assault enemy fortifications arise.

Chariots are the first depicted products of military engineering, as seen on the famous Standard of Ur, *c.* 3500 BC. But there is no evidence of a sophisticated obstacle-crossing operation until 511 BC, when **Darius I**'s Persians built a bridge of boats across the Bosporus. All the early, successful military nations – Hittites, Egyptians, Assyrians, Persians, Greeks and Romans – depended upon military engineers to achieve their conquests and secure their bases. And the **Crusaders**, returning to Europe, would copy Byzantine forts – until the revolution caused by **explosives, artillery** and firearms in the 14th century.

Simply because engineers were virtually the only technically trained soldiers available, they became explosives experts and the first gunners – and were rewarded with twice the pay of the highly skilled British longbowmen. In **siege warfare** during the **Hundred Years War**, gunners such as Jean **Bureau** created much new work for military engineers by systematically making high-walled forts and **siege engines** obsolete. By the end of the 17th century military engineers, such as the pre-eminent Marshal Sébastien de **Vauban**, had created scores of star-shaped forts and, at the same time, evolved standardized ways to overcome them by a combination of digging, bombardment, demolition and infantry assault.

Thus the military engineer has from the first needed to combine general engineering knowledge with a sound military education. France had specialist engineer officers as early as 1690, for example, though her sappers and miners at that time belonged to the artillery. By 1715 all engineering was combined in a single corps; in 1868 they acquired responsibility for the electric **telegraph**; in 1876 military **railways** came under their wing; and in 1904, aeronauts. It is interesting to see how closely this parallels the British Army experience – the Royal Engineers can fairly claim to have had a hand in the beginnings of most British military technological developments.

In the United States the Engineer Corps has traditionally had a civil function as well and many major civil engineering projects have been pushed forward under military leadership – the Panama Canal is a typical example.

In most modern armies, where separate technical corps exist to deal with specific aspects such as **communications** or vehicle repair, the field engineer has reverted to the many and varied tasks of his traditional role. In the battle zone a prime engineer function in support of mobility is the overcoming of obstacles. Of these, perhaps the most important is gap crossing which can be undertaken in many ways, from fascines carried on the front of a **tank** and dropped into a ditch to provide a quick way forward, to more permanent bridge structures.

Other counter-obstacle work involves clearing lanes through enemy minefields, employing techniques that range from sappers searching for **mines** with electronic detectors and digging them up manually, to various forms of mechanical mine-clearing equipment. Until **World War II**, mine-clearing was entirely a manual operation, though there were one or two experiments carried out using **World War I** tanks. The first practical mechanical mine-clearer, a rotating drum carrying lengths of chain which flailed the ground, exploding the mines in its path, mounted on the front of a Matilda tank and known as Scorpion, was used at the Battle of El **Alamein** in 1942. Successful, but very slow and cumbersome, the idea was developed into the much more efficient Sherman Crab, widely deployed in the later stages of World War II. Other devices, such as rollers pushed in front of a tank and ploughs to dig the mines up, have also been used; the British developed a new flail system, as a result of the **Falklands** War, which can be mounted on various vehicles. Yet another solution tried is the explosive hose, projected over a minefield and then detonated, in turn detonating the mines by overpressure, again pioneered by the British and known in its modern version as Giant Viper. A combination of flail, mine plough and Giant Viper appears to offer the most promise: but no system is foolproof.

During the **UN–Iraq War** engineers again improvised gapping devices attached, often, to armoured vehicles. For their part, Iraqi engineers were faced with the need for rapid repair of airfield runways after cratering by special munitions (*see* **Bombs**); and should have been used for counter-mobility tasks, as first practised on a large scale in World War I, to deny routes and key points to the UN by demolitions at bottlenecks, by the destruction of stores and by the setting of booby traps.

Not so glamorous perhaps, but equally vital engineering tasks are the provision of adequate and safe water-supply points in military operational areas, and a general engineering requirement to provide military works of all kinds wherever they may be needed.

Finally, in British service, the Royal Engineers take on the task of bomb disposal (though not generally the dismantling of terrorist devices), a relic of their World War II involvement and also, by a curious anachronism which demonstrates their versatility, the provision of postal facilities for all three services.

English Channel, Battles of the. Since Roman times the narrow seaway dividing the British Isles from France has assumed vital importance as a strategic and commercial key to Europe as well as Britain's moat. King Alfred's building of a navy to **blockade** the **Vikings** in their ports was vital to the defence of southern England in the latter half of the 9th century. But for lack of a navy King Harold was unable to repulse the Norman invasion in 1066, whereas at the start of the **Hundred Years War** in 1337, construction of a fleet by **Edward III** and his victory off **Sluys** in 1340 put a stop to French **amphibious** raiding and ensured Britain's unimpeded access to France in the centuries to come.

Not until 1588, when the Spanish Armada entered the Channel with the aim of escorting an army to invade Britain, was there another major battle in these waters. Then a week-long harassment by 50 nimble English **carracks**, armed with the latest cast-iron **artillery**, outfought the clumsy, ill-armed Spanish **galleons** and obsolete **galleys**, and culminated in a fireship attack on the Armada in Calais Roads, which forced the Spaniards to put to sea, to face further attacks. Driven northwards, they lost numerous vessels from shipwreck off Scotland and Ireland on their way home.

The three Anglo-Dutch wars of 1652–4, 1665–7 and 1672–4 were mainly of a naval nature and produced numerous encounters and major battles in the Channel in which the superior British **gunnery**, broadside firing, tactics and blockade, dictated by Admiral Robert **Blake**'s *Fighting Instructions*, eventually proved decisive. These superior methods would be applied by the British against the French in the subsequent **wars of the 17th and 18th centuries** and the **French revolutionary wars**, when the French were practically excluded from the English Channel. Indeed, not until **World War I**, when the new German Navy threatened to interfere with cross-Channel traffic and gain access to the Atlantic, was there a recurrence of fighting in these waters.

From the outbreak of war in August 1914, the Anglo-French Allies determined to prevent the Germans interfering with cross-Channel routes and from using these waters for access to the Atlantic. The German seizure of Antwerp, Zeebrugge and Ostend, and their failure to capture the other French ports as the prize for the Battle of Ypres in 1914, set the scene for the action of the next four years. While Britain based **fighters** near **Dunkirk** to intercept German Zeppelins and aircraft heading for England, German U-boats entered the Channel and, on 22 September 1914, sank three British **cruisers** with the loss of nearly 1,400 men. Throughout the war the Royal Navy regularly bombarded Ostend and Zeebrugge docks, hoping to prevent their use, and **convoyed** ships to

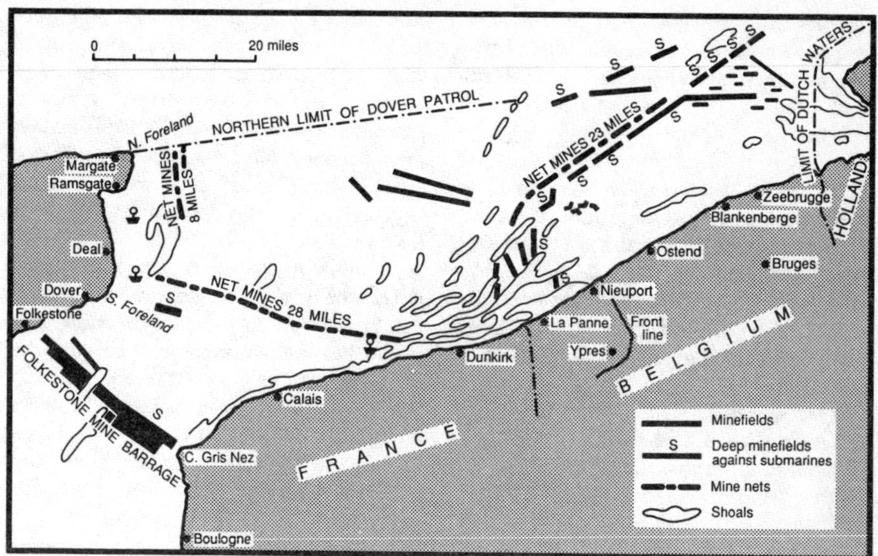

Straits of Dover: minefields and net barrages at the end of 1917

Holland through waters infested with **mines** and **submarines**.

As the German submarine campaign got into its stride, a mine and net barrage was laid to shield the Straits of Dover in April 1916 and patrolled day and night by armed drifters and aircraft, supported by **destroyers**. The Germans took to raiding the barrage. For example, on the night of 26/27 October, 12 of their **destroyers**, without loss, sank a destroyer, a transport and six drifters, damaged other vessels and bombarded Folkestone. Destroyer sought destroyer in combat, sometimes by ramming. Gradually the German threat was curbed – notably after the **amphibious** hit-and-run raids against Zeebrugge and Ostend in 1918.

Throughout **World War II** the struggle for the Channel was intense, commencing with anti-submarine measures in 1939 plus attempts to prevent the laying of mines from the air, but escalating to endless action after May 1940 as the Germans seized the entire coastline. During the **Dunkirk** evacuation and throughout the Battle of **Britain** a pattern emerged. Both sides' convoys would be attacked by aircraft by day and night and, off Dover, by **artillery**. At night, too, **torpedo-boats** would stalk each other and the convoys they guarded. Overhead, the struggle for air superiority was ceaseless and the bombing of ports a regular feature of amphibious raiding and invasion countermeasures. In fact, although the British made their first of many amphibious hit-and-run raids in June 1940, the Germans never copied them. Nor was there a real German threat of invasion after September 1940, disbelieve it as the British did.

Occasionally, prior to the **Normandy** invasion in 1944, there would be unusually sensational events – such as the passage up-Channel from Brest on 11–13 February 1942 of the **battle-cruisers** *Scharnhorst* and *Gneisenau* and the **cruiser** *Prinz Eugen*. They had been out of action in French ports since March 1941, kept under close surveillance and repeatedly bombed. Yet, despite ample warning of what was intended, thick weather and a series of accidents and cumulative human errors enabled the Germans to steam half-way up the Channel in broad daylight on the 12th before the British were alerted. They were tackled by a few torpedo-bombers, at great cost, and torpedo-boats after passing Calais. To great public indignation, they escaped without damage until both battle-cruisers struck mines laid in their path from the air off the Dutch coast.

Other events were the costly **Dieppe** raid on 19 August, against which German naval forces played only an accidental role in repulsing the attackers; and the attack, kept secret until long after the war, by torpedo-boats against an American invasion rehearsal off the Devon coast in May 1944 when several **landing-craft** were sunk with heavy loss of life.

These sensations apart, it was the British and their Allies whose small ships and aircraft fixed an unbreakable grasp on the narrow seas from 1941 onwards. This enabled them to sail convoys almost as they chose while, at the same time, deterring the enemy from venturing out by day. Because of this unremitting effort, at no small cost, it was possible to raid and invade almost with impunity.

Ericsson, John (1803–89). A Swedish soldier who came to England in 1826 and designed the steam **railway** locomotive *Novelty* which was rejected in favour of the **Stephenson** *Rocket*. In 1837 Ericsson's steam-driven screw-propelled *Francis B. Ogden* towed a Royal Navy ship on the River Thames at 10 knots, but the Admiralty rejected it as 'impractical'. Emigrating to America, he built *Princeton*, a steam warship for the US Navy. On 9 March 1862, during the **American Civil War**, his *Monitor*, the first armoured-turret ship, fought ***Merrimack*** to revolutionize naval warfare. He also experimented with **torpedoes** and solar engines.

Eugene of Savoy, Field Marshal Prince (1663–1736). As a keen student of warfare, Eugene joined the Austrian Army in 1683 after rejection by the French Army. He distinguished himself during the relief of Vienna in September 1683; commanded a **cavalry** regiment in pursuit of the Turks in 1684 (*see* **Ottoman Empire, Wars of**); was promoted major-general in 1685 and field marshal in 1688 at the age of 25! Between 1690 and 1715 he was almost constantly at war and, though frequently outnumbered by superior forces, showed genius as well as immense bravery. In Italy in 1690 he outmanoeuvred a larger French army; and at Zenta in 1697 routed and ejected from Hungary a Turkish army three times the size of his own.

At the start of the War of the Spanish Succession in 1701 he again outmanoeuvred the French in Italy, although checked in 1702. But with **Marlborough** at Blenheim in 1704, at Oudenarde and the siege of Lille in 1708 and **Malplaquet** in 1709, he brought the French to the verge of collapse, having already thrown them out of Italy in 1706. Finally he negotiated a favourable peace with France in 1714 and, the next year, joined forces with the French to rout the Turks in the Balkans. He used this victory to good advantage in the ensuing years to negotiate treaties very much to Austria's advantage. As the greatest soldier-statesmen of the period he and Marlborough rang the changes between the somewhat limited **wars of the 17th** and more total **wars of the 18th centuries**.

Explosives. Many are the claims as to when and who invented gunpowder, but there seems to be general acceptance that probably the Chinese or Indians were first, before or during the 12th century. Beyond doubt, however, the black powder consisting of, approximately, 75 per cent saltpetre, 12 per cent sulphur and 13 per cent charcoal, was used in the first cannon (*see* **Ammunition** and **Artillery**) in the 1340s. Though not until the 16th century, apparently, was it employed for blasting.

Gunpowder burns rapidly, mostly on the powder's surface and, in a confined space, produces an explosion with much smoke. By the 15th century it had been discovered that mixing the ingredients with alcohol or urine formed a paste which could be 'corned' into chunks to slow the burning and increase energy output; a process further refined by Thomas Rodman in the 1840s.

The first effective challenger to gunpowder as a propellant was created by Christian F. Schönbein in 1846 when he added nitric acid to cotton, producing gun-cotton (nitrocellulose). It was adopted by Austria in 1860, though unpopular elsewhere after a serious explosion in an Austrian factory. Nevertheless, interest continued as it was much more powerful than gunpowder, and a version known as Poudre B, developed by Paul Vieille, was adopted by the French Army in 1885. Meanwhile, in 1875 the Swede Alfred **Nobel** successfully used nitroglycerine as a gelatinizing agent with gun-cotton and produced ballistite, which was smokeless. Frederick Abel and James Drewer from Britain were working on the same lines but extruded their propellant in the form of cord – hence cordite.

These smokeless propellants not only increased the effective range of artillery and small arms, but also opened a new era in shooting techniques since guns would no longer be shrouded in smoke, giving away their position, and gunners would the more easily be able to observe the fall of shot.

The cordite taken into British service, though efficient, caused serious barrel wear owing to the very high temperatures developed on firing. But it was not until flashless propellant was demanded during **World War I**, to avoid blinding the gunners and giving away gun positions, that serious efforts and improvements were made; success was not finally achieved until after **World War II**.

The requirement for shell fillings was quite different, a detonation some 100,000 times the burning rate of propellant being needed to burst the shell casing and generate sufficient blast at the target. Picric acid (known as lyddite in British service) was originally demonstrated in France in 1885. Though used for many years it had a high melting-point which made filling difficult; a tendency to form dangerous compounds with heavy metals; and over-sensitivity for use in armour-piercing shell.

Germany meanwhile found a safer alternative in Trinitrotoluene (TNT), which was adopted in 1902. In the USA a less sensitive replacement for picric acid (known as 'Explosive D') was taken into service.

By 1907 the British too had adopted TNT in place of lyddite. However, though safer it was also less powerful and work between the wars and subsequently has led to the development of RDX (Research Department Explosive) for use in British shell and to the development of various plastic explosives, mostly developed by Imperial Chemical Industries for demolition and similar tasks – much favoured by terrorists in the form of the Czech-made Semtex, safer to handle until primed, easily moulded to fit any container and hard to detect.

In 1944 the demonstration of **nuclear** explosives threatened such disastrous consequences that they became a factor in **deterrence** and, so far, they have only twice been exploded during hostilities. More recently, highly efficient liquid fuel–air explosives have been developed to detonate land**mines**. An explosive vapour, dispensed by projectile or low, slow-flying aircraft, is detonated to produce an over-pressure which explodes the mines and, incidentally, causes severe damage to people in the open.

F

Falkenhayn, General Erich von (1861–1922). An unusual Prussian infantryman who instructed the Chinese Army and served as a General Staff officer in the Boxer War. As a favourite of the Kaiser, he became Minister of War in 1913. His disagreements with General Helmuth Johannes von Moltke, the Chief of the General Staff, led to his replacing von Moltke on 14 September, as the Army retreated from the **Marne**. Successes against Russia in 1915 were to his credit, but his disastrous decision to attempt to defeat France by attrition at **Verdun** in 1916 brought his transfer in August brilliantly to command the German invasion of Romania. Sent to command Turkish forces at Gaza in October 1917, he was too late to prevent defeat and was replaced in January 1918.

Falkland Islands, Battles of the. *World War I.* Following his victory at Coronel, Admiral von Spee dallied until 26 November 1914 before, after instructions from Germany, making for home via Cape Horn. Meantime the British had sent two **battle-cruisers**, under the command of Vice-Admiral Doveton Sturdee, to the Falkland Islands to join the old **battleship** *Canopus* and five **cruisers** to catch von Spee's squadron of two armoured and three light cruisers. Von Spee closed to attack Port Stanley early on 8 December, 24 hours after Sturdee's arrival for coaling. When *Canopus* opened fire at extreme range, von Spee withdrew. Not until Sturdee left harbour in pursuit did he recognize the presence of battle-cruisers which were both faster and better armed than his ships. Von Spee manoeuvred to get within range, but inflicted only minor damage. Sturdee aimed to engage from long range and, using the latest **gunnery** director techniques, overwhelmed his enemy. All the German ships, except one cruiser, were sunk, with the loss of 1,540 lives.

War of 1982. During **World War II** the island base was extensively used. Afterwards it was of declining importance to Britain while Argentina increasingly pressed claims to what she calls 'the Malvinas'. On 2 April 1982 substantial Argentine forces under General Benjamino Menéndez landed near Port Stanley, as well as a small group on the island of South Georgia. After overcoming brief resistance, they assumed control of all the islands and began to prepare a defence many, wrongfully, thought might not be called upon.

Immediately, Britain began dispatching ships and **submarines** as the advance guard of an **amphibious** task force, under Rear-Admiral John Woodward, which moved via Ascension Island (that became a half-way main staging base 4,000 miles from the Falklands) to recapture South Georgia. On 25 April, after **helicopters** had crippled a **submarine** and after a brisk fight ashore, South Georgia was retaken to become an advanced base for the approaching main task force. The establishment of a **blockade** and the buildup of British forces off the Falklands were conditioned by **logistic** restrictions; the time it took to prepare, sail and assemble the 31 warships (including two **aircraft-**

carriers and six submarines) and 80 support vessels involved; and the weather, which in the South Atlantic was appalling.

It was a considerable feat that the naval blockade was in place by **nuclear** submarine on 12 April and complete on 1 May when the main force arrived in what was known as the (200-mile) Exclusion Zone. On the 1st, Port Stanley airfield was bombed by Harrier fighter-bombers from the aircraft-carriers and a Vulcan bomber, which was seven times refuelled in flight, from Ascension. Thereafter, until the last night of the war, only supply by air from Argentina for the garrison of about 11,000 was possible. Furthermore after 2 May, when the cruiser *Belgrano* was sunk by a British submarine on the 2nd and Argentine naval forces (including an aircraft-carrier) withdrew to port, their best strike aircraft were compelled to operate at their extreme range of endurance. Thus, after the failure of the Argentines to sink the British carriers, the struggle for air superiority went the British way, helped by a successful hit-and-run amphibious raid by SAS against aircraft on Pebble Island on 15 May.

On 21 May, under the command of Major-General Jeremy Moore, British troops landed at San Carlos Water, unopposed except for suicidal attacks by Argentine fighter-bombers which sank warships, but without harm to the assault ships, including the cruise ship *Canberra*. Nevertheless loss to a **missile** of the container ship *Atlantic Conveyor*, with three out of four heavy-lift helicopters, placed a strain on logistic support as the land forces marched on Goose Green and over 60 miles of marshy, roadless terrain to attack the hilltop perimeter strongholds of Port Stanley.

The British had ascendancy over the poorly led Argentines from the outset, but were always operating on a logistic shoestring. It was of great assistance when, on 6 June, Bluff Cove was captured unopposed. Thereafter, although two logistics ships were hard hit on the 8th, this advanced base greatly boosted the buildup to the assault on Port Stanley. Phase 1, the taking of the outer defences, was complete by 12 June. Phase 2, the seizing of the inner ring dominating the town, was carried out against demoralized opposition by the 14th and compelled General Menéndez to agree to a cease-fire.

Faraday, Michael (1791–1867). The son of a blacksmith, Faraday studied science when an apprenticed bookseller, before becoming a laboratory assistant in 1813 to the eminent scientist Humphry Davy. In 1820 he developed Hans Oersted's observation of the relation between magnetism and electric current into his own discovery of electromagnetic rotation. Henceforward his studies of the **electromagnetic effect** were to give physics, **communications** and **power generation** a revolutionary boost. As Director of the Royal Institution his experiments in 1831 showed how an electric current was produced by passing soft iron through a magnetic field, a demonstration of induction which led eventually to the invention by others of the **dynamo** and much else besides.

With an uncommon ability to express complex matters clearly, Faraday stated the laws of electrolysis that were to inspire James Maxwell's mathematical electrical theories (for Faraday was no mathematician) and the invention of practical **batteries**. Faraday was the father of **electronics**, modern technology's most powerful force, which comes close to turning the military art into a science.

Fifth column. A term used for **propaganda** purposes by General Emilio Mola in 1936 to describe Fascist collaborators operating disruptively, by rumour or sabotage, behind the Republican lines (initially in **Madrid**) during the **Spanish Civil War**. It was in common use by all sides throughout **World War II** to denote almost any form of subversive activity, real or imagined. As such it had significance as a demoralizing or unsettling influence upon civilians as well as combatants. For example, the followers of Vidkun Quisling, the Norwegian traitor who seriously undermined **Norway**'s resistance in 1940, were classified as fifth columnists. And the fear of a fifth column was very successfully exploited by the Germans both during the Battle of France in 1940, and their subsequent victorious campaigns, when reports of its presence were frequently grossly exaggerated to the detriment of morale and resistance, as well as often causing dire consequences for innocent people accused of such activity.

To the present day it is still occasionally used for journalistic purposes.

Fighter aircraft. Before **World War I** the feasibility of arming **aircraft** had been established. A **rifle** was successfully fired from one in 1910, a **machine-gun** in 1912 and a 37mm cannon in 1913. The long-held expectation of aerial combat was fulfilled on 5 October 1914 when a French Voisin V89 brought down a German Aviatik with its machine-gun, an event which coincided with an acknowledgement that special fighting aircraft had to be developed to counter hostile **reconnaissance** and **bombing**.

It was also clear that the best way of shooting was by pointing an aircraft with a forward-firing weapon at the target. To begin with only rear-engined, pusher types were suitable. But on 1 April 1915 Roland Garros, flying a tractor Morane monoplane with a machine-gun firing through the propeller (which was armoured to deflect bullets), shot down an unsuspecting German aeroplane in flames. Within a few days of the Morane itself falling into German hands, a Dutchman, Anthony **Fokker**, had invented a device which activated firing of the gun only in moments of safety – and had sold it, mounted in his monoplane, to the Germans who scored their first victory with it on 15 July. Henceforward, the struggle between so-called fighters to dominate **air warfare** was the key to the winning of air superiority to enable reconnaissance, bombing and so on to be carried out.

Since it was quickly realized by the fighter pilots, above all the aces, that their success and survival depended upon machines which could outfly the enemy's, competition flourished to produce fighters which were faster, more manoeuvrable, higher climbers, sturdier and better armed. Between 1914 and 1918, speeds increased from the 87mph of the Fokker E1 monoplane to the 138mph of the French Spad-XIII; operational ceilings rose from 12,000ft to over 20,000ft (where oxygen deprivation reduced performance of both man and machine); biplanes (twin-winged) were much preferred to monoplanes because they were sturdier and often more manoeuvrable; and armament could be two or three machine-guns. Indeed, machines such as the Bristol Fighter F2B had a single forward-firing gun and one or two rearward-firing guns manned by a gunner/observer.

In practice nearly all fighters were multi-role, employed for reconnaissance, ground attack and bombing (usually low level), and **bomber** and **airships** interception, as well as fighter versus fighter combat. Fundamentally, fighter tactics were based upon a dive from above, shooting to kill the victim at ranges rarely in excess of 100 yards and then a scramble for height or dart for safety and home. But the duels which often occurred between fighters frequently were tail-chasing contests as pilots circled tightly and attempted to turn within the turning circle of the enemy plane to bring guns to bear. Manoeuvrability was crucial.

In the 1930s, as engine performance improved and all-metal construction began to supersede wood and fabric, the biplane fighter, which was no longer able to catch faster monoplane bombers, was replaced by sturdier, faster and better-armed monoplanes whose poorer manoeuvrability proved no great handicap. The very sturdy and reasonably agile German Messerschmitt Me109, which was improved upon throughout **World War II**, in 1939 had a top speed of 357mph, a ceiling above 30,000ft, and one 20mm cannon (firing through the propeller boss) plus two wing-mounted machine-guns, making the fitting of **armour** for vulnerable places essential for the enemy. They and their reciprocating-engined opponents would carry out all the traditional roles and, at war's end with long-range fuel tanks, operate freely, like the 437mph North American Mus-

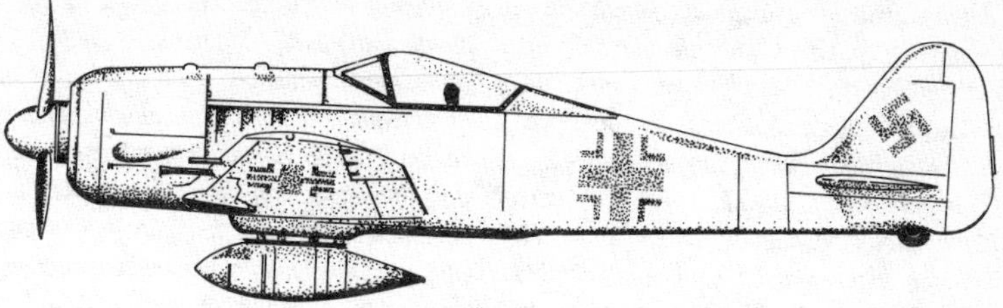

Fighter aircraft, 1941 (German FW190)

tang, at combat ranges in excess of 500 miles to Berlin and back. At the same time, fighters were built not only larger, but also, like the Bristol Beaufighter, for long-range and night fighting, with two engines and a second crewman to navigate and work the air-to-air **radar**.

The appearance of the **jet engine**, which almost coincided with the reciprocating engine reaching its peak of performance, brought about a new breed of fighter. The German Messerschmitt Me262 was the first combatworthy type. With a speed of 541mph, a ceiling of nearly 35,000ft and four cannon, it would have won Germany outright air supremacy in 1944 had not **Hitler** insisted on it being developed primarily as a bomber. Yet it was aerodynamically unable to make full use of its twin-engine power to overcome compressibility and fly above the speed of sound (Mach 1) – as could the next generation of swept-winged machines, such as the Russian Mig15, with its speed of 683mph in level flight, and a ceiling of around 50,000ft.

Nearly all jet fighters, with their variety of wing configurations (some variable), have a multi-combat role. Most have speeds in excess of Mach 2, are fitted with sophisticated **electronic** equipment (including **computers**), are greedy for fuel and benefit from in-flight refuelling, and are capable of carrying a variety of cannon, bombs, **rockets** and **missiles**, mostly externally. They place extreme physical and cerebral loads on their crews and call for sophisticated direction from ground controllers, plus support by complex technology based on **aircraft-carriers** or

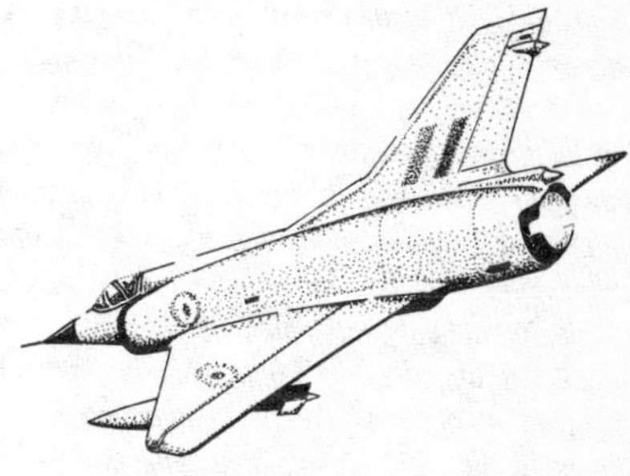

Mirage III fighter-bomber (French)

airfields with high-grade runways. Indeed, it was the desire to escape from such vulnerable airfields which prompted the invention of the British subsonic Vertical Take-Off and Landing (VTOL) Harrier, which can operate efficiently from very small spaces and from ordinary ships' decks.

The technology which enables aircraft to engage targets with fire-and-forget missiles well beyond visual distance has revolutionized combat tactics. On-board **electronics** engineers, supplied with intelligence of the enemy from a variety of ground and airborne sources, now spend as much time arranging defensive electronic countermeasures (ECM) as striking at the enemy. Stand-off engagements in which neither pilot sees the other are as likely as visual, shorter-range shots with heat-seeking homing missiles. Cannon fire is virtually a last resort.

Fisher, Admiral John (later Lord) (1841–1920), saw active service in the **Crimean War**, in China in 1858–60 and Egypt in

1882. Although he commanded many ships, it is his tenure at the Gunnery School, his strong influence on naval **gunnery** and his welcome to new technology in a conservative age for which he is most celebrated. 'Jackie' Fisher thrust the Royal Navy into the 20th century ahead of its rivals when the German Navy was challenging for supremacy. As First Sea Lord, from 1904 to 1910, he is credited with pushing through reorganization of **dockyards** to cope with the latest vessels (including the revolutionary ***Dreadnought* battleships**), the **Director Fire Control System**, **submarines** and the conversion from **coal** fuel to **oil**. He regarded submarines as a deadly weapon and ruthlessly scrapped 154 obsolete vessels. On retirement in 1910 he had created a modern fleet ready for war.

Called back by Winston **Churchill** as First Sea Lord in October 1914, it was he who ensured victory at the **Falklands** and who resigned in May 1915 because he disapproved of the **Gallipoli** campaign.

Flame warfare. The use of fire in support of military operations has been an option available to commanders since ancient times. Timber-built sailing ships were always notably vulnerable to it, and captains dreaded igniting their own vessels. Nevertheless it was occasionally used at sea and first recorded as Greek Fire at the siege of Constantinople in AD 672. To begin with, a compound of pitch, sulphur and, probably, saltpetre, invented by Callinicus, was hurled in pots with a burning fuse. Next pumps and pipes were installed in ships to project the flaming liquid at short range – the first flame-thrower.

The secret of Greek Fire was kept until 812. But thereafter various flaming liquids were used in **siege warfare**, either to set buildings alight or to pour from fortress walls on to assailants below. Fireships continued to be used, most famously by the British when they sent eight among the Spanish Armada at its anchorage at Calais, drove it to sea in panic and thus prevented the invasion of England in 1588.

Flame warfare became more sophisticated, in **World War I**, when used by the Germans in trench warfare in 1915. Their crude *Flammenwerfer* was a steel tube on a base plate through which burning **oil** was squirted by compressed air. A cumbersome British projector was an equally dangerous device to operate, especially since the compressed air still contained oxygen. Ranges never exceeded 70 metres (some manpack kits only 30 metres). And because so much fuel was burnt during projection, target effect was low.

Despite the dubious value of these early weapons the idea emerged again in **World War II**. Short range was again a problem but the projectors themselves had been made more reliable and easier to operate; the Germans, especially, had an effective manpack set in use during the Battle of France in 1940. The United States, too, found flame-throwers, both manpack and mounted in **amphibious** tractors, particularly effective in dealing with the Japanese, deeply dug-in on the Pacific islands they so tenaciously defended.

British attention turned to larger equipments. Crocodile, a system based on the Churchill tank, towed an armoured trailer full of jellied petrol (**napalm**). The fuel was fed through pipes running under the vehicle's hull to the flame gun in the hull gunner's compartment at the front. Pressurized by nitrogen, the fuel was ignited by a petrol jet and an electric spark as it left the projector. The gunner could fire 100 shots to a range of about 80 metres without replenishment, a system used with considerable effect against the concrete defences of the Atlantic Wall and throughout the campaign in North-Western Europe in 1944/5.

The British have now abandoned the flame-thrower but a system mounted in the American M60A2 tank was used in **Vietnam** and the Soviet Army had a system in service mounted on T55. The latter displaced only the coaxial machine-gun, so that the main armament could still be used; in both systems the fuel was carried on the vehicle so the gunners had only a limited number of flame rounds available before replenishment was needed.

The **Korean War** saw the use of air-dropped napalm as a flame weapon, also used extensively in Vietnam, and this is now the more usual application of the concept.

Fleming, Professor Sir John (1849–1945). A student under James Maxwell who in 1881 took the chair of mathematics and physics at University College, London. In 1883 he joined the Edison Company as a consultant and began work on photometry, potentiometers and carbon-filament electric-light lamps. He next joined Guglielmo **Marconi** in developing transatlantic **radio** transmissions and in 1904 invented the thermionic (vacuum diode) valve which was to revolutionize **electronic** (a word he coined) **communications**. As a brilliant lecturer he followed in the footsteps of Michael **Faraday** in propagating science and technology for peace and war.

Fletcher, Admiral Frank (1885–1973). One of the US Navy's most experienced airmen, who in 1942 won the most exacting **aircraft-carrier** battles against the best Japanese carrier forces, but who was underrated and denied due credit. In February 1942 he led aircraft-carriers in raids on Japanese bases in the Pacific Ocean. As commander Task Force 17, with two carriers, he succeeded in May, despite misleading **intelligence** and errors, in defeating the Japanese strategic plan in the drawn Battle of the **Coral Sea**. A month later, when in command of carrier Task Forces 16 and 17 at the decisive Battle of **Midway**, his dive-bombers massacred the Japanese carriers, though his own *Yorktown* was also sunk and he was wounded. But when commanding Task Force 61, with three carriers in support of the Battle for Guadalcanal, he was faulted for abandoning the **amphibious**-force transports to the mercy of the Japanese. He lapsed again in the initial battles for the **Solomon** Islands between 22 and 25 August, when poor intelligence and misjudgements again appeared to let opportunities slip. On 18 October, an exhausted man, he was replaced, but later given a command in the Okinawa battles.

Flexible response, Strategy of. A philosophy of **deterrence** regarded as a means to match threats with graduated measures. Although conceived in the 1950s with **nuclear deterrence** in mind, it has been widened in scope by the desire to maintain forces strong enough to resist non-nuclear aggression. Therefore it contributes to expansions of military might.

Fokker, Anthony (1890–1939), taught himself to fly in 1911 and the next year set up a factory in Germany producing military **aircraft**. In 1915 he invented the first practical interrupter gear to permit a **machine-gun** to fire through the propeller's arc, and fitted it to his E1 monoplane **fighter**. The machine gave Germany a temporary lead in the struggle for air superiority that continues to the present day. During **World War I** he designed and built 40 different types of aircraft (mainly biplanes) for Germany and afterwards in America built monoplane multi-engine **transport aircraft** which led the way in the development of airlines.

Food, Military. Until scientists developed the technology of food preservation beyond the salted, spiced, dried or smoked-meat stage, armies, like many civil communities, were at the mercy of **logistics**. Tenuous sources of supply and transport usually made it impossible for armies to maintain large forces without causing famine by a scavenging which denuded the countryside. Likewise, on long sea voyages, ships' companies were dependent on what livestock they could carry to supplement an otherwise unappetizing diet. In effect, navies and armies had need to keep on the move in search of fresh food to survive (*see also* **Logistics**).

The development of modern food technology may be dated to the discovery in mid 18th century that scurvy was caused by an unbalanced diet lacking fresh fruit and vegetables. Warships' crews were among the first to benefit. Research begun in France in 1795 inspired Nicolas Appert's bottling of heat-sterilized preserved foods, which the Royal Navy used on a polar expedition in 1825. Then, in 1839, an Englishman, Peter Durand, used lighter, stronger, tin-plated canisters, but without ensuring prevention of contamination during the canning process. Indeed, not until Louis Pasteur's research into fermentation showed scientifically in 1864 how control of preservation could be achieved, were food manufacturers able to supply the military with food which could be stockpiled safely for long periods. By this time refrigeration by immersion in ice, which had been invented in 1842 by H. Benjamin, was vastly improved upon by mechanical methods by 1870 to revolutionize the transport and storage of frozen food.

Foods which were 'pasteurized', dehydrated or refrigerated became commonplace in the 1860s. To their immense relief logisticians henceforward could base their calculations upon reliable sources of food; the **medical** authorities were less concerned about food poisoning; the transport services economized by moving concentrated foods on wheels instead of on the hoof; and all this was approved by those whose diet was assured and improved as well as made more attractive. In fact, the standard of military food was never better nor worse than the logisticians and cooks who supplied, prepared and served it. It usually reflected a nation's eating standards; the more so whenever cooking was carried out by sailors and soldiers in their own messes, instead of by better-trained cooks working in well-equipped kitchens.

As always, **siege** conditions, when large bodies of men lodged for long periods in trenches, made feeding difficult. Starvation causes obsession with feeding. During the Siege of **Port Arthur**, the Japanese suffered badly from beriberi, caused by eating little but polished rice, and the Russians had scurvy from the lack of vitamin C (as yet undiscovered) from fresh vegetables. The morale and health of those who dwelt for days on end in damp, cold trenches during **World War I** suffered because it was difficult to provide regular hot meals. Eating out of tins was not always appetizing, and the variety of tinned food was limited and monotonous. Fresh hot food in insulated containers helped.

In **World War II** national standards of diet varied greatly. Gradually strangled by **blockade**, the Axis powers' diets declined as substitute (*ersatz*) food became commoner. On the other hand, the Western Allies made strenuous and by no means unsuccessful efforts to feed their troops with a high-calorie, varied diet. Moreover in a war when small groups often had to cook for themselves, improvements were often achieved by an exchange of rations between allies. The standards reached since then, as the latest additives and freeze-drying have been introduced, have improved in quality as technologists have found better ways to compress and pack combat rations.

Ford Motor Company. Formed at Dearborn in 1903 by Henry Ford I, who had built his first car in 1896, the company took the lead in **motor vehicle** technology with its innovative Model T car. By the introduction of the first moving assembly line in 1913, Ford was able to raise production and cut the Model T's price to $500. During **World War I** Model Ts were converted to military use and in 1918 a prototype light tank based on it was designed. After the war Ford acquired many other companies, including truck-builders such as Dodge. It also built tri-motor **transport aircraft**. Prior to **World War II** it supplied engines for **armoured fighting vehicles** and subsequently tanks (including the engines for the many Sherman M3 tanks), in addition to vast numbers of other military vehicles.

Fortifications (*see also* **Siege warfare**). From the earliest times fortifications of some kind or another have been constructed, usually in peacetime, as the basis of defensive positions against the threat of war. More often than not they would be built on commanding ground and, if possible, supplement existing obstacles, such as rivers and ditches, with earthworks, **stone** barricades and **wooden** stockades.

A Roman legion on campaign would throw up field fortifications wherever it halted for the night – and probably improve on these as time went by with a wall and stockade, especially if it was decided to garrison the place as a settlement. Indeed, as leaders in the development of fortifications, the Romans have to their credit many fortified walls, including Hadrian's Wall and ditch shielding England from Scotland, a stone and earthwork barrier 117 kilometres long which provided a footpath between regularly spaced forts. But these were dwarfed by the Great Wall of China, 4,100 kilometres long and 7·6 metres high, which was wide enough to take **chariots** and was constructed between the 4th century BC and the 16th century.

In its defence against Muslim invasions, the Eastern Roman Empire considerably strengthened and extended existing fortifications to protect cities, including Constantinople, by doubling or even trebling the number as well as raising the height of walls and making weapon slits for archers, ideas which were subsequently brought back to Europe by the **Crusaders**.

Before firearms were introduced, the medieval castle was able to hold out under siege for long periods, often for years. The situation did not change dramatically as guns began to appear, since they were neither sufficiently accurate nor powerful enough to pierce the walls. However, as guns became more efficient in the mid 15th century the traditional moated castle became an anachronism, though the need for permanent fortifications to protect cities and vital centres remained. Designers gradually introduced more complex bastion structures with as little as possible rising above ground level. The influence of the Frenchman Marshal Sébastien de **Vauban** in the 17th century is probably the most significant, his ideas governing fortress design until the late 19th century. Vauban retained the traditional plan of inner enclosure, rampart, moat and outer rampart – already refined to counter the effects of gunfire; however, he extended the outer works so as to compel the enemy to start his siege operations much farther away. Every face of a Vauban fort was supported by other works to flank and rear and his fortresses became vast polygons featuring great bastions interspersed with smaller ones, all capable not only of providing mutual support but also of surviving independently, supplied by underground magazines and covered walkways.

Most European countries followed Vauban's lead and his fortresses may still be seen today. However, across the Atlantic another form of fortification began to appear as the United States was drawn into civil war in the 1860s. Few permanent masonry forts had been built in the New World, though the familiar wooden frontier post provided a basis for the defence of troops and settlers during the American Indian Wars in the far West. It was the **American Civil War** which provided a portent for the future with its extensive use of entrenchments in the defence of vital areas.

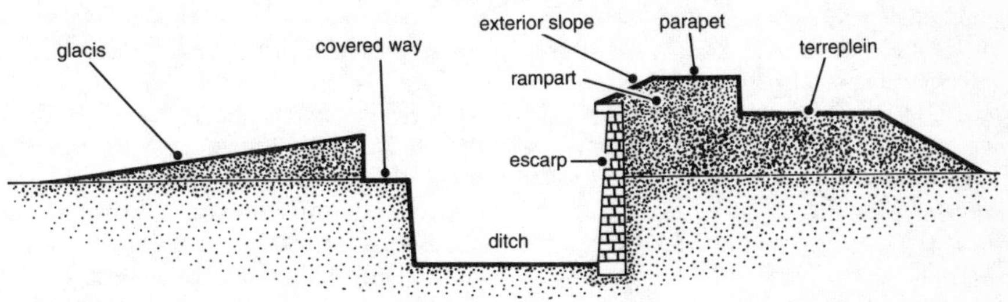

Cross-section of a Vauban fortress

In the late 19th and early 20th centuries the age of the permanent fortification was by no means dead and the Belgian General Henri Brialmont emerged as a worthy successor to Vauban with the fortress complexes he designed to protect major cities such as Liège (surrounded by no less than 12) and Antwerp. Similar massive defences were erected at Verdun, Belfort and Przemyśl – forts built of reinforced concrete, with much of the construction underground and the armament mounted in steel rotating turrets, many of Brialmont's 'disappearing cupola' design, only emerging above ground to fire. These complexes were so vast they were described as 'land battleships' – a real misnomer since, unlike their maritime namesakes, they were immobile, a basic weakness in a world of increasing mobility. There was invariably a way round, even if the forts themselves were impregnable – which few actually were. Such defences are never, in the end, impervious to modern weapons. Nevertheless the heroic defence of **Verdun** by the French during **World War I** is a classic of its kind.

Though permanent fortifications played their part in World War I it was field fortifications which were such a significant feature of that conflict. Elaborate entrenchments stretched from the Swiss border to the North Sea, consisting of lines of fire trenches in depth to several kilometres, with communication trenches running forward to link them together. Numerous dugouts – often very deep – provided shelter for frontline and supporting troops, for stores and for the treatment of casualties. All of this created a world of its own below ground level which left an indelible impression on those who experienced it; a defensive system when supported by **artillery** and protected by belts of **barbed wire** and the quagmire that was no man's land which the infantry of neither side could breach. The war in France and Flanders could be described as a perpetual siege.

Despite the arrival of the **tank** and general **mechanization** of armies during the 1930s, the European powers still saw the need for permanent defences. The best known is probably the **Maginot Line**, which consisted of an elaborate network of large fortresses, connected by tunnels and with its own **railway** system and electrical supply, the whole provisioned for several months. Starting at the Swiss border the Maginot Line ended on the Belgian border at Montmédy; a fact the Germans were quick to grasp in 1940, turning the French flank with ease as they poured through Belgium, making the line useless.

The Belgians themselves had built a line of forts along the Albert Canal, the most famous and supposedly impregnable of which was Eben Emael; it fell in a few hours to a combined **glider** and parachutist attack which landed on top of it.

There were others, but none so elaborate as the Maginot Line and none in the end was any more effective, though the Stalin Line facing Poland did manage to delay the German invasion of Russia in 1941 for about two days. Others worthy of mention are the Mannerheim Line in Finland; the Czech Little Maginot Line; the Greek Metaxas Line facing Bulgaria; Germany's own **Siegfried Line** opposite the Maginot Line and the **Atlantic Wall** to oppose Allied landings in Europe; and Corregidor, America's fortress in the Philippines, which fell to the Japanese in a month. Since 1945 the only comparable system has been the Bar Lev Line, held by the Israelis along the eastern bank of the Suez Canal. The false sense of security which this gave them allowed the Egyptians to breach it with remarkable ease when the opportunity arose.

The defender behind fixed fortifications has invariably put himself at a disadvantage by surrendering the initiative to his opponent and has often, as a result, lost the war.

Franco-Prussian War. The outbreak of hostilities on 15 July 1870 may be deemed the result of **Napoleon III**'s delusion that the French Army was superior to Prussia's, and his underrating of Otto von **Bismarck**'s diplomatic ability in unifying the German states. As a result a French force of 224,000 men, in a state of utter **mobilization** and **logistic** confusion, was faced by 475,000 well-organized Germans, admirably directed by General Helmuth Karl von **Moltke**. The assembly of German formations and stores at the frontier proceeded with smooth precision. Frenchmen wasted immense time and effort wandering from place to place trying to find units which, as often as not, were concentrating in the wrong place with only a minimum of stores and equipment.

Lacking a plan and **intelligence** of the enemy, Napoleon ordered an advance on Berlin. But the Prussians advanced too, colliding with the French near Saarbrücken on 2 August in the first of several frontier battles. Casualties were heavy, but marginally in the German favour. German tactical handling was not all it might have been, but it was more than a match for the French, whose **artillery** was outmatched, and who were forced remorselessly backwards as their morale faded. At Fröschwiller and Spichern on the 6th and Mars-la-Tour on the 16th, the two sides blundered into each other, making charge after countercharge – but leaving the Germans in possession of the field.

On the 12th, Napoleon, a sick man, handed command of the Army of the Rhine to Marshal Bazaine (who was retreating under constant pressure on the fortress of Metz) and to Marshal MacMahon with an army at Châlons he was trying to reorganize. At Gravelotte on 18 August von Moltke, with 200,000 men, brought Bazaine's 100,000 to battle – and suffered appalling losses storming the village of Saint-Privat, at the same time seeing his right wing collapse in a panic which he personally helped check by leading up reinforcements. Bazaine, who was on the verge of a significant success, failed to follow up his temporary advantage and withdrew supinely into Metz, where he and 173,000 men remained almost impassive in the hope of relief by MacMahon's army.

Accompanied by Napoleon and with 100,000 men, MacMahon moved to the rescue by an indirect route, close by the Belgian frontier, whence von Moltke skilfully manoeuvred him into the trap at **Sedan** where on 1 September he was crushed and Napoleon taken prisoner.

Belatedly, as the Germans surrounded the frontier fortresses of Strasbourg, Belfort and **Verdun** and began to march on Paris, the French proclaimed the Third Republic and began a last ditch resistance by **guerrilla warfare** as they tried to rebuild their armies in the field. **Paris** was surrounded on 19 September, but held out despite dire shortages of food and munitions and the mutinous behaviour of garrison elements. At the same time the Germans had great difficulty restoring the **railway** to supply their troops and bring up the siege train with which to batter the defences and city, starting on 5 January 1871.

Meanwhile Bazaine surrendered Metz on 27 October and desultory engagements were fought in the vicinity of **Amiens**, Coulmiers, Orléans and Le Mans, battles which offered no greater benefit to the French than a salving of pride, and simply tried German patience while peace negotiations dragged on. Paris sued for an armistice on 26 January, but Belfort's garrison fought on until ordered to capitulate on 15 February, marching out with full military honours two weeks before the German Army entered Paris in triumph.

Frederick II (the Great), King (1712–86). Prior to studying the art of war, Prince Frederick of Prussia focused his considerable intellect on the cultural arts, science, philosophy and politics. When acceding to the Prussian throne in 1740 he had evolved a comparatively liberal outlook without this in any way weakening his iron resolve as C-in-C of a well-trained army combined with astute political sense. As a stern disciplinarian, he acknowledged that his flexible military philosophy of rapid manoeuvre by combined arms was based upon a study of Prince **Eugene**'s campaigns. These methods were enhanced by application of fire-power from faster-moving **horse artillery** and overhead fire from howitzers.

His lightning Silesian campaigns (1740–42 and 1744–5) during the War of the Austrian Succession revealed his genius, even though he came close to disaster in 1744 when ejected from Bohemia by superior Austrian forces. For in four battles, between 4 June (Hohenfriedberg) and 25 November 1745 (Görlitz), he outmanoeuvred and defeated the Austrians and Saxons to retain possession of Silesia. After this campaign he wrote his celebrated *Military Instructions for the Generals*, a work which laid down the strategic and tactical procedures used so effectively during the **Seven Years War** (1756–63) and raised the art of **war of the 18th century** to a new zenith ahead of the changes to be wrought by the **French revolutionary wars**.

French revolutionary wars. The outbreak in 1789 of the French Revolution (the latest among others elsewhere) was followed in 1792 by a French declaration of war against Austria which was caused by internal political pressures and a belief that, to survive, the revolution must be spread wider. As a result, the purged and disorganized Navy and Army were thrust into a series of wars against most European nations over the next 23 years. Nevertheless revolutionary fervour and good use of **artillery** enabled the French at the battles of Valmy and Jemappes to repel invasion. Indeed, after the shock of Louis XVI's execution in January 1793, it usually was fervour and artillery which repeatedly held off the combined forces of Austria, Britain, Holland, Piedmont, Prussia and Spain, whose monarchies and aristocracies were both outraged and badly scared by events.

Blockade was imposed by the Allies and maintained with some effect at sea (except when a vital food **convoy** from America got through in June 1794), despite a British victory over the escort at the four-day Battle of the Glorious First of June. The naval war of 1794 established a pattern of blockade incidents throughout the Mediterranean, the Atlantic, the West Indies and the Baltic, interspersed by the major British victories of St Vincent (February 1797) over the Spanish fleet after Spain's alliance with France in 1796; Camperdown (October 1797) over the Dutch fleet; the Nile (1798), when Admiral **Nelson** destroyed the anchored French fleet which had carried **Napoleon**'s army to invade Egypt; Copenhagen (1801), where Nelson smashed the Danish fleet to deter the Baltic states from breaking the British blockade; and **Trafalgar** (1805), when Nelson routed the combined French and Spanish fleets in what was to prove the last major battles between sailing warships of the line, and would establish British naval supremacy for the next 100 years (*see also* **Naval warfare**).

On land, however, the French were more successful. Reinforced in 1793 by **conscription** and benefiting from the earlier reforms of General de **Gribeauval** and the introduction in 1794 by General Lazare **Carnot** of the flexible divisional system (first proposed by the Duc de Broglie in 1759), the Army began to win significant victories, notably in Italy and Austria in 1797 under Napoleon **Bonaparte**. These victories, which ruined the Alliance by 1797, were offset by the reverses at sea and the formation of an Anglo-Russian Alliance in December 1798, until, that is, Napoleon deserted his marooned army in Egypt to return to France and become First Consul in November 1799. He thereupon raised a new army and, in May 1800, took it to Italy to retrieve failing French fortunes and win a crushing victory over the Austrians at Marengo on 14 May – at the same time as the French troops in Egypt were being rounded up by an Anglo-Turkish force.

In this war of constantly fluctuating fortunes and shifting alliances, the peace signed between Britain and France in March 1802 lasted barely fourteen months before Napoleon used his dictatorial powers to extend the revolutionary conflict into an attempt to dominate Europe by military force – a force which, nevertheless, relied but little on technical innovations at a time when the Industrial Revolution was changing the world. For his fertile military mind concentrated mainly on improved organizations, methods and tactical innovations – such as the creation of army corps composed of all-arms divisions consisting of cavalry, infantry and mixed artillery brigades, and the rapid manoeuvring of those formations, supported by the latest, concentrated **gunnery** techniques, to envelop and overwhelm outmoded armies.

Lacking mechanization, Napoleon's **logistic** systems could only remain locked in the limbo of the past, dependent in the final analysis on living off the country. He benefited greatly from Claude Chappé's semaphore-signal **communications** network, but tended more to reject than encourage the development of new technology and weapons. For example, he was unimpressed by **rockets**, rejected Robert **Fulton**'s **submarine** and proposals for a canal steam paddle-boat to enhance logistics, and denied the Navy the support it so urgently needed to break the blockade and make feasible the projected invasion of Britain in 1803. The invasion had to be postponed when Napoleon was

Napoleon's German and Austrian campaigns, 1805–13

compelled to react in 1805 against the latest European alliance of disenchanted nations – and abandoned after the defeat of Trafalgar.

But the stunning campaigns which overran Europe in the years to come with such mighty victories as Austerlitz (1805), Jena (1806), Eylau (1807) and Wagram (1809), failed to impose Napoleon's so-called Continental System. Added to the drain of resources after the French invasion of Portugal in 1807 and occupation of Spain, followed by British intervention in 1808, the logistically flawed invasion of Russia in 1812 and the ensuing disastrous winter retreat from Moscow proved fatal. The growth of Allied forces in Germany in 1813, defeat at Dresden and **Leipzig** and the invasion of southern France by a British-led army in 1814 as the collapsing French armies withdrew, induced Napoleon's abdication after defeat in the Battle of Paris on 30 March.

Napoleon's return from exile the following year and final defeat at Waterloo in June was simply the last throw of an arrogant demagogue whose methods and technology had been overmatched by superior forces.

French wars, 1635–1783. The resurgence of French national pride and outwards urge in the aftermath of victory in the **Hundred Years War** was relatively slow to develop until Cardinal Armand Jean Richelieu became virtual dictator in 1619. Try as France might in the 16th century (*see* **Wars of the 16th century**) to gain ascendancy in Italy and over Spain (**Spanish Wars**), she was thwarted by strong opponents (including the British) and various wars of **Religion** which in 1635 culminated in the **Thirty Years War**. Richelieu's military aspirations (he personally conducted the successful 14 months siege of La Rochelle (1627–8)) stimulated French military ambitions. The army which Cardinal Giulio Mazarin took over on Richelieu's death in 1642, and which eventually came under control of the glory-seeking King Louis XIV, was modelled on **Gustavus Adolphus**'s Swedish army. Under Vicomte Henri Turenne, however, the **tactics** of manoeuvre with the aim of economically achieving the **strategic** aim without fighting a major battle were highly developed.

This doctrine, hinging on the developments of **fortifications** and **siege warfare** devised by Marshal Sébastien de **Vauban**, would serve Louis XIV and France well in wars with Spain (1653–9, 1667–8), Britain (1668), Holland (1672–8), the German states (1674–9) and the Great War (1688–97). Yet in a period when the more scientific organization, methods and logistics of the Minister for War, the Marquis François Louvois, were being introduced, weapon development was somewhat neglected. France was slow to adopt the flintlock **musket**, the ring **bayonet** and the latest tactics which were to be used so effectively by the Duke of **Marlborough** and Prince **Eugene** to outmanoeuvre and defeat France in numerous confrontations, including the decisive battles of Blenheim, Ramillies, Oudenarde and **Malplaquet**, during the War of the Spanish Succession (1701–14).

Notwithstanding these setbacks, a weakened France was at war again, defending herself against Spain in the War of the Quadruple Alliance (1718–20) and once more (this time in alliance with Britain) in an almost bloodless contest between 1727 and 1729. This was followed by participation in the War of the Polish Succession (1733–8) to win Lorraine. Retaining almost insatiable ambitions, she was engaged under Louis XV in expansion of her colonial empire in competition with Spain, Britain and the other maritime nations, and forfeiting few opportunities to meddle with European affairs. Inevitably in 1756 she became involved with the **Seven Years War**, with the result that she forfeited many colonies, including Canada and India, largely owing to naval inferiority.

Since the Battle of **Sluys** the French Navy had been inferior to Britain's and, in due course, to Holland's. Combined Anglo-Dutch supremacy in numbers and techniques (*see* **Naval warfare**) was a decisive factor in the **wars of the 17th and 18th centuries**. Although the **blockade** was never absolute, France's trade suffered. In the War of the

Spanish Succession she lost supremacy in the Mediterranean; and in the Seven Years War many colonies, because of an inability to reinforce them. Nevertheless, attempts to restore her Navy paid off during the **American War of Independence** (1775–83) when she declared for the rebellious colonists and was able, with Spanish collaboration, significantly to evade the blockade and reinforce the American Army.

Defeat in the Seven Years War prompted important reforms among the French forces. General Jean de **Gribeauval**'s sweeping changes to **artillery** were chief among them, along with proposals concerning **army organization**, which would lead to adoption of the divisional system and improved **logistics**, which would have their impact on the **French revolutionary wars**.

Frigate. The name given to small, fast open-decked Mediterranean Sea sailing vessels, it was applied in the late 16th century to Portuguese and Spanish merchant and warships and was copied by the French in the 18th century. By 1756 and the **Seven Years War**, they were recognized as a class of three-masted, single-deck 5th- or 6th-rate warships mounting 30 to 40 guns and, therefore, rarely included in the line of battle. During the **French revolutionary wars**, however, the British built a 4th-rate class of 44-gun frigates which proved excellent maids of all work in the exercise of sea power and commerce protection.

With the arrival of steam the name went out of use towards the end of the 19th century as **cruisers** and **destroyers** took over the frigate's function; but it was revived during **World War II** to denote an enlarged **corvette** with a displacement of a little over 1,000 tonnes and a speed of 20 knots. In 1943 the US Navy also readopted the name, but to describe vessels of similar speed but 2,400 tonnes displacement – a sort of slow **destroyer**. Mostly these vessels were employed on escort and **anti-submarine** tasks, fitted with **sonar**, **radar** and suitable weapons.

Following World War II the name has become almost meaningless since it has been used to describe vessels with displacements between 1,000 and 3,500 tonnes and speeds well over 30 knots. Once more, however, they are maids of all work, fitted with sophisticated weapons and equipment for anti-submarine and anti-aircraft roles, escort and scouting duties, plus all manner of other tasks deemed unsuitable for smaller craft and less expendable destroyers.

Frogmen. A name given in **World War II** to specially trained free divers wearing thin waterproof clothing, flippers (dating from the 16th century) and self-contained breathing apparatus to enable them to operate at shallow depths for protracted periods of time. Reliable breathing apparatus and techniques were developed between the World Wars, notably in Britain by Sir Robert **Davis** and in France by Jacques Cousteau. The Italians too were very active and used such equipment in 1941 for men riding **torpedoes** against shipping in harbour. Frogmen carried out many vital roles in **amphibious warfare** such as beach **reconnaissance** (usually coming and going in **landing-craft**, canoe, or **submarine**), pilotage for approaching assault craft, and demolition of underwater obstacles. They were used extensively for the **Normandy** invasion in 1944 by so-called Combined Operations Pilotage Parties (COPP) and for all the US Navy's **Pacific Ocean War** landings by Underwater Demolition Teams (UDT) after heavy casualties to the assault parties at Makin in November 1943 caused by reefs and obstructions. Sabotage is carried out by frogmen against shipping and bridges, usually by their laying limpet **mines** against targets or placing charges beneath vessels in shallow water. They also take part in clandestine operations to examine, for example, ships' underwater, **anti-submarine** equipment; and are in common use for inspection and maintenance of marine equipment of all sorts.

Fuller, Major-General John Frederick Charles (1878–1966). A British light infantryman and arguably the most important and prescient military **philosopher** of the 20th century. He led an anti-**guerrilla** force in the Boer War and in **World War I** developed instruction techniques for senior commanders before, in December 1916, becoming chief staff officer of the Tank Corps. To his brilliantly analytical and far-seeing mind may be traced the genius of the battles of **Cambrai** and **Amiens** and the basic concepts of today's **armoured** forces. After World War I he was instrumental in creating the climate for experiments with **tanks** and armoured forces. Owing to his acerbic temperament and forthright speech and writings, he made many enemies and, after 1927, was diverted from the main military stream of promotion. Nevertheless his books, notably *Lectures of FSR III*, which was about modern war, had a profound influence on German and Russian armoured and guerrilla warfare doctrine, while in journalism he was a gadfly whose pronounced Fascist sympathies ensured that, in **World War II**, he was not offered further employment. However, he provided to his own country useful intelligence of German preparations.

Fulton, Robert (1765–1815). An American gunsmith who in 1779 designed a steam paddle-boat and in 1796 proposed improvements to canal transport with steam vessels. These were rejected both in America and France in 1797, as later was his **submarine** *Nautilus* (copied from **Bushnell**'s *Turtle*) when offered to **Napoleon** in 1800 and subsequently the British and Americans. In America in 1807, however, his paddle-boat *Clermont* with a Boulton & Watt engine proved a commercial success, and he developed it in 1813 into the centre-paddle-wheel **frigate** *Demologos*, with 24 32-pounder guns, which found little favour with admiralties.

G

Galleass. A vessel of the mid 16th century, built with the intention of combining the obsolete **galley**'s ability to manoeuvre, in calm or against the wind, with the advantages of the **galleon**. Like so many compromises it failed because it retained the galley's inherent clumsiness and, with some 300 oarsmen, **logistic** shortcomings without being able to match the galleon's mobility and fire-power. The six included in the Spanish Armada were outclassed and the type abandoned.

Galleon. The product in 1570 of Sir John Hawkins's modifications to improve the **carrack**'s handling, by the removal of the forecastle which tended to push the bows to leeward and inhibit the steering of a straight course. The Spaniards copied it 17 years later, called it galleon and had 20 ready for the Armada a year later. The British never used the Spanish name for these warships which, until 1860 and with few great changes except in armament, soon were regarded as the dominant ships of the line of battle – in modern parlance, capital ships.

Galley. The name for vessels powered by oarsmen with assistance from sails, which must have existed prior to 3000 BC. They were tactically important because they could be propelled against the wind and in conditions of calm. As warships they first were

Galleon

used as carriers of fighting men bent on boarding for hand-to-hand combat. Or, in Phoenician practice from 1100 BC, to sink ships by holing them below the water-line with a **ram**.

Top-heavy Mediterranean galleys, which were unstable in rough seas, were made larger and faster by increasing the number of oars and mounting them in banks. Biremes (two banks) would be rivalled by **triremes** (three banks). There is evidence of some galleys with 17 banks and short bursts of speed up to nine knots. But the hundreds of enslaved oarsmen required had only limited stamina and imposed a considerable security risk and a **logistic** burden on captains and combatant crews. Handier were the smaller, more seaworthy **Viking** longboats, whose oarsmen also were committed fighters.

The galley's days were numbered with the coming of **artillery**, which could be mounted only with limited traverse on platforms above the oar banks. Nevertheless the Swedes used gun galleys with considerable success in 1790 at the second Battle of Svenskund, when they surrounded an inshore Russian fleet and destroyed 64 out of 151 ships for the loss of only 4.

Gallipoli, Battles of. It was a Russian request for aid to Britain and France in 1914 which initiated the misconceived attempt, after Turkey's entry into the war on 29 October on the side of the Central Powers, to strike at Constantinople. It was doubtful if the presence of an Allied fleet in the Bosporus would knock Turkey out of the war, as intended by Winston **Churchill**, and in addition the feasibility of forcing the Dardanelles Narrows was uncertain. Uncertainty was augmented when, as an experiment without Cabinet approval, the Royal Navy advertised intentions by bombarding the outer Dardanelles forts on 23 November.

Thus the subsequent drift into a full-scale operation in 1915 to force the Narrows by ships alone forfeited surprise, the one factor which might have bought success. Even with surprise, in this grossly mismanaged campaign, the chances of forcing and then keeping the Narrows open without occupying the Gallipoli peninsula – as attempted with the loss to **mines** of three **battleships** and damage to other ships between 19 February and 18 March 1915 – were dubious in the extreme.

The landings by the Army, then inevitably called for, were delayed until 25 April because there had been no prior consultation between the two services. When it came to launching **amphibious** attacks on the Asian shore, and at the neck as well as the tip of Gallipoli, the Turks, under the supreme command of the German General Liman von Sanders, were ready with strengthened defences on both shores. The result was that General Sir Ian Hamilton's troops, who came ashore for the most part in rowing boats and without preliminary naval bombardment, landed either unopposed in wrong and inhospitable places (as at Anzac Cove, where the skill of General Mustafa Kemal prevented expansion), or elsewhere, against emplaced opposition, to be massacred. Instead of on the first day cutting the peninsula at the neck and forming a four-miles-deep beachhead at the tip, only four toe-holds were obtained, lodgements which, in the weeks to come, were never fully linked up.

Conditions ashore were appalling, supply precarious and **health** undermined by sickness which sapped the troops' morale. Though the French had secured temporary possession of the Asian shore at Kum Kale, passage of the Narrows remained closed to surface vessels. British **submarines** penetrated into the Sea of Marmara but could not stop Turkish supplies. However, the depredations of German submarines, which sank three battleships among other vessels, caused severe curtailment of naval support at Gallipoli.

In a renewed effort to outflank the unyielding defences of the neck and seize the Narrows, a surprise, unopposed landing at Suvla Bay, in conjunction with a night attack at Anzac Cove, was made on 6 August. At first it made excellent progress. But once more an over-optimistic plan, difficult terrain, supply

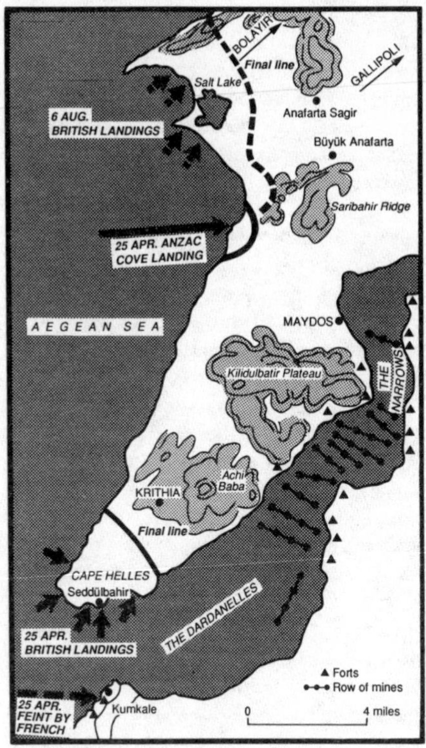

The Gallipoli campaign, 1915

and water shortage, indecisive British command and stubborn Turkish resistance stopped the advance, this time in sight of its final objectives.

In November a political decision to withdraw was made on military advice. Heavy losses were expected. Yet, as the only brilliantly executed plan of the entire campaign, it was carried out with secrecy and surprise in two phases with vast loss of stores but few men. In total each side had lost about 251,000 men since the start.

Gas, Industrial. There are several claimants to the discovery of flammable **coal** gas. Usually the credit goes to William Murdock with his production plant of 1795 and the lighting of Boulton & Watt's high-production Soho factory in 1808. Soon gases were increasingly used not only for lighting but also heating, power and innumerable industrial and military purposes. These included hydrogen for **balloons** and **airships**; ammonia for **explosives** and fertilizers; oxygen for metals production, cutting and human resuscitation; chlorine, phosgene and cyanide for poison gas (*see* **chemical warfare**); and nitrogen for fertilizers and **flame**-throwers. After 1821 (when the first natural-gas well was drilled in America), as supplies of gas and **oil** became more generally available, these also became the staple feedstocks of the vast gas industry.

Gee. A radio **navigational** aid devised by the Decca Company during **World War II** to help **bombers** find their target. Three (later four) master stations' signals when picked up by a receiver gave a fairly accurate position, though it was not good enough for precision bombing. In various forms it was (and still is) used for navigation at sea and was invaluable in navigation for **airborne** and **amphibious** landings. However, it was vulnerable to jamming.

Generators, Power. For military purposes, until electricity could be generated in commercial quantities, the means of propulsion (*see* **Propulsion, Means of**) was the principal power requirement. But, as the **electromagnetic effect** came to be understood, interest in electricity supply led to the invention by Alessandro Volta in 1800 of the chemical–electric primary cell, a development which provided power for the **telegraph** systems – which navies and armies, as well as commerce, seized upon.

It therefore followed that when in 1859 Gaston Planté discovered the lead–acid battery system there was a demand for machines capable of generating electricity for recharging Planté's storage batteries. This requirement was first rather inadequately satisfied in 1868 by charging from dry-cell batteries, but far more efficiently by the **dynamo** invented by Thomas **Edison** *c.* 1879 as a way to power his recently invented electric-light bulbs.

Since then military authorities have been

to the forefront in the development and uses of all kinds of battery and generator for service in ships, land vehicles and **aircraft**.

Geneva Conventions. In the shadow of the excesses of **World War II**, an effort was made to draw up international agreements regulating the conduct of nations at war. In 1949 the Holy See and 58 governments signed four Conventions dealing with: (1) the amelioration of wounded and sick in the field; (2) armed forces at sea; (3) prisoners of war; and (4) protection of civilians in time of war. In many instances the Conventions confirmed what was already generally practised by many countries, such as registration of and recognition of military dress (*see* **Clothing, Military**) and respect for prisoners' rights. But they also reflected the findings of international courts dealing with war crimes and thus established more precisely the human rights of civilians as well as the military.

The Conventions have been only partially effective, and often virtually unrecognized in many **guerrilla**-type conflicts. Prisoners, hostages and the wounded continue to be maltreated by fanatics who probably neither know nor care about humanitarianism or the Conventions. Nevertheless, the existence of rules which threaten penalties through international law, or the observance of which offer material or psychological and political advantages to those who comply, have made the exercise worth while, as many repatriated people discovered, for example, in **Korea, Vietnam**, the Middle East, the **Falklands**, Iraq and Yugoslavia. Compliance with the Conventions is at least civilizing.

Genghis Khan (*c.* 1162–1227). Named Temujin (meaning ironworker), this great Mongolian leader became Genghis Khan in 1206 and cunningly unified the local nomadic tribes and created an efficient army. He was even more a ruthless dictator and statesman than a clever strategist, who chose only to lead in person the toughest campaigns. In 1211 he began the conquest of the Chin Empire but left its completion to others while he moved westwards, reaching the Caucasus mountains in 1222 before entering southern Russia and then returning in 1225 to conquer the Tangut Kingdom. His empire and armies proved remarkably stable, as was proved when the **Mongol** invasions continued long after his death.

Germany, Battles of. Germany's central location in Europe has made it a battleground, compelling its people to form strong defence forces. Successively the land was invaded by, among others **Romans**, Huns, **Vikings**, Magyars and **Mongols**. It was at the heart of the Wars of **Religion**, severely damaged by the **Thirty Years War**, the War of the Spanish Succession and the struggles in which Prussia asserted her dominance in the **wars of the 18th century**, especially during the **Seven Years War**. And it was repeatedly fought over during the **French revolutionary wars**.

Germany's good fortune and aggressiveness post 1813 spared her combat on her own soil until **World War I**, when the Russians invaded East Prussia and were ejected after the Battle of **Tannenberg** in 1914. Light **bombing** of targets in the west did little damage and the post-war encounters between Communist and Freikorps factions were of short duration.

World War II was a very different matter owing to technology and **air warfare**. At the start enemy machines overflew the Fatherland mostly dropping **propaganda** leaflets, but, occasionally, **bombs** on naval and coastal targets. Yet no sooner had her armies invaded the west on 10 May 1940 than the Allies began the **strategic bombing** campaign against transport and industrial targets, a campaign which, although notable at first for the inability of bombers to find and hit their targets, or the defences to shoot down many bombers, compelled the Wehrmacht to divert very large resources to home defence. For although the populace soon got used to raids and were encouraged by the gradual improvement of the defences and relatively minor damage inflicted, the government feared political repercussions. The menace assumed terrible dimensions as the scale and accuracy of British attacks increased dramatically in June 1942 with 1,000-bomber raids on Cologne, Essen and Bremen.

Furthermore, the occasional low-level daylight attacks by the British, when supplemented on 27 January 1943 by American Boeing B17s bombing Wilhelmshaven in daylight, gave notice of forthcoming accurate, round-the-clock bombing. In consequence, **fighters** had to be withdrawn from the battle fronts to strengthen the defences against day and night area attacks which devastated cities and vital industrial targets. Nevertheless, by December 1943 the fighter and **radar**-controlled system of defence had curbed the daylight bombers and was getting the upper hand at night. It was a short-lived victory only, once American long-range fighters appeared, even over Berlin, to defeat the German day fighters in spring 1944, clearing the way for almost unopposed attacks as the Allied armies approached the frontiers in the summer.

In both west and east the Allied armies reached the German frontier in September but were stopped by **logistic** deficiencies and a revived German Army. In the west the **Siegfried Line** provided the shield behind which the Ardennes counterstroke was prepared and launched in December. However, the losses incurred when that ill-judged attack was repulsed, deprived the East Front of forces badly needed to defend East Prussia and Silesia. The result was that when the Russians launched massive offensives in January 1945 the Front soon gave way, despite fanatical resistance by young and old. By 31 January it had backed to the River Oder and southwards through Silesia into Austria.

At that moment the British, Canadians and Americans were engaged in clearing territory to the west of the Rhine in operations which destroyed the best of what remained of the German Army. On 7 March, at Remagen, the Americans made the first of several subsequent Allied crossings of the Rhine, presaging the overrunning of the country as far as the River Elbe, leaving the eastern half, including Berlin, to fall to the Russians in their last offensive starting on 16 April. All that remained on 7 May were a few recalcitrant pockets in a ruined country which proved incapable of the **guerrilla warfare** that **Hitler** had called for.

Giap, General Vo Nguyen (b. 1912). A North Vietnamese history teacher who joined the Communists in the 1930s to resist the French in Indo-China. In 1944 he organized a **propaganda** unit for the Vietminh and thereafter played a leading role as **guerrilla** leader in the Indo-China War of Independence. He achieved fame with his brilliant handling of **logistics** and the Battle of Dien Bien Phu in 1954, when his crushing defeat of the French forced them to withdraw from **Vietnam**.

As Minister of Defence and C-in-C of the North Vietnamese Army (NVA), Giap prepared the military action against the South which began in January 1959. But his initial strategy of infiltration by the Vietcong guerrillas was too slow-acting and gave time for the Americans effectively to reinforce the South, compelling Giap to attempt to entice a major enemy force to destruction, as at Dien Bien Phu. This led to the commitment in 1964 of NVA units to reinforce the Vietcong guerrillas as a prelude to the series of disastrous offensives which culminated in the utter failure of the 1968 **Tet offensive**. Giap nevertheless pursued the same strategy again in 1972 when the Americans withdrew ground units, only to be defeated once more by the much improved South Vietnamese Army. He was then replaced by General **Dung**.

Gilbert Islands, Battles of the. In January 1942 the Japanese took unopposed possession of the islands with a small garrison. On 17 August, as a distraction from the **Guadalcanal** landing, 222 US Marine Raiders, carried 2,500 miles by **submarine**, landed on Makin. Though the island was eventually taken, the raid was a fiasco with unnecessary loss of life.

The Japanese now fortified Makin and

Tarawa, which the Americans, under Admiral Kelly **Turner**, assaulted on 20 November 1943 with heavy loss. For although Makin was easily overwhelmed by army infantry, beach obstacles caused difficulty and an escort **aircraft-carrier** was sunk by **submarine**. At Tarawa, which was very strongly held, crossing the reef seriously held up the **landing-craft** and Amtracs. As a result casualties to the 4,700 **Marines** finally amounted to 985 killed and 2,193 wounded. On the other hand, only 17 Japanese out of the garrison's 5,000 survived. Turner admitted the failures to reconnoitre thoroughly, to 'gap' obstacles and to provide sufficient supporting fire, lessons already learned and passed on by the British after the **Dieppe Raid**.

Gliders. Before suitable power plants were invented, heavier-than-air craft were of necessity gliders. Their first indirect military use was by the Germans in the 1920s to train pilots for the day when they would again be permitted an air force. It was the Russians who, in the 1930s, built an assault glider (to carry 18 men), without making much use of it since they preferred **parachutists**. As did the Germans who, nevertheless, formed a secret glider-borne unit and built a few 10-man DFS230 gliders (fitted with three braking **rockets** and a tail parachute to arrest landing), supported by fewer, still larger, cargo DFS242s.

On 10 May 1940 the Germans committed 11 DFS230s, towed by Junkers Ju52s, to land on the Belgian fort of Eben Emael. Two went adrift when towing cables broke, but the rest swiftly seized the fort from its thoroughly surprised garrison. Meanwhile nearby bridges over the Albert Canal had also been taken by glider-borne troops. At various times throughout **World War II** the Germans again would use gliders for small operations, such as the *coup de main* Corinth Canal crossing in April 1941, anti-**guerrilla** raids and the rescue of Benito **Mussolini** in 1943. But never again after **Crete** in May 1941 (where about 80 gliders were landed) would they use them on a large scale.

In the summer of 1940 the British (still unaware of German gliders in Belgium) preferred glider-borne troops to parachutists for large-scale landings. They built two main operational types; many 29-man, side-loading Horsas (which could also carry **jeeps** and small **artillery** pieces); and a few enormous nose-loading Hamilcars which could carry a light **tank**. Similarly enthused, the Americans built the 13-man, nose-loading Waco Hadrian. The first British glider operation, in Norway on 19 November 1942, failed when two gliders crashed attempting to land **Commandos** on a glacier to attack the heavy-water plant at Vermork.

Large-scale Anglo-American employment began with the invasion of Sicily on 11 July 1943 when several gliders, out of 144, cast off in the wrong place. Some landed in the sea, with the loss of 250 men; the rest were scattered. In **Normandy** on 6 June 1944, however, they played vital roles by capturing strategic bridges and gun-sites at night and in daylight, bringing in large numbers of infantry, tanks and artillery to reinforce parachute formations dropped ahead of them. In Holland, in September, and the Rhine crossings, in March 1945, they landed to plan in daylight in well-concentrated mass – although too far from the bridge at **Arnhem** to help the parachutists.

Meanwhile, in **Burma**, Waco gliders had played a vital role by landing the first wave of General **Wingate**'s Chindits behind the Japanese lines in March 1944.

After World War II gliders fell out of favour, partly because parachutists were regarded as operationally more flexible and economic but very largely because the **helicopter** offered a far better means of transport which could get in *and out of* much smaller landing-places.

Göring, Reichsmarschall Hermann (1893–1946). A larger-than-life Bavarian politician and colleague of **Hitler** who won fame in World War I as an air ace and as later leader of the Richthofen squadron. Although responsible for organizing the training of pilots and the supply of aircraft for the secret air force, and becoming Air Minister and C-in-C of the German Air Force in 1935, his military talents were regarded as little better than those of a battalion commander. As Hitler's right-hand man, he ably and ruthlessly eliminated political opposition, yet his claim that the Air Force could single-handed prevent the evacuation from **Dunkirk** and win the Battle of **Britain** were fundamental errors in letting Britain survive in 1940. The wasteful assault on **Crete** in 1941 was another mistake, which deprived the Army in Russia of vital resources; and claiming ability to supply Stalingrad by air in 1942 was as irresponsible a boast as saying Germany could not be bombed. Worst of all was his permitting the Air Force to fall behind in the technological race. He very much contributed to Germany's defeat and was condemned to death for war crimes at the Nürnberg Trials.

Gorshkov, Admiral Sergei (b. 1910). As a Rear-Admiral during **World War II**, Gorshkov commanded 400,000 Russian sailors in land and river operations in the Ukraine in collaboration with Nikita Khrushchev, under whom he became C-in-C of the Soviet Navy in 1957. He inherited the fleet Josef **Stalin** had built up to challenge the world, including a programme for 1,200 **submarines**. No submariner himself, Gorshkov nevertheless gave high priority to their construction – above all **nuclear**-powered, missile boats over surface vessels. By his retirement in 1985, he had vastly expanded the fleet and established it worldwide. Yet, technically advanced as its vessels were, they suffered from serious technical deficiencies stemming from poor workmanship and his dictum that 'better is the enemy of good enough'.

Greek–Persian wars. The unification and militarization of Persia under **Cyrus the Great** (559–530 BC) and the westwards drive by **Darius I** into Europe in 511 and subsequent seizure of Thrace and Macedonia, inevitably produced friction with the other Grecian states, under the rising influence of Sparta and Athens. An unsuccessful anti-Persian revolt by the Ionian city of Miletus in 499 was supported by Athens, with the result that Darius decided to conquer Greece. But his fleet was wrecked by a storm in 492, delaying the invasion until 490. Darius probably committed two out of his five well-organized and equipped divisions (each of 10,000 men) under General Datis to a diversionary landing at **Marathon**, which drew off Miltiades's 10,000 strong Athenian Army while the rest of the Persian Army sailed for Athens. But the defeat of Datis on the beach at Marathon led to the campaign being called off.

Ten years later the Persians, now under Xerxes, tried again, this time with a march down the east coast supported by a fleet estimated at some 1,200 **galleys** (mostly **triremes** and 50-man penteconters) and 3,000 transports. The army overcame famous Spartan resistance at Thermopylae, the fleet defeated the Greeks off shore and the army occupied Athens. But when the Persian fleet attacked the Grecian fleet of 325 triremes under Themistocles at **Salamis**, they were routed. This victory enabled the Greeks to go on the offensive in 479 to destroy the Persian fleet on the beaches at Mycale in August. Meanwhile, in July the Persian Army under Mardonius had drawn into battle at Plataea the outnumbered Greek Army under the outgeneralled Pausanias. But in a fluctuating struggle the Persians were routed by the superior fighting qualities of the Greek **infantry** phalanx.

The Battle of Mycale put an end to Persian sea-power and Plataea started Persian decline. Meanwhile Greek strength rose to a peak in 336 BC when **Alexander the Great** came to power, determined to destroy the Persian empire. He firmly believed in the superiority of a smaller, well-disciplined and better-equipped army of all arms over much larger but inferior forces. And he proved it by stunning victories at Granicus in 334 (35,000 Greeks versus 40,000 Persians); Issus in 333 (30,000 versus 100,000); **Arbela** in 331 (47,000 versus, very approximately, 200,000); and Hydaspes in 326

(11,000 versus 34,000). These triumphs virtually eliminated Persian power for centuries to come.

Grenades. Probably by mid 15th century the invention of gunpowder had led to the making of hand-thrown **explosive** charges fused to burst among the enemy. These unreliable grenades (so named, possibly, from the pomegranate they resembled) required careful handling by so-called grenadiers who, by the mid 17th century, were often grouped in specialized **infantry** companies that frequently led the assault in **siege warfare**. By the mid 19th century, however, the popularity of grenades had declined, only to be revived during the Siege of **Port Arthur** and by later adoption as standard issue by the German Army prior to **World War I**. As trench warfare developed, with often quite short distances between the opposing armies, the grenade became an essential weapon. The few that the British had, known as 'jam pots' and 'hair-brushes' from their shape, were crude devices lit by an ordinary match. Early in 1915 a Belgian company developed a new type containing a striker, detonating cap and fuse. A lever retained the striker against a spring which, when released, struck the detonating cap. The explosive filling was ammonal.

William Mills, a British foundry owner, saw how the Belgian grenade fusing could be improved and produced a successful version which became known as the No. 36. Some 75 million were produced in the UK during **World War I** and the pattern remains in service, essentially unchanged in concept, to this day.

The limitations of a hand-thrown grenade led to demands for a longer-range projector. The first idea was to fit a cup over a rifle muzzle, from which the grenade was discharged by the firing of a special ballistite cartridge; normal ammunition could not be used while the cup was fitted. A typical refinement is that of the Belgian FN Company's Telgren telescopic rifle grenade which can be fired with any type of rifle ammunition. The grenade fits directly on to the rifle muzzle and the tail telescopes into the head for stowage so that the infantryman can carry several on him. The Soviets in their turn produced a number of magazine-fed grenade-throwers giving a rapid-fire capability.

In addition to their original high-explosive filling, grenades have now been developed specifically for **anti-tank** use, with **smoke** fillings and with irritant gas fillings for riot control.

Gribeauval, General Jean de (1715–89) was appointed Inspector of Artillery in 1776 tasked, in the aftermath of the **Seven Years War**, to reform France's **artillery**. He concentrated on greater mobility with lighter carriages, pieces and limbers drawn by paired **horses**; improved accuracy and rate of fire by the adoption of better **ammunition**, tangent scales and elevating screws; and efficiency by standardization with 4-, 8- and 12-pounder field guns and 6-in howitzers, manned by soldiers in place of hired civilians. These measures, benefiting from the introduction to France in 1782 of British **iron**-making processes, revolutionized **gunnery** in time for the **French revolutionary wars** and **Napoleon**.

Ground-effect machines. Also known as air-cushion vehicles or hovercraft, capable of travelling across water and reasonably obstacle-free terrain. It was Sir John **Thorneycroft** who, in the 1870s, first proposed the theory that if the hull of a ship was shaped to form a plenum chamber and then filled with pressurized air, the ship would rise out of the water and hence create less drag, allowing greater speeds with fuel economy. He was unable to suggest a system to prevent the air escaping round the sides of his design and it was Christopher Cockerell in the 1950s who produced the first practical hovercraft. He included a slot round the circumference of his vehicle which jetted the pressurized air inwards towards the centre, thus saving most of it to provide the necessary lift. Subsequent development showed the advantage of using a flexible skirt to enhance this effect if high loadings were to be achieved and this is the configuration to be seen on most modern vehicles.

The layer of pressurized air on which the hovercraft rides has two functions: to reduce surface friction, thus requiring less power for forward motion, and to provide a suspension system for the vehicle. The effect is comparable to a saucer sliding on ice and is equally difficult to control without **aircraft** techniques; the result is a seagoing vehicle using aircraft-construction methods and tending to be very expensive in consequence.

With only about 30 per cent of the world coastlines able to accept the beaching of a conventional landing-ship, while it has been estimated that about 70 per cent would accept a hovercraft, the concept has an obvious attraction for military **amphibious** operations. One study suggested that a large hovercraft capable of carrying four main battle tanks would theoretically be able to embark its vehicles in Malta and land them in the vicinity of Baghdad without meeting any serious obstacles in its way.

Though Britain led the way in hovercraft design and construction, it abandoned the military application chiefly because of cost, though it has in service some large cross-Channel ferries carrying upwards of 400 passengers and 60 vehicles at about 60 knots. It did, however, supply the Imperial Iranian Navy with BH7 hovercraft, though little has been heard of them since the revolution there. The United States made very successful use of the SRN5 (US designator SK5) during the **Vietnam War**, operating in the marshy area known as the Plain of Reeds. The US Marine Corps are now operating large seagoing LCAC – Landing Craft Air Cushion – built by Bell Aerosystems and capable of carrying some 70 tonnes of cargo across open beaches and as far inland as the terrain will allow. The Soviets too operated their large AIST220 in a similar way and the French are building a 200-tonne NES2002.

Guderian, General Heinz (1888–1954). A light infantryman who, prior to **World War I**, specialized in **radio**. Between 1914 and 1918 he served mainly with the **staff** on the Western Front. In 1922 his task was to help develop the **mechanization** of the German Army; by 1929 he had become convinced that **tanks** in all-arms, armoured (Panzer) divisions would in future dominate land warfare. With **Hitler**'s support, but obstructed by traditionalists, he promoted the creation of the German armoured forces which spearheaded the invasion of Poland in 1939. As an armoured corps commander then and at the head of a corps and a group in the invasion of France in May 1940, he proved his theories right by brilliant leadership. Likewise in Russia in 1941 he led the drive towards **Moscow** and Kiev, but was sacked when he disobeyed Hitler in the winter defeat.

In the aftermath of Stalingrad, in February 1943, Hitler recalled him as Inspector to restore the armoured force from ruination. He laboured effectively but only to see the armour repeatedly squandered by Hitler. On 21 July, after the attempt on Hitler's life (of which he was aware), he was made Chief of the General Staff, principally directing operations on the Eastern front, a task made impossible by Hitler's follies – to which Guderian repeatedly and tempestuously objected until sacked once more in March 1945.

Guerrilla warfare. From the earliest times until the appearance in more general use of uniformed troops in the 17th century, a great many small armed groups, living off the countryside and merging with the populace, fitted today's notion of the guerrilla band. The name was coined from the Spanish *guerrilleros* waging a small war against the French from 1809 to 1813, meaning lightly armed irregulars operating behind the enemy lines in the sort of underground struggle which then, as now, gave rise to terrible excesses among participants and to civilians caught up in what usually amounted to civil war.

Most formal **wars of the 17th and 18th centuries** were accompanied by some kind of behind-the-lines combat, including sabotage and ambushes, while insurrections, such as the **American War of Independence**, have often been the essence of guerrilla warfare. Giuseppe Garibaldi's independent forces in Italy's Wars of Independence were guerrillas who suffered their worst set-backs when attempting to stand and fight against unshaken regular troops. The Frenchmen in plain clothes who cut lines of communication during the **Franco-Prussian War** in 1870, though called *francs-tireurs* by the Germans and shot out of hand, were guerrillas. Many were the bands at large during the **American Civil War** – some in uniform, some not. The situation was the same in most of China's struggles (including the Boxer Rebellion), as well as in the South African Boer War. Generally it was the wearing of a recognized military uniform (*see* **Clothing, Military**) which, as often as not, distinguished regulars from irregulars bearing arms – although recognition of an armband as a uniform sometimes stretched imagination.

World War I was fought, for the most part, by uniformed armies. Even the so-called German guerrillas under General von **Lettow-Vorbeck** in East Africa were in uniform, as were British soldiers fighting in the Arab revolt against the Turks. On the other hand, the Irish patriots who rose against the British in 1916 wore plain clothes, and some were shot after trial. During the Russian revolutionary wars, when guerrilla warfare was rife and identification usually uncertain, injustices and excesses in the name of freedom and liberty often were quite barbaric. Indeed, what happened then opened a new chapter in guerrilla warfare through development of systematic, large-scale support and supply by outsiders (for example of the Whites by the Americans and British).

In the Chinese Civil War it was the defeat of the Communist revolt by the Nationalists in the 1930s (their so-called 'bandit suppressions' were really counter-guerrilla operations), that compelled the famous Long March – in effect a large-scale, evasive migration by irregulars and their camp followers. China, in fact, was infested by guerrillas of all kinds. So too was Spain during the **Spanish Civil War** with the so-called **fifth column** behind Republican lines and prevalent use of **aircraft** and **radio** for supply and control. These, indeed, were rehearsals for **World War II**'s guerrilla groups (sometimes called resistance fighters and sometimes partisans) which sprang up among the invading Germans and Japanese.

Guerrilla operations evolved to a pattern. Bands of various sizes were recruited from the local populace, initially organized by agents sent in by air or sea from bases such as Britain, Russia, North Africa and India, and then supplied by air with arms, ammunition and other warlike stores. Usually the bands assumed local political affiliations – some pro-government, others not, many Communist – and fought (like the Greeks and Yugoslavs) as much among themselves as against the Axis. Amid Russian forests and swamps, large bands hid prior to striking at the Germans in support of major offensives. Only in Yugoslavia, Russia, parts of France and Italy did bands in the West engage in overt resistance long before the arrival of approaching Allied troops, and usually with disastrous consequences, as in northern Italy and the Vercors in 1944, when bravery did not compensate for lack of heavy weapons and inadequate training against German regulars. Prudence dictated that guerrillas limited their attacks to sabotage, as ordered by **radio** messages from the so-called Special Operations Executive (SOE); made full use of well-trained Allied troops sent in to support them; and did not come out in force until the moment was ripe – when the enemy was departing.

After World War II many guerrilla organizations remained in being and hid their arms for future use in political struggle. Notably in China, the Philippines, **Vietnam**, Malaya and Indonesia, in the Far East and in Palestine, Greece and elsewhere, cadres remained for the prolonged struggles to come. In addition to these hot spots there were countries that had not been major battlefields but whose nationalist aspirations or internal conflicts prompted resort to arms and guerrilla warfare. Their belligerents, like the majority throughout Africa, in Cuba, Central and South America, used political affiliations with the Great Powers, or such oil-rich ones as revolutionary Libya, to advise, train and supply their bands in readiness for the day of action.

Modern guerrilla warfare, ranging from coercive terror bombing (as in Ireland) to overt resistance by large bands, is more commonplace than ever before. Although guerrillas have won total successes (as in Algeria and Cuba), the vast majority have either been defeated (as in Malaya) or contained in struggles which continue for years, without decision, to the desolation of peoples and places – notably in East Africa and **Afghanistan**. Technique becomes more sophisticated as **communications**, weapons and **explosive** technologies improve. Struggles now take place in urban surroundings as well as rural areas. Counter-guerrilla precautions and operations demand an immense outlay in **intelligence** networks, **psychological warfare**, skill in execution and precautionary measures. It is an unending struggle which, materially and socially, consumes almost as much as full-scale war.

Guibert, Comte de (1743–90), fought in the **Seven Years War** and in 1772 made his name in high Parisian society with his *Essai général de tactique*, a popular work which was more influential for its proposal to form a citizens' army based on **conscription** than for its emphasis on mobility as practised by **Frederick the Great**. But in 1779, in *Défense du système de guerre moderne*, he reverted to the orthodox concept of manoeuvre by disciplined forces in the Prussian manner. Three years after his death the *levée en masse* in France began the era of national conscription which continues to this day.

Gulf War. *See* **UN–Iraq War**.

Gunnery. The art of gunnery is concerned with the operation of equipment and the control of gunfire (*see also* **Artillery** and **Ammunition**). From the 14th to mid 19th century, on sea and land, there was no effective fire-control system largely because the **electromagnetic effect** had not been mastered. But with the invention of the **telegraph** and **telephone**, and **power-generation equipment**, gunnery became more than an art as indirect fire control became possible during the **American Civil War**.

Up to the 1890s, however, gunners mainly laid their pieces by direct line of sight, well knowing that, owing to indeterminate and inferior charges and projectiles, windage between barrel wall and projectile, problems of rangefinding, lack of hydraulic recoil systems and, among other natural causes, rolling, pitching and wind drift, the zone of dispersion of shot was so large as to make the chances of hitting a small target or making effective corrections, even at close range, low. Indeed it was the lengthening ranges of guns which made it essential to impose better gun control in order to take advantage of their capabilities beyond 7,000 metres on land, let alone 12,000 at sea.

At sea in sailing-ships, guns were either fired independently by the gun captain or in broadsides under command of a deck officer, according to the captain's requirements, concentrating, very likely, on such vital parts as masts and rigging. The design of fast warships, with their guns housed in several spaced-out turrets, made these practices obsolete. Henceforward the centre of mass was usually the target for centrally controlled gunfire. In 1905 Captain Percy **Scott** RN proposed that control of a ship's guns should be from a fire **director** placed high up on the foremast. The director contained an observer officer, a gunlayer and a gun trainer, the latter two equipped with instruments indicating elevation and azimuth respectively, transmitted electrically to the gun turrets below. This system was full of errors but it did significantly improve gun control and was adopted by the Royal Navy by 1913.

For fire control, the bringing of fire to bear and subsequent correction of the fall of shot, changes of range and bearing had to be integrated, together with the effect of the ship's motion, and passed continuously to the gun-sights. The Dreyer fire-control table, adopted prior to 1914, met the basic requirement, despite many shortcomings. The

German system, designed by the electrical firm of Siemens & Halske, incorporated a mechanism for automatically adjusting the sight so that all the layer had to do was align his sight with the target and fire. Neither side, nor the French who adopted a system similar to the Dreyer table, had at this stage adopted **gyroscopic** control, though the Germans were about to do so when the outbreak of war in 1914 brought trials to an end.

The Battles of **Dogger Bank** and **Jutland** exposed the flaws of these systems (many were already noted even in peacetime exercises) when haze plus funnel and gunsmoke obscured vision, and the marking of fall of shot to assist correction was severely impaired when several ships were engaging the same target at once. The problems of night firing with the help of **searchlights** had also to be tackled more seriously, and had been partly solved prior to 1939 when **radar** ranging and assisted gun control were already being practised by the Germans.

World War II saw an increase in naval air power which tended to make direct engagements between surface ships something of a rarity. It is probably true to say that the larger ships generally practised their gunnery against land targets in support of **amphibious** operations during this war. The requirement for accuracy was just as urgent but at least the targets were stationary.

Since World War II the surface-to-surface **rocket** missile has largely superseded the gun for ship-to-ship engagements, though most warships of any size will have at least one gun of up to 105mm calibre, rapid-firing and automatically controlled by **computer**. Such weapons were used effectively against shore targets during the **Falklands** War.

On land. As gun ranges and accuracy improved it became feasible to withdraw the guns from the main battle line and conceal them behind natural features. This in turn required the guns to fire indirectly, where the layer could not see his target and hence fire had to be corrected by a forward observer. The introduction of Morse code, transmitted by flags or **heliograph**, provided the vital link between observer and guns and made indirect fire a practical proposition during the **American Civil War**, especially where tethered **balloons** were used to lift the observer aloft.

In 1885 the Germans used a forward observer with a telephone to control artillery fire during trials against **fortifications**. Balloons were soon in use with the same instrument. Yet although in **World War I** the telephone and, by 1915, **radio** made it possible for ground and air observers to give precise and detailed corrections to the guns, the difficulties of judging range still made it essential to fire a number of establishing rounds at the target – thus alerting the enemy to the effective concentrations to follow. By 1916, however, this tactical disadvantage was overcome to some extent by predicted fire (i.e. firing for effect from surveyed gun-sites and making use of scientifically ascertained ammunition data and meteorology to hit targets whose location was known) and first used decisively for counter-battery tasks in 1917 at Riga, **Caporetto** and **Cambrai**.

By the end of World War II communications and survey methods had become so accurate that the artillery of an entire corps, firing without registration, could be at the disposal of a single forward observer communicating simultaneously by radio with all gun positions, either observing from the forward infantry or from a spotter **aircraft**. Post-1945 indirect fire techniques have been even further refined with the use of **radar** and **computers**; many of the variables previously limiting accurate fire have been eliminated or much reduced.

Two tasks remain for direct fire: for use as **anti-aircraft** and **anti-tank weapons**. Both are concerned with shooting at fast-moving targets, with the advantage that, unlike the Navy's weapons, they are usually fired when stationary. The problem of achieving hits on fast-moving aerial targets was not easily solved, even with the introduction of fire-control radars, which improved the kill rate from 20,000 shots per aircraft in 1940 to 4,000 per aircraft in 1941. Modern high-flying jet **bombers** are beyond the reach of any gun system and surface-to-air missiles have now taken over against high-altitude targets. However, most countries have retained low-level air-defence gun systems of 35–40mm calibre. Usually vehicle-mounted, computer-controlled and incorporating surveillance and tracking radars, these are sophisticated and expensive equipments which, like the German Gepard, can claim a chance of a hit of about 80 per cent against a transonic crossing target – the worst case.

The anti-tank gunner is concerned with visible targets at 3,000m or less, small targets which he must either hit first time or perish himself from retaliatory fire. Modern technology has now largely overcome the **rangefinding** problem but that in turn has revealed other variables which need to be corrected: wind speed, ambient conditions, muzzle velocity, barrel bend and wear. A modern tank fire-control system integrating all these problems is computer-controlled with electronically stabilized guns and sights, leaving the gunner with the simple task of aligning his sight on the target and firing when the time is ripe. With such a system, an 80–90 per cent chance of a first-round hit is to be expected.

In the air. During World War I the **machine-gun** quickly became the **fighter aircraft**'s weapon and on the single-seater biplane it was mounted pointing forwards above the upper wing or, later, firing between the propeller blades by means of an interrupter gear. The pilot pointed the aircraft at the target and manoeuvred to obtain a hit. Results were not impressive, though it was not before 1916 that much was done to improve matters, when the Frenchman Yves Le Prieur demonstrated a wind-vane corrector. It was very complex and Major Geoffrey Norman designed a simpler system which altered the setting of a machine-gun foresight to allow an aim-off related to aircraft and bullet velocities.

The introduction of multi-gun fighters in the 1930s did not materially alter the means of fire control. As late as 1937 it was felt that air combat would be at such short range that special sighting equipment would be unnecessary. However, increased aircraft speeds and the need to engage attacking fighters accurately from the power-operated gun turrets being fitted to bombers soon called for improvements; gyrostabilized gunsights were in service with the Royal Air Force by 1943.

The air attack of ground targets became an important feature of **air warfare** during World War II and this called for the fitting of larger guns (though Westlands had fitted a 37mm gun to a fighter as early as 1927). The use of **armour** on fighters and bombers also had its effect and 20mm cannons were specified for German fighters before 1939, with Hawker's 40mm anti-tank gun being on trial in 1941; by 1943 a six-pounder anti-tank gun had been successfully mounted in a Mosquito. However, in the later stages of World War II rockets gradually replaced guns for ground attack, paving the way for their introduction for air-to-air combat in the **jet** era. Nevertheless, the US A10, specifically designed for ground attack, carries a six-barrelled 30mm Gatling gun as part of its weapon fit and it is in this role that air gunnery chiefly survives, now controlled by computer and with head-up displays allowing the pilot to aim his weapons without being distracted from control of his aircraft.

Gustavus II, King, Adolphus (1594–1632), who came to the Swedish throne in 1612, was of considerable intellect and a great civil reformer. His fame, however, rests on his military reforms and campaigns in the **Thirty Years War**. Exploiting the fire-power increased by the invention of the wheel-lock **musket** (half its predecessor's weight) and the first practical fixed cartridge, he was able to thin out and extend the **infantry**'s battle line. The **cavalry** was taught to integrate with the infantry and to charge knee-to-knee to produce shock. And the **artillery** was subdivided into siege, field and regimental branches, the latter equipped with the famous light gun for close support.

Supported by sound **logistic** services, he

conducted indecisive campaigns in Russia (1614–17) and in Poland (1617–29), prior to intervening in the **Thirty Years War** in 1630 on the Protestant side. After the Battle of **Breitenfeld** in 1631 he occupied north-west Germany and in 1632 invaded Bavaria where he won the battles of the Lech in April, Alte Veste in September and, on 16 November, Lutzen where he was killed rallying shaken cavalry. His influence would long be felt.

Gyroscope. A word coined by Léon Foucault in 1852 for a rotating wheel whose axis is free to turn retains its direction unless displaced. The first practical gyroscope, which waited upon a suitable electric motor, was made by Ludwig Obry in 1895 and soon after incorporated in the Whitehead **torpedo** to predetermine its course and depth. Since then gyroscopes, which can be north-seeking, have been applied to **navigation** systems; as **artillery** gun and gun-sight stabilizers; for automatic pilotage and stabilization of ships, **aircraft** and **space vehicles**; and in **rockets and guided missiles**, besides many other military and civil uses.

H

Hadfield, Sir Robert (1858–1940). A brilliant metallurgist and founder of the Hadfield Company, who published more than 220 scientific and technical papers. Among his most important contributions to military technology were manganese **steel** in 1882 and silicon steel in 1885 – the former of immense importance for **armour**, **ammunition** and **tank** tracks; the latter for the burgeoning electrical industry. Hadfields became one of the principal armaments firms during and after **World Wars I** and **II**.

Haig, Field Marshal Sir Douglas (later **Earl**) (1861–1928). A British cavalryman who believed in the tactical soundness of the knee-to-knee charge. He commanded I Corps on the outbreak of **World War I** in the retreat from Mons, at the Battles of the **Marne** and the First Battle of Ypres before taking command of First Army in February 1915 and becoming C-in-C British Expeditionary Force (BEF) in December.

He was conscious of the need for the BEF to collaborate fully with the French and remained on good terms with Marshal **Joffre**, in trying to relieve the strain on the French Army at **Verdun** with his own strategically controversial offensive on the Somme in 1916; but sceptical of General Nivelle in 1917. At **Arras** in April and May the BEF played its part and, for the rest of the year in Flanders and at **Cambrai**, took at dreadful cost the full strain in the aftermath of the French Army's mutiny. In March 1918, he placed the BEF under Marshal Foch to see them through the crises of the **Hindenburg** offensive, and, at **Amiens** in August, he opened the way to the war's end in 1918. Often maligned as anti-tank in outlook, in fact he was ready to use any tools made available to achieve victory.

Halder, General Franz (1884–1972). A Bavarian artillerist and one of the best brains in the General Staff, he specialized in operational and training matters. As Army Chief of Staff from August 1938, he willingly carried out the planning and execution of **Hitler**'s campaigns in Poland, **West Europe**, **North Africa**, the Balkans and Russia. But when things began to go seriously wrong in the Caucasus in September 1942, he was removed after a final disagreement with Hitler. Yet, indignant as Halder sometimes was at Hitler's interference in military matters, and flirt as he did with the notion of deposing him, he was no political resister, simply a soldier who stuck to the military last and took pride in winning victories – until awakening too late to a hopeless situation.

Halsey, Admiral William (1882–1959). A naval officer's son of charismatic dimensions, Halsey was in command of US **aircraft-carriers** at **Pearl Harbor** in December 1941, though he had not become an aviator until 1935. Highly rated for his aggression by Admiral **Nimitz**, Halsey's carriers raided Japanese shipping in 1942 and launched the air attack on Tokyo in April. As a result he missed the Battle of the **Coral Sea** and was relieved of command with a skin complaint prior to **Midway**. In October 1942 he relieved the weak Admiral Ghormley, Commander South Pacific, and immediately inflicted heavy losses on the Japanese in the attritional battles of the **Solomon Sea**. In June 1944 he took command of 3rd Fleet, an immensely powerful force of **battleships** and carriers, and smashed Japanese air power in the battles preceding **Leyte Gulf**. At Leyte Gulf, however, Halsey's obsession with the sinking of Japanese aircraft-carriers led him, to Nimitz's disgust, to uncover the **amphibious** force in order to chase Japanese carriers, whose role was known from intelligence to be diversionary. But his fame remained undimmed.

Hannibal (247–183/182 BC). The most celebrated of Carthaginian generals of that name who became C-in-C of the Army at 26 and two years later precipitated the Second **Punic War** (219–202 BC). Lacking sea power, he gambled on invading Italy by a march from Spain through the Alps in October. During the ensuing campaigns he lost only three out of sixteen major battles, besides the victory at **Cannae** in 216 BC. A brilliant strategist and tactician, who made much use of **elephants**, he often fought against adverse odds. But in the end the Roman **blockade** fatally weakened Carthage and led to his crushing defeat by Scipio at the Battle of Zama in 202 BC. He eventually committed suicide.

Hanoi, Battles of. Since early times the capital and principal port of the Vietnamese region of South-East Asia, and often fought over, Hanoi became the capital of French Indo-China from 1887 until 1945. The city grew into a vital industrial and communications centre linked by **railway** to Haiphong and the hinterland.

Occupied by the Japanese in 1943, it was often bombed by the Americans from 1943 until 1945 during **World War II**. Immediately after that war, the Vietminh made it their main objective at the start of the war for **Vietnam**'s independence from the French. Vietcong guerrillas fought within its urban sprawl. When the French were defeated in 1954 it became the capital of North Vietnam.

Throughout the struggles of the 1960s and 1970s, the city's outlying communications, along with Haiphong port, became targets for US bombing. Transport, plus airfields and anti-aircraft batteries, were the principal targets, the city itself receiving relatively slight damage at first. In 1966 **oil-fuel** targets, power stations and **railways** in the so-called Iron Triangle of Hanoi, Haiphong and Thanh Hoa were hard-hit – forcing the building of pipelines and delivery of oil in drums instead of bulk. Targets in urban areas were attacked accurately by smart missiles. In 1967 the Americans attacked bridges and railways to seal off Hanoi and Haiphong from the rest of the country, causing a major evacuation of both cities and the dumping of stores in urban areas which pilots were forbidden to attack. Not until December 1972, in an attempt to bring the procrastinating North to the conference table subsequent to the collapse of their offensive in the south, did President Richard Nixon at last permit attacks on all targets regardless of location. Hanoi city was bombed heavily and 1,318 people reported killed. When American anti-war activists asked the mayor of Hanoi to condemn the 'Christmas Offensive' and claim 10,000 had been killed, he refused in order to protect the North's political credibility. A month later the Paris Accords (which the North never intended to keep) were signed, bringing the war to a close.

Harris, Air Marshal Sir Arthur (1892–1984). Took part as a British Home Defence **fighter** pilot and then **artillery** observation pilot in **World War I**, developing an enthusiasm for General **Douhet**'s theories in the 1920s. In key command and staff appointments in the 1930s, he pressed the development of heavy RAF **bomber** forces as an independent war-winning instrument, by advocacy that lacked absolute scrupulousness. In **World War II**, when made Commander Bomber Command in 1942, the bombers at his disposal were incapable of fulfilling his task of 'knocking Germany out of the war'. Unable to hit small, military targets, he was compelled to adopt 'area bombing', in which urban areas were devastated without achieving the main aim. Bomber Command contributed mightily to final victory, but Harris's reputation was tarnished by his seeming insensitivity and unwillingness to abandon area bombing in favour of more accurate and economic methods.

Health services, Military. Until people began to move beyond their immediate locality they had but a limited knowledge of the health afflictions of distant communities. Exploration accompanying greater mobility, above all by sea, not only led to discoveries of new maladies but also carried them far and wide to infect and ravage unimmunized peoples. Chief among the carriers and often the worst sufferers were navies and armies. Ships' crews on long, worldwide voyages were not only exposed to almost every kind of plague and ailment but, for example, suffered from scurvy owing to inadequate diet (*see* **Food**). Armies landed in unknown territory often became victims of insect-borne infections such as malaria and yellow fever, for which **medical services** had no remedy. When hygiene was rudimentary and empirical it was common for the death and incapacitation roll to be far higher from sickness than from wounds.

Yet although before the second half of the 19th century there was only an uneducated understanding of the causes of disease, an awareness of ways and means to promote hygiene existed. **Roman** invaders brought with them sewage systems and controls over water supply, and also water-borne diseases such as typhoid and cholera. They also noticed the connection between swamps and malaria and regulated food purity. One major disaster that followed the collapse of the Roman Empire was the uncontrolled spread of disease and plagues in the Middle Ages, which depleted populations and produced a severe manpower shortage. Another was the spread of venereal disease by seafarers, starting in the 16th century, and the ill-effects of excess drinking of alcohol.

Military health services improved with the advances of humanitarianism and science. The awakening public concern about overcrowding in urban and industrial areas, linked to the gathering of health statistics and advances in medicine, led to vital 19th-century social reforms. A turning-point in health care came with the publicity about the suffering of sailors and soldiers in the **Crimean War** and the work of Florence **Nightingale**. Henceforward far greater attention was paid to **logistic** services, which provided more nutritious **food**, sensible **clothing** and shelter, not only to economize in lives and improve combat fitness, but also to enhance **morale**. But it would be many years before there was adequate awareness of what was implied or required, and many campaigns, such as the Spanish–American War, the Boer War, the **Russo-Japanese War** and **World War I**, before the main causes of ill-health and the worst of its problems, abuses and defects had been discovered and eradicated by **research**, technical developments and organization.

In the latter half of the 19th century the medical authorities began to pay stricter attention to prevention than to cure. The implementation of legislation in civilized countries raised standards and concentrated the minds of surgeons and doctors on the need to care for sailors and soldiers by supervising their living conditions, feeding arrangements and physical well-being. Proper sanitation was imposed upon the military, who were compelled to study and enforce sanitation and hygiene in the field by specialists, particularly when operating in underdeveloped countries, waterless **deserts** and frozen wastes where standards and facilities were primitive. It was realized before 1900 what caused most water-borne disease, but it was not until the 20th century that insects and parasites, such as flies, lice and mosquitoes, were discovered to be bearers of many intestinal complaints as well as such mass killers as malaria and yellow fever. From the identification of cause to the discovery of vaccines to immunize people against such scourges as smallpox, cholera, typhus and tetanus, was a relatively short step, one which contributed significantly to reduced sickness and mortality rates in **World War I** and almost total extinction of some afflictions in **World War II**. Likewise, as the reduction in insect population was tackled in campaigns to eliminate the static water in which they bred, so new drugs emerged to suppress or prevent malaria and yellow fever in World War II, even more effectively when insecticides, such as DDT, rid humans of lice and typhus. Indeed, one decisive advantage of the Allies over the Japanese was their possession of the latest health technology and techniques, enabling them to operate in environments which were fatal to the unprotected.

The present day's efficient sophisticated fighting men are as dependent as ever upon physical and mental fitness, complemented by comprehensively enforced health regulations and procedures, as well as by balanced diets and clothing that is compatible with good health as well as combat function. They can make the difference between victory and defeat.

Heinkel, Professor Dr Ernst (1888–1958), was the designer of important German **aircraft** in **World War I**, including the unarmed Albatross B types, the Brandenburg **fighters** and seaplanes and the Austrian Hansa–Brandenburg Ç1 biplane **reconnaissance** and light **bomber**. In 1922 he built a factory at Warnemünde and another in Sweden. When the secret German Air Force was sanctioned by **Hitler**, Heinkel, who had already designed a fast monoplane civil airliner, was ready with designs for reconnaissance seaplanes, the He51 biplane fighter and the highly successful He111 twin-engine monoplane bomber (5,656 built). And in August 1939 he produced the world's first **jet**-propelled aircraft, the He178. During **World War II** his factories produced, among other aircraft, the troubled He177 bomber and the first twin-engine jet aircraft, the inferior He280.

Helicopter. Interest in Vertical Take-Off and Landing (VTOL) **aircraft** reaches as far back as the 15th century and was stimulated in 1907 when a man-lifting machine made a brief hop. But it was not until 1923 that Juan de la Cierva's Autogiro (with unpowered rotor blades) came close to vertical take-off. Yet it was 1936 before the first practical, piston-engine Focke Wulf 61 flew at a speed of 76mph to an altitude of 11,000ft, without stimulating much further development by the Germans; and 1946 before the little American Bell 47 met a military requirement.

After 1946 technical and military development was rapid. The French Alouette was the first military version with a **jet engine** and a speed of 110mph. Speed limitation to about 200mph is, however, a handicap, caused by stalling of the rotor blades in forward flight, although the latest, wider chord blades with large tips to some extent overcome the problem. With increased power bigger machines became feasible and were used operationally during the Malayan War of Independence and the **Korean War**. To begin with they carried men, stores and wounded in and out of small jungle clearings and to and from the front line. They undoubtedly saved many lives by speedy evacuation of casualties to the **medical services**. Used by the French for counter-**guerrilla** warfare during the Algerian War of Independence and by the British for **amphibious** assault at Suez in 1956, they reached a level of maturity during the **Vietnam War**. Helicopter units were, in 1961, the first US air units to serve in Vietnam. To begin with mostly single-rotor types with payloads of between 600lb and 5,000lb and speeds in the region of 100mph were employed. But soon they were joined by twin-rotor cargo machines, such as the Chinook, which could lift 33 troops or about 20,000lb of cargo. Machines of these types were principally under development in Britain, France and Russia but it was in both technology and strategic and tactical techniques and concepts that the USA thrust itself into the lead in Vietnam.

To begin with great progress was made

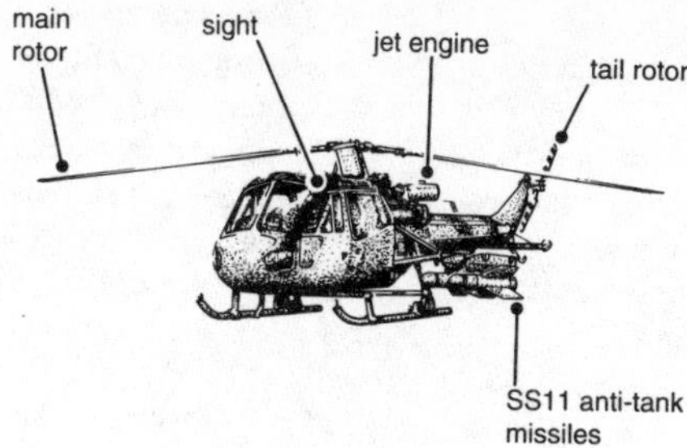

Helicopter (*Westland Scout*)

with support and maintenance services for machines which required a great deal of each through normal wear and tear and frequent battle damage. Cost-effectiveness was always under scrutiny by the critics, but to begin with were well on the helicopter's side. In undeveloped terrain it introduced vital mobility by its ability to move large numbers of troops, with their **logistic** support, into normally inaccessible places. As a result existing **airborne forces** of **parachute** and **glider**-borne troops became outmoded. But as the helicopter engaged more closely in the combat zone and counter-fire increased, it had also to be armed, initially as a defensive measure. **Machine-guns, rockets and guided missiles** were successively fitted and used for suppressive fire against landing zones and, in due course, offensively against **armoured fighting vehicles** (AFVs). Alongside these developments emerged the revolutionary concept of *air mobility* (sometimes designated *air cavalry*) and the suggestion that the armed helicopter might be a substitute for AFVs, instead of just a complement. The vulnerability of helicopters to ground fire, above all guided weapons, has nevertheless been frequently demonstrated, especially in **Afghanistan** (where Russian-armed Hind machines lost heavily), the **Arab–Israeli wars** and the **Falklands** campaign.

It is in **naval warfare**, usually operating from ships, that the helicopter is most comfortable since it is not so easily ambushed as over land and can play a leading role in most operations. In **anti-submarine** and anti-**mine** warfare it is particularly effective as well as vital in **amphibious warfare** and for logistic support, as repeatedly shown over the Falklands and the Persian Gulf between 1980 and 1988. Armed with guided missiles it also proved deadly against a surfaced Argentine submarine in the Falklands War and Iraqi vessels in 1991.

Helicopters depend heavily upon technology for tactical effectiveness, above all when operating in close country where they depend upon sophisticated **navigation** and **night-vision** equipment in order to extend their round-the-clock versatility; and for their chances of survival upon **armour** for crews and vulnerable areas. Consequently modern battlefield helicopters, such as the American Apache, have increased significantly in complexity, weight and cost to the classic point at which questions are asked as to cost-effectiveness. These questions were partially satisfied by day and by night over Iraq in 1991, when in addition to front-line missions, infiltrations into the enemy rear were launched with insignificant losses, a success story attributable in some part to the fact that these operations were across open **desert** against an electronically blinded opponent.

Heliograph. A signalling system invented by Sir Henry Mance (1840–1926), consisting of two mirrors reflecting sunlight in any required direction, the beam being interrupted by a key-operated shutter. Thus messages could be sent to a distant observer using Morse code. Mance's design was first used in 1878 during the Second Afghan War (*see* **Afghanistan, War of**) and was soon generally adopted throughout the British Army, though a version known as the heliotrope had also been used during the **American Civil War** of 1861. Since the idea required bright sunlight it was of limited value in Europe, though very successfully used for many years by the Indian Army and during the Boer War.

Henry the Navigator, Prince (1394–1460). A soldier son of King John I of Portugal, whose unsuccessful **amphibious** expeditions against the Canary Islands inspired his interest in exploration and maritime ventures. In 1418 he initiated the numerous voyages of discovery along the West African coast which were to reach De Sintra at the time of his death. He directed these operations from Sagres, where he built a school staffed by navigators, astronomers, cartographers and **shipbuilders**. They significantly advanced **navigation**, charting and long-sea-voyage techniques, besides developing the caravel, which was to be progenitor of the **carrack** and all great sailing ships into the 19th century.

Hertz, Heinrich (1857–94), was a physicist who, between 1885 and 1889, developed the **electromagnetic** theories of James Maxwell to demonstrate the production and reception of **radio** impulses. These were vital experiments which, before his premature death, enabled the measurement of radio waves and proved their relationship to heat and light.

Hindenburg, Field Marshal Paul von (1847–1934). A Prussian infantryman who distinguished himself in both the Austro-Prussian War of 1866 and the **Franco-Prussian War** of 1870. He retired in 1911 as a general but was recalled in August 1914 to command the retreating Eighth Army in East Prussia. Teamed most happily with General **Ludendorff** as his Chief of Staff, they won the Battle of **Tannenberg**, defeated all enemy offensives in 1914 and threw the Russians back in Galicia and Poland in 1915.

On 29 August 1916 he became Chief of the General Staff, authorized to exercise the Kaiser's powers as Supreme Commander. With Ludendorff as his chief executive (First Quartermaster-General) he went on the defensive in the West and successfully knocked Russia out of the war in 1917. At the same time he approved unrestricted **submarine** warfare and rigorously tightened up the war economy. Following the failure of the Hindenburg offensive of 1918, it was Hindenburg who instructed the Kaiser to abdicate and who brought the defeated troops home. Even his great prestige, for which he owed much to Ludendorff's dynamic ability, was insufficient to stave off revolution. Nevertheless, as President from 1925, he managed, in tumultuous times, to retain a measure of control until appointing **Hitler** as Chancellor in 1933.

Hitler, Adolf (1889–1945). An Austrian who served as a corporal in the Bavarian Army throughout **World War I**, was several times decorated and wounded – and who loved war. He formed a hatred of Communism (among other things, which included Jews), was lucky to survive the Revolution and in July 1921 took over the Nazi Party. The failure in 1923 of his first attempt, in league with General **Ludendorff**, to seize power also alienated his trust of high-ranking officers. Nevertheless, when made Germany's Chancellor, in 1933, it was at the call of Field Marshal von **Hindenburg** (the President) and with army support.

From the outset Hitler was bent on war against Russia. Rearmament and repossession of territory lost through the Versailles Treaty were steps in that direction. Yet, although naval and army expansion and re-equipment, along with creation of an air force (the Luftwaffe), were central to his aims as Supreme Commander, he seems to have been confused in mind as to how they should be employed – except, initially, as a card in a political, bluffing hand in a game in which he won all the tricks he contracted for. Nevertheless, by March 1939 when virtually all lost German territories had been reassimilated without war, he gambled. The occupation of Poland, which might well have been achieved without resort to arms, was an outright expression of Hitler's megalomanic desire for war.

As a charismatic bully, who divided to conquer and who often tried to be all things to all men, Hitler's handling of his commanders and his conduct of war was that of a gifted amateur. Mentally wedded to his experience of the trenches, he had a capacious memory that could absorb every detail of the military art and technology which fascinated him, but his lack of a sound military and technological education prevented him putting it to skilled use. None of this mattered while he followed the directions of the expert staffs who, sometimes, were too content to admit the soundness of his appreciations. Unfortunately for Germany, Hitler not only increasingly imposed unsound decisions upon the General Staff, but also indulged in systematically undermining their authority, converting them to lackeys. At the same time he failed to give the Navy adequate support and also permitted his favourite, Hermann **Göring**, to ruin the Luftwaffe.

Hitler reached his peak of power at the start of **World War II**. The story of his decline and destruction is reflected in many entries here.

Hobart, General Sir Percy (1885–1957). A tempestuous British Sapper of great ability and courage who won distinction in **World War I** and became a leader of armoured forces prior to **World War II**. In 1939 his career was side-tracked by enemies and he was forced out of the Army. But he was recalled by Winston **Churchill** and given command of armour. As Commander 79th Armoured Division, he was given the task of developing and training the specialized armoured forces which would lead the **amphibious** assault on **Normandy**. To a considerable extent he ensured success with economy of lives both on 6 June 1944 and in many subsequent operations, including landings in Holland and the River Rhine crossing in 1945.

Ho Chi Minh Trail. A strategic supply route which North Vietnam began constructing through Laos in 1959 to carry the **logistic** burden of its projected invasion of South Vietnam. Built by 30,000 troops under General Vo Bam to follow mountainous jungle tracks and connect with routes eastwards towards the coastal centres, it was developed during the **Vietnam War** to become a main artery with roads up to nine yards wide in stretches. It was 625 miles long, and had more than 10,000 miles of subsidiary trails and a 3,125-mile pipeline. In due course trucks were used but in the early days most stores were man-packed or pushed on bicycles of French manufacture.

The existence and significance of the Trail was soon realized by the South, but it was not until 1965 that the Americans tried cutting it by air attack. Over the years they expended enormous efforts on this, compelling the North to divert great resources to repairs and reducing capacity and flow, but never shutting it down. Indeed, only physical occupation could have done that, with an invasion of Laos, which America rejected for political reasons.

Holland, John Philip (1841–1914), was an Irish–American schoolteacher who, in 1879, built a streamlined, 16ft **submarine** (Holland No. 1). It was followed in the next 12 years by eight petrol-engined experimental submarine **torpedo-boats** for the US Navy, of which the 105-tonne No. 9 became prototype for seven production models. The **Vickers Company** ordered Holland boats for the Royal Navy, but had to redesign them before acceptance into service in 1902 of the first of a series of 63; much safer diesel engines were adopted in 1911.

Horses. The origins of the horse are speculatively vague. Seeming to have spread throughout the mainland masses by natural and predatory migrations, the horse did not reach the American continent until taken there by the Spaniards in the 16th century. Unavoidably, it was an essential combat and logistic factor in land warfare until the mid 20th century when, happily, it was superseded by **motor vehicles**.

Early horses appear to have stood 48 inches (twelve hands) at the withers. Selective breeding to satisfy military requirements has evolved riding animals standing 80 inches and heavy draught horses of great girth and strength, a trend stimulated in the **Middle Ages** by the demand for stronger horses to carry heavily armoured soldiers into battle. The trend was gradually reversed when **armour** was made obsolete by the long**bow** and firearms, resulting in calls for faster horses carrying less weight.

Regardless of their carrying capacity and speed, horses always are afflicted by fundamental weaknesses which restrict their military usage. They are big targets, easily frightened and difficult to protect since they cannot dig trenches, nor carry armour or an economic load. Furthermore they are finicky, requiring three regular meals and at least eight hours rest a day if they are to maintain condition and resist disease. Many a campaign or battle has faltered or been lost because horses could not be maintained, and they died in large numbers. During the Boer War, for example, 300 British Army horses a day were expended, most from lack of care or hard driving; few from combat; and First German Army's 84,000 horses on the march to the **Marne** in 1914, which required 924 wagons of fodder per day, declined so badly from **logistic** inadequacy that its **cavalry** was all but made inoperative.

Hundred Years War. The war between France and Britain, which began in 1337 and continued sporadically until 1453, was initiated by **Edward III**'s claim to the French throne in 1327 and exacerbated by territorial disputes and commercial rivalry between the two nations. To begin with, Edward concentrated upon forming alliances and enlarging his navy to combat French raiders in the English Channel. The complete naval victory off **Sluys** in 1340 won the British assured freedom to transport troops to France to back the fomentation by **propaganda** of internal dissent within that country, leading to the moment in 1346 when Edward felt strong enough with his smaller but better organized, equipped and trained army, to challenge the French to a decisive battle.

The defeat at **Crécy** of the French **armoured cavalry** by the intense fire-power of longbowmen, **artillery** and dismounted **infantry** won Edward a **strategic** advantage which a team of great commanders, such as the Duke of Lancaster and the Black Prince, were able to exploit until a favourable peace was dictated at Brétigny in 1360. In the years between and for the rest of the war, technology in the form of **explosives**, artillery and the gradual introduction of plate **armour** revolutionized land warfare. Existing **fortifications** proved vulnerable to a new kind of **siege warfare** in which gunpowder charges and shot quite easily demolished the walls of old-fashioned castles, making them obsolete. Most impressive of all was the manner in which Edward controlled strategic movements through foresighted operational instructions; and the precision of those moves made with the help of rudimentary maps (*see* **Charts and maps**). This ascendancy was emphasized in 1356 by the manner in which two widely separated British armies coordinated their activities before the Battle of Poitiers when, once more, a smaller British army (6,000 strong under Edward the Black Prince), in a prepared defensive position, routed a poorly coordinated French one of 20,000, under King John, by fire-power and superior tactics.

For the next half century the peace was frequently disturbed by political offensives and broken by five British peace-enforcement *chevauchées* in 1369, 1370, 1373 (15,000 strong under Lancaster which travelled 600 miles), 1375 and 1377, which hunted elusive French armies, laid siege to castles and waste to the countryside until a halt was called in 1380 and a new peace signed in 1396.

The war was renewed in 1415 by King Henry V to reassert his claim to the French throne at a time when the French were in political disarray. His capture of Harfleur and march towards Calais, while a much larger French army closed in on his emaciated force, culminated in the victory at Agincourt, where the French showed they had learnt nothing from the past. In 30 minutes, on a frontage of 900 yards, 5,000 archers, backed by 900 armoured infantry, slaughtered 6,000 out of 25,000 Frenchmen.

In the subsequent expulsion of the British from France, highlighted by the raising of the Siege of Orléans in 1429 by the inspirational Jeanne d'Arc and the rise in French

morale, it was the decline in British leadership and the decisive part played by the French artillery arm under Jean **Bureau** which were of greatest significance. Time and again the British were outgeneralled by leaders such as Bertrand Duguesclin and Arthur de Richemont. In the culminating siege and Battle of Castillon it was the concentrated fire of, perhaps, 200 to 300 guns and the original tactics of their commander Jean Bureau which settled the issue. The war was brought to an end by the routing of the relieving army under the redoubtable John Talbot, Earl of Shrewsbury.

Hydrogen bomb. The idea for a thermonuclear bomb, based on the fusion of deuterium, was first mooted in 1942 but, though work was done to show the idea was feasible (the US Super Project), there was no further development till 1950, when President Truman authorized work to proceed. It was decided to use tritium, an isotope of hydrogen, mixed with deuterium to keep down the otherwise very high ignition temperature: the name 'hydrogen' bomb stems from this 'burning' of the hydrogen isotope. On 1 November 1952 the first test device of between four and eight megatonnes was exploded in the Pacific, obliterating a small island and leaving a mile-wide crater. The device relied on an atom-bomb trigger, releasing neutrons to bombard a lithium liner which in turn started the fusion process generated by the reaction of the tritium/deuterium. The device had a uranium jacket and it was the fission of this, from the neutrons emitted by the reaction, that produced the radioactive fission materials.

Practical warheads and **bombs** soon followed the early tests and hydrogen bombs now form a significant proportion of the major powers' nuclear armoury.

Hydrography, the scientific study of the seas, lakes and rivers, became vitally important as an adjunct to **navigation** when seafarers began voyaging beyond familiar waters. Early ocean navigators, such as the Chinese, Phoenicians and **Vikings**, began keeping records (often secretly) of prevalent winds, currents, sea depths and bottoms, coastlines and harbour entrances which, over the years, were transferred to **charts and maps**. Before the 15th century Arab traders were to the fore in ocean survey. But it was navies which, in the interests of warlike operations, took the lead in the gathering and recording of hydrographic detail. In 1720 the French Navy established a Hydrographic Department to collate, update and publish a mass of information, followed belatedly by Denmark in 1784, Britain in 1795, Spain in 1800, Russia in 1827, the USA in 1830 and others in the 19th century. An international organization was established in 1921.

Nevertheless, despite some assistance from aerial survey, there remained many poorly charted and recorded areas up to **World War II**, as those engaged in **amphibious** operations soon discovered. Clandestine inspections of beaches and soundings of approaches were therefore undertaken by such organizations as the British COPP (Combined Operations Pilotage Parties) before raids and major landings. Operations were intensified and sophisticated by the use of **sonar** during the **Cold War** and as **submarine** warfare demanded new information about the profoundest depths for which, previously, there had been no requirement. These investigations have provided information of greatest value to scientists, fishermen and in various commercial activities, such as the search for minerals.

Hydrophone. The invention by Alexander Bell of the **telephone** in 1876 brought with it the electromagnetic loose-contact microphone (the word coined by David Hughes in 1878) and subsequently improved upon it in many ways for speech **radio**. In 1914 the Royal Navy experimented with so-called hydrophones (simple, waterproofed microphones) on the seabed and in ships as part of a passive **anti-submarine weapon** system. But they were found unreliable and in due course active **sonar** proved a far more effective detector. Nevertheless elaborate hydrophone arrays can and do secretly monitor the passage of ships and submarines in strategic waterways. (*See also* **Sound Locators**.)

I

Identification Friend or Foe (IFF). In the heat of battle there is always the chance of friend harming friend and therefore a need for some form of quick recognition or challenge by password. In hand-to-hand combat, especially between dense **infantry** and **cavalry** formations this was rare. But stand-off weapons such as the **bow and arrow** and firearms significantly added to the chances of error.

On land IFF has usually depended upon the wearing of distinctive **armour** and uniform **clothing**, notably in the first instance by the Greeks and Romans. In regular armies helmets often provided the best recognition features until the increased adoption in the 17th century of brightly coloured tunics: reds, blues, whites, yellows etc., a fashion which was abandoned in the 19th century as the need for camouflage to reduce casualties in face of heavy fire became paramount. **Guerrilla** bands usually preferred to avoid identification and often, contrary to custom or international Conventions, wore nothing plainer than a removable badge or armband.

As the range of **artillery** increased the chances of error multiplied; while **mechanization** and the invention of **armoured fighting vehicles** introduced IFF problems with vehicles of similar configuration. To avoid accidents painted symbols or pennants were sometimes carried, but mostly there was an optimistic dependence upon recognition training.

At sea, as on land, there was also a reliance on ship recognition, colour schemes, signalling challenges and the display of ensigns, always with the possibility of deceit by captured vessels flying false colours.

In the air, too, recognition, colour schemes and markings were the normal IFF until the arrival of **radio** and **radar** held out a promise of IFF by **electronic** means. The first example of this was the fitting by the British of a discrete pulse repeater in friendly aircraft which would differentiate friend from foe on a ground controller's visual display unit. Many sophisticated variations have followed but, like most electronic devices, they are vulnerable to exploitation by the enemy.

Image intensifiers (II) were developed to satisfy the need for a **night-vision device** which would help the human eye to make use, passively, of ambient light in the dark. Early II devices enhanced light by 10^5 or more through a bulky three-stage 'cascade' cathode-ray tube with a large-diameter lens to gather maximum light. Miniature 'channel' tubes have since been developed for goggles worn by **helicopter** crews and **infantry**.

Independent Air Force (IAF). In October 1917 the British created 41st Wing in France to bomb industrial targets in Germany as a reprisal for air raids against England. Up to 6 June 1918 its light and heavy **bomber** squadrons carried out 55 day and 87 night raids (not all on Germany) with heavy losses and only minor effect. Then, reinforced, renamed the IAF and put under the command of **General Hugh Trenchard** it prosecuted this experiment in strategic bombing (*see* **Bombing, Strategic**) until the end of **World War I**, but without decisive results from 205 day and 373 night raids, the dropping of 665 tonnes of **bombs** and the loss of 458 aircraft.

Industrial development. From prehistory man's survival has depended upon some form of industry; growth activities which, initially, were designed for the procurement and processing of food, and for weapons for hunting and defence. These enterprises, as time went by and particularly with the 18th-century Industrial Revolution, tended to conglomerate and therefore become of vital military as well as economic **strategic** importance. *See*, therefore, **Armaments industry** and related entries, especially **Aircraft construction industry; Dockyards and naval bases; Electronics; Explosives; Food, Military; Gas, Industrial; Mechanization, Military; Metallurgy; Motor vehicles; Nuclear energy and weapons; Ordnance services; Power-generation equipment; Propulsion, Means of; Research and development; Shipbuilding, Naval; Technology, Effects of; Training, Military and technical.**

Infantry are soldiers who fight on foot (or from within **fortifications**), no matter if they are transported into battle by **ship**, **horse**, **chariot**, **motor vehicle**, **tank** or **aircraft**. At first, armed with clubs and **edged weapons**, they have evolved in striking power and personal protection under the impulse of technical effects (**Technology, Effects of**) and to satisfy current **tactical** requirements. After the Assyrian soldiers, armed with **swords**, short **spears**, **bows** and wicker shields, came the Greek hoplites, with longer spears and bronze **armour**, fighting in the compact, close-ranked phalanx, followed by the even better-armed and more flexibly effective Roman legion.

Infantry's survival in face of the threat from **cavalry** and other more mobile, shock arms has depended on its capability to project sufficient fire-power, commensurate with its ability to hold ground, without being destroyed *in situ* by rapid-fire weapons such as bows and arrows, small firearms and **artillery**. Failing that, it has had to rely on mobility to take evasive action, to disperse and, when the time was ripe, concentrate again at some vital spot. To act, in fact, like **guerrillas** and **Commandos**. **Viking** raiders were little more than **amphibious** guerrillas, Hun and **Mongol** cavalry, when dismounted, simply a more flexible and highly mobile infantry formation. And the lightly armoured British men-at-arms and longbowmen of the **Hundred Years War** were a fine example of self-contained infantry formation, with integrated weapons capable of hitting the enemy at long range while holding prepared positions or during the assault on fortifications.

Roman infantryman

The development of hand-held firearms from the 16th century onwards simply supplied weapons for more sophisticated versions of the bow-and-arrow/man-at-arms organization. But because the early **muskets** and **pistols** were incapable of developing the longbow's volume of accurate fire, it was found necessary to introduce **pikemen** (modernized spearmen) and mix them with musketeers. The measure became unnecessary in the late 17th century with the introduction of the flintlock **musket** and the **bayonet**.

Increasingly, from the 17th century onwards, infantry had to take account of the greater destructive power of the **grenade** (with the creation of élite grenadiers who often led the assault on fortifications), and of **artillery**; as well as the fire-power of enemy infantry and cavalry formations, as demonstrated in the **Thirty Years War** by **Gustavus Adolphus**. More and more the construction of protective breastworks and digging of trenches (in addition to formal **siege** works) became crucial; the Battle of **Malplaquet** in 1709 demonstrated, for example, how important were breastworks.

Nevertheless, close-ranked formations prevailed until the **Crimean War** in the 1850s, when the latest muskets and **rifles** made it possible for a thin line of infantry, without artillery support and deploying into square, to defeat cavalry. This was when the **empty-battlefield theory** emerged, as combat arms not only thinned out but, for survival, were compelled to disappear below ground in deeply entrenched positions.

The **American Civil War** and the Boer War indicated not only the wane of cavalry as the arm of shock action, but also the feasibility of mounted infantry. Vital as it was (and is) for infantry to be fit and capable of marching long distances, they always rode when they could. They were thus to be found in armoured **railway** trains, so it was logical for **armoured fighting vehicles** to appear within a few years of the invention of the petrol engine. And equally logical that, when infantry were massacred and bogged down in **World War I**, the suggestion they should travel inside **tanks** was acted upon. Infantry transport has been developed ever since, in the shape of the specialized armoured personnel carrier, lorries and **helicopters**, besides motor cycles.

Since the invention of the **machine-gun** and the trench **mortar**, infantry have been forced to adopt the first for close combat and both to give longer-range, heavy-fire support for movement and in static defence. Likewise, the appearance of the tank at once made infantry call for their own **anti-tank weapons**; while the latest signals **communication** systems, particularly **radio**, were adopted once weight and reliability had been got right. This accumulation of technology has turned some infantry into all-arms units with sophisticated **tactics**, besides calling for skills far beyond those of the simple marching rifleman and grenadier. By the beginning of **World War II** the average infantry unit was packed with specialists whose average intelligence ratings had to be much higher than their predecessors', a trend which made it necessary for units, basic rifle companies each of three or four platoons, to acquire an additional special support company, as well as a **logistics** company which cared for transport, supply and administration.

Infantryman, c. 1942 (*German, equipped with sub-machine-gun and, on hip, respirator*)

World War II also brought forth any number of élite infantry units, such as **airborne** and the Commando, which were often criticized for swallowing the best men to the detriment of leadership in ordinary units. Be that as it might, wise commanders have learnt to conserve their infantry in the knowledge that they are indispensable in most circumstances and terrain and all too easily squandered through their vulnerability, no matter how good their fieldcraft, their accurate fire in support of fluent movement or their endurance in exacting conditions. Without infantry it is almost impossible to operate efficiently in tank-proof terrain, such as **mountains**, swamps, woods and built-up areas. On the other hand, infantry depend greatly upon artillery, tanks and engineers to operate almost anywhere and are at a distinct disadvantage in wide open **desert** or steppe which lack natural barriers.

Finally, it should be remembered that **guerrillas** and armed civilians, as well as any sailors, soldiers in general or airmen if so redirected are potential infantrymen.

Infrared (IR) devices. The discovery by Heinrich **Hertz** and others in the late 19th century of the **electromagnetic** spectrum of radio, thermal and light frequency bands introduced to science the existence of infrared emissions, which were invisible to the human eye, in the 10^6 to 10^8 gigahertz bands. At the higher end is the so-called Far IR band (*see* **Thermal imaging**); at the lower end, Near IR, which was first used in so-called photoelectric cells, producing 'invisible rays' for detection of moving objects. This, during **World War II**, was developed by the Germans and the Allies as **night-vision** and **navigation** devices and, later, for night **photography**.

The early night-vision equipment depended upon illumination of the target by a **searchlight** with an IR filter over the lens. The viewer, a somewhat cumbersome monocular or binocular, worked on the principle of converting the photonic flux of the incoming light into an electron flux displayed on a luminescent screen. Such devices are called active because they depend upon an external IR source. They suffer from the disadvantage that an enemy observer with an IR viewer can detect the source and take appropriate action. Nevertheless, from the 1950s until superseded by passive **image intensifiers** and thermal imagers in the 1970s, IR kits were widely used on Russian, American and British **tanks**, usually with the searchlight coupled to the gunner's and commander's sights while the driver had a system of his own.

IR devices continue in many applications, notably as **sensors** for **surveillance** on land and from **space**.

Intelligence gathering. In **Sun Tzu**'s celebrated *Art of War* (*c.* 500 BC) he laid emphasis on the need for spies from the local populace, the enemy's officials, double agents and agents who managed to escape from the enemy encampment. Since the first authenticated Sumerian handwriting, dating from *c.* 3100 BC, a quantity of verbal, written and encoded intelligence (*see* **Codes and ciphers**) has been gathered. The sources and form of reports available to political and military leaders changed but little for many centuries; but the quantity and speed of dissemination increased with the gradual spread of literacy, with better **charts and maps** and signal **communications** systems, which were linked to the conquest of the oceans, exploration and the opening up of land routes.

By the time of the **wars of the 18th century**, and notably the voluminous and accurate intelligence gathered by the Duke of **Marlborough** during the Wars of the Spanish Succession, elaborate and fundamental techniques had been developed in European theatres of war. He and his successors were well briefed as to geography, local economies, political situations, state of forces and commanders' intentions. The most successful operations nearly always benefited significantly from foresighted provision of finance to create the sources and organizations needed for the discovery, synthesis and dissemination of information. The worst disasters often were the result of false or inadequate intelligence or counter-intelligence.

The types of intelligence are many and various: political, economic, **strategic, tactical, logistic, industrial** and so on. Their variety and complexity have increased in pace with the advance of technology, particularly **electronics**. Since the beginning of the

Industrial Revolution in the 18th century, intelligence-gathering has had an ever-increasing effect on commercial as well as military developments. It is as helpful for a manufacturer as it is for a government or military commander to obtain advance knowledge of a competitor's future products and weapons. Savings in time, effort and money with **research and development** to counter, for example, a new type of sea-**mine** and also design an improved model is then possible.

Covert sources include spies (agents of variable trustworthiness who are not nearly so romantic as often portrayed), patrols, decrypted codes and ciphers, intercepted mail and **telephone** messages, and the pattern and volume of traffic using communication networks. Overt sources include reports in journals, ground and aerial (including **space**) **photographic reconnaissance** missions, plain-language messages, industrial catalogues, transport timetables, overheard careless conversations and disclosures during routine diplomatic exchanges. The establishment of reliable sources and the collection of data are time-consuming and complex, involving much subterfuge. So too is the transmission of information, which is liable to interception by hostile counter-intelligence organizations. Double agents, deception from the planting of spurious material and the inadvertent disclosure of sources are among the perils involved when setting up channels of communication. Many a thriving network has been shut down or destroyed because of organizational or technical defects, an astute enemy, carelessness or sheer bad luck.

The synthesis of information received is, of course, vital and dependent upon expert recognition of the evidence's truth and falsehood (duly corroborated), the rejection or filing of dubious clues, and the presentation of clear, concise reports and summaries sufficient for assured planning of operations. **Computers** help accelerate the analysis of vast volumes of material and avoid overlooking relevant material. Today, as in the past, the most disconcerting errors can be made in the transmission of crucial information, as for example when a vital report about German night dispositions was not sent to Admiral **Jellicoe** at **Jutland** in 1916 and made all the difference to his chances of victory; or the inadequate exploitation and interpretation of available political intelligence, which deluded the Germans as to Russian vulnerability in **World War II**; or the misinterpretation of information about the movements of the battleship *Tirpitz*, which led to disaster for **Convoy** PQ17 in 1942. Errors such as these were mainly due to typical human and organizational breakdowns, which delayed reports, induced incompetent synthesis, allowed poor presentation or highlighted commanders' unwillingness or inability to credit the evidence and the conclusions drawn from what, to the uninitiated, could appear as a black art. Indeed, the secretiveness which is central to the functioning of **secret services** and associated, numerous intelligence agencies, is itself an enemy of credibility. As a major example, the vital security needed strictly to limit the access of Allied commanders in **World War II** to awareness of the deep penetration of high-grade enemy codes and ciphers prevented those in the know emphasizing to subordinates, who were not in the secret, the authenticity of startling news – a dilemma which led to misunderstandings over **Pearl Harbor** in 1941 and many another incident.

Intelligence of quality is vital to economy of effort and national survival. It is dearly bought or obtained, potential political dynamite (particularly during **cold wars**) and vulnerable to misinterpretation. It can be abused and misused, but ignored at peril. Well organized and run, it pays dividends beyond price. There is a vast **bibliography**, much of which, owing to inherent secrecies and deceptions, contains faults and misconceptions of the sort common to intelligence matters. Instances of intelligence effect will be found among many entries in this Encyclopedia.

Iran–Iraq War. Disputes between Iran (Persia) and her Muslim neighbours reach far back in history. Border incidents in the 1960s owing to Iraq taking action against dissident Kurds in the north were only stifled

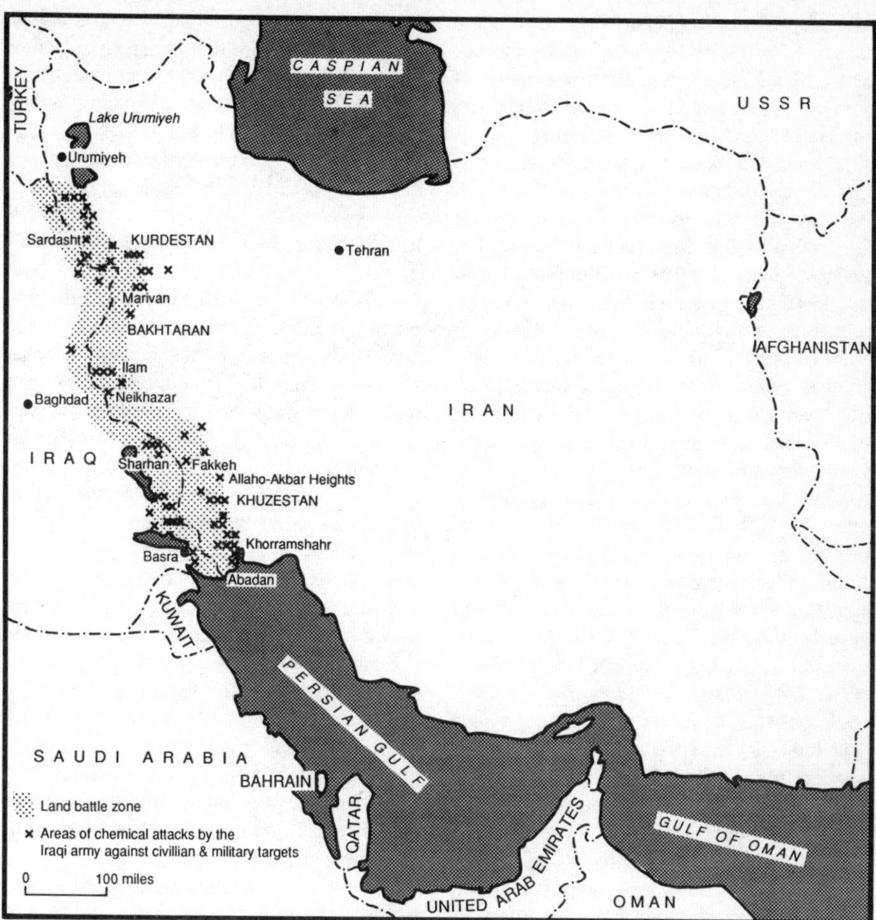

The Iran–Iraq War, 1980–88

by agreements in Iranian favour in 1975. The chaos in the aftermath of the Shah of Iran's deposition in 1979 gave Iraq under President Saddam Hussein an opportunity to reassert her authority and also gain possession of the important waters of the Shatt al Arab at the head of the Persian Gulf. With armed forces roughly equal in number in September 1980, Iraq (mainly equipped by Russia) invaded Iran (mainly equipped by the USA and the UK), whose forces had lost some 60 per cent effectiveness with the Revolution.

Initial thrusts towards strategic and economically vital Abadan and Khorramshahr misfired in bad weather against unexpected resistance. Iranians, inspired by the religious zeal of Ayatollah Khomeini's followers, mounted suicidal counter-attacks which gradually produced stalemate along the length of the 750-mile frontier. Iran continued to attack, despite a debilitating equipment poverty following arms embargoes by America and Britain. She traded men's lives in vast quantity to wear down the numerically smaller Iraqi population. Besieged Khorramshahr was relieved in May 1982 as the Iraqis, facing exhaustion, were forced back across their own border and as Basra came under threat. When Iraq, in effect, sued for peace, Iran demanded the overthrow of the Iraq government as a condition. Iraq reverted to a sullen defensive posture.

Iran went to great lengths to obtain arms in order to prosecute a war of attrition which cost an immense number of lives (precise figures have not been substantiated). Between September 1981 and January 1984, 13 offensives, of which five were at corps and four at divisional strength, had been counted, with claims of over half a million Iranian and 250,000 Iraqi casualties. By then gas, supplied by Russia and dispensed by Russian weapon systems, had been used for the first time since 1936 in Ethiopia. The collapse in February 1984 of a promising attack by 100,000 Iranians to capture Basra, was put down largely to the use of **chemical weapons** – mustard and nerve gases – which contributed to loss of **morale** and 40,000 casualties to virtually unprotected troops. Immediately Iran was compelled to purchase protective clothing and attack in the knowledge that the slightest success on their part was liable to provoke chemical attack. In 1985 further disastrous repulses occurred as Iran's recruiting fell off from lack of enthusiasm. Even the zealots could only raise 12 per cent volunteers for the slaughter. In 1986, however, there was an improvement, once protective clothing was widely issued, and attacks made ground. Again Iraqis had their backs to the wall as Iranian morale improved.

Yet stalemate on land followed in 1987 and 1988. Iraq opened air attacks on **oil** installations as Iran began pinprick attacks against and mining (**Mines, Naval**) of international shipping in the Gulf. Iran, realizing she could not break the enemy, went in for limited ground attacks, **rocket** bombardment of cities and **guerrilla warfare**, with armed Kurdish collaboration, threatening Iraq's oilfields. Iraq responded in kind and killed thousands of unprotected Kurds caught down wind of a gas attack. All at once in 1988, it seems, Iran's military morale momentarily collapsed. Her forces withdrew with heavy losses when Iraqi troops advanced on a wide front across the frontier, taking vengeance on the Kurds with massacre by gas attacks which drove the survivors into Turkey, where they were disarmed. In May 1988 Iran took advantage of UN mediation to accept an armistice and a UN supervisory presence.

Iraq–UN War. *See* **UN–Iraq War**.

Iron. Iron smelting is the process of intensely heating a mixture of iron ore and charcoal until it becomes an incandescent sponge, separated from the residual clay and ash which form sinter or slag. It has been suggested that smelting occurred by accident in many different parts of the world, *c.* 3500 BC, when **wood** fires were lit in the vicinity of iron ore (sometimes known as red-paint rock). A science of iron **metallurgy** existed, it seems, by *c.* 1500 BC as, gradually (by AD *c.* 400), this precious metal supplanted **bronze** for weapons and **armour**.

Refinement by the removal of slag and other impurities is the key to production of iron or **steel**. Initially this was done by hammering the heated sponge to produce wrought iron in tough, malleable form suitable for forging or rolling. But always there was a tendency to buckle under stress and deteriorate by rusting.

The early laborious methods could manufacture only small quantities of metal at very high cost. It was rising demand for improved weapons and armour which sponsored the invention of the more sophisticated, cheaper, mass-production casting (*c.* 1000 BC) and refinement by air blast furnace in AD 1380. And then, in the latter half of the 19th century (owing to iron's propensity to shatter when struck hard) the need for very dense and tough armour to resist projectiles from more powerful naval **artillery** called for special armoured steel. Robert **Hadfield**'s invention of manganese-steel armour plate in 1882 was revolutionary. Iron armour and armour-piercing shot became obsolete by the end of the 19th century, and soon too was decreasingly used for **bombs**, shells and other projectiles.

J

Jackson, Admiral Sir Henry (1855–1929). A Royal Navy specialist in **navigation** who in 1881 turned to **torpedoes** and, in 1892, began experiments with Hertzian waves (*see* **Hertz, Heinrich**). In 1895 he transmitted a 'spark' **radio** signal from one end of a ship to the other and then one ship to another. After he met Guglielmo **Marconi** in 1896 and discovered they were working on the same lines, they collaborated. But in 1900 it was Marconi's instead of Jackson's more advanced radio set which the Admiralty adopted.

In 1905 he became Third Sea Lord and thereby, in a technologically revolutionary period, controller of the building of **dreadnought battleships** and **battle-cruisers** as well as the development of **submarines**, torpedoes, radio, and aerial navigation. He founded the Portsmouth War College in 1911, became Chief of War Staff in 1913 and was First Sea Lord from May 1915 to December 1916, and thus in charge at the Admiralty during the Battle of **Jutland**. After **World War I** he became Chairman of the Radio Research Board and deeply involved with experiments concerning the ionosphere (the Heaviside layer).

Japan, Battle of. During **World War II** the first direct attack on Japan took place on 18 April 1942 when 16 B25 Mitchell medium bombers, launched from the US **aircraft-carrier** *Hornet*, bombed Tokyo. Material damage was slight, but the impact on Japanese and American morale considerable.

From the outset the Americans determined to wreck Japan with a campaign of strategic bombing (*see* **Bombing, Strategic**), using their chosen instrument, the B29 **bomber**, with its 30,000ft ceiling, 3,250 miles range and 20,000lb maximum bomb-load. It was intended to base the aircraft at Chengtu, in China, and on Saipan Island in the Marianas (once it had been secured) which, respectively, were 1,550 and 1,400 miles from Japan. The Chengtu B29s opened their attacks on Japan with a night raid on 15 June 1944, but in the months to come would also hit targets in Manchuria, Formosa and Hangchow. Overall they were not a success, bedevilled as they were by target-finding difficulties, frequent engine failure and **logistic** difficulties which reduced the number of missions. Losses also were heavy – about 18 per cent, including a few from deliberate ramming.

Operations from Saipan began on 24 November, aimed at the Japanese aircraft industry. They were only moderately successful even though the Japanese air defences, lacking **radar**, made only a patchy impression. But so effective were seven Japanese air attacks from Iwo Jima on Saipan's airfields that occasionally the B29s were diverted to attack that target as a defensive measure. Gradually the Americans improved their techniques, but the most damaging raid prior to 9 March 1945 was by naval aircraft-carrier bombers against the Tokyo area on 16 and 17 February, which destroyed 200 Japanese aircraft against the loss of 60 machines to much improved Japanese air defences. On 9 March, however, the B29s began night area attacks with incendiary bombs on Tokyo and other major cities and with devastating results. Several B29s were lost to gunfire but the Japanese night fighters made no interceptions. A month later, once Iwo Jima began operating fighters, the offensive was pressed home also by day, supplemented by carrier-borne aircraft and, towards the end, by medium bombers from Okinawa. They laid waste to Japan as her defences, deprived of sufficient fuel for outclassed fighters, crumbled. The dropping of **atom bombs** on Hiroshima and Nagasaki in August merely underlined the hopelessness of Japan's condition and made it easier for her to sue for peace.

Jeep. Supposedly originating from 'GP' (General Purpose), the word has passed into the language as describing any small four-wheel-drive vehicle. The Jeep emerged as the result of a US Army requirement in 1940 for a small vehicle seating three passengers and with room for a 0·30in machine-gun mount. Willys-Overland won the contract, producing a four-seat vehicle which saw service in all theatres of **World War II**, being used for **reconnaissance**, raiding, casualty evacuation, signals and many other tasks. Some 600,000 were built by 1945 and no subsequent equivalent has quite matched the simple ruggedness of this unique vehicle.

Jellicoe, Admiral Sir (later **Earl**) **John** (1859–1935). A Royal Navy **gunnery** specialist who played an important role in developing long-range shooting at sea. Shortly after the outbreak of **World War I** he assumed command of the Grand Fleet, which he led at the Battle of **Jutland**. It was his misfortune that, having brilliantly lured the German High Seas Fleet into a deadly trap, he was denied the fruits of victory by inadequate transmission of vital **intelligence** and the shortcomings of his fleet in relaying signals and in **night** fighting. Made First Sea Lord at the end of 1916, he obdurately refused, against the evidence of operational analysis (*see* **Analysis, Operational**), to institute the **convoy** system to combat the **submarine** threat. He had to be overruled by the Prime Minister, who removed him from office at the end of 1917.

Jet engines function by sucking in air from the atmosphere into a centrifugal or axial combustion chamber, where fuel is injected and ignited; the hot gas then being blown at high pressure by a **turbine** through a nozzle to produce thrust as well as drive the compressor (*see also* **Propulsion, Means of** and **Rockets**). The idea was conceived in the 1920s by Frank **Whittle** and shortly afterwards (and quite independently) by Hans von Ohain. Both initially opted for a reliable centrifugal compressor but von Ohain, who was well funded in Germany, achieved the first jet flight in August 1939 with the axial Junkers 004 engine in a Heinkel He178; while the underfunded Whittle had to wait until May 1941 before his centrifugal Rolls-Royce engine flew in the Gloster E1.

The high thrust-to-weight ratio of the jet engine (Junkers 004 1,980/1,630lb, Rolls-Royce 850/623lb) gave it an enormous advantage over more complicated propeller reciprocating engines, in addition to providing an aerodynamically cleaner shape and greater operational efficiency at higher altitudes. First mounted in **fighter** and light **bomber aircraft**, the jet's revolutionary potential to raise the performance of heavy bombers and large **transport aircraft** was at once realized, even if the problem of high fuel consumption at lower altitudes was also recognized. But during **World War II**, from 1944 onwards, it was in British and German fighters that they made their mark – and might have changed the course of the air war had the Germans not squandered their initial lead.

After the war the development of the *pure jet* went ahead in parallel with the *turbo-propeller* engine. The latter made use of excess power to drive a propeller, a more efficient and economical arrangement at speeds below 500mph at lower altitudes – but with far fewer military attractions, except in **helicopters** to drive the rotors.

Military development naturally concentrated on very high thrust to achieve greater speeds and load-carrying. The axial-flow engine gradually superseded the centrifugal type. Size was increased for larger air intake to satisfy multi-stage compressors, whose casings and turbine blades had to withstand operating temperatures in the region of 3,000 degrees Celsius, a requirement which called for very advanced **metallurgy**, new lightweight **steel** alloys and, in due course, man-made materials.

The demand for additional power in combat aircraft for short and uneconomical bursts of power led to the *afterburner* – a device to increase thrust by 50 per cent by raising the temperature of the gas to about 4,000 degrees Celsius in a second combustion chamber situated before the nozzle – producing as much as 27,000lb thrust. While, to increase payload for transport aircraft, the ducted *turbofan* engine with a fan driven by the turbine to improve efficiency at take-off was developed.

Additionally, and for military use only, there appeared the *pulse jet* used in the German V1 flying bombs – a very noisy but simple device in which fuel and air were mixed in a combustion chamber and the gases fed straight to a nozzle. And the very simple *ram jet*, a jet engine without a compressor which depends upon high speed to function and therefore needs assisted take-off. It was used in the temperamental German Messerschmitt Me163 fighter and a few air-launched guided missiles.

Joffre, Marshal Joseph (1852–1931). A Sapper who took part in the Siege of **Paris** prior to serving in many parts of the French Empire. In 1911 he was made Chief of Staff over the heads of others because he was rated a safe Republican. He was responsible therefore for the faulty Plan XVII adopted at the outset of **World War I**, and when that was defeated magnificently rallied the Allied Armies to win the Battle of the **Marne**. When the stalemate of **siege** and trench warfare set in along the Western Front in October 1914 he was tasked to clear France of the Germans, but the job proved beyond him with the resources at his disposal. Only his high prestige kept him in place. He was none too receptive to such new ideas as **tanks**, preferring the traditional use of firepower and sapping to achieve a breakthrough – and exhausting the French Army in the process. As the enormous casualty list increased throughout 1915 and 1916 his authority waned, and he was replaced in December 1916.

Jomini, General Baron Antoine de (1779–1869). A bank clerk whose *Traité de grande tactique* concerning **Frederick the Great** drew him to the the attention of **Napoleon**. As a senior staff officer he was at many battles, including Jena and the advance to **Moscow** in 1812, before deserting to the Russians in 1813 and taking part on their side in the Battle of **Leipzig**. Meanwhile he had begun his *Histoire critique et militaire des guerres de la Révolution* which became the foundation of his celebrated *Précis de l'art de la guerre* (1838) which succinctly analysed **strategy**, **tactics**, **logistics**, **engineering** and diplomacy and clearly foresaw the influence of burgeoning technology and of **railways** on future war.

Jones, Professor Reginald Victor (b. 1911) whose research with **infrared** detection and **electronics** led him into **radar**. At the outbreak of **World War II** he became involved in scientific **intelligence** for the Air Ministry and was crucially responsible for detecting German navigational **beams** and devising countermeasures. Appointed Assistant Director of Scientific Intelligence, he was engrossed with **air warfare**, the **bomber** offensive and the struggle against the German V weapons. After 1945 he continued as a government scientific adviser in various important capacities.

Jungle warfare. Combat techniques among dense forests are as old as mankind, and mainly have been of the essence of **guerrilla warfare** with ambush and hit-and-run raiding. Only since the Spanish–American War of 1898 has special attention been paid to it. Nevertheless, tribal conflicts and numerous small colonial campaigns had invariably emphasized the tactical problems of fighting at extremely close ranges against a completely concealed enemy; the menace to **health** in regions where disease was rampant; and the **logistic** and **engineering** difficulties, which could be crippling. Indeed, the main differences between the enervating jungle wars of the 20th century and those of the past was the increase in scale and, as a result of the latest demands of humanitarianism associated with higher valuations of life, the need to take greater care of soldiers. The American soldiers sent in haste to fight in Cuba were ill-clad for the conditions and unprotected against the ravages of malaria and yellow fever. But, as experience was gained in their successive involvements in Central America, they were better prepared to cope with the fundamental problems.

World War I provided only one major example of jungle warfare, in East Africa, where for four years the Germans waged a guerrilla war in which the jungle provided a vast hiding-place. The campaign absorbed some 160,000 British troops at a cost of about 18,000 casualties, plus some 50,000 dead black Africans at a price to the Germans of 18,000 soldiers and 7,000 labourers. In the struggle disease accounted for many more deaths than did wounds.

World War II was abundant in jungle warfare, once the Japanese expanded into South-East Asia and the Pacific islands. For an urban nation, too, they demonstrated remarkable adaptability to the conditions and tactically outmanoeuvred their opponents by outflanking movements, closely supported by engineering effort, through deep jungle. Their conquests of **Malaya**, the Dutch East Indies and **Burma** in 1942 were largely the result of these psychologically unsettling methods against opponents who were mentally unprepared for the circumstances. These threats were only thoroughly overcome when, in 1943, Allied troops had been taught how to move and live in the claustrophobic jungle; and when reliable supply by air drop into clearings did away with the need to react to outflanking – a logistic answer to a tactical problem, one that exploited a campaign of long duration in hostile terrain for which the Japanese were less well prepared medically than their enemies.

The lessons of World War II were carried forward into the **cold** and **limited wars** to come.

All the contestants who survived prolonged periods of combat in Indonesia, **Malaya**, **Vietnam**, the Philippines, East Africa and Central America drew on well-recorded past experience for survival. There are few other military activities which require more careful and thorough **training** and acclimatization.

Junkers, Professor Dr Hugo (1859–1935), was a technically educated innovator who in 1895 began developing and building gas engines at Dessau. In 1907 he collaborated in the design of a tail-first monoplane and thereafter concentrated on the design and construction of revolutionary cantilevered, thick-wing, all-steel or duralumin **aircraft** as well as Jumo diesel engines (*see* **Propulsion, Means of**). Throughout **World War I** the Junkers Company built 15 types of metal monoplane, of which only two (the J9 and J10) went into production. The J1 biplane, with an **armour**-protected engine, was the only successful combat model he built.

In 1919, however, his single-engine, six-seater F13 metal airliner was a great success and led to the tri-motor G23 airliner which eventually became the famous Ju52 **bomber/transport** of the early 1930s. In the meantime he had built airframe and engine factories in Sweden and Russia, with considerable influence on future Russian, as well as German military designs. Before his death the German factory had taken a lead in designing four-engine airliners (the G38) and dive-bombers (the Ju87) and was on the eve of involvement with the first **jet**-propelled machine.

Jutland, Battle of. The Germans' realization at the start of **World War I** that they were outmatched at sea plus the failure of their limited **submarine** campaign in 1915, pushed their naval command to the conclusion that only by a strategy of attrition might they wear down the Royal Navy and perhaps loosen the Allied **blockade**. A new C-in-C, Vice-Admiral Reinhard Scheer, planned in February 1916 to lure a portion of the British Grand Fleet into a trap against the High Seas Fleet. At the same moment Admiral **Jellicoe** was prepared to seek battle, provided the conditions were not dangerously adverse. In taking risks, Jellicoe had the advantage of superior **communications** and **intelligence**, with the capability of reading the German intentions through possession of their **codes**. Thus when German light forces, and later the High Seas Fleet, began sweeps across the North Sea. Jellicoe was always able to have overwhelming, undisclosed forces at sea to meet them – though several times missing contact.

Despite **radio** deception measures, overall British Intelligence told them the Germans were preparing for sea on 30 May, with the result that their ships left port two hours ahead of the Germans. Both sides led with their **battle-cruiser** squadrons (nine British ships against five German). When making contact in mid-afternoon 31 May, Admiral David Beatty was not aware that the High Seas Fleet was at sea and Scheer was unaware that the British were near, since his zeppelins had not spotted them and he had discounted submarine sighting reports. Admiral Hipper ran before Beatty to lure him on to the High Seas Fleet and, by superior **gunnery** with better **ammunition**, caused two battle-cruisers to blow up. At 1630hrs Beatty's scouting **cruisers** found the High Seas Fleet, causing Beatty to reverse course and speed northwards, pursued by the Germans – straight into Jellicoe's 28 **battleships**. By 1735hrs, when the Grand Fleet was almost in sight, Beatty had inflicted considerable damage on all but one of Hipper's battle-cruisers (including a **torpedo** hit from a **destroyer**). But his reporting to Jellicoe was lax and it was only at 1800hrs that Jellicoe received accurate information of the enemy, enabling him to form line and steam across

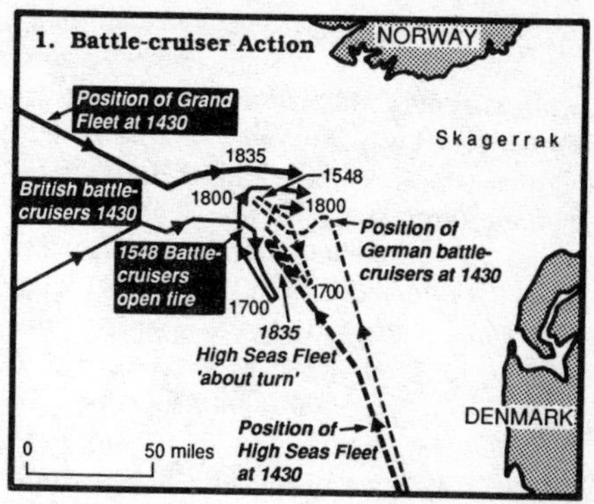

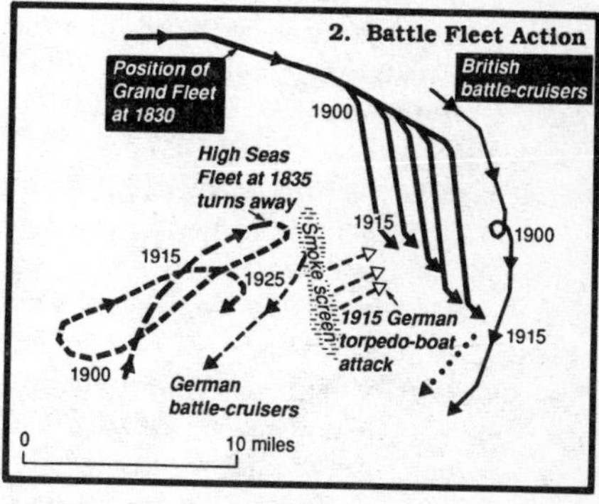

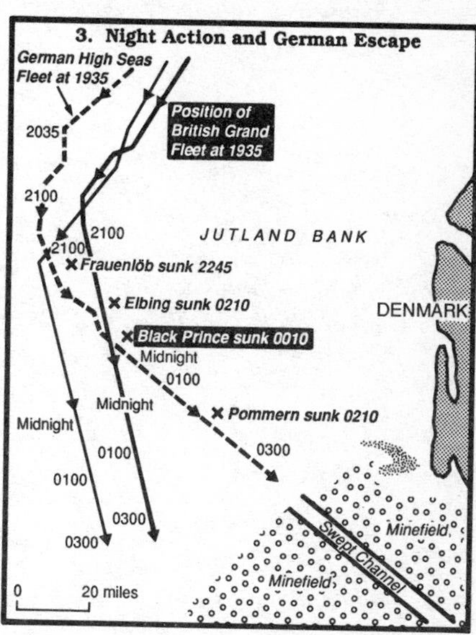

The Battle of Jutland, 31 May 1916

the approaching High Seas Fleet (crossing the T) to deluge it with shells. Scheer turned away to avoid annihilation of his 22 battleships, permitting Jellicoe to manoeuvre between the Germans and their ports as darkness drew near. When Scheer turned intuitively back against Jellicoe, to inflict further losses (including another battle-cruiser) and damage, it was only once more to run into a storm of fire from a better-placed Grand Fleet, which now had superior light conditions for gunnery. Again he was forced to turn away westwards, his retreat most valiantly covered by a charge from the four remaining battle-cruisers and by 17 destroyers. Only twice in this phase of the battle did the Germans score hits, whereas the British scored 19 as the Germans retreated into the gloom.

Jellicoe (who had not trained the Grand Fleet for night action) adopted a defensive formation, confident that the Germans had been cut off and could be overwhelmed the next day. During the night, despite numerous indications by sight and sound of engagements that the Germans were working their way past the rear of the Grand Fleet towards the safety of their minefield's swept channel, Jellicoe continued on a course which allowed Scheer to escape. In the dark only light British forces engaged. Lacking rehearsed and sound night-fighting techniques, they were fooled by the Germans (who were aware of the British signal challenges) and failed adequately to report what was happening around them. Furthermore, radio intercepts by the British Admiralty confirming Scheer's course were not passed on to Jellicoe. The result was that the Germans escaped with trivial losses.

Germany, whose losses in ratio to the British were in capital ships two to three, in cruisers four to three, and destroyers five to eight, claimed a victory, though their battle-cruisers, with one exception, were in a very poor way. It would be August before they were ready again, whereas the British had merely to refuel and rearm to be at sea to continue the unrelenting blockade. When Scheer did make further break-out attempts on 18 August and 3 November 1916, and again in April 1918, it was to no real avail, and two battle-cruisers were torpedoed by British submarines on 19 August 1916. Inactivity therefore doomed the German Navy to that steady decline in morale which ended in mutiny in November 1918, after one more foray had been mooted.

K

Kamikaze operations. The Japanese word 'kamikaze' means 'heavenly wind' (after a typhoon in 1570 which wrecked a Chinese invasion fleet). In June 1944 it described Japanese pilots recruited by Captain Jyo Eiichio to crash-dive on to enemy ships, particularly **aircraft-carriers**, as the most likely way of destroying the vastly superior Allied navies. Although the method complied with the martial spirit of *bushido*, it was not adopted by the Japanese until the aftermath of the Battle of **Leyte Gulf** – thus forfeiting a golden opportunity to inflict possibly decisive damage on the American Fleet during that crucial encounter. For when the kamikaze pilots struck at carriers on 25 October, they sank one and seriously damaged four more within four hours – a higher success rate than achieved throughout the foregoing, massed air attacks on the Seventh Fleet. In the months to come they caused immense damage to material and **morale**. By 12 December seven heavy carriers and 16 other ships were damaged and seven others sunk. On 9 January 1945 34 hits from 100 kamikaze aircraft put down one carrier and damaged four more; there were similar results off Okinawa when 355 suicide sorties launched on 6 April alone scored 28 hits and sank six ships, despite intense, improved defensive measures. To these can be added a few B29s brought down by ramming during the Battle of **Japan**. In effect, kamikazes were relatively economic human guided missiles, impervious to **electronic** jamming.

Kesselring, Field Marshal Albert (1885–1960). A brilliant artillerist and staff officer who won a reputation for colossal energy and powers of command and organization in **World War I** and afterwards was selected as a member of the élite **Truppenamt**. He played an important role in reorganizing the Reichswehr and laying the foundations of an Air Force (Luftwaffe) once Germany again was permitted one. When the secret Luftwaffe was formed it was Kesselring who, in 1933, was its Chief of Administration and who built it up until June 1936 when the Chief of Staff designate was killed in a crash. Kesselring took his place in August – 15 days after the Luftwaffe's existence was announced. It was he who shaped the new force for its wartime role as a tactical air force, shorn by him of **Douhet**'s theories and a strategic bombing (*See* **Bombing, Strategic**) role. He, too, steered it through the early days of combat involvement in the **Spanish Civil War**, before asking Hermann **Göring** for command of First Air Fleet in June 1937.

In this appointment he was responsible for the principal air effort in the Polish campaign of 1939, frequently debating matters with **Hitler**. Given command of Second Air Fleet in January 1940, he led it in the invasions of Holland and France, the Battle of **Britain** and in the first five months of the invasion of Russia – operations in which it always was the predominant air element. But on 28 November 1941 Hitler made him C-in-C South, tasked to control German interests in the **Mediterranean** theatre where the war was not progressing in the Axis' favour. Henceforward his difficult task was to supply the **North Africa** battles and attempt by diplomacy to resolve German–Italian wrangles. In 1943 he increasingly took control in the withdrawal from Africa and the defence of Sicily and mainland Italy, until given overall command in Italy in October, where he was to stay until appointed Commander Army Group West in March 1945 as the war drew to a close. He was made titular head of state in the south when Hitler was cornered in Berlin.

Tried for ordering the execution of partisans in Italy in 1944, the sentence of death was commuted to imprisonment. He was released in 1952.

Keyes, Admiral Sir Roger (later **Lord**) (1872–1945). A pugnacious sailor who won fame during the Boxer Rebellion in 1900 and, in **World War I**, as naval chief of staff at **Gallipoli** in 1915. He planned the spectacular but unsuccessful **amphibious** raids against Zeebrugge and Ostend in 1918 and retired in 1931. But, through Winston **Churchill**, he engineered his recall in 1940 to become Director of Combined Operations, charged with raiding the enemy coast after the collapse of France and developing amphibious forces and techniques for the future invasion of the Continent. The foundations he laid, including the building of **landing-craft and ships** and the formation of **airborne forces**, were far-sighted and of immense importance. He also saved the **Commandos** from disbandment. But there were few raids and he made many enemies until superseded in October 1941 by Admiral Lord Louis **Mountbatten**.

King, Admiral Ernest (1878–1956). A US Navy officer who was rather more distinguished for his anglophobia than his intellect. As C-in-C Atlantic Fleet prior to **Pearl Harbor**, he was responsible for defence against German **submarines** when the USA was involved in the Battle of the **Atlantic**. His unwillingness to adopt the **convoy** system, a refusal he persisted in after becoming Chief of Naval Staff in December 1941, had disastrous consequences for Allied shipping in the West Atlantic. A bully whom President Roosevelt treated with kid gloves, King fought tooth and nail against the US Army and the British to obtain priority for Navy operations in the **Pacific Ocean** campaign. To his credit goes the decision to invade Guadalcanal 'on a shoestring' in August 1942.

Kitchener, Field Marshal (Lord) Horatio (1850–1916). A Sapper who won praise for his performance during the Sudan Campaigns of 1884–5 and fame as C-in-C the Egyptian Army in the eventual crushing of the Mahdi's forces at Omdurman in 1898. As Chief of Staff and, subsequently, C-in-C British Army in South Africa during the Boer War, he won notoriety for an obdurate ruthlessness, as demonstrated by his ill-considered and disastrous overturning of the **logistics** organization, and the construction of badly administered concentration camps as a counter-**guerrilla** measure which so antagonized public and Boer opinion. His dictatorial behaviour nevertheless found favour in Britain. It was to popular acclaim that he was made Minister for War in 1914 on the outbreak of **World War I**.

Characteristically he upset current War Office planning by insisting the war would not be short but of at least three years' duration; and refusing to make use of the existing reserve force Territorial Army organization as the foundation of the great citizens' army he demanded – without careful study. Indeed he assumed dictatorial, overcentralized powers not only to direct military **strategy**, but also for the **mobilization** of the nation's industrial resources. The results were states of administrative, equipment and training chaos which were long in being overcome. Nevertheless, such was the people's adulation for this politically naïve man, but whose leadership of recruitment and national purpose was dynamic, that it was politically impossible to remove him. Instead the government sent him on a fact-finding journey to **Gallipoli** in May 1915 to advise on continuance of a campaign he disapproved of. It took the opportunity in his absence to strip him of many functions and much power, stopping short of sacking him. On 5 June 1916 he was drowned in a mined cruiser on a mission designed to help Russia out of her difficulties.

Korean wars. Korea has had her share of conflict throughout the ages. Her involvement with higher military technology began when she was caught in the crossfire of Japanese and Chinese rivalry and thus became, after occupation by Japan in 1894, a battleground in the Chinese–Japanese War of 1894–5. Likewise she was marched over during the **Russo-Japanese War** of 1904–5 and finished up under the Japanese heel until the close of **World War II**.

After she was overrun in 1945 (up to the 38th Parallel) by victorious Russian troops from Manchuria, she was partitioned into two zones. To the south US forces took the surrender of the Japanese; to the north the Russians, who converted their zone into a Communist state with Pyongyang as capital. On 15 August 1947 the South declared itself a republic, with Seoul as capital. No sooner had the US Army withdrawn in 1949 than the North exerted political, **propaganda** and **guerrilla** pressure upon the South, a campaign which culminated in a full-scale, Russian-equipped invasion with seven infantry divisions and one armoured brigade of the North Korea Army (NKA) on 25 June 1950.

The Republic of Korea's (ROK) Army was newly formed, under-equipped and inadequately trained. The US troops who, under a United Nations resolution, began arriving piecemeal by air and sea on 30 June were unready for immediate combat. Seoul had fallen on the 28th when the ROK Army withdrew in confusion. US troops came into action on 5 July near Osan and were almost annihilated when ROK troops gave way. There followed a none-too-orderly retreat towards the port of Pusan. On 7 July General **MacArthur** was appointed C-in-C United Nations Command and Eighth Army, consisting of American and ROK troops, plus a British brigade, was formed under General Walton Walker. The farther North Korean forces advanced the stronger the resistance they met from air attacks and the continually reinforced UN armies, until stalemate occurred around the Pusan perimeter.

Then came the Inchon landing in September, the UN advance to the Yalu River, the entry of China into the war in November and the UN retreat to the 38th Parallel, along with the epic Battle of Hungnam and the evacuation from Wonsan in December. In this period the UN forces again plumbed depths of demoralization, but in January 1951, when they bent once more under the weight of a renewed Chinese offensive and again lost Seoul, they reasserted themselves. Under General Matthew Ridgway they checked and threw back, almost to the 38th Parallel, a very brave opponent, who lacked sophisticated weapons and techniques.

Once more, on 22 April, the Chinese tried to break through on the west flank, but this time the UN forces had their enemy's measure and, except where a ROK division broke and the British 29th Brigade narrowly was saved from engulfment as it stolidly filled the gap, yielded ground grudgingly. Fighting continued across the front. On the east flank on 15 May, a ROK corps cracked but American troops restored the position by the 22nd and the UN launched a general counterstroke next day to roll the Chinese back to their start lines where both sides entrenched. The war's mobile phase was over. Until an armistice was arranged on 27 July 1953 the two sides glared at each other from deep emplacements, occasionally indulging in outbreaks of aggression which frequently were timed to coincide with critical stages of the peace negotiations, accompanied by endless propaganda, including unproven North Korean accusations of **biological warfare** by the UN.

Meanwhile sporadic naval warfare accompanied the UN **blockade** of North Korea and North Korean mining (**Mines, Naval**) operations (under Russian direction and often by sampan). Carrier-borne aircraft regularly struck the North, and **helicopters** were used

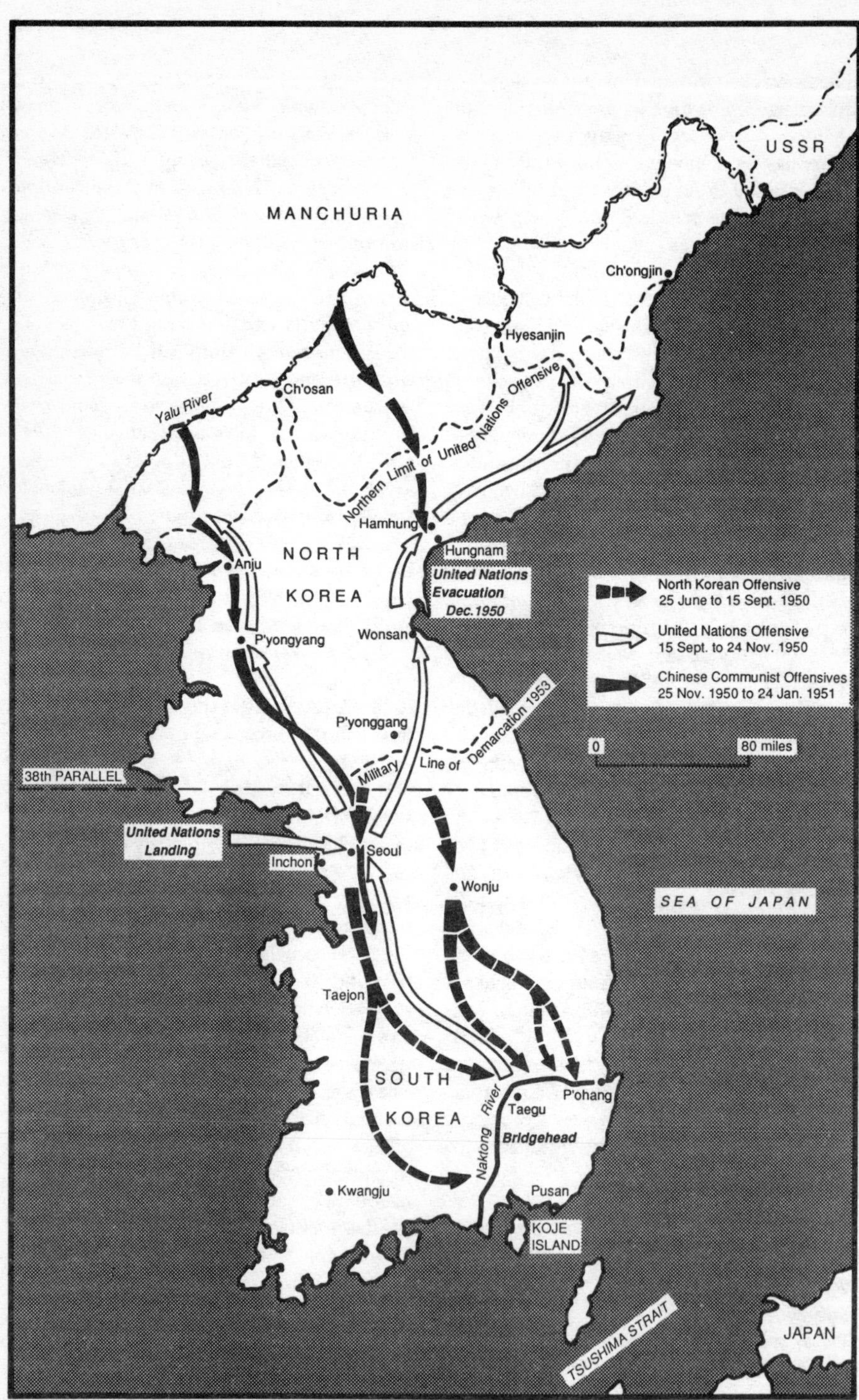

The Korean War, 1950–53

to detect mines. Indeed, it was now that the helicopter demonstrated its versatility in rescue missions of airmen shot down behind the enemy lines; for command and control; and, above all, resupply and the evacuation of wounded. The principal air activity, however, was over the North where attacks on **logistic** and tactical targets during the mobile phases of the war were continuous. After 1950 the North Korean Air Force was wholly on the defensive and dog-fighting between Russian **jet fighters** (mostly piloted by Russians, Chinese and East Europeans) and US jets escorting bombers was confined to the region of the Yalu River. In these engagements the superiority of US Air Force and Allied training was repeatedly shown, with a claimed ratio of ten to one in their favour. But the UN's losses of ground-attack aircraft were much heavier. From all causes UN losses were 1,213 aircraft between June 1950 and July 1953.

Korolev, Sergei (1906–65), was a pilot and **glider** designer who in 1933 built a liquid-fuelled **rocket**, was backed by the Russian Army and narrowly escaped Josef **Stalin**'s purges. During **World War II** he experimented with rocket-propelled **aircraft** and afterwards copied a German **V2** for a rocket which flew in 1947. For security reasons known only as the Chief Designer, he led the Soviet rocket and **space** programme to provide transatlantic **nuclear** warheads and launch satellite vehicles for **propaganda** and scientific purposes. With ample funding and high priority, he gave the USSR the lead in space in 1957, though within narrow margins of safety. After his death the USSR steadily fell behind the USA in its space and technological programme.

Krupp. The Krupp company was founded in Essen in 1801 by Friedrich Krupp, whose **research** made possible high-quality **steel** casting ahead of British competitors. His son Alfred expanded the foundry to produce, among other things, rails and wheels for **railways** and in 1851 the first successful breech-loading cannon. Krupp's then became big makers of **artillery**, warships and other weapon systems, coming to be known as 'the Arsenal of the Reich'; and was first to introduce the **Bessemer** and open-hearth steel-making processes. During **World War II** many of its factories were laid in ruins by **bombing**, but have since been rebuilt.

Kursk, Battles of. This important Russian route centre on the main railway line from Moscow to Kharkov and the Crimea fell into German hands without much resistance in October 1942 at a time the Russians were hard pressed to stay in **World War II**. It was retaken early in February 1943 as the Germans fell back precipitately after defeat at **Stalingrad**, but held as the centre of a salient formed in the aftermath of Field Marshal von Manstein's counterstroke in March, after the wet season put a stop to movement.

Von Manstein's aim was to pinch out the Kursk salient as soon as possible, before the Russians could fortify so obvious an objective. But when Hitler decided to augment this relatively limited operation to a major offensive with political overtones – and employ the latest, untried Tiger and Panther **tanks** in mass – delay was inevitable and surprise forfeited. Not until 5 July was a start made. By then the Russians, under General Georgi **Zhukov**, were forewarned and had assembled 1,330,000 men, 3,600 tanks, and nearly 20,000 guns behind deep defences to meet 900,000 Germans with 3,700 tanks and 10,000 guns. Each side also had about 2,500 aircraft. Predictably the Germans attacked the flanks of the salient to pinch it out. Always in control, the Russians snuffed out the northern pincer within hours and by 16 July had absorbed the southern thrust after an intense armoured struggle. On 10 July the Americans and British had landed in Sicily and on the 12th the Russians had begun their summer offensive at Bryansk. Hitler's decision to adopt defence was therefore inevitable as evidence of a massive Russian counterstroke at Kursk accumulated. From the very heavy losses in this, the last German offensive in the East, the German Army never recovered.

Kut-al-Amara, Siege of. When General Sir Charles Townshend's 6th (Poona) Division was prevented by the Turks from capturing Baghdad in 1915, he withdrew to Kut where, with 15,000 diseased men, he allowed himself to be besieged. For two months he made no effort to conserve supplies or send out 6,000 Arab non-combatants. At the same time he falsely reported that he was out of supplies. The result was that premature attempts at relief, like forlorn efforts to send supplies up the River Tigris, failed. Two weeks before he surrendered on 29 April 1916, the first attempts were made to drop supplies from **aircraft** which, in the heat, were barely capable of taking off. Some loads fell into Turkish hands; only 16,800lb arrived, but many flour bags burst on impact.

L

Laager, Wagon. Since the recorded inception of wheeled vehicles, *c.* 3500 BC, they have been utilized as mobile **fortification** systems. For example, at the Battle of **Adrianople** in AD 378, the Goths and their allies formed a wagon circle on high ground as a decisive tactical and **logistic** pivot to defeat the Roman Army. Even more effective were the wooden wagons of the Hussite leader John **Zizka** who, in 1419, armed them with **artillery** and, on vital ground, linked them with thick boards to form a continuous, protected circle. And at the night Battle of Kutna Hora in 1421 they fought in close formation while on the move.

More elaborate versions of the Hussite laagers, with an inner ring of 'soft skinned' wagons, subsequently operated in Europe until modern fire-power made them untenable. Settlers in the plains of such countries as South Africa and America, however, were able to revive the laager against ill-armed tribes. And with the advent of **armoured fighting vehicles** in the 20th century it again became possible in **World War II**, especially in the **desert**, to form night laagers of soft-skinned vehicles surrounded by **tanks**.

Landing-craft and ships. Until the 18th century **amphibious warfare** was chiefly carried out in whatever shallow-draught craft and ships were available. River barges and ships' boats were most commonly used, either to beach or come alongside quays. In 1758, however, the Royal Navy began construction of flat-bottomed craft, some with ramps, which were carried aboard troop transports. Until World War I there were few innovations. For the major landings at **Gallipoli** mainly ships' rowing-boats were used – though in the initial landing a converted tramp steamer (the *River Clyde*) was modified for an assault landing under fire after beaching. At the Zeebrugge raid in April 1918 troops were disembarked under fire on the mole from an adapted **cruiser**. Adaption of craft such as barges and fishing-boats for beach landing was the Germans' chosen improvisation when preparing to invade Britain in 1940; but, when Britain and the USA were faced with no alternative other than large-scale amphibious operations to defeat the Germans, Italians and Japanese, they were compelled to develop a considerable fleet of specialized vessels.

As a model, only faintly recognized, they had the Japanese, who had used special landing-ships carrying mechanized, shallow-draught craft during their landings in China in the 1930s. Both the Americans and the British adopted such ships, mostly by conversion of seagoing ferries with strengthened davits, known as LSI (Landing Ship Infantry). Fast motor-launches were initially used by the British in 1940 for hit-and-run coastal raiding, and ramped, mechanized lighters to carry vehicles and stores ashore (LCM – Landing Craft Mechanized) – such as intended at Dakar. But it was 20-man, wooden, 5-tonne Higgins LCP (Landing Craft Personnel), speed 10 knots, and steel, partially armoured, 13-tonne LCA (Landing

Landing-craft assault (LCA)

Craft Assault), speed 6–10 knots, which were used in very large numbers for raiding and by assault infantry. They were followed by 384-tonne, seagoing LCI (Landing Craft Infantry), speed 14 knots, which could embark about 200 men.

To get **tanks** and other vehicles ashore, LST (Landing Ship Tank) and LCT (Landing Craft Tank) were built. The first LSTs were converted cargo ships, but these could not function on sloping beaches in tidal conditions. The British developed a special ship whose draught could be adjusted by the flooding of ballast tanks – a method since employed by modern roll-on roll-off ferries, which is what LSTs (with a displacement of anything between 1,600 and 5,000 tonnes) were. LCTs displacing some 500 to 600 tonnes were intended for use well forward in an assault, carrying five 40-tonne or eleven 30-tonne tanks, or ten 3-tonne lorries. Only partially armoured, it was often desirable for LCTs to launch amphibious tanks when well off shore. The Japanese used British-designed 3-tonne swimming tanks, but the British produced 30-tonne Sherman tanks (called DDs, from Duplex Drive), kept afloat by a collapsible screen and powered by twin screws driven by the tank's engine (via the rear idler).

Numerous variations of these basic landing vessels were produced, to perform all manner of roles: beach survey and obstacle clearance; close fire support by guns and rockets; assistance with navigation; and a host of **logistic** tasks once a foothold had been secured ashore. Highly adaptable was the American amphibious wheeled DUKW, which carried a useful load and could drive on land without adjustment after swimming ashore, as too could the American Amtrac assault vehicles which were propelled through the water by their tracks and served as light tanks or armoured personnel carriers once ashore.

Since World War II landing vessels have been improved upon for both military and civil purposes, although to some extent the **helicopter** has supplanted them. To their number has been added **ground-effect machines** (often known as hovercraft), which can skim over almost any flat surface without changing mode and have many military uses.

Laser (Light Amplification by Stimulated Emission of Radiation). When atoms are stimulated to a higher level of activity than normal they emit light as they revert to their usual level. In ordinary light sources the atoms are randomly stimulated and consequently emit light independently. However, if all the atoms emit radiation in step with the stimulating wave, and if this effect can be sufficiently multiplied, the resulting beam of coherent light (i.e. at a single frequency) can be extremely powerful. This is the laser effect

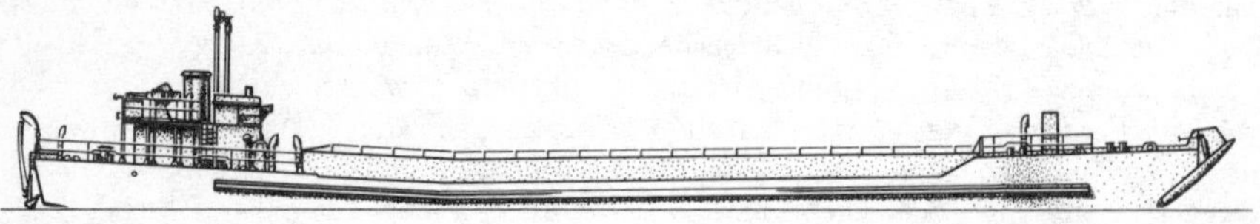

Landing-craft tank (LCT) vessel

and the phenomenon was recognized by Professor Dr Albert Einstein in 1917, though it was not until 1960 that the first practical laser was built by the American T. H. Maiman.

It was quickly recognized that the laser beam was able to burn through many materials and could cause serious injury to human tissue. There was much talk of its use as a 'death ray', but this was not a seriously pursued option; the first practical military application was as an extremely accurate **rangefinder**. In the field of **tank gunnery** this made a dramatic impact since the problem of judging distance correctly had always far outweighed all other system difficulties. Lasers have likewise been used as markers to indicate the target for homing missiles.

The laser has also found a ready military application as a training aid. The British Simfire system uses a low-power ruby laser to trigger sensors mounted on vehicles, and even on infantrymen's helmets, when an 'enemy' weapon was correctly aligned on a target and the trigger pressed, the sensor in turn activating a smoke generator to indicate a hit. Simfire produced a realism in training previously unimagined and such systems are now widely used.

Lawrence, Colonel Thomas Edward (1888–1935). Born in North Wales, Lawrence spent most of his youth in Oxford, where he was educated. As a young archaeologist before **World War I** he worked in North Syria. When war started Lawrence was first employed in military intelligence in Cairo, subsequently being sent to the Hejaz to work with Sherif Feisal, leading the main Arab forces fighting against Turkish rule. Here Lawrence soon took a prominent part in the Arab revolt, leading raids against the Turks and advising Feisal in his support of the British drive through Palestine to Damascus.

After the war Lawrence took part in the discussions about the future of the former **Ottoman Empire**, at the Versailles Conference. He then sought obscurity in the ranks of both the RAF and the Tank Corps. He died due to a motor-cycle accident in Dorset.

Laws of war. Even primitive people have sought to limit the excesses of war by the adoption of codes of practice and law. The Greeks and Romans were to the fore in making laws, as were Christian leaders from St Augustine onwards, particularly in their endeavours to make war a last resort for a just cause – always with suitable provisos. The medieval rules of **chivalry** were war laws framed mainly for a privileged, belligerent feudal minority and typical of all such societies worldwide. The **Crusades** and Wars of **Religion** were blighted by acts of ungoverned barbarity. The reluctance of European armies in the **wars of the 17th and 18th centuries** to engage in major battles was only a transient economic reaction in which the alternative conventions of **siege warfare** only marginally limited mayhem.

Not until the upsurge of humanitarianism, accelerated by the horrors of the **Crimean** and Austro-Italian wars in the 1850s, were strenuous attempts made to codify and enforce international laws. The Declaration of Paris in 1856, which clarified the rules of **blockade**, and the **Geneva Conventions** of the 1860s, which created the **Red Cross Society** and gave better treatment to prisoners of war and the wounded, were milestones on the way to The Hague conferences of 1899 and 1907, which in turn established rudimentary but almost unenforceable rules of war. After **World War I**, improvements were made to the Hague rules regarding **submarine** warfare, the outlawing of **chemical warfare**, the rights of neutrals and attacks on open cities – most of which were broken at some time or other during **World War II**. However, the flagrant atrocities of the latter war, well publicized and, to some extent, codified during the post-war war crimes trials, did produce significant changes of approach to the conduct of war by individuals. The acknowledgement, *ex post facto* though it might be, of crimes connected with starting an aggressive war and against humanity found their way on to the statute books of many nations and into military law, with a deterrent and educational effect not to be underrated.

A fairly comprehensive set of laws now exists. Unfortunately the means of enforcement are not always available, particularly since 'might', with the threat of force, has a tendency to be 'right'. But the mobilization of public and international opinion against those in breach of the laws has been effective many times, even if not immediately; against the superpowers in **Vietnam** and **Afghanistan**, for example, and in Yugoslavia.

Le Creusot foundries were started in 1782 by the English **ironmaker** William Wilkinson further to exploit the deposits of ore in that part of France for arms manufacture using British plant. They did well during the **French revolutionary wars**, but prospered wonderfully after Adolphe and Eugene Schneider took over in 1837 and concentrated more on the manufacture of **railway** equipment and river steamers. In 1874 they began the manufacture of **armour**, **artillery**, other weapons and **motor vehicles**. The famous 75mm gun was among their products and, in **World War I**, **armoured cars** and the first French **tank**, the Schneider M16. Much diversified, they supplied both sides in **World War II** and continue to manufacture arms. (*See also* **Schneider-Creusot Company**.)

Lee, General Robert E. (1807–70). A Virginian and West Point Military Academy graduate who saw active service in Mexico in 1847 and 1848 and willingly took command of Virginia's forces in the **American Civil War**. He was less than enthusiastic when made military adviser to Confederate President Jefferson Davis in 1861, responsible for the direction of military operations. Initially his generalship was diffident. Yet Davis retained him and, after 1862 when his masterly conduct of the **Seven Days Battle** saved Richmond and the Confederacy, made him (in modern terms) Chief of Staff as well as Commander of the Northern Army.

An exponent of manoeuvre, Lee realized that Davis's politically orientated defensive strategy was likely to lose the war for the South. Yet on both occasions when he invaded the North to threaten Washington, defeat was the outcome: at Antietam in September 1862 and Gettysburg in July 1863. The latter exposed flaws in his method of command – and also indicated that the South's chances of survival were remote against a powerful opponent who always outnumbered him. Nevertheless it was Lee's thrifty utilization of ever-weakening armies which prolonged the war until 9 April 1865, when he surrendered in person to General Grant.

Leipzig, Battle of. Driven back from Moscow in 1812, but defiantly blocking the Allied advance towards France in 1813, **Napoleon**, with 200,000 dispirited men, tried on 16 October to hold Leipzig against an Allied army of 320,000 men (Prussians, Austrians, Russians and Swedes) under General Prince Gebhard von Blücher. Heavy fighting throughout the 16th and 17th pressed the French back, but Napoleon, though threatened with encirclement, refused to withdraw. By the 18th the Allies, fully concentrated, launched a massed, frontal attack and threatened to encircle the French, who were deserted by the Saxon corps. Next day ferocious fighting in the city was climaxed when the bridge over the River Elster was prematurely blown up by a corporal. Nevertheless most of the French, panicked by **rocket** fire, were fortunate to escape but left behind 325 guns, 40,000 muskets and 900 wagons. Each side lost about 60,000 men in the longest battle of the **French revolutionary wars**, with tremendous portent for the future. For **artillery** had dominated and Napoleon's aim of a unified Europe under French hegemony was ruined.

Lepanto, Battle of. The **Ottoman Empire**'s expansion in the Mediterranean Sea, which had been checked in 1565 during the Great Siege of **Malta**, resumed in 1570 when the Turks invaded Cyprus. A combined fleet of 209 **galleys** from Spain, Venice, Malta and Rome, under the command of Don Juan of Austria, met 235 less well-built and armed Turkish galleys under Ali Pasha off the Greek coast on 7 October 1571. Both sides deployed in three divisions with a small reserve. Allied **artillery** and small arms caused losses before the fleets grappled for hand-to-hand combat. At first the Turkish wings had the advantage, but the triumph of the strong Allied centre proved decisive. The Allies lost 12 galleys and 8,000 men; the Turks some 20,000 and about 170 galleys (117 captured) from which 15,000 enslaved Christian oarsmen were rescued. In this last major engagement between galleys, Ottoman power remained unbroken in the short term. But the effects on **morale** were eventually decisive.

Lettow-Vorbeck, General Paul von (1870–1964), was the German commander in East Africa at the outbreak of **World War I**. Grossly outnumbered, he fought a skilful **guerrilla** campaign, living off the country and captured enemy material. Progressively the Allies were compelled to commit disproportionate resources to a very damaging campaign. German casualties were lighter than their opponents'. Sickness took the greatest toll in raiding which was governed by **logistics**. Desperately short of weapons, supplies and medicines, von Lettow-Vorbeck's army was the last to surrender on 23 November 1918. Post-war he remained an advocate of irregular action in war and politics.

Leyte Gulf, Battle of. After defeat in the **aircraft-carrier** battle of the Philippine Sea in June 1944 and ahead of the correctly anticipated invasion of the Philippines in October, Admiral Soemu Toyoda concluded that Japan's last hope lay in a surprise naval victory aimed at destruction of the enemy's **amphibious** fleet. He devised a complex plan conditioned by lack of strong carrier air forces and the inescapable **logistic** imbalance which made him concentrate his Fleet at Brunei, where oil was plentiful, instead of Japan where, owing to **blockade**, it was not. Admiral **Nimitz** was fully informed of the plan through access to Japanese **codes**. He therefore knew that the battleships and carriers (under Vice-Admiral Jisaburo Ozawa) sailing from Japan, with few **aircraft**, were intended as a decoy from the main **battleship** force which, under Vice-Admiral Takeo Kurita and supported by two subsidiary fleets and land-based air attacks, were tasked to destroy the invaders disembarking in Leyte Gulf.

The landings began on 20 October, but already the Japanese were (incorrectly) satisfied they had won a great victory off Formosa on 13–16 October against Admiral **Halsey**'s carriers. They were thus unaware, as they approached San Bernardino Strait and Surigao Strait, that Halsey's Third Fleet was guarding the former and Admiral Thomas Kinkaid's Seventh Fleet the latter. At the same time as Ozawa's fleet had been allowed, through American error, to slip through unreported, Kurita's fleet was detected on the 23rd, hammered in the Sibuyan Sea by **submarines** and Halsey's aircraft (with the loss of the super-battleship *Musashi* and two heavy cruisers, plus other ships) and forced to turn back on the 24th at the same time as Japanese aircraft were annihilated. Meanwhile the two subsidiary Japanese forces converging on Surigao Strait were running into a hot reception from Kinkaid's **torpedo-boats** and battleships. Halsey, hearing belatedly of the approach of Ozawa's fleet and disregarding the fact that it was only a decoy, abandoned the San Bernardino Strait to charge northwards to attack it.

However, Kurita had reversed at night to his original course and passed through the strait at dawn, headed for Leyte Gulf. There only a few escort carriers and light forces protected the invasion fleet because Kinkaid's battleships were engaged in Surigao Strait, albeit to excellent effect with the virtual annihilation of battleships and cruisers. So Kurita had Leyte Gulf at his mercy, attacks by aircraft from a few light carriers manoeuvring for their lives before Japanese guns being little more than harassment. But he lost tactical control, then his nerve and finally turned for home, his mission unaccomplished.

Farther north, off Cape Engano, Halsey had found Ozawa and was in the process of sinking all four carriers and several other ships, but not the two battleships. These

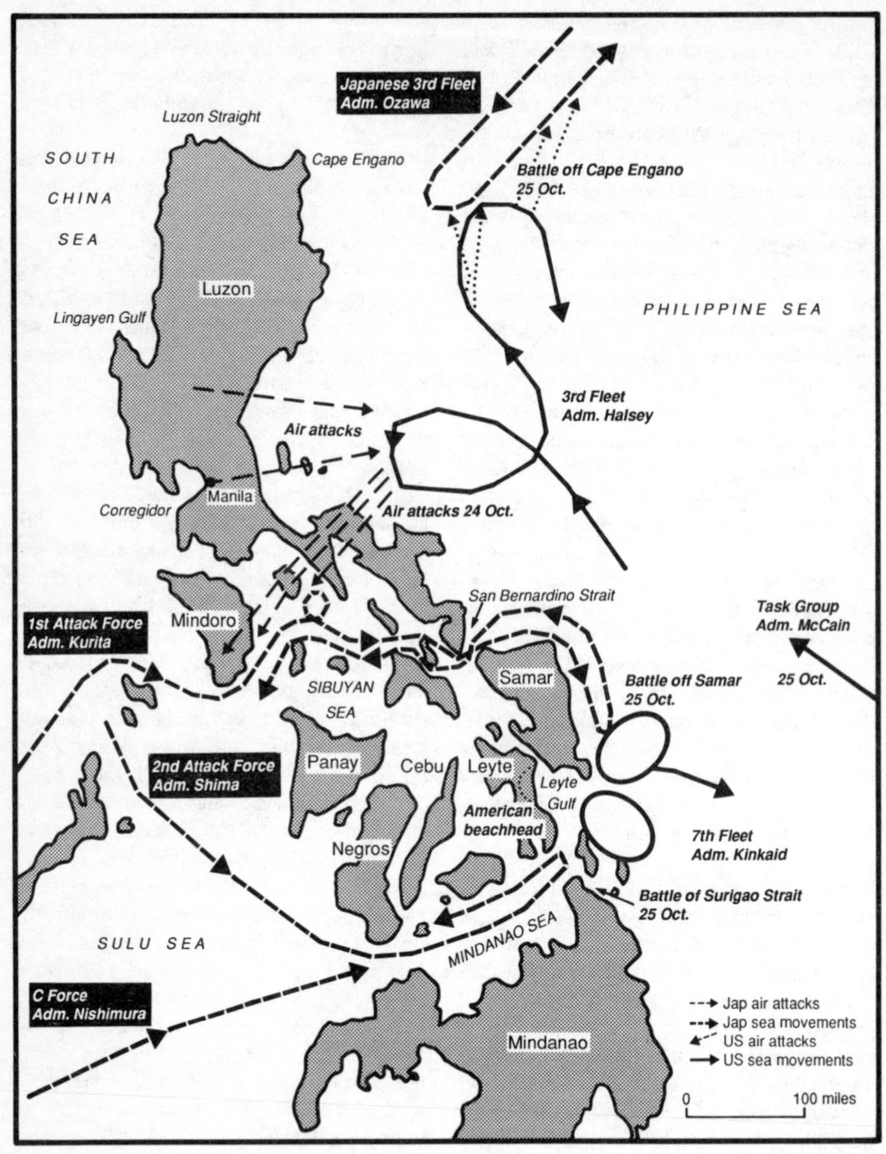

The Battle of Leyte Gulf, 23–26 October 1944

would escape since, in response to a terse signal from Nimitz, Halsey tardily sent back his own battleships to deal with Kurita – only again to be robbed of his prey by Kurita's premature withdrawal. Despite some staggering errors on both sides, it was a crushing victory for the Americans, who destroyed about 500 aircraft and sank four carriers, three battleships, six heavy and four light cruisers, eleven destroyers and one submarine, at a loss to themselves of three light carriers, three destroyers and 200 aircraft.

Liddell Hart, Sir Basil (1895–1970). A defence correspondent of the London *Times* newspaper who philosophized copiously on war and cleverly rewrote history in support of his theories. His nebulous theory of the 'indirect approach' was an effort to dissuade Britain from land warfare on the Continent linked to a reversal of earlier support for armoured forces on grounds of the **tank**'s vulnerability. With the ear of the Prime Minister (Neville Chamberlain) and as adviser to the War Minister (Leslie Hore-Belisha) he carried great weight. **World War II** utterly discredited him. The manner of his corrupt, post-war recovery of credibility is documented in Professor John Mearsheimer's *Liddell Hart and the Weight of History* (1989). Liddell Hart was ill at ease with technology and probably terrified by its potential impact on war.

Limited war. A term in popular use since the 1950s to describe wars with limited aims conducted in well-defined areas; also to the exclusion of **nuclear weapons** and, until the 1980s in the **Iran–Iraq War**, the **chemical** kind too. But efforts to limit war to save life, maintain the survival of hierarchies and minimize economic damage are as old as time; some kinds of pragmatic measures are mentioned in the **laws of war, chivalry, siege warfare** and various **wars**, notably of the **17th and 18th centuries**.

Not until after the **Crimean** and Austro-Piedmontese wars of the 1850s, however, did humanitarianism become a prime motivation. Vast though its dimensions were, the **American Civil War** remained limited because other nations could not influence the actual fighting. The **Franco-Prussian War** stayed within bounds because neighbouring states saw no political purpose to be achieved by involvement in a struggle which did not immediately threaten the balance of power in Europe. Nations sympathetic to the Boers did not intervene in their war against Britain in 1899 because they were in no position to do so effectively. But **World War I** spread far and wide and became unlimited because so many nations had founded relationships upon treaties with secret clauses, based on the popular assumption that war was a glorious, unavoidable, even desirable natural activity – provided the initiators felt they could win quickly and cheaply. The tragedy of the unpopular **World War II** lay in the inability of people to dispose of irresponsible and greedy demagogues – the Japanese military clique (including Emperor Hirohito), **Mussolini** and **Hitler**.

Although the threat of global nuclear war has been contained since 1945, **cold war's** many ramifications have perpetuated a state of global tension which has made outbreaks of limited war unavoidable. The various wars for independence have all generated limited conflicts, often with outbreaks of civil war in their aftermath. Controlling such wars has preoccupied the major powers who, with varying degrees of success, have applied diplomatic, economic or coercive measures to stop them or restrict their spread – the **Arab–Israeli**, Indo-Pakistani and **UN–Iraq wars** are prime examples of relative success. Limiting the wars in which the major powers are directly involved, notably the ones in **Vietnam** and **Afghanistan**, was much more difficult, though achieved in the long run.

The 20th century has seen more fighting than any other in history. Two global wars and ceaseless cold and limited wars are nothing to be proud of. Nor can there be much confidence in preventing nuclear war in the future if **atom bombs** and the most sophisticated chemical weapons fall into the hands of megalomaniacs as irresponsible as those who started World War II.

Lincoln, President Abraham (1808–65). A lawyer with experience as a soldier (without seeing action) who, because he was President of the USA throughout the **American Civil War**, was C-in-C the Federal forces. His strength in this role lay in understanding that only by a strategy of offensive action by superior forces could the North win the war. Hence his adoption of **blockade** and insistence upon invasions of the South. But the key to the North's victory was his willingness to give commanders wide powers of initiative within the broad strategic plan; and to dismiss those who failed his expectations, despite the political consequences.

Lissa, Battle of. On 20 July 1866 when the Austro-Prussian War was almost over an Italian fleet, under Count Persano, covering a **convoy** invading the island of Lissa, was intercepted by an Austrian fleet under Count Tegetthof. A few ships were armoured but only the Italian *Affondatore* had a **ram** and a turret. The Italians interposed as the Austrians charged with the intention of ramming. Some managed to do so (though not *Affondatore*), resulting in one seriously damaged ship and the Italian flagship sunk with 381 hands. The Austrians sank the biggest Italian ships and claimed a victory which led tacticians to false conclusions concerning the ram's potency, though to correct ones about the need to locate masts astern of turrets in order to fire straight ahead.

Lloyd George, (Lord) David (1863–1945). A radical, reforming politician of outstanding eloquence who became Britain's Prime Minister in December 1915. His reorganization of government through a Cabinet of only five, in permanent session, served by a strong secretariat; his unrelenting and by no means irrational pressure on admirals and generals, linked to strong dissatisfaction with their strategies and ways of waging **World War I**; his management of industrial production and manpower shortages through **conscription** and reorganization; and his handling of Britain's allies, producing in March 1918 an effective unified Allied command, were major performances during a war-winning period in office.

Logistics. It seems that the first use of the word logistics to describe the 'practical art of moving armies', in every respect except strategy and tactics, appeared in Antoine **Jomini's** *Précis de l'art de la guerre* in 1836. It appears to have been rescued from obscurity by Captain Alfred **Mahan**'s influence on the US Navy in the 1880s, helping it assume modern usage in what is now defined as ways and means of financing and procuring war resources, in conjunction with the supply and transport of men and materials.

Until the mid 19th century, indeed, logistic restraints changed but little. The time ships spent at sea was limited by the supplies they carried and their distance from sources of replenishment – a relatively simple equation compared with that facing land forces. For armies were strictly bound by main communication systems – rivers, tracks and roads – and the necessity to take food on the hoof or in animal-drawn carts while living off the country. Mobility improved as the Romans built special military roads and as forests were gradually cleared and swamps drained. But it was often essential for a large army, locust-like, to keep on the move from one feeding area to another, often leaving famine in its train. The demand, post 14th century, to make and transport **artillery ammunition** exacerbated the problem, for new technology tends to increase the logistic load.

Frequently the blame for unsuccessful campaigns is laid upon logistic failures; while the successful ones are credited to the commander, despite any of his logistic shortcomings. For example, one hears far less about the excellent logistic arrangements made by the ever-victorious Duke of **Marlborough** than about **Napoleon**'s and **Hitler**'s supply breakdowns when defeated in their invasions of Russia.

The introduction of steam-powered **railways** and the **telegraph** in the 1840s were logistically revolutionary, especially when associated with long-term storage of **food** by bottling and canning. The **Crimean War** of the 1850s was almost the last British campaign attended by logistic consequences sufficient to bring down a government as the result of public indignation at the ill-treatment of the forces. For although the transport of men and stores to the theatre of war by rail and steamship had worked well, the provision and handling of supplies and their delivery to the combat zone had been almost medieval because the Commissariat had fallen into decay since 1815. The soldiers' food and **medical** care provided were a scandal brought quickly to public attention by the more rapid signals **communications** of the day – which also expedited identification and implementation of the remedial measures required, as in the sending of nurses under Miss Florence **Nightingale**. At the same time it was discovered that the **engineering** services required to support the handling of equipment and provisions themselves created a more complex logistic load that could best be dealt with by **mechanization**. The unloading of ships and docks clearance work at Balaklava, for example, were in due course partly carried out by steam cranes, a light railway and tractors.

The accelerating development of mechanization and signal communications fundamentally conditioned logistics to the point at which, provided adequate forethought was given to the matter (based on sound **intelligence** and administrative practice), armed forces need never want. Pasteurization, refrigeration and dried foods (all of which were available by the 1870s) solved the problem of the long-term storage of **food**. The invention of practical, liquid-fuelled internal combustion and **turbine** engines in the 1880s led to much faster and more reliable means of propulsion and created a demand for a network of greatly superior roads as well as enormously simplifying the bunkering of ships and, in due course, railway locomotives. The conquest of flight by heavier-than-air machines in 1903 opened the way to the greatest transport revolution of all. Discoveries of the cause of such diseases as malaria and yellow fever, which ravaged armies in the Spanish–American War, led to prevention and cure, at the same time as disciplined hygiene and the latest **medical** practice marginally reduced human wastage.

World War I stimulated demand for all the above, including the first air delivery of supplies to besieged land forces at **Kut-al-Amara** in 1916. By 1918 a major switch from animal to mechanized transport on land had taken place; a significant reduction in fatalities to the wounded; a considerable increase in the size and scope of logistic services, with consolidation of the best basic organizations and methods required; and development of communications networks without which response to rapidly changing conditions and circumstances was impossible.

World War II not only benefited from the lessons of World War I, but also introduced numerous sophisticated improvements, of which the most important were the conquest of sickness in disease-ridden places (such as **Burma**, South-East Asia and parts of the **Pacific**) and **air transport** (**Airborne forces**) on so large a scale that, given the resources, mobility was vastly enhanced, **blockade** made more difficult and siege surmountable. Indeed, it was the calls of war which created the infrastructure (such as airfields in the most inaccessible places) and the latest means of transport (such as the **landing-craft** which were forerunners of roll-on roll-off ferries) which made far easier the supply of wars to come, besides enhancing everyday life and the mobility of peoples worldwide. In the **electronic** epoch, the diversity of communications made possible extremely swift, flexible and well-calculated reactions to the most surprising circumstances. No better example is that of the improvised, overnight British response in the mounting of the **Falkland** Islands campaign in 1982, done without previous planning at a range of 8,000 miles from Britain and with the nearest base 4,000 miles distant.

Lôme, Stanislas Dupuy du (1816–1885). In 1842 Lôme was sent to study **ship building** in England, and in 1848 designed and built the first steam-driven French warship. Interest in **armoured** warships led to his building the armoured steam **frigate** *La Gloire* which, in 1859, made wooden sailing-ships obsolete. As Inspector-General of Naval Material from 1861 he not only developed steamships, including an efficient horizontal three-cylinder compound engine and a project for a cross-Channel **railway** ferry, but also during the **Franco-Prussian War** designed a steerable **balloon** to fly out of besieged Paris. Finally he conceived Gustave Zédé's **submarine** *Gymnote* with fore-and-aft control by hydroplanes.

Louvois, Marquis François de (1639–91), became Louis XIV's Secretary of State for War in 1662. A brilliant and ruthless administrator, he reduced the power of the nobles in the Army, enforced honesty, improved equipment (by the introduction of better **artillery**, the flintlock **musket** and the **bayonet**), besides encouraging **Vauban**'s construction of modern **fortifications**, the creation of well-supplied magazines and a hospital for disabled soldiers. Thus, in collaboration with Jean Colbert, who modernized and fortified the **dockyards and naval bases** and increased the fleet from 20 vessels in 1661 to 270 in 1677, he gave his king the military power to dominate Western Europe, which was broken by the combined genius of **Eugene** and **Marlborough** in the War of the Spanish Succession.

Ludendorff, General Erich (1865–1937). A brilliant member of the German General Staff with a large share of the credit for rapid capture of the Liège **fortifications** in August 1914. Later that month, as Chief of Staff to General **Hindenburg**, he helped win the Battle of **Tannenberg**. From then onwards, he dominated the partnership controlling the Eastern Front until appointed to supreme command of Germany's war effort in August 1916. As Quartermaster-General, he executed the defensive strategy in the West and the undermining of Russia in 1917, followed by the Hindenburg offensives in 1918. Arguably, his loss of nerve after the Battle of **Amiens** shortened a war already lost. He was sacked on 26 October. In 1923 he was at **Hitler**'s side in the abortive Munich *Putsch*.

M

MacArthur, General Douglas (1880–1964). An infantryman who took part in the Mexican War in 1914 and was a divisional commander in **World War I**, followed by command of American forces in Germany. He was Chief of Staff of the Army, 1930–35, at its nadir, and military adviser to the Philippines Army 1935 to 1937, when he retired. Called back to command the US and Philippine forces in July 1941, he was defeated by the Japanese and forced back to Bataan. From there, on President Roosevelt's order, he was evacuated to Australia to take command of the Allied forces, tasked to advance through New Guinea. MacArthur's relations with the Navy and Allies were not always easy. He managed to have overruled the Navy's intention to bypass the Philippines in order to satisfy his ego by conquering the islands in 1944–5. The Japanese surrendered to him on 2 September 1945. He remained in Japan as Supreme Commander to democratize the nation and restore its economy.

Upon the outbreak of the **Korean War** in 1950 he was made Supreme Commander of UN forces. While Eighth Army withdrew into the Pusan perimeter, MacArthur planned and, in September, launched the Inchon landing and then invaded North Korea. In November the Chinese stopped Eighth Army short of the Yalu River and forced it back to the 38th Parallel, where the position was secured. On 11 April 1951 MacArthur was sacked by President Harry **Truman** after a disagreement over his insistence upon bombing enemy bases in Manchuria.

Machiavelli, Niccolò (1469–1526), was a profound thinker, powerful statesman and military philosopher who in 1505 promoted the Florentine militia. In 1514 he completed his celebrated political treatise *The Prince* and in 1520 the complementary *On the Art of War*. The latter was regarded as the 16th century's most influential military work. It visualized fewer mercenaries, larger armies, the decline of armoured **cavalry** and the value of fire and movement by lighter, more agile forces. Machiavelli's otherwise sound tactical analysis was flawed, however, by an underestimation of **artillery** fire-power and an objection, because they caused wars, to **fortifications**, though in doing so he implied the significant effectiveness of such weapon systems.

Machine-guns. In all probability the first clumsy, mechanical repeater weapon was a multi-arrow crossbow, *c.* AD 1200 (*see* **Bows and arrows**). When firearms appeared multi-barrel systems comprising bundles or rows of small guns were introduced, *c.* 17th century, in a similar method of increasing fire-power. From these evolved, in 1718, James Puckle's practical, lightweight, hand-cranked, revolving-chamber, single-barrel flintlock gun, designed to repel boarders at sea, firing nine bullets in succession. Progress with machine-guns (MGs) was delayed, however, until Alexander Forsythe invented the mercury-fulminate percussion cap, in 1807; Johann von Dreyser's reliable breech-loading **rifle** of 1827, which was adopted by the

French Reffye mitrailleuse

Maxim machine-gun

Prussian Army in 1848; and Samuel **Colt**'s revolving **pistol** of 1835. After 1856 various manual machine-guns appeared of which the American Williams was first into action, as an **infantry** weapon, on 31 May 1862 at the Battle of Fair Oaks.

Eleven types of revolving-chamber or multi-barrel guns were in use during the **American Civil War**. The most successful was Richard Gatling's revolving-chamber gun with a paper (later brass) cartridge (*see* **Ammunition**). This was followed in France by the Reffye, a 25-barrel *mitrailleuse* which fired at 75 to 125 rounds per minute (rpm) and was used by the French as **artillery** during the **Franco-Prussian War**. However, the invention by Alfred **Nobel** in 1885 of ballistite, smokeless powder, made manual guns obsolete. For Hiram **Maxim** at once exploited the recoil produced by this more powerful propellant to actuate a belt-fed, semi-automatic, water-cooled gun with sustained fire of 500rpm.

The 7·69mm Maxim was first used during the Sudan campaign, the Boer War and with a decisive increment to the defensive in the **Russo-Japanese War**. Most European armies adopted it, but it was elaborately modified by the **Vickers Company** for the British who, in 1915, formed a Machine-Gun Corps for specialized support instead of integration with infantry or **cavalry**. At about the same time the French introduced a gas-operated Hotchkiss which used some of the propellant gases, diverted via a barrel port, to operate a piston connected to an extractor and ejector which disposed of the empty cartridge cases, while at the same time a spring was compressed which, on reasserting itself, allowed a new round to be fed into the chamber on the breech mechanism's forward stroke. The Hotchkiss was air-cooled, but with a very-heavy-finned barrel to dissipate the heat generated by firing.

A third type of MG, which emerged from Austria in the early part of the 20th century, was a version operating on the blow-back principle, in which barrel and breech block were not locked together, there was no piston, the action taking place purely on recoil energy. The **World War II** Sten sub-machine-gun was a similar idea and was equally unsafe and unreliable, depending as it did on consistency of cartridge-filling techniques that was not always achieved.

These earlier cumbersome weapons, all weighing around 70kg, soon generated a need for a lighter MG which one man could carry and fire. The best known in World War I were probably the British (though American invented) 7·69mm Lewis gun and the US 7·62mm (0·30in) Browning automatic rifle; both weapons, gas-operated, magazine-fed and weighing about 7kg, survived to be used during World War II. They could be fired like a rifle or on a bipod. Naturally, for **aircraft** armament the lightweight, air-cooled Hotchkiss, Lewis and German Parabellum guns, firing tracer and armour-piercing bullets, were preferred to the heavier water-cooled Maxim types; while machine-guns of various types also were used against low-flying aircraft.

Developments in the 1930s to meet the

needs of the infantry section and of **tank** secondary armament produced some very successful designs. On the British side the 7·69mm Czech-designed Bren became the universally used infantry light machine-gun (LMG), a typical feature being the ease with which a hot, and hence inaccurate, barrel could be removed during action and replaced by a new one carried by the crew. The Germans introduced two highly effective LMGs: the 7·92mm MG34 (800–900rpm) and the later MG42 (1,200–1,300rpm). Both could be fired from a tripod, giving them an MMG capability. For armoured vehicles the British had developed another Czech-designed MG: the 7·92mm Besa, while the Americans relied on the World War I 7·62mm Browning.

All these weapons fired rifle ammunition but there was also a need for a larger-calibre MG for longer-range and more substantial targets such as light armoured vehicles. Both Vickers and Browning produced a 12·7mm (0·50in) MG and the British also developed a 15mm Besa mounted in some **armoured cars**.

The trend in recent years has been to reduce the infantryman's burden and this is reflected in the design of MGs. The British introduced a 7·62mm General Purpose Machine-Gun (GPMG) to replace both the Bren and the Vickers MMG. It can be magazine- or belt-fed, fired from bipod or tripod and, with an easily fitted heavy barrel, is capable of sustained fire; the US M60 MG has a similar versatility. The larger-calibre (12·7mm) US M85 provides two rates of fire: 450rpm for ground targets and 1,000rpm for air defence – an interesting reflection of Maxim's original design, which offered the same facility.

Madrid, Sieges of. During the War of the Spanish Succession and the **French revolutionary wars**, Madrid repeatedly was taken but escaped any prolonged siege. At the beginning of the **Spanish Civil War**, however, General Franco's first attempt and subsequent bombardment by **artillery** and **bombing** failed to enable him to break in. Fighting also went on in the surrounding countryside to complete the city's isolation, but the Italian troops involved were routed; indeed, Madrid was never completely cut off, men and materials got through. In July 1937 the garrison launched a sustained counter-offensive – which was thwarted by the besiegers after nearly three weeks' grapple.

Not until February 1939, when the Republican cause was collapsing, was the battered city – by now on the verge of starvation – in peril again. On 28 March the enemy walked in unopposed. Madrid had been a symbol of government determination. But through the cloud of propaganda generated about its resistance, commentators failed to note that General **Douhet**'s theories of air power were suspect. **Bombers** had failed to destroy the defence.

Maginot Line. An enormously expensive, 200-mile **fortifications** system, named after the French Minister for War, André Maginot, built after 1931 along the Franco-German frontier between Basel and Longwy. It consisted of underground, air-conditioned forts sprouting **artillery** from retractable, rotating, armoured turrets, bristling with **machine-guns** and shielded by ditches and **barbed wire**. It was narrow in depth and flanked by neutral countries. In May 1940 the Germans outflanked it via Luxembourg and Belgium and in June broke through it quite easily. Maginot-mindedness remains a byword for tactical/technical misconception and stagnation.

Mahan, Admiral Alfred (1840–1914). A US Navy officer who fought in the **American Civil War**, but won his great reputation as a military philosopher when a lecturer at the Naval War College, of which he became president in 1886. From the study of history he formulated celebrated theories concerning the influence of sea power on history, published in three classic volumes prior to 1897. When not at ordinary naval duty (he commanded a **cruiser** from 1893 to 1895), involved with the Spanish–American War or engaged in international affairs, he was studying **strategy**. He concluded that commerce and **logistics** were key issues and that sea power was decisive. He believed that, as a deterrent (*see* **Deterrence**), Anglo-American sea power should guarantee peace.

Malaya, Battle of. Vital to Japan's war of expansion were the simultaneous attacks on Hong Kong, the Philippines, **Pearl Harbor** and the Malay peninsula. The Twenty-fifth Army under General Tomoyuki Yamashita landed in southern Thailand and at Kota Bharu on 8 December 1941 as British air power was smashed and, on the 10th, the Royal Navy's two capital ships, based on Singapore, were sunk at sea by **torpedoes** and **bombs**.

Their bases and logistics secure, the Japanese force of 100,000 well-trained and strongly supported men could advance through **jungle** and rubber plantations much as they chose. For the British had deployed in strength far to the south, shielding Singapore, without counting on a major attack so far north. In consequence the Indian brigade defending Kota Bharu had to be withdrawn, beginning the long retreat down the west coast of the peninsula. Everywhere the British, unprepared for jungle warfare, gave way

The invasion of Malaya, 1941–2

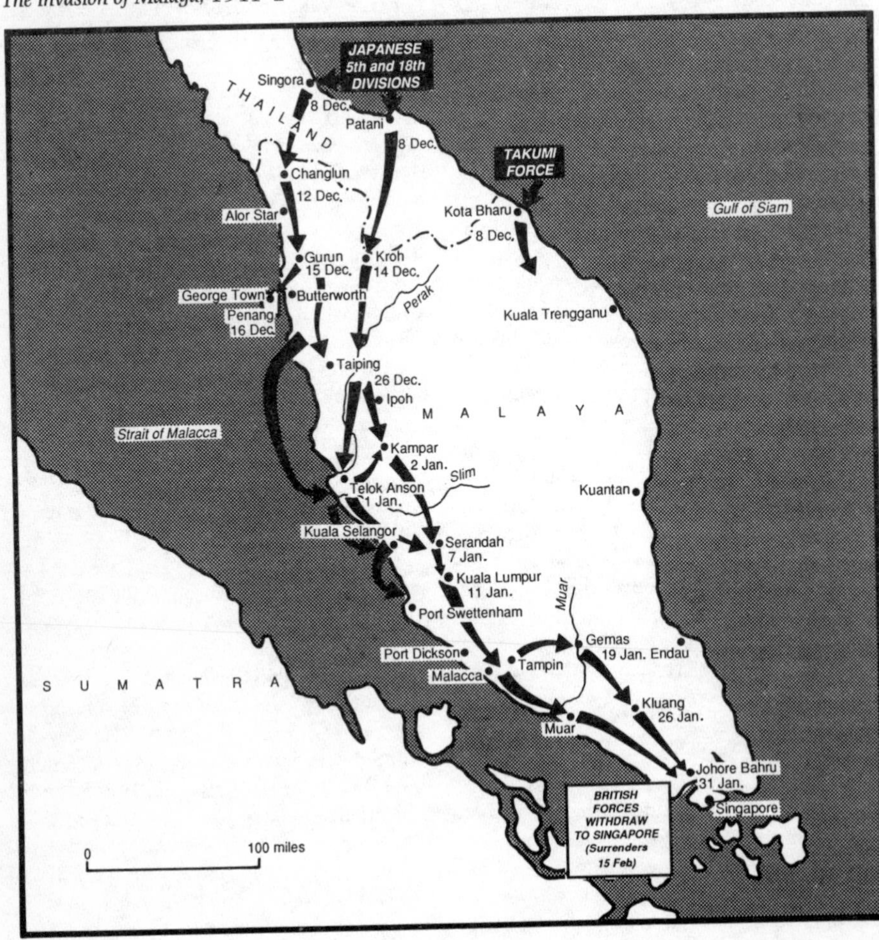

before the Japanese, whose tactics of infiltration were far superior. Successive, hastily prepared river-line defences collapsed when the Japanese found weak spots, crossed and unexpectedly appeared in strength far to the rear.

Penang was captured on 16 December, Port Swettenham on 9 January after the Slim River line was broken by **tanks** on the 7th. The withdrawal was virtually unchecked since the original Johore Line on the River Muar was only briefly defended by Indians and newly arrived Australian troops, who acquitted themselves well. By the 15th it had been penetrated and by the 31st those of Malaya's shattered defenders who were left had retreated across the causeway into Singapore, which soon fell.

Malplaquet, Battle of. When the Princes **Eugene** and **Marlborough** besieged Mons on 4 September 1709, a French army of 90,000 under Marshal Claude Villars threatened the besiegers by entrenching a 7,000-yard gap between woods to the south-west of the city. The princes attacked both French wings with **infantry** on the 11th. They sent their right wing, most unusually and at risk of confusion, through a large wood, compelling Villars to reinforce his hard-pressed wings by stripping his entrenched centre. Whereupon the Allied infantry and **cavalry** reserve, supported by **artillery** cross-fire, punched through the centre to bring on a great clash with the French cavalry counter-attack. Marshal Louis de Boufflers, who had assumed command from the wounded Villars, then ordered a withdrawal, which the Allies were too exhausted to pursue. French casualties were 12,500; those of the Allies 20,500 in a portentous indication of the multiplying defensive effects of fire-power and entrenchments.

Malta, Sieges of. Strategically located and naturally defensible in the central Mediterranean, Malta has inevitably endured many long sieges, apart from numerous invasions by **Greeks**, **Romans**, **Muslims**, **Normans** and **Crusaders**. The first siege of note took place between May and 8 September 1565 when the Knights of St John defied 60,000 Turks, who were forced to withdraw when a Spanish fleet arrived. And the second when, after **Napoleon** occupied the island in June 1798, a two-year siege by the Maltese, assisted by the British, Portuguese and Neapolitans, starved out the French garrison – leaving the British in possession.

The third and longest siege began on 10 June 1940 when the Italians declared war on Britain and France and launched air raids on the island. With only meagre defences and far distant from the nearest friendly bases, in Egypt and Gibraltar (after France capitulated on 21 June), the island was nevertheless never completely cut off. Indeed, throughout 1940 the Italians were unable to prevent the passage of **convoys** through the **Mediterranean** and the reinforcement and supply of the garrison. Hurricane **fighters** arrived in August, along with more guns. Offensive action by surface warships, **submarines** and **bombers** so hampered the movements of Italian convoys to **North Africa** that the Germans were compelled to intervene. On 10 January 1941 they launched the first of many furious and extremely damaging attacks on ships, the **dockyards** and airfields. British bombing was curtailed and surface warships withdrawn. Yet, to complete a pattern repeated throughout, no sooner were Axis air attacks diminished than British offensive operations increased.

The populace lived in caves and shelters when under attack. Casualties were not excessive, but damage considerable and **food** and fuel supplies short, desperately so when only cargo submarines could get through. The scale of air attack coincided with the intensity of operations in **North Africa**, rising to crescendos during the spring campaign of 1941 and in the period of the November to January 1942 offensives; it reached a peak when Field Marshal **Kesselring** opened a massive pre-invasion air offensive from March until May 1942, and again in mid-August 1942 during the passage of the last convoy before the siege was lifted. The invasion never came, because **Hitler** feared the outcome.

Convoy battles were decisive. The garrison was in trouble when all the merchant ships were sunk in a February convoy and when, in April, 30 out of 47 Spitfires flown in from **aircraft-carriers** were destroyed in two days. It felt better when 64 more, flown in during May, produced a daylight air victory over a convoy which, nevertheless, got only two out of six ships through. It drew courage when, as in mid-August, a tanker and four other ships out of fourteen arrived, even though the escorts had suffered terribly, losing a carrier, two cruisers and three destroyers.

Once French North-West Africa had been invaded in November 1942, the passage of convoys became much easier and Malta reverted entirely to the offensive against southern Europe. By 30 June 1943, 1,436 Maltese civilians had been killed and 3,415 wounded.

Manila Bay, Battle of. Within six days of the start of the Spanish–American War in April 1898 a US Navy squadron, under Commodore George Dewey, had arrived in Manila Bay from Hong Kong. With six cruisers armed with 8in and 6in guns, he outclassed the ten run-down, largely unarmoured wooden warships (including four **cruisers**), under Rear-Admiral Patricio Montojo, which he found at anchor in the bay. In two phases, totalling three hours of none-too-accurate **gunnery** (14 hits out of 157 8in shells fired), the Americans (with eight wounded) destroyed the Spanish squadron (with 381 dead and wounded). Dewey then bombarded the shore forts and sent for the Army, which took possession of Manila in August.

In **World War II** the bay was to witness the arrival of the Japanese on 2 January 1942 and the ensuing sieges of Bataan and Corregidor. Followed in 1944 by numerous attacks on shipping in the harbour and the eventual fall of the badly damaged city to US troops in February 1945 as a culmination of the Philippines battle.

Mao Tse-tung (1893–1976). A mostly self-educated peasant who was a co-founder of the Chinese Communist Party in 1921. In 1927 he began staging unsuccessful peasant revolts against the Kuomintang in the latest round of Chinese civil wars, but by 1931, as Party chairman, had established a reputation and a foothold, with Russian help, in the south-east. By 1934 he had demonstrated that, for the time being, **guerrilla warfare** alone offered long-term success and that short-term survival demanded the Long March to the north-west to win respite from Kuomintang pressure. By 1937, at the start of the Chinese–Japanese War and **World War II**, he had won control of the party and created a military doctrine which dovetailed his own style of Marxist dogma. Until 1946. Mao collaborated with General Chiang Kai-shek in resisting the Japanese, but was careful not to exhaust his followers in the process.

Resumption of the civil war was not long delayed, followed in three years by the complete overthrow, with Russian help and numerous captured weapons, of the Kuomintang. In this period his doctrine of guerrilla warfare as the way to create a secure base for ultimate seizure of absolute political power was fully vindicated, to become the model for most other Chinese-motivated struggles for independence, e.g. in Malaya and **Vietnam**. Thereafter Mao concentrated more on consolidating Chinese power, only resorting to force of arms in **Korea** and India when it seemed China's frontiers were directly threatened.

Marathon, Battle of. The outbreak of the **Greek–Persian Wars**, **Darius I**'s delayed invasion of Greece in 491 BC and the landing in the Bay of Marathon of two Persian divisions under General Datis, as was hoped, drew off Miltiades' Greek Army from Athens, which was the objective of the other three Persian divisions. Miltiades found some 20,000 Persians drawn up in three ranks on the shore with their flanks in the air and the fleet beached. The Greek **infantry** phalanx attacked with reinforced wings which overlapped the Persian flanks while its centre held Datis' attention. The Persian flank guards panicked and ran for the boats, carrying the centre with them. Datis' rearguard covered embarkation but he lost 6,000 men before sailing away. The Greeks lost 192. Darius abandoned his campaign.

March to the Sea. After General **Sherman** had advanced from Chattanooga to capture Atlanta on 31 August 1864, he conceived a deep penetration aimed at the coast and Savannah to tear the heart out of the Confederacy and perhaps end the **American Civil War**. Sending the Army of the Cumberland back to Chattanooga eventually to defeat a Confederate army at Nashville on 16 December (and also reduce his **logistic** load), Sherman struck out on 15 November in three self-contained columns amounting to 68,000 men. To economize in manpower guarding lines of communication, he destroyed the **railway** behind him. Taking only 600 ambulances and 2,500 **ammunition** wagons, he lived off and laid waste the country as he went. With barely 15,000 troops, the Confederates fell back to hold Savannah. But Sherman seized Fort McAllister at the mouth of the Ogeechee River on 13 December, made a junction with the Navy and occupied the city on the 21st when the enemy withdrew. Then he turned northwards to seize Columbia and Wilmington, won the Battle of Bentonville on 20 March 1865 and entered Goldsboro on the 23rd, thus poised to join General Grant at Petersburg for the war's last act.

Marconi, Guglielmo (1874–1937). An Italian physicist who, in 1894, began experimenting with **electromagnetic effect** and spark **radio** waves. In England in 1896 he made a series of transmissions of ever-longer range which, in 1901, culminated in the spanning of the Atlantic and gave proof that signals were not blocked by the earth's curvature. Henceforward the Marconi Company, keenly encouraged by the military, was a world leader in the development of signal **communications** and **electronics**. (*See also* **Fleming, Professor Sir John** and **Jackson, Admiral Sir Henry**.)

Marines. In naval warfare there has always been a requirement for men trained to fight both at sea and on land, although until the 17th century the distinction between sailors and soldiers often was indistinct. In Greek, Persian, Carthaginian and Roman warships the majority of those who hurled missiles and engaged in hand-to-hand combat after boarding were soldiers sent to sea for the occasion; as were, for example, the longbowmen and men-at-arms in the English fleet which defeated the French at **Sluys** in 1340. Such men could also help maintain the sailors' discipline and adapt more easily to **amphibious** operations.

It was the British who, in 1664, formed the first modern Marine corps, the so-called Admiralty regiment, to be copied in 1665 by the Dutch who, in 1667 during the Second Anglo-Dutch War, used them most effectively in a raid on the Medway dockyards. And in 1775 the Americans formed a Marine Corps, which fought against the British during the **American War of Independence**.

The variety of tasks on ship and shore performed by Marines developed in line with technology and its impact on tactics. When ships' companies disappeared under **armour** in steamships, and there no longer was a role on deck to snipe enemy marksmen or take part in boarding operations, Marines were often allocated to man at least one gun turret per ship. The latter role led in 1862 to the formation of the Royal Marine Artillery, to differentiate them from the ordinary Royal Marine Light Infantry, which had been formed in 1855. Whatever branch they belonged to, Marines were used ashore as an arm of sea power; seizing or defending forts and **dockyards** as the vanguard of invasions, and carrying out hit-and-run raids.

Both the British and the Americans in **World War I** formed Marines into units as part of field-force formations in France, but their most significant fight was as assault landing-parties in the Zeebrugge raid on 23 April 1918. Between the wars, while all countries (not Japan) neglected amphibious warfare the US Marines began developing the technology and techniques they foresaw might be necessary in a **Pacific War** against Japan. This process led to the creation of a self-contained organization that included armoured vehicles and aircraft as well as special **landing-craft**. Like the British, they did little more than formulate ideas and make a few prototype craft – just enough to have something in mind when, suddenly in 1940, the German victories in Europe and mounting Japanese aggression in the East made it plain that large-scale amphibious warfare was unavoidable. When the British were compelled to take the lead in 1940, however, the Royal Marines were unready, owing to other commitments, to adopt the **Commando** role at once. That would wait until 1942, in time for the **Dieppe** raid.

Since then, Marines the world over have spearheaded nearly all the major amphibious operations, besides helping develop the latest assault techniques with surface vehicles and **helicopters**. As a result they have augmented their élite status by reaching such high professional standards, in most phases and natures of war, that they are automatically considered for immediate employment in dealing with so-called 'brushfire' outbreaks in the **Cold War** and in the opening stages of unexpected small wars such as that of the **Falkland Islands** in 1982. At the same time they also pioneer or keep alive raiding techniques and the special skills needed in arctic, **jungle**, **mountain** and **desert warfare**.

Marlborough, John Churchill, Duke of (also **Prince**) (1650–1722), was from an impoverished family. He made his fortune at court and won distinction while fighting alongside the French in the Third Anglo-Dutch War. At the Battle of Sedgemoor in 1685 he was responsible for the defeat of the Monmouth Rebellion against his mentor, James II. But in 1688 he deserted James to help put William of Orange on the throne. William III, however, mistrusted Marlborough and only reluctantly gave him leading roles in government and military operations against the Jacobites. But, shortly before his death in 1702, on the eve of the War of the Spanish Succession, he appointed Marlborough C-in-C of the British and Dutch forces opposed to the French, thus, as the favourite of Queen Anne, making him prime minister in all but name.

When leading a fractious alliance, Marlborough proved a political genius as well as employing brilliant **strategy**, **logistics** and **intelligence gathering** from the highest enemy sources. Never was he defeated in many sieges, nor at the battles of Blenheim (1704), Ramillies (1706), Oudenarde (1708) and **Malplaquet** (1709). At the same time he saw to the defeat of French naval power and the **blockade** which brought France to her knees. He fully exploited the flintlock **musket** by controlled platoon, instead of company, firing; welcomed the **bayonet**; and, as a shrewd tactician, practised cooperation between all arms in battle.

When political opponents intrigued and engineered his dismissal in 1711, the chance of enforcing a favourable peace disappeared. But, following the Queen's death in 1714, he finally was restored to royal favour.

Marne, Battles of the. The German plan devised by General Count Alfred von **Schlieffen** before **World War I** was dominated by the concept of a wide swing through Belgium, west of Paris, with a strongly reinforced right wing aiming to envelop the French Army between the capital city and the eastern frontier forts. By August 1914, however, the Chief of Staff, General Helmuth von Moltke (nephew of Field Marshal von **Moltke**), had not only reduced the number of formations in the right wing but launched them with an inadequate system of signal **communications** as well as a defective **logistic** plan.

The First Battle. On 3 August the Germans stormed into Belgium, pushing the Belgian Army back into Antwerp and falling upon the French Fifth Army and the British Expeditionary Force (BEF) between Namur and Mons. Obligingly the French assisted by launching strong attacks with their reinforced right wing in Alsace and Lorraine on the 4th, thus denying resources to their threatened left in Belgium. This offensive was

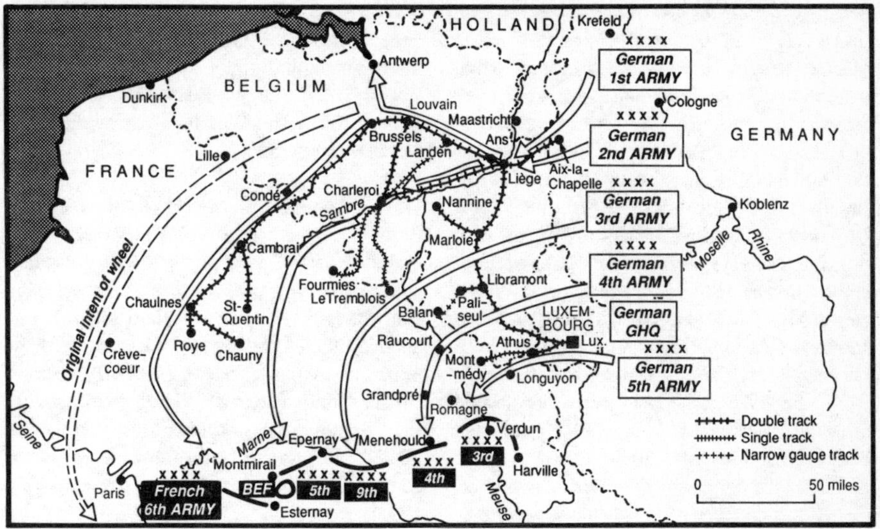

The Battle of the Marne, 1914

checked by the 18th with heavy loss and thrown back by a violent German counter-offensive which, nevertheless, diverted further strength from their own right wing. Heavy attacks at the battles of the Sambre and Mons on 22 and 23 August at last warned the Allies under General **Joffre** of their peril. They began a long retreat towards Paris, pivoting on **Verdun** and gradually reinforcing their left wing while the French defeated the German offensive on the eastern frontier.

By 30 August the Germans were in some difficulty because, owing to inadequate planning of **telephone** communications and much reliance on a defective **radio** system, their C^3I was beginning to fail. Their First Army therefore made several erratic movements and swung east of Paris (instead of west). At the same time signs of exhaustion appeared among men and **horses** when the **railway** system failed to keep pace and supply dwindled. Seeing an opportunity to strike the German right flank while also protecting Paris, Joffre created a Sixth Army north-east of the city, a presence which compelled First German Army to stop advancing in order to form a strong flank guard. A diversion which opened a large gap between the First Army and the Second on its left, into which the BEF and Fifth French Army entered, largely unopposed and only belatedly noticed by the Germans. Indeed, at this very moment on 8 September a worried German staff officer with plenipotentiary powers arrived at HQ Second Army to find it about to retreat. Appreciating that the troops were exhausted he ordered a general withdrawal, which saved Paris and, perhaps, First German Army – a retirement which ended in an entrenched river-line position for the Battle of the Aisne on 15 September.

The Second Battle was fought between 15 and 19 July 1918 as part of the Fifth and last **Hindenburg** offensive. It was intended to exploit the Third Offensive, which had been brought to a halt by the French and Americans at the battles of Cantigny and Château-Thierry in May and June respectively. On this occasion the German left wing, attacking to the east of Reims in an effort to reach Épernay in conjunction with the right wing attacking from Château-Thierry, met a well-forewarned and prepared defence. Directed by General Foch, it gave but little ground. To the west of Reims, however, where the defences were weaker, a large penetration was made to allow a strong crossing of the Marne, a success soon obliterated when heavy air and artillery bombardments were directed against the bridges and their approaches, while reinforcements blocked further advances by German troops, who at last saw the hopelessness of their situation.

General **Ludendorff** ordered a withdrawal of the Soissons–Reims salient, but was pre-empted on the 18th when French and American troops launched Foch's long-prepared counterstroke aimed at its elimination. Caught on the wrong foot, the dispirited Germans were hurled back by well-coordinated **tank** and **artillery** attacks, which inflicted heavy casualties and indicated to Ludendorff that the Hindenburg offensive was over. He cancelled the next offensive in Flanders, evacuated what was left of the Soissons–Reims salient and reverted to the defensive on the eve of the Battle of **Amiens**, which broke his resolve.

Marshall, General George (1880–1959). A brilliant infantryman who was First US Army's Chief of Staff in France in **World War I** and went on to become Chief of Staff of the Army on 1 September 1939. It was his task therefore to lift the Army from a low point in its history to readiness for a war he never doubted was coming, a task he carried out with enormous success. Yet it is for his great influence on President **Roosevelt**, Winston **Churchill** and many another top Allied leader during **World War II** that he is most renowned. The way in which he shaped American policy to obtain priority for tackling Germany before Japan (and thus securing preference for the Army over the Navy) and his dogged resolve to invade Europe at the earliest moment had immense impact, for better or worse, on the post-war world. He helped shape this world when, as Secretary of State from November 1945 to January 1949, he put through the European Recovery Programme (the famous Marshall Plan) to save Europe from economic chaos and Communism. In this period he helped obtain recognition for Israel and the founding of **Nato**. Later, as Secretary of Defense from September 1950 until September 1951, he implemented the Nato agreements and strengthened the armed forces during the **Korean War**. In 1953 he was awarded the Nobel Peace Prize.

Martel, General Gifford le Q (1885–1958). A Royal Engineer who was the British Tank Corps' first staff officer in 1916 and wrote a paper proposing **tank** armies. As a technical staff officer between the **World Wars** he developed obstacle-crossing vehicles; cheap light 'tanks', which later became **infantry** weapon carriers; and was instrumental in introducing **Christie** cruiser tanks into British service in 1936. He was out of his depth when commanding the **Arras** counterstroke in May 1940.

Marx, Karl (1818–83). A renowned radical philosopher who, through publication of his famous *Das Kapital* between 1867 and 1894, created a worldwide political/economic revolution. In collaboration with Friedrich **Engels** he studied military **strategy**, with a belief in war's totality through intense class, economic and **psychological warfare** besides actual combat. The Russian-led Communist socialist states, which appeared during **World War I**, drew selectively on Marxism; later to contribute to the causes of **World War II**, the **Cold War** and many other conflicts, including the subsequent collapse of the USSR in 1991 with all its ramifications.

Matapan, Battle of. In an endeavour to interfere with British reinforcement of Greece in March 1942, the Italian Fleet, with a **battleship** and eight **cruisers** commanded by Admiral Angelo Iachino, attempted to intercept **convoys** off **Crete**. Aware of the move, Admiral **Cunningham**'s cruisers made contact on the 27th and lured the Italians towards his three battleships and an **aircraft-carrier**. An abortive **torpedo** attack on the Italians persuaded Iachino to turn for home. A second attack stopped a cruiser. The delay inflicted led to a night action aided by **radar** and **searchlights**, resulting in the sinking of three Italian cruisers. But British signalling errors let the battleship escape.

Maxim, Sir Hiram (1840–1916). An American inventor who simply was quickest to realize that the smokeless ballistite **explosive** produced by Alfred **Nobel** in 1885 made feasible the automatic, belt-fed **machine-gun**, using a brass cartridge case, upon which he was working. Having demonstrated the gun and cartridge's effectiveness, his major successes thereafter were marketing them to most of the world's armies. His brother, Hudson, was a scientist who studied atomic theory and subsequently, from 1888, developed various explosives for the US Navy and Army.

Mechanization, Military. The necessity for military mechanization has been apparent since the first use of the lever and the invention of the **wheel**. It was applied to **chariots**, to move many early **siege engines** and to enhance fire-power. Block and tackle was vital in sailing ships and for the lifting of heavy equipment. There was an element of mechanization in the crossbow, notably with Leonardo da **Vinci**'s giant, wheeled model of *c.* 1485. Indeed nobody, prior to the 18th century, did more than da Vinci to promote mechanization with such imagination. From his fertile mind came mobile **artillery** and a mechanical flail, a sort of **tank**, flying machines and water-powered machinery.

Mechanization in the modern sense is often dated to the invention of the Daimler internal-combustion engine as a means of **propulsion** in 1883. But the development of steam power, culminating in James **Watt**'s relatively efficient engine of 1763 and followed in 1769 by Nicolas Cugnot's steam-driven tricycle of 1769, is more appropriate. From that came the evolution of steam-driven ships (leading to the final phasing out of sail by the 1880s) (*see* **Shipbuilding**), artillery, **railways**, traction engines and cranes. Leading to the first steerable **balloon** in 1852, the petrol-engined **motor vehicle** in 1885, **airships** in 1888 and **aircraft** in 1903, along with electro-oil-driven **submarines** and petrol-engined **torpedo-boats**.

By 1900 the automobile was in general use, an **armoured car** had been built and European armies were trying out trucks and motor tractors for **logistic** purposes. In 1911, for example, the British War Office devised a scheme encouraging manufacturers to build trucks to military standards – making it easy to requisition a standardized fleet in wartime. Throughout **World War I**, however, mechanization became essential because animals could no longer cope with the vast logistical loads of 20th-century warfare; and because motor transport often proved far more economically flexible than railways. At the same time, tractor-towed artillery and the **tank** introduced a new operational dimension of mobility and protection by making the **horse** obsolete in battle – a horse population which, in any case, was greatly reduced as civil carriers found motor vehicles easier and more cost-effective to run.

Immediately after the war the three services of many nations embarked on studies and trials which progressively led to the development of more reliable and efficient mechanical equipment; along with new factories to make it, the training of drivers and mechanics, the creation of base and field workshops, the means to store and move fuel, labour-saving handling equipment, earth-moving machinery, mobile **bridging** and much improved roads for rapid transit. And at sea mechanized **landing-craft** gave a boost to **amphibious warfare**.

The enormous cost of scrapping obsolete equipment, and the procurement of new, retarded change until **World War II** loomed closer; and usually there was a tendency to buy cheaper vehicles, such as four-wheeled, two-wheel drive (4 x 2) trucks, with poor cross-country performance, inferior artillery tractors and unreliable tanks. Indeed, in that war only the British and American armies became completely mechanized (and even they had to employ animals in **jungle** and **mountain warfare**). The Germans and Russians relied upon horses to the end; and the latter would have been in trouble had not the

Americans and British supplied them with some 400,000 lorries, 51,000 **jeeps**, 2,000 railway locomotives and 12,000 wagons.

Since 1945 the three services of every reputable power have become fully mechanized – often creating new operational and logistical problems besides solving old ones, and giving rise to the need to satisfy immense fuel consumption and meet the demands of complex manufacture, repair and maintenance of machines that are capable of performing reliably in all climates and terrains. The recruitment and training of many managers and specialists to make full and efficient use of ever-diversifying ranges of equipments, often when under operational stress, have become essential.

Medical services, Military. Until the latter half of the 19th century the plight of wounded and sick sailors and soldiers was desperate, to say the least. From the earliest days treatment and care had been blighted by scientific ignorance, hit-and-miss methods, magic and bigoted religious dogma. Often the wounded might, for lack of evacuation services, be left to die where they lay or deprived of care for lack of surgeons, trained nurses, dressings and medicines. Their chances of survival were less than 50 per cent; the sick, owing to poor **health services**, often fared far worse.

The first military surgeons are said to have been Greek, namely Podlarius and Machaon, *c.* 1200 BC. The father of modern medicine, however, probably was another Greek called Hippocrates, *c.* 460 BC. True or false, there is no doubt that the Greeks initiated the scientific approach to medicine which gradually was spread by the medical officers of the Roman Army and was connected to knowledge from Arab and oriental countries, culminating in the founding of the first medical school at Salerno in the 10th century. Henceforward there would be slow progress during the **Crusades**, but after the 14th century faster acquisition of knowledge and experience came in the spirit of the Renaissance and through such investigators as Leonardo da **Vinci**.

The urge of the military medical authorities to save life was stimulated by the manpower shortages of the **wars of the 16th, 17th and 18th centuries**. But although numerous fundamental discoveries were made by distinguished members of the **Royal Society** and the like, the wounded benefited but little. Moreover, the appearance of **explosives** and firearms in the 14th century had made their plight far worse because it was far more difficult to deal with severe burns and gunshot wounds, rather than relatively clean cuts and abrasions. Lack of anaesthetics, sterile instruments and dressings and so on ensured that at least 50 per cent of the wounded who were fortunate enough to reach improvised medical stations alive died from shock, sepsis and other infections, besides inept surgery, treatment and care and incorrect diet.

The demonstration in 1846 of ether as a general anaesthetic by the American Thomas Morton was a major turning-point in surgery. An event soon followed by the scandalous treatment of the wounded in the **Crimean War** and the reforms forced on the military by Florence **Nightingale** and humanitarians after the fall of the British government. Her *Notes on Matters Affecting the Health, Efficiency and Hospital Administration of the British Army* and her *Notes on Nursing* became bibles for all reformers including administrators on both sides during the **American Civil War**.

Medical technology advanced at the same time. Anaesthetics, often objected to by doctors on religious grounds, were used for the first time for surgical operations. In parallel with a drive to cleanse and adequately ventilate hospital wards went campaigns to improve public, as well as military, sanitation and raise health standards with water-borne sewage systems everywhere. At the same time improvements to **food**, plus Joseph Lister's discovery of antiseptics in 1865 leading to sterilization of medical instruments and wounds, were revolutionary. Mortality rates were reduced dramatically. Populations rose enormously, producing vast pools of manpower to expand industry and fill the ranks of enlarged armed forces. A process improved upon after the Spanish–American War in 1898 when the causes of malaria and yellow fever, along with means of their suppression and prevention, were discovered.

Nevertheless, mortality rates in **World War I**, though lower than before, remained high, notably from disease in undeveloped countries, such as those of East Africa. Even though it was discovered that Lister's carbolic acid had serious defects as an antiseptic, the latest generation of surgeons (against the objections of their seniors) were able to show in practice that it was thorough cleansing of a wound, with the removal of infected tissue, which best controlled the deadly gangrene. In addition, medical organizations which evacuated casualties rapidly and treated them early saved the most lives.

Mechanization and the use of motor ambulances, which had hastened evacuation of casualties in World War I, were put to even better use in **World War II**. Evacuation by **balloons** had first been tried in the **Franco-Prussian War** and sporadically, with France setting the example, in World War I. It was much more widely used in World War II, principally in such underdeveloped theatres of war as **Burma** and the **Pacific**. The greatest revolution, however, was through increased use of vaccines against disease and tetanus, together with blood transfusion and use of plasma. Most dramatic of all was Alexander Fleming's antibiotic **penicillin** to control infection and strikingly improve the wounded's chances of survival. Such chances were further enhanced by the introduction of DDT insecticides to control the carriers of disease (particularly malaria), in addition to suppressive drugs such as mepacrine. Armed forces lacking these benefits (as did the Japanese in particular and the Germans to some extent) risked the loss of battles and significant decline in **morale**.

Commanders, above all those in action, have a duty to care for their men's health: in the latter half of the 19th century it became normal for medical practitioners to act as their advisers on such matters, giving advice which was ignored at one's peril, with the risk of serious consequences. Naturally the medical authorities had no difficulty in obtaining priority for **helicopters** to evacuate the wounded with unprecedented speed to medical units; they helped raise survival rates, in the **Korean War** for example, to 98 per cent of those who arrived alive at a surgical station. More controversial was the employment of psychiatrists to deal with cases of **combat fatigue** once a more enlightened view of this phenomenon was taken after World War I. Improvements continue as knowledge increases and prejudices are overcome.

Mediterranean Sea, Battles of the. This large inland sea can lay claim to have been at the heart of the evolution of **naval warfare** since the earliest recorded time. The proximity of Europe to Africa, plus the exits to the boundless eastern lands, tempted seafaring invaders and stimulated the development of warships (*see* **Shipbuilding**) to protect trade. The first known reference to a naval engagement dates from *c*. 2700 BC when the Pharaoh Sahu-reh sent eight ships to rescue prisoners from Phoenicia. In due course battles between fleets of over 100 vessels occurred in line with the growth of population and changes in technology. As examples of this trend the battles of **Salamis** (about 1,100 ships in 480 BC), Aegospotami (about 300 in 405 BC), Ecnomus (about 680 in 256 BC), Actium (800 plus in 31 BC), **Lepanto** (about 570 ships in 1571) and the Nile (30 ships in 1798) are worthy of study. The Battle of **Lissa** (38 in 1866) marks the tactical transition from sail to steam and the strange revival of the **ram**.

In **World War I** there were no major sea battles in the Mediterranean. The Allied navies were challenged by only a few Austrian and German **submarines** emerging from the Adriatic and the Bosporus. But Italy's entry into **World War II** on 10 June 1940 immediately caused action as she attacked **Malta** and attempted to prevent the passage of **convoys** through the Sea while reinforcing her forces in **North Africa**. On 3/4 July, however, the British Fleet (under

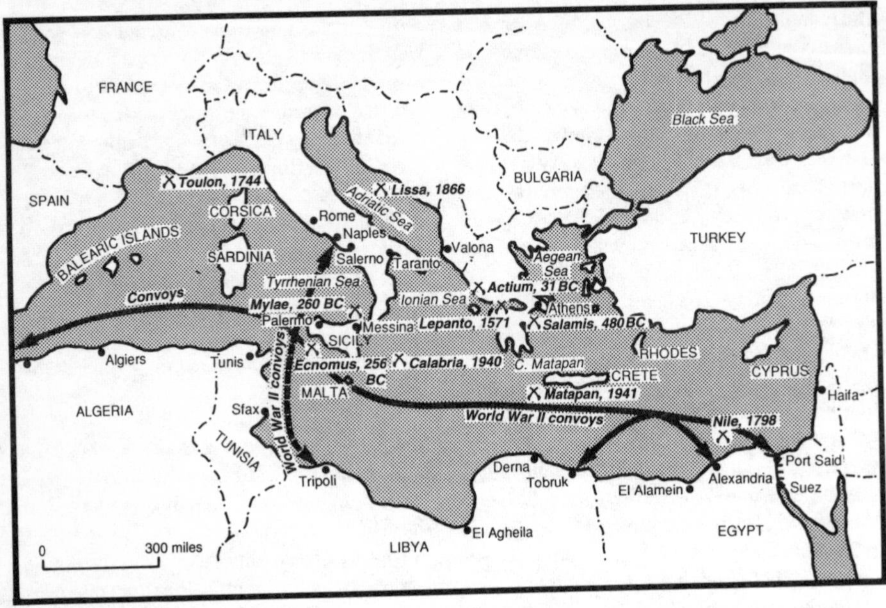

The Battles of the Mediterranean Sea

Admiral **Cunningham**) attacked French warships in Oran and seized others in Alexandria. It then appeared off Messina to damage and drive into port the Italian Fleet at the Battle of Calabria on 9 July. There the Italians mostly remained until 11 November when British **aircraft-carrier**-borne torpedo-bombers in a celebrated night attack at **Taranto** severely damaged three **battleships** and two **cruisers** for the loss of two machines, a blow from which the Italians never recovered.

Before the Germans attacked **Malta** in January 1941, Britain managed to fight through a few **convoys** with relative ease. But from then on convoys were sent mainly to sustain **Malta** and its interceptions by air, surface vessels and submarines of Axis shipping to North Africa. Every convoy provoked a battle, in which Axis aircraft and submarines played the leading roles. Occasionally the Italian Fleet put to sea, usually to be driven off, as at **Matapan** in March 1941. Off Greece, **Crete** and **North Africa**, however, the Royal Navy suffered heavy losses in 1941, mainly from submarines and aircraft. In December 1941 it was almost crippled by Italian two-man midget submarines which entered Alexandria harbour and severely damaged two battleships. In addition to suffering the sinking at sea of another battleship by German submarine, and the loss in a minefield (**Mines, Naval**) of three cruisers and a destroyer, thus making it easy for the Axis armies to be reinforced for the major offensive in May 1942.

This was the nadir of British fortunes, even though, with American assistance, Malta was kept in action. Soon the boot was on the other foot when the British, amply supplied by aircraft, strongly reinforced and invariably well posted by **intelligence**, were able to play havoc with the Axis supply lines. This was the overture to victory at El **Alamein** and the Allied landing in North-West Africa (*see* **North Africa, Battles of**). As a prelude to the relief of Malta, it guaranteed the clearing of the North African shore, thus reopening the Sea to convoys, with enormous benefits to Allied **logistics**.

For the remainder of the war the Allies had free use of the Mediterranean to strike where they chose against the enemy in

Sicily, Italy, the Adriatic and Aegean Seas and France. Such freedom was largely denied the Italians (until they surrendered in September 1943) and Germans who were hunted relentlessly whenever their submarines, small craft and aircraft ventured forth – the surviving major elements of the Italian Fleet having either surrendered or been destroyed.

Since 1945 sporadic naval actions by light forces have taken place in the Eastern basin in connection with **Arab–Israeli wars**, plus the major amphibious invasion of Suez in 1956.

Megiddo, Battles of. Megiddo was a **fortification** guarding a strategic bottleneck in northern Palestine between Egypt and Mesopotamia (Iraq). It was the scene in 1479 BC of the first recorded battle, a clash between the Egyptian Army under Thutmose III and the Hyksos Army under the King of Kadesh. Each had about 1,000 **chariots**, but Egyptian dash and the flank leverage they applied was decisive and led to the fall of the fort with immense loss to the Hyksos of men and material.

On the same ground in 609 BC the Pharaoh Necho scored another decisive victory (sometimes known as Armageddon, meaning 'the ultimate') over a Jewish army under King Josiah. And in the years to come it would frequently be disputed by contending forces.

The nearby Allied attack against the Turks in 1918 in **World War I**, however, was the largest ever fought there. After defeat at Gaza in 1917 the Turks and Germans fortified a strong line with three armies numbering 36,000, under General Liman von Sanders, between Jaffa and the River Jordan. The Allied commander, General Sir Edmund **Allenby**, had 57,000 men, 12,000 **cavalry**, plus **armoured cars**, and the Arab **guerrilla** force operating in the desert on his right flank. By elaborate deception he deluded von Sanders into believing the attack would fall on the Turkish inland flank. Instead he attacked near the coast, where only 8,000 Turks with 130 guns stood against 35,000 British with 400 guns. The attack started on 19 September, overran the Turks on the coastal sector and pivoted inland. Those Turks not enveloped were thrown into rout, harried by **aircraft**, mobile forces and Arabs. Damascus fell on 1 October, Aleppo on the 25th. Five days later Turkey signed an armistice.

***Merrimack* versus *Monitor*.** In the **American Civil War**, *Merrimack* was a wooden, screw steamship which the Confederates converted into an ironclad, fitted with a ram, armed with six 9in, two 7in and two 6in guns firing broadside. *Monitor* was a Federal low-freeboard, steam raft mounting a rotating turret with two 11in guns. *Merrimack* raided Federal shipping blockading the James River on 8 March 1862, sank a sloop by ramming and a **frigate** by gunfire and forced another frigate ashore. But when next day she appeared again, *Monitor* was waiting. *Merrimack* tried to ram but was evaded by the more manoeuvrable *Monitor*. For three and a half hours they bombarded each other at close range, the former firing more rounds and scoring 23 hits, the latter firing 53 rounds at a much slower rate, because the turret had to be hand-traversed inboard to reload, and scoring 20 hits. Each retired with slight damage and weary crews. Next day *Merrimack* was back, but *Monitor* declined to engage in case damage would leave her opponent a free hand. A few days later *Merrimack* was burned by the Confederates, to prevent her falling into Federal hands, and *Monitor* sank in rough water.

Mesopotamian campaigns. Situated between the rivers Tigris and Euphrates and the land masses of Europe, Asia and Africa, it is no mere chance that Mesopotamia was not only a place to trade and a strategic crossroads, but also a vital cultural and technological, urban centre. New **metallurgy** and weapon systems, such as the **chariot**, sprang from a land which, perforce, survived in a state of belligerence. Over the centuries its invaders included Persians, Phoenicians, **Greeks, Romans**, Arabs (*see also* **Muslim wars**),

Mongols and Turks (*see* **Ottoman Empire**) all of whom desired possession of the vital overland routes of a country difficult to defend.

The construction of a **railway** from Constantinople to Baghdad and Turkey's siding with Germany in October 1914, doomed Mesopotamia to involvement in **World War I**. On 6 November a small British–Indian force landed at the mouth of the Shatt-al-Arab, captured Abadan and, over the ensuing year, advanced slowly up the two rivers. But on 22 November 1915 the outnumbered, **logistically** weak British were defeated at Ctesiphon and thrown back into **Kut-al-Amara** where they surrendered to the Turks on 29 April 1916. In December the British, greatly reinforced and now well administered under the command of General Sir Frederick Maude, advanced again, defeated the Turks, took Baghdad on 11 March 1917 and proceeded, by the war's end, to overrun the rest of the country, including its **oil**fields. This was a **mechanized**, **desert** war employing numerous river craft, **motor vehicles**, **armoured cars** and **aircraft**.

As an Arab state in 1918, under British mandate until 1932, Mesopotamia came to be known as Iraq. The British developed airfields and used aircraft and armoured cars to police the country; and returned by force during **World War II**, in May 1941, when the Iraqis, with German incitement, threatened vital oilfields.

Since 1945 Iraq has undergone several revolutions, culminating in the coming to power of Saddam Hussein, the **Iran–Iraq War** (1980–88) and the **UN–Iraq War** (1990–91).

Messerschmitt, Willy (1898–1978). A **glider** designer who, after **World War I** began to design **transport** and small **aircraft**. He became chief designer of the Bayerische Flugzeugwerke in 1927 and later its head, when it was renamed after him. His all-metal Me109, with its single-spar wing, proved one of the outstanding **fighters** of **World War II**. Less successful, though playing important roles in the German air force, were the twin-engine Me110 fighter, the gigantic six-engine Me323 transport (adapted from a glider), the Me163 (Comet) **rocket**-propelled fighter and the **jet**-propelled, twin-engine Me262.

Metallurgy. When **bronze** was discovered *c.* 6500 BC and came into common use *c.* 4000 BC, the art and science, chemistry and technology of metallugy found its beginnings. Henceforward, and predominantly driven by military requirements, the study of metals and alloys – their composition, structure, strengths and weaknesses, manufacture and uses – became vital to civilization and the military art. Time and again inventors carried out **research and development** to induce what they believed was a military requirement, or in order to satisfy a previously stated demand. Over and over again the basis of new technology is rooted in a new metal or alloy. The powerful threat of a steel-tipped arrow from a long**bow** prompted manufacture of top-grade **steel** for improved **armour** protection – which in turn called for higher steel production. Without **aluminium** the development of **aircraft** would have been retarded; if Hans Oersted had not isolated it as a precious metal in 1825 somebody would have needed to do so in order to give Alfred Wilm a suitable metal to alloy with other metals to produce duralumin in 1909. Modern **communication** systems depend on a variety of conductors based on copper. And aircraft, **space vehicles**, **rockets and guided missiles**, their means of **propulsion** and **electronic** controls would be impossible without special alloys, including those with **titanium**, which perform under conditions of extreme temperatures.

The many references to it in this Encyclopedia would suggest that metallurgy is fundamental to progress in many fields as well as in military technology.

Middle Ages, Wars of the. The period of European cultural, economic and technical stagnation between the decline of the **Roman** Empire in the 5th century and the **Hundred Years War** in the 14th to a large extent may be ascribed to the overthrow of one sophisticated, though decaying, military system by another simpler, predatory one. Invading Huns, Goths and **Vikings** and their like were bent mainly on destruction with basic weapons; they contributed virtually nothing to improve what they found. The **Muslim wars of expansion**, commencing in 632, initiated wars of **religion**, of which the **Crusades** (1096–1226) were but one phase, fought with weapons and tactics that were only marginally superior to those of Rome in her heyday.

At sea the **galley** remained the basic fighting ship until the Battle of **Sluys** in 1340; with ramming followed by boarding, supported by relatively ineffectual missile-throwers, as the principal tactical methods. On land there was only a slight shift in tactical emphasis caused by the combatworthiness of the more lightly equipped but well-armed, more mobile armies of the East and Islam. Although the use of a wagon **laager** at the Battle of **Adrianople** in 378 was of some future significance. The barbarians tended to win because their morale and belligerence were superior to those of decadent opponents. Battles continued to be head-on and hand-to-hand, with only a nod to **strategy** and few attempts at tactical outflanking. From **Belisarius** and **Edward III**, few great innovative captains appeared on sea or land. The historian **Vegetius** merely extolled the Roman military system.

For 900 years the atrophy of society and the decay of **communications** – both **logistic** and signal – retarded intellectual and technical changes. To preserve what remained of Roman civilization, men tended to spend vast effort and treasure on the construction of **fortifications**, without keeping pace with countervailing **siege engines**. This resulted in the growth of bigoted, self-interested organizations, as in **chivalry**, bent on retaining the status quo and resisting progress. The principal military hallmarks of the Medieval Age are the extremely gradual improvements made to missile-throwing weapons and sea and land communications. Indeed, it is no coincidence that the end of that period and the start of the Renaissance coincide with the appearance of the longbow, the invention of gunpowder, the development of firearms and the first, rudimentary **charts and maps**. For in little more than a century after the battles of **Sluys** and **Crécy** the caravel was revolutionizing discovery, **naval warfare** and commerce; castles were obsolescent; land communications were being restored and new approaches to **strategy** and **tactics** were being formulated by thinkers such as Niccolò **Machiavelli**. While inventors like Leonardo da **Vinci** were designing better sources of power and weapon systems. Paradoxically, the restoration of strong military methods with advanced technology began to create a more civilized society upon the ruins of the old.

Midway, Battle of. Midway Island, which the US Navy began developing as a naval and air base in 1940, was the main objective of the Japanese in June 1942 to consolidate their perimeter defences by extending their earlier conquests and at last destroying the American **aircraft-carriers**. Having in May failed at the Battle of the **Coral Sea** to seize Port Moresby, Admiral **Yamamoto** still attacked the Aleutian Islands, as a diversionary operation. Against Midway he sent the carrier fleet as advanced guard to the **battleship** fleet and the **amphibious** invasion force. Amply forewarned of the Japanese plans, Admiral **Nimitz** assembled three carriers with 250 aircraft (under Vice-Admiral **Fletcher**), plus 109 machines on Midway, against Admiral Chaike Nagumo's four carriers with their 275 aircraft.

A complicated plan, overconfidence and sloppy staff work attended the Japanese approach to Midway on 4 June. They were unaware of Fletcher's presence, whom their **reconnaissance** failed to locate. The initial air strike against the island was complete and a second being prepared when Nagumo received news of the American carriers. A **bomber** attack by the Americans went astray and the **torpedo** aircraft, attacking unsupported, were annihilated. A moment later, however, the dive-bombers arrived, caught

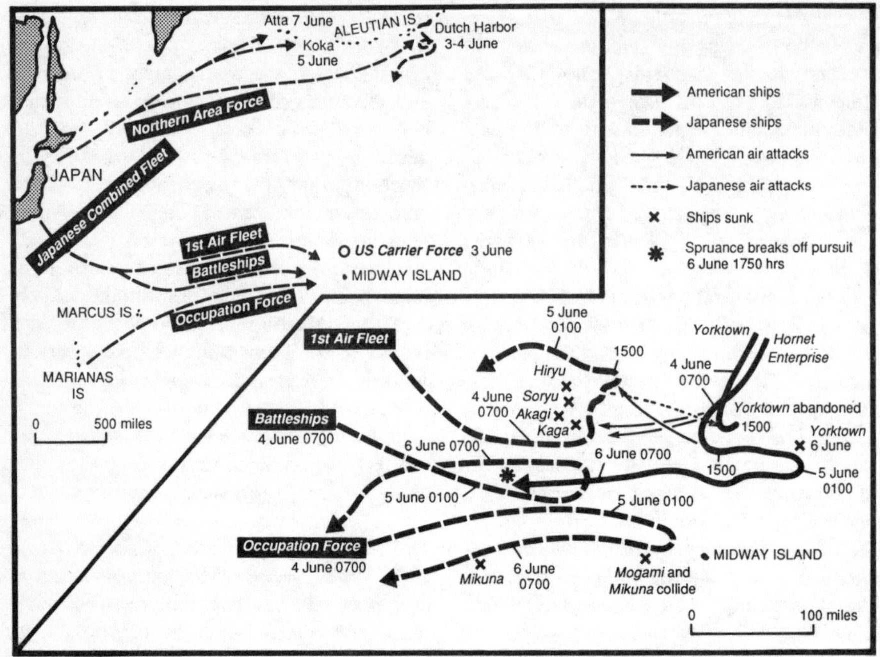

The Battle of Midway, 1942

the Japanese in the midst of rearming their machines and set three carriers ablaze. The fourth carrier, however, escaped and it was her aircraft which crippled the US carrier *Yorktown* (finally sunk by **submarine**) before herself being found and sunk by the Americans. Realizing there was no hope of taking Midway, Yamamoto withdrew, having lost the cream of his naval airmen and Japan's hope of ultimate victory.

Mines, Land. With the construction of strong, stone **fortifications** arose the need to breach their walls as the crucial phase of **siege warfare. Siege engines** had their uses but their effect was often slow to materialize. Demolition of a wall could be swifter and more complete, especially if the foundations were not on rock-solid ground, by undermining the wall via a tunnel and then burning away the props holding up whatever was above, a process which was expedited in due course after the invention of gunpowder.

The appearance of heavy **artillery** led to the gradual disuse of mining, although the technique was never completely abandoned, not even with the development of Marshal **Vauban**'s star forts in the 17th century. They were occasionally employed during the **American Civil War** and extensively during **World War I** when, on **Vimy Ridge** as an extreme example, mining and countermining on a very large scale went on for months on end to win control of the ground beneath the crest line. And on Messines Ridge in 1917, 500 tonnes of ammonal practically wiped out part of the German front line prior to a British assault. Towards the war's end a few buried **anti-tank** mines had been made by the Germans after the Battle of **Cambrai**.

However, it was in **World War II** that the buried mine came into its own. By 1939 the need for two basic types had emerged: the small anti-personnel mine designed to wound; and the larger anti-tank mine designed to break the track of a **tank**. Both categories were cheap to produce and easy

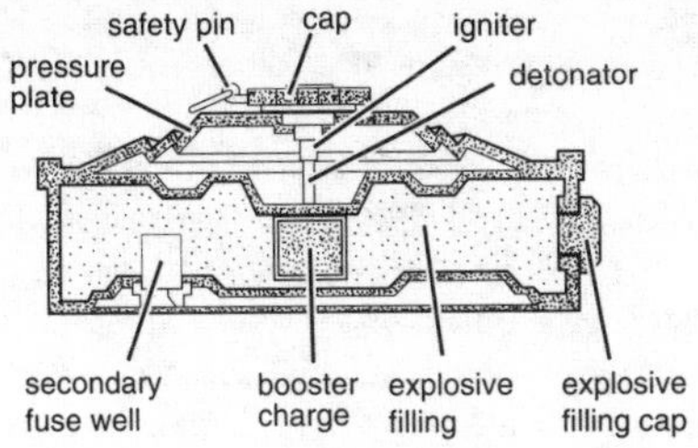

Anti-tank mine

to handle, but much more difficult to detect and remove once buried in the ground. Extensive minefields were laid in every theatre of war, usually sited to channel the enemy into a defile where he could be suitably attacked.

Both anti-tank and anti-personnel mines were detonated by pressure fuses and usually encased in metal boxes. One of the first improvements was to replace the container with a wooden one to defeat the metal detectors then in use. The anti-tank mine was comparatively insensitive to handling and, since the only certain way to clear a minefield (as it still is) was to dig up the mines by hand, it was not long before explosive anti-handling devices were fitted. Anti-personnel mines designed to jump up and spray pellets at body height, when triggered remotely, also began to appear.

Minefields became such a hazard, and could be laid so much more quickly than they could be cleared, that mechanical clearing devices were soon being developed, particularly by the British. Probably the most successful idea was the flail tank: a set of chains mounted on a rotating boom extended from the front of the vehicle; the chains beat the ground as the tank moved forward, exploding the mines as it went. Other ideas have included heavy rollers and ploughs.

Since 1945 mines have become even more complex. In the anti-tank field they have been designed to attack the more vulnerable belly, or even the side, as well as breaking the track. This has led to much more sophisticated fusing, such as by fuses actuated on contact with the tank hull, or influence fuses actuated by heat, noise, vibration or magnetic effects. It is now common to encase mines in plastic to outwit the metal detector.

Modern anti-personnel mines are made in innocent shapes to look like small stones or other casually discarded objects and then scattered widely and rapidly from **helicopters** or vehicle-mounted projectors. Deployed over large areas by the Argentines during the **Falklands** War, they have continued to be a major hazard long after the war has ended. In Afghanistan similar mines, distributed far and wide by Soviet helicopters, have caused innumerable casualties among the civilian population.

Modern mine clearing has not really advanced a great deal from World War II; the British-designed Aardvark flail is merely an updated and more efficient version of the successful Sherman Crab. Similarly, Giant Viper – an explosive hose carried across a minefield by rocket and then fired, exploding the mines by sympathetic detonation – is the modern development of a wartime device. The unseen mine will continue to be a major hazard in all land-warfare operations.

Mines, Naval. The idea of underwater (as opposed to waterline) attack with **explosives** was first made feasible by David **Bushnell** in 1775 and demonstrated in 1776 by a mine with a timing device (called a torpedo until the 1870s), laid by **submarine** against a British ship. But until Samuel **Colt** exploded a *controlled* mine by electricity passed through a five-mile cable in 1843; and the **Crimean War**, when the Russians used *contact* mines with a sulphuric-acid fuse in a glass tube (though lacking the ability to tell friend from foe), there was little progress.

Subsequently mines were used with **strategic** and **tactical** effect in nearly all naval wars, especially after the incorporation in *buoyant* mines (tethered to a sinker on the sea bed) of the Herz horn, with unlimited life until its acid fuse was struck to form electrolyte for a primary cell with enough current to detonate the main charge.

Buoyant mines had a significant effect upon the **Russo-Japanese War** and **World**

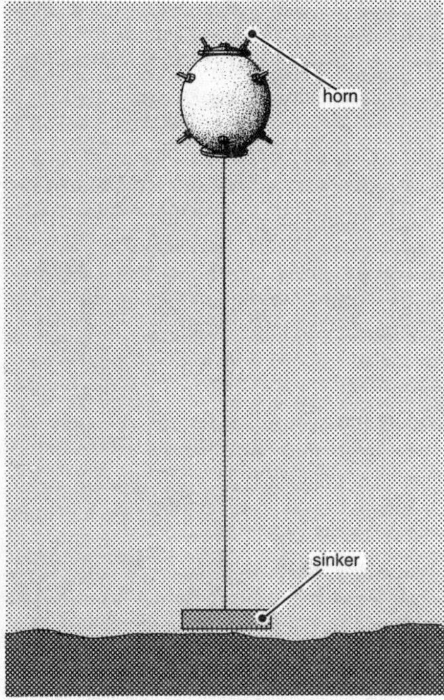

Bouyant sea-mine

War I when various *influence*-detonated types were invented, actuated by magnetic, pressure or acoustic means. The magnetic mine is triggered by the approaching ship's own magnetic field; the pressure mine relies on the reduction of pressure beneath a ship in shallow water to deflect a diaphragm which triggers the detonator; the acoustic mine originally used **hydrophones** to detect a passing ship – a somewhat unreliable device which in its modern form using microelectronics, is once again popular.

Modern technology has improved all these types of mine in current service with the world's navies and has also introduced the atomic mine. Extensive minefields were laid by surface ships and submarines during both World Wars, especially in confined waters, where they had some success as an anti-submarine weapon. Techniques have changed over the years and during **World War II** the availability of larger **aircraft** meant that mines could be laid from the air. This was used to best effect by the Americans during the **Vietnam War** when much of the coastline was denied to Vietcong supplies by this means; in particular, Haiphong harbour was quickly and effectively closed from the air in 1972 by the dropping of thousands of magnetic, acoustic and pressure mines, thus denying to North Vietnam its major port and causing her a major supply crisis. However, it must be said that mine warfare is perhaps more effective in the threat it poses than in the actual damage done. In World War II the Axis lost 1,316 ships to mines and the Allies 1,118; comparatively small numbers in proportion to the total shipping losses of that war.

Minesweepers have become an important element in the naval armoury. By streaming a paravane attached to a long cable from the side of the ship, the aim was to snag the mine's anchor chain, which would then slide along the cable into the jaws of a cutter, allowing the mine to bob free to the surface. The ship's crew would then engage it with rifle fire until one of the horns was struck and the mine exploded.

Magnetic mines are dealt with by degaussing, the passing of an electric cable round a ship's hull to neutralize its own magnetic field. Acoustic mines are exploded by simulating a ship's presence with discrete underwater noise. Pressure mines, however, are virtually unsweepable and require very special measures.

Missiles. *See* **Artillery** *and* **Rockets and guided missiles**.

Mitchell, General William (1879–1936). An officer in the US Army Signal Corps who learnt to fly in 1916, flew as an observer in **World War I**, but by September 1918 was commander of a Franco-American air force of 1,500 machines. He used the force in mass (sometimes with formations of 200 aircraft) in the Saint-Mihiel battle and the Meuse–Argonne offensive. By then he was a fervent champion of air power, proposing the parachuting of **airborne infantry** behind the German lines in 1919, and of strategic bombing (*see* **Bombing, Strategic**) by independent air forces on the British model.

In the early 1920s, as Assistant Chief of Air Services in the Army Air Corps, he tried to prove, by a somewhat spurious demonstration against an anchored **battleship**, that warships were outmoded by aircraft. His vigorous campaign on behalf of the Air Corps made many enemies. It led to his conviction by court martial and suspension from duty in 1925 for accusing the Army and Navy Departments, in the aftermath of the *Shenandoah* **airship** crash, of 'incompetency, criminal negligence and almost treasonable administration'. He resigned from the Army in 1926 and devoted the rest of his life fighting for air power.

Mitsubishi Company. The Mitsu Bishi (*sic*) Company was formed in 1870 by Yataro Iwasaki (1835–85), a petty samurai who was a financial adviser to the Meiji regime and specialized in shipping. In 1884 he acquired the new state-financed shipyard at Nagasaki. The firm expanded, with vital help from the **Vickers Company** and others, into a sprawling engineering conglomerate with factories at Kawasaki and elsewhere. In 1909 it built engines for a **cruiser**. From then onwards it was crucial to Japanese armament manufacture, warships and **aircraft**, including the Type 10, shipborne **fighter** of 1921, the Ki30 light **bomber**, Jiro Horikoshi's A5M and famous A6M Zero-Sen fighters and the G4M and Ki46 bombers.

Mobilization. Prior to the 15th century the raising of armed forces was conjoint with **recruiting** and **conscription**, based on tribal loyalties and feudal systems, and usually adjusted to accord with agricultural requirements and available manpower, frequently including mercenaries. On orders from his monarch or overlord, a chief or lord was obliged to train, arm and call out his sworn followers for service either as long as required or until the money ran out or the harvest had to be gathered in. From the 16th century onwards, owing to the decay of feudalism in Europe, manpower shortages and the increasing complexity of ships and weapons, these rather local arrangements had, perforce, to be supplemented by the creation in peacetime of reserves of committed key, skilled sailors and soldiers to be the cadres of expanding forces when war threatened. This system became essential in the 17th century when France, for example, resorted, through the medium of improved **communications**, to a national *levée en masse*. Henceforward, and notably during the **French revolutionary wars**, mobilization became very systematic, geared to the national economy in order to raise the largest supportable reserve forces with stockpiles of weapons and materials and an industrial plan to satisfy foreseeable **logistic** requirements.

Rapidly proliferating **technology** and the impact of humanitarianism vastly complicated planning and procedures besides creating highly significant political/strategic issues.

Permanent staff (**Staffs, Military**) corps were formed to cope with the military problems of the 19th century. The Prussian General Staff was to the forefront with organization and method of mobilization for the Austro-Prussian and **Franco-Prussian wars** of 1866 and 1870 respectively. In these struggles the relatively smooth movement of the Prussian permanent and reserve forces to their pre-invasion assembly points was sometimes undone by inefficient control beyond railheads – but it was undeniably superior to that of their enemies. The mobilization of the French armies in 1870 was a monument to poor planning and inefficient control and execution. Recalled reservists roamed in search of their units which, as often as not, lacked equipment and supplies and were not located where required. **Morale** sank. Defeat and revolution ensued.

Orders to mobilize are expensive expressions of national policy, usually actuated by proclamation. They can be selective or general in scale of application. Either way they have political meaning since they demonstrate militant intent. The European nations' mobilizations in advance of **World War I** were not only managed with commendable efficiency, their implementation also fell little short of declarations of war. There were instances of the military authority informing politicians that cancellation was impossible and war inevitable. The long-prepared mobilization schemes of 1914 were closely dovetailed with **strategic** plans and economic policies, knowledge of which by an enemy could forfeit surprise. It assisted the Germans to be aware of France's Plan XVII and its intention to invade Germany via the Thionville–Metz area; and that the French were unaware of the Schlieffen Plan.

All these plans were based on the concept of a short war which would be 'over by Christmas'. Britain was among the few nations to plan for a long war, though she had made no preparations for such an event. After Christmas, therefore, the major participants suffered from shortages, principally of **explosives** and **ammunition** which, in cases, took over two years to rectify. Likewise the unexpected development of a war of material made demands on weapons and equipment, such as **motor vehicles**, **aircraft** and **tanks**, which had hardly been envisaged. This required the creation of brand-new production facilities, which placed such immense demands upon manpower that the armed forces began to suffer and women had to be mobilized to a far greater extent than ever before.

Mobilization for World War II reflected the lessons of its recent predecessor. Control of manpower and resources was applied beforehand. Allowances, often excessive, were made for the disruptive effects of **air warfare**. China had long been in a state of civil war before Japan, with large permanent forces, struck in 1937. Most European nations had begun a partial mobilization before 1 September 1939. Several of Poland's combat troubles occurred because she delayed mobilization for political reasons – though air attack caused disruption too. It was a defect of German policy that, for political popularity as well as economic reasons, **Hitler** restricted industrial production until far too late; and resisted the **recruiting** of women to the end. Instead, Germany lived off her victims and employed unreliable slave labour.

Since 1945, in an almost universal environment of **cold** and **limited war**, many nations have retained large forces, supplied by conscription, and large reserves. The overshadowing **nuclear** threat suggested the need for large pools of manpower and resources for **civil defence** and to replace the wholesale destruction of global war. The massed Communist forces, which were bankrupting their nations, were faced by anti-Communist nations which, to defend their economies, depend on sufficient warning to mobilize their cheaper reservists and move them long distances to threatened places. It has become an element of **deterrence** to demonstrate an ability effectively to mobilize at speed to match any variety of threats. The fact remains that the slightest indication of a hostile mobilization from **intelligence** sources acts as a warning for counter-mobilization and for diplomatic activity to defuse the political situation.

Moltke, Field Marshal Count Helmuth Karl von (1800–91). A member of the Prussian General Staff with a bent for history who eagerly embraced the **railway** as a revolutionary factor in war. He envisaged the Prussian Army playing a leading role in the unification of Germany and, with royal backing, was appointed Chief of the General Staff in 1857. He then joined Chancellor Otto von **Bismarck** and Minister of War, General Albert von Roon, in the foundation of the German Empire through the war with Denmark in 1864, the Austro-Prussian War of 1866 and the **Franco-Prussian War** of 1870–71. These conflicts were quickly won within economic bounds through von Moltke's brilliant modernization of the Army and his own firm, if not faultless, handling of combat operations.

The key to modernization was his strengthening of the General Staff corps by the inculcation of intensive professionalism together with a broad outlook that encouraged this élite to take decisions within a framework of general directives (as opposed to precise orders). The sense of initiative evolved created expert planners and commanders who were given advanced promotion to the highest appointments. He envisaged prolonged battles by armies moving on broad fronts, poised to concentrate at decisive battlefields to overwhelm the enemy by heavy fire-power with the latest weapons. His Staff officers and the **telegraph** controlled carefully devised movements in which the railway was the vital instrument. The effectiveness of his ideas was expressed in victory, though he never entirely managed to coordinate operations in the combat arena and was often embarrassed by **logistic** shortcomings.

Mongol invasions (*see map overleaf*). The irresistible and vast outwards surge of the Mongol armies from the Gobi Desert was initiated by tribal warfare in 1190 and the coming to power of **Genghis Khan** in 1206. Dependent upon the **horse** for their mobility, the Mongol warriors were protected by leather and mail **armour**, and armed with lances, **bows** and short **swords**. As time went by they assembled a **siege** train of slings and trebuchets, along with **flame** projectors and a tubed, gunpowder-propelled, spread-shot weapon. Signal **communications** were by shouts, drum beat and flags. A serviceable **intelligence** system along with **propaganda** warfare shaped **strategy** and spread rumours, alarm and despondency. **Logistically** their hardened warriors mainly lived off the land.

A Mongol army was composed of *toumen* of 10,000 men, each of ten battalions. Probably the largest force ever assembled was about 240,000. Tactically they were superior to many of their opponents in diligently employing ingenuity to achieve surprise and swiftly to direct their thrusts round flanks on to the enemy rear. Occupied territory was firmly governed by indigenous officials. Nevertheless, their intentions were entirely predatory; they brought little or no culture, technology or other benefits in their train.

The campaigns of Genghis Khan, from 1206 to 1227, conquered northern China, swept through central Asia into Persia and thence into Anatolia, the Caucasus mountains and Russia. His successors continued their wanderings, launching forth in 1236 from the Gobi Desert into northern Russia to reach Kiev in 1240 and invade Poland and Germany in 1241, where they won a shattering victory at Liegnitz, prior to overrunning Hungary (after the battle of Mohi) and then withdrawing through the Balkans after the death of their leader, the Khan Ogedie.

In 1258 a Mongol army under the Khan Mongke extended control over China and in the Middle East, but after his death factional disputes weakened Hulagu's army in Arabia, enabling a Mameluke Muslim army to win a decisive victory at Ain Jalut in 1260. Henceforward, despite incursions into South-East Asia and India at the end of the 14th century, the Mongol empire, under Tamerlane, began to disintegrate and quickly lose its identity.

Montgolfier brothers, Joseph (1740–1810) and **Jacques** (1745–99). Scientifically educated by their paper-manufacturer father, they experimented with paper-made hot-air **balloons**. In June 1783 the heat from wool and straw lifted an unmanned balloon to 1,500ft; and in November took François de Rozier aloft over Paris with a passenger, thus initiating the age of manned flight. Both continued to sponsor flight while pursuing other scientific and industrial projects, including a hydraulic ram.

Montgomery, Field Marshal (Lord) Bernard (1887–1976). An infantryman who won a reputation as a Staff officer in **World War I** and, in 1940, commanded a division with flair in the retreat to **Dunkirk**. By 1942 he had been an army commander in Britain, but it was when given command of Eighth Army in Egypt in August that he came to public attention. His conduct of the second and third Battles of El **Alamein** restored that formation's **morale** and won Montgomery a reputation for care, without taking risks, in his conduct of war. He was to lead Eighth Army in pursuit of the Axis armies to Tunisia, to Sicily and into Italy – a period in

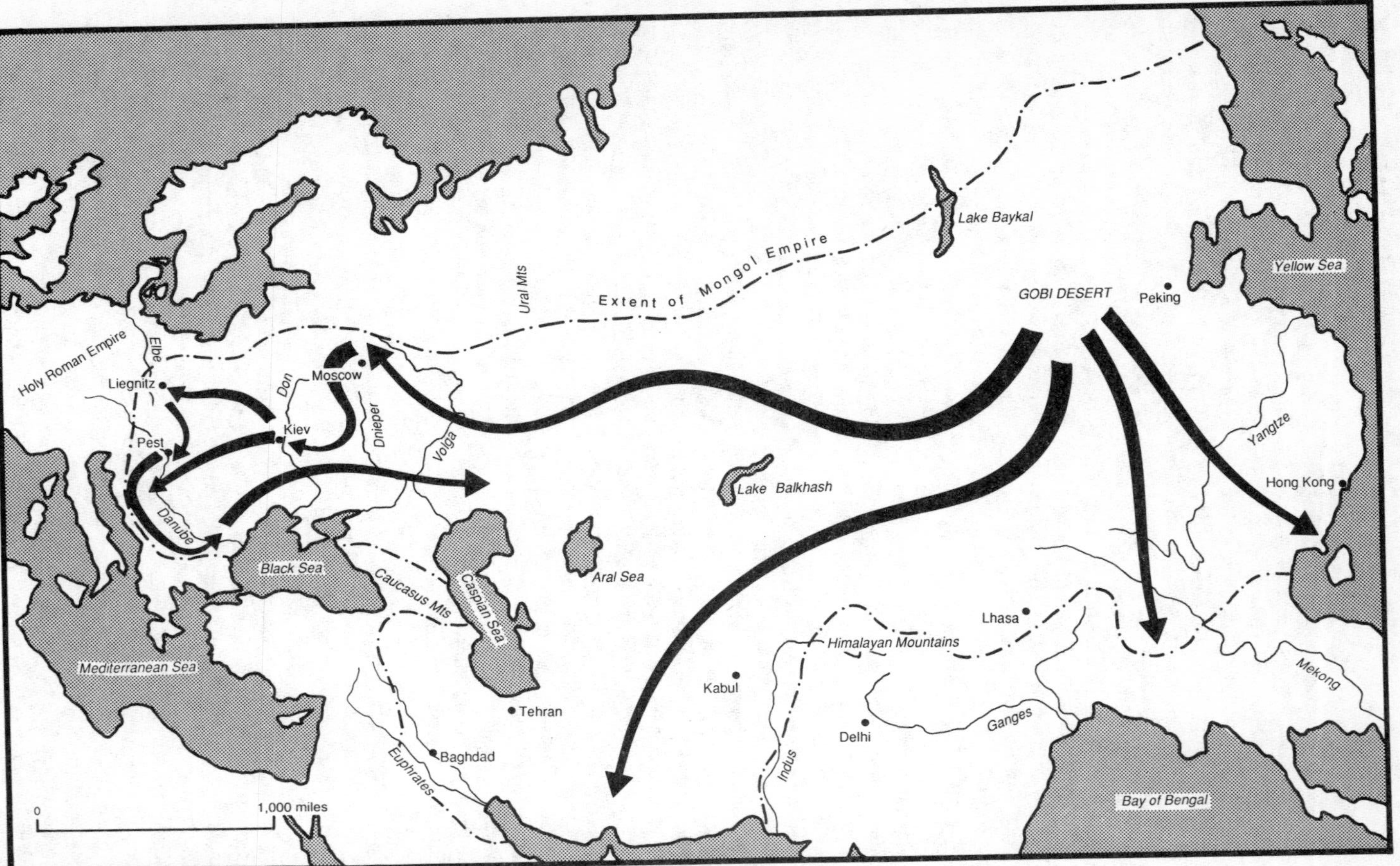

The Mongol expansion and invasion of Eastern Europe, 13th century (see entry page 223)

which he managed to sow mistrust of his methods with American allies. Yet he commanded the Allied armies brilliantly in the **Normandy** battles in 1944 and if he had been allowed his way by General Eisenhower, might have finished **World War II** that autumn with a narrow-front advance through **Arnhem** into Germany. He had to be content with commanding 21st Army Group on the Allied left flank in the final Battle of **Germany**, the culmination of which was his acceptance of the German surrender on 4 May 1945.

As Chief of the British Imperial General Staff after the war he was not a success, but as Chairman Western Europe's C-in-C's Committee from 1948 until becoming Deputy Supreme Allied Commander in Europe in 1951 until 1958, he played an important role in establishing **Nato**'s forces.

Morale. There is some complexity in defining what morale really is, let alone determining how to create and maintain it. The *Oxford English Dictionary* refers to it as 'conduct, behaviour, especially with regard to confidence, hope, zeal, submission to discipline'; and records its appearance in this context in 1842 in relation to military matters. But of course it had always been vital to almost any human activity under stress. For example, it is not to be taken for granted, as sometimes supposed, that people who are 'cheerful' or in 'high spirits' have high morale; for sometimes it is disguised by silent, dour resolution.

The military usually place maintenance of morale next to top among the principles of war (*see* **War, Principles of**) and go to immense trouble enhancing it. Chiefly the aim is to provide strong senses of purpose and motivation, clear, sensible leadership, excellent weapons and equipment, sound **logistic** services, adequate pay and sustained successes in all activities, including the ultimate one of battle. In which one of the most demoralizing factors can be a sense of real, or imaginary, inferiority to the enemy in weapons and technology.

Every skirmish, battle, campaign and war provides examples of the effects of morale. At all levels individuals have demonstrated varying spiritual qualities which accounted for only part of a group's overall morale. A leader of high morale may very well compensate for several followers with low morale. In moments of extreme duress or euphoria there can be wild swings of morale – often only passing reactions if a corrective is sensibly applied by leaders, who need not necessarily be appointed officers. But (as will be seen below) in total war the maintenance of civil morale has also to be assured.

History contains many examples of morale breaking down in units and sub-units. When a few men or small groups abandon the battlefield it is as likely as not due to collapse of morale – perhaps only temporarily, for any of a number of reasons, and some quite insignificant. In such instances, loss of or failure of a leader was often the cause. It is the far fewer, battle-losing collapses that start a rot which catch greatest attention, sometimes blotting out the numerous occasions when, in appalling conditions, men have persevered and triumphed. *See* for example **Alexander the Great**, **Attila**, **Belisarius**, **Dönitz**, **Edward III**, **Frederick the Great**, **Genghis Khan**, **Giap**, **Guderian**, **Hitler**, **Kesselring**, **Lee**, **Marlborough**, **Montgomery**, **Mountbatten**, **Napoleon Bonaparte**, **Nelson**, **Sherman**, **Slim**, **Stalin**, **Togo**, **Trenchard** and **Zizka**.

Since morale became a word to conjure with in the 18th century, numerous campaigns have illustrated the conflicting effects of high and low morale. Often suggesting that it is an index of operational and logistic merit and emphasizing how, in war, it gives people something to look forward to – most of all survival. *See* for example the following wars and their related campaigns and battles: **French Revolutionary**, **Crimean**, **American Civil**, **Franco-Prussian**, **Russo-Japanese**, **World War I**, **Spanish Civil**, **World War II**, **Vietnam** and **Iran–Iraq**.

Mortars are short-barrelled (10 calibres or less) smoothbore high-angled weapons for dropping projectiles on top of and behind the enemy defences. It is likely they have been in use since the siege of Constantinople in 1453 when Mohammed II built one to lob a heavy stone to sink a ship anchored off the Golden Horn – ranging being by observation of fall of shot and variation of the propelling charge. By the 17th century projectiles of between 30 and 200lb were fired to ranges of 2,000 metres, and although their **ammunition** still was sometimes solid shot, high-explosive shells (often called bombs) have usually been preferred, especially against **fortifications**. Mainly an army weapon, they were sometimes mounted in specially adapted warships for attack on land targets – for example during the **Crimean War**. Not until the 1860s, however, was a fixed propelling charge thought of, plus the idea of varying the mortar's elevation to adjust range.

Circa 1908 the Germans built a *Minenwerfer* with adjustable angles of elevation as an infantry-support weapon capable of being carried by two men on a wooden base; and they also developed a 305mm weapon. These became models for the mortars, with ranges up to 400 metres, which were so important in the trench warfare of **World War I**. They were soon copied by the French and British. Typical of the simple, rapid-fire mortars produced was the British 102mm (4in) tube mounted on adjustable legs and resting on a base-plate weighing 32kg. It threw its bombs to about 350m, at a rate of 30 rounds a minute. Firing was a simple matter of dropping the bomb down the tube on to a fixed striker which detonated the percussion cap. This remains the broad principle upon which most modern mortars operate.

In the 1930s German development continued and by 1939 two extremely effective weapons were available: an 80mm with a range of 2,000m, and an 81mm reaching out to over 4,000m. The devastating effect of the German mortars exposed the inadequacy of the British equipment, rectified in time by improvements to the 76mm and by the introduction of the heavy 107mm (4·2in) mortar with a range of about 4,000m.

The need for mortars in mobile operations was fully appreciated by the end of **World War II** and, though basic designs remain generally unaltered, much effort has gone into improving bomb ballistics and sighting equipment. Larger mortars, with calibres up to 240mm, have been produced, notably by the Russians, some breech-loading, some rifled, some automatic, like the 82mm Vasilyek, and some vehicle-borne. However, most armies have found the 81mm calibre satisfactory for close infantry support, backed up by the 107mm weapon and with a light section mortar such as the British 51mm, hand-held and firing a 2kg bomb out to about 750m.

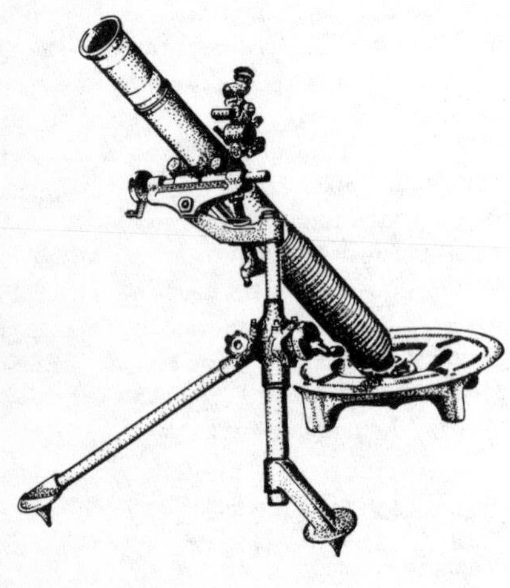

Mortar (81mm)

Moscow, Battles for. Ten times Moscow has been the objective of foreign armies, although, because of persistent internal disturbances, it ceased to be the Russian capital between 1703 and 1918. The **Mongols** captured and destroyed it in 1237, 1293 and 1382, but in 1408 another raid was repulsed. In 1521 Tartar forces again reached the fortifications before being bought off; and in 1571 plundered the city for the last time. It was captured by Polish forces in 1610 (they withdrew in 1612) and threatened in 1614 by **Gustavus Adolphus**'s Swedish–Polish forces during the Russo-Polish War (1609–18). Thereafter it developed as an industrial and cultural centre until June

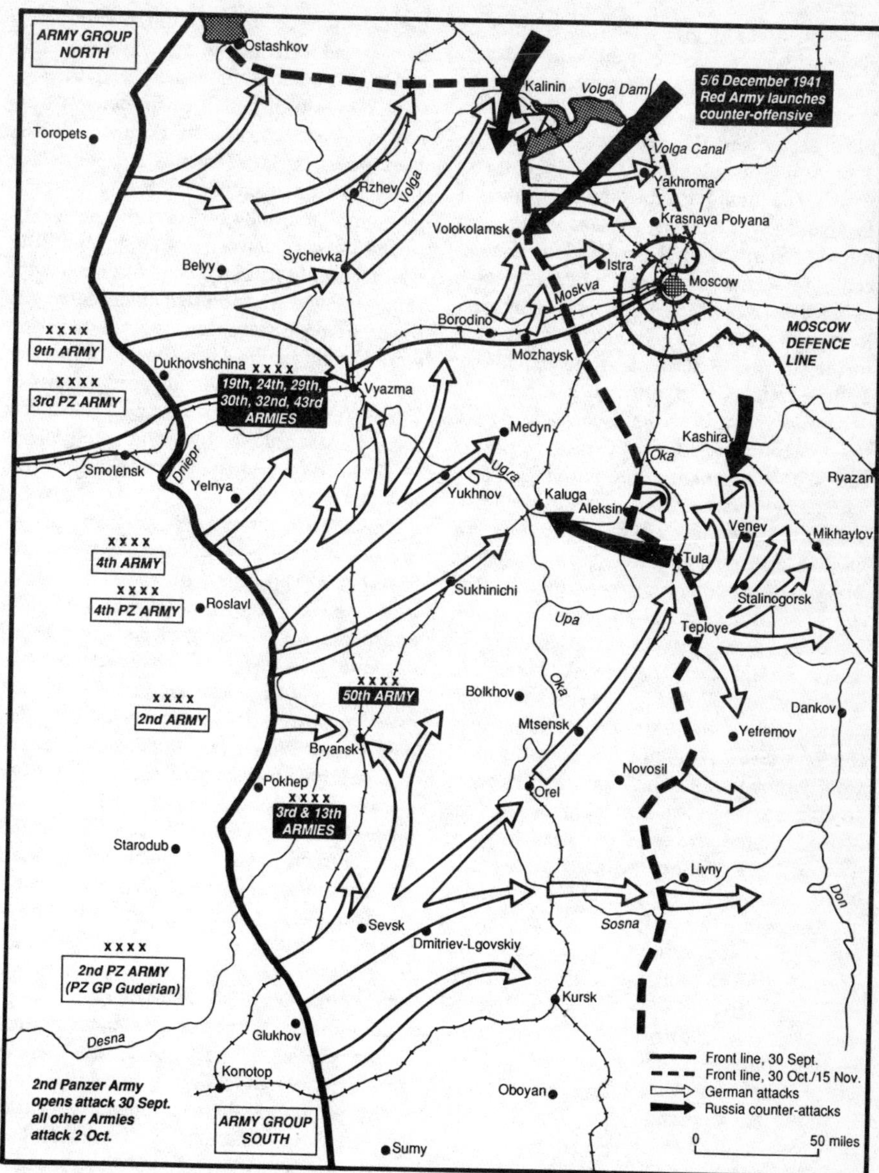

The Battle of Moscow, 1941

1812 when the French, under **Napoleon**, invaded Russia.

Napoleon, with an allied army of 440,000, hoped to impose his will on Russia by defeating her 600,000-strong armies before reaching Smolensk. But he chose Moscow as his objective because it was a principal **logistic** base – in the knowledge that his own logistic and **medical services** were totally inadequate for such a gamble. The Russians, however, refused battle and laid waste the country as they withdrew. Before Napoleon's centre forces of 301,000 in seven corps brought the Russians to the drawn Battle of Borodino on 7 September, they had wasted away to 130,000. This partially demoralized force, which eventually reached Moscow on 14 September, was only 95,000 strong. So, even if the Russians had not burned down 75 per cent of the city, the French already were doomed to retreat, in a celebrated disaster from which, in hard winter conditions and owing to the desertions of allies, barely 15,000 escaped death or capture.

During **World War II** Moscow was one of three of **Hitler**'s objectives in the invasion of Russia in June 1941, although the Russians, who always assumed Moscow was the main objective, made no attempt at evasion. The 55 divisions (including five panzer corps) of Army Group Centre under Field Marshal Fedor von Bock routed the frontier forces, capturing some 600,000 men and 5,000 **tanks** at Minsk and Smolensk, before, in August, being delayed by Hitler's vacillation and by logistic difficulties. When eventually the final thrust was launched on 6 October, via Vyasma and from Tula, the Germans again routed a depleted enemy. But the onset of winter rains and mud, followed by deep frost, delayed the **mechanized** forces and brought logistic collapse. Within 25 miles of Moscow the halt, imposed by exhaustion, preceded a desperate Russian counter-offensive under Marshal **Zhukov**, which threw the Germans back and put an end to their intentions. Never again would they seriously threaten the city.

Rule One of the Book of War, according to Field Marshal **Montgomery**, is: 'Never invade Russia.'

Motor vehicles. The search for a means of **propulsion** on land in lieu of animal power began in the 17th century with attempts to make compressed-air piston engines – 200 years before Karl Benz's first internal-combustion engined, horseless carriage ran in 1885. In the interim there had been many experiments: Nicolas Cugnot's steam-driven carriage had run in 1769 and steam tractors existed and were used in the **Crimean War**.

The first Benz motor vehicles were tricycles with power transmitted to the wheels by a chain. Those of Gottlieb **Daimler** initially were based on horse carriages, without the shafts, with the engine at the rear and a four-speed capability. Thereafter developments followed apace, so that by the 20th century front-engine vehicles with transmission of power via a clutch and gearbox to the rear wheels were almost standard; brakes were greatly improved to cope with increasing speeds and weights; and sprung suspensions, shock absorbers and pneumatic tyres were introduced. At the same time efforts to improve mechanical reliability were paramount.

The advantages of motor vehicles, including motor cycles, for military use were soon recognized. This was particularly so when, in the 1890s, load-carrying lorries and passenger omnibuses were built for commercial purposes and the first **armoured car** was proposed. **Mechanization** loomed ahead, bringing with it a vast programme to dispose of obsolete wagons and animals. Small wonder that the Germans shied away from such expensive construction and thus exacerbated the **logistic** shortcomings which, at the start of **World War I**, contributed to their defeat at the **Marne** when **railways** and **horse** transport failed to deliver sufficiently. Meanwhile the British profited from lorries with three-tonne capacity and the French moved reinforcements in Parisian taxi-cabs.

As World War I proceeded the demand for motor vehicles and their supporting services multiplied in manifold ways. Indeed, without such expansion the requirements of the massed armies and, above all, the **artillery** would have been impossible to meet. There were too few animals to haul the guns and supply their vast appetite for **ammunition**. These were tasks for which motor vehicles had not previously been envisaged, which now prompted industry to design a variety of types, among which, very significantly, the **Ford** and Fiat companies' small vans were found to be very useful in the **desert** campaigns of the Middle East and the undeveloped terrain of East Africa.

The need for extensive road construction and maintenance became mandatory, especially because the narrow-tyred, four-wheeled, two-wheel-drive (4 × 2) vehicles had a poor cross-country performance and only a handful of tracked vehicles were utilized for transport tasks. Prior to **World War II**, however, specialized multi-wheel drive, as well as full-track and half-track vehicles (the latter pioneered by Adolphe Kegresse), were available. Yet in 1939 most armies mainly had 4 × 2 trucks which were a handicap on inferior roads, such as the Germans found in Russia. Later, multi-wheel drive became common and load-carrying (sometimes enhanced by trailers) increased – subject to the restrictions imposed by **bridging** limitations.

The variety of special vehicles (many with diesel engines) increased also: the **jeep**; fuel bowsers; **engineering** plant; artillery tractors; and articulated tank transporters capable of carrying 50 tonnes of supplies as well as armoured vehicles. The trend was pursued after 1945 to satisfy requirements for better engines, transmissions, suspensions, cross-country and **amphibious** performance.

Mountain warfare. Since commanding heights are the embodiment of vital ground, mountains create extreme conditions for warfare. Not only do they dominate terrain by providing naturally strong **fortifications** to reinforce defensive operations, they make **logistics** and endurance more difficult than in any other kind of terrain. Every route has to be cleared of enemy observation, which as often as not means struggling to every peak in sight, peaks which, if much over 10,000ft, cause anoxia to men or animals which have to carry equipment and supplies, in cloud and cold, where every ridge can hide an ambush. Mobility is at a premium in warfare which is inherently slow-moving – calling for the **infantry** specialist and much patience. It demands climbing equipment and techniques and is usually unsuitable, to a large extent, for **mechanized** forces – though, as in **jungle warfare**, the presence of a single **tank**, dragged to commanding ground by ingenuity and **engineering** toil, can be worth a squadron in an open plain.

Mountains, particularly forested ones, make ideal hide-outs for **guerrilla** forces. That is why such campaigns have been prolonged and effective in, for example, **Afghanistan**, northern India and Pakistan, Italy, parts of France and Spain, Greece, Yugoslavia and parts of China and **Burma**. It does not follow that hill people make the best mountain soldiers, but their feel for the uplands is an obvious advantage. It was no fluke, for example, that Gurkha and French North African troops did so well in the Italian mountains before Rome in 1944, or that battles among the Himalayas have taught so many **tactical** lessons, such as the need to picket and fortify all heights overlooking a line of march – and the need to retire quickly, with fire support, when abandoning them in face of anticipated enemy reoccupation.

Fire and movement is as much the essence of mountain warfare as anywhere else. It simply calls for more exacting aiming techniques in ranging on targets at varying heights above or below gun positions; the most skilful siting of forward observers; and the use of pieces, including **mortars**, which provide plunging fire from the upper register. For a shrewd enemy rarely deploys on a forward slope to face direct fire, but sites behind crest-lines where shots in the lower register cannot easily reach and are difficult to observe. On the other hand, construction of positions, which also provide essential shelter against weather, can be difficult in rock. The effect of the defensive is thus emphasized and has rarely been better demonstrated than by the Germans in Italy, from 1943 to 1945, when their systematic route denial by demolition and booby trap placed an even heavier work-load than usual on Allied engineering resources.

Aircraft, needless to say, have changed the conduct of mountain warfare, even though flying among peaks and steep-sided valleys in unstable weather and against enemy **anti-aircraft** artillery can be daunting and where accurate delivery of **bombs** is difficult. They still provide **intelligence** and can be of immense assistance delivering ammunition and supplies to isolated positions, as well as evacuating wounded, thus saving them prolonged and painful journeys by stretcher down rough tracks and precipices, tasks which, since the **Korean War**, have been made immensely easier with the arrival of the **helicopter**.

Mountbatten, Admiral Lord Louis (1900–79). A Royal Navy signalling specialist whose royal connections made him a social celebrity prior to **World War II**, in which he displayed his great talents as a leader. After commanding a **destroyer** flotilla from 1939 until the Battle of **Crete** in May 1941, when his ship was sunk, he was appointed Chief of Combined Operations in October. This post made him responsible for development of **amphibious warfare** and hit-and-run raiding and, in 1942, obtained for him a seat with the Chiefs of Staff and the ranks, in addition to Admiral, of General and Air Marshal. The **Dieppe Raid** was his responsibility and the lessons learnt disseminated by him to the immense benefit of commanders in all theatres of war. In October 1943 he became Supreme Commander South-East Asia Command (SEAC), charged with the reconquest of **Burma** and **Malaya**, tasks accomplished by September 1945 with the collaboration of Generals **Slim**, Stilwell and others, though throughout with only the minimum of resources.

In 1947 he was sent to India as the last Viceroy, to transfer British power to what would be India and Pakistan. From 1955 to 1959 he was First Sea Lord, involved with the Suez operations in 1956. As Chief of Defence Staff from 1959 to 1965, he brought the three armed services under the control of a single Ministry of Defence. He was assassinated by the Irish Republican Army while on holiday in Eire.

Movement control. The control of movement by military forces throughout the world is fundamental to operations by sea, land and air. It is a major General **Staff** responsibility starting at ministries and working through formation HQs to sub-unit level. It aims to make efficient and economic use of available facilities to move men and material to their right destination in good order, on time. It depends upon foresight, the availability and use of full information about resources and conditions, and the execution of control through clear instructions at every phase of an operation.

The invading Persian **amphibious** forces in the **Greek–Persian Wars**, for example, needed strict movement control in their assembly and deployment. There had to be a semblance of it when **Hannibal** crossed the Alps through a series of defiles which, in modern parlance, were critical points. Gradually, with the appearance of reliable **charts and maps**, improved inland waterways and roads, greater precision of movement planning became feasible by such great generals as **Marlborough** and **Napoleon Bonaparte**. And usually in the trenches, during the assault phases of **siege warfare**, a measure of control was essential to mitigate traffic chaos.

The introduction of fairly reliable steamships and **railways** in the 1840s made it possible to schedule movements with a higher degree of certainty than when dependence was upon wind conditions and animal power. Once it became possible to calculate quite accurately the rate and volume of delivery at any of a number of destinations on the world's surface – a reliability which improved with the development of **mechanization** and **aircraft** – chance was reduced in **strategic** and **tactical** planning. From the **Crimean War** onwards the integration of civil handling and transport agencies under military control was increasingly formalized. At the heart of most **mobilization** schemes were to be found detailed movement plans, allied to defensive measures against enemy interdiction or disruption. The German military, indeed, were insistent upon having a say in the development of their railway and communications system in peacetime. Later the British began subsidizing shipping companies to provide, for example, adequate lifting gear for heavy equipment to enable handling at ports lacking adequate shore facilities.

As control became more necessary and complex in line with new **technology** in both **World Wars**, armed forces recruited specialist movement-control staffs, many of whom were civilian transport and supply experts. Spaced along the lines of communication, movement control strands (very often joint service) reached out from HQs to ports, railway stations, airfields and all manner of depots where movement orders were executed in collaboration with civil agencies. Closer to the front, the operational naval, army and air staffs worked together on beaches, at maintenance centres and airheads directing men, equipment and supplies into **logistics** channels. Military police played a prominent role manning traffic-control centres and in route direction closer to the front where tactical movement and defensive measures become much more acute and have to be dovetailed into operational plans. Indeed the closer to the front the more likely that movement becomes a controlled activity, geared to time of day, the situation and future plans. For signs of movement can often indicate forthcoming intentions and have to be scheduled astutely.

History provides many examples of the impact of movement control on operations. Examples are provided in the preparations for war. Lack of control by the French during the **Franco-Prussian War** was notoriously disastrous. Failures by the Germans to integrate rail and road movements during the Austro-Prussian War lost them opportunities after the Battle of Königgrätz and contributed to their armies' exhaustion at the Battle of the **Marne** during **World War I.** In both campaigns railway stock stood idle failing effectual provision to offload them into road vehicles. On the other hand, skilful, high-density rail movements of reinforcements from one place to another during World War I were major factors in the stabilization of threatened trench fronts.

The exceedingly widespread **World War II**, with its large theatres of war threatened by **blockade** and shipping shortage from naval and air action, created demands for world-wide control of shipping to minimize waste. Intricate communication networks helped prevent ships being tied up in ports which could not accept their cargoes. **Convoy** routeing took account of priorities of movement as well as avoidance of attack. The rapid development of air supply called for special air traffic control compatible with handling arrangements at base and front – the Germans set an example of this during the **Spanish Civil War**. The finest immediate postwar example of the techniques involved is, of course, the **Berlin Airlift**, with its precisely integrated systems of deliveries to airfields in the West, rapid loading of aircraft, scheduled flights along **radar**- and **radio**-controlled corridors, and swift unloading and turn-round at Berlin, where reception was geared to a non-stop service. These practices are repeated with still greater sophistication for most **cold** and **limited war** operations – and are on constant stand-by.

Mulberry harbours were a combination of blockships and prefabricated, concrete-and-steel breakwaters sunk off **Normandy** in June 1944 to ensure the large-scale **logistic** support of the invading Allied forces across sheltered beaches. Two were prepared in Britain, but the one allocated to the Americans was abandoned before completion after severe storm damage. Floating, anchored pierheads were provided to handle 6,000 tonnes per day. Yet twice the quantity of vehicles and cargo (respectively 332,645 and 1,602,976 tonnes by 29 July alone) came across open beaches from beached and 'dried out' ships at low tide.

Museums, Military. For the student of history, museums provide rich sources of information through visual display, archives and libraries. Since the Renaissance, collections of weapons and the like, many of them privately owned, have been made. Not until the 17th and 18th centuries were many public collections formed, usually in capital cities, such as Paris, Rome, Vienna, St Petersburg and London, where arms and armour were displayed in historic buildings. The Industrial Revolution and the growth of **technology** widened the need for and scope of military museums, which became repositories for machinery, records and illustrations that inventors and manufacturers could study for ideas and methods. At the same time, as a result of improving travel arrangements, an increasing number of people were able to visit the central museums with relative ease for educational purposes, nostalgia and as a leisure activity.

Some military museums try to cover every aspect of specific periods, such as the Imperial War Museum in London which deals with the **World Wars** and subsequent wars only. Others, like the Smithsonian Museum in Washington, include military equipment among other technical exhibits. Most major nations have their own naval, army and air museums, often displaying working examples of ships and machines in **dockyards**, army camps and airfields. On a smaller and simpler scale are the collections of artefacts and uniforms belonging to ships' companies, corps, regiments and squadrons which have found homes in local depots, separated from national collections. Inevitably there is duplication; often a researcher has to hunt far and wide for some information.

Display techniques have evolved over the years from the static kind to the far more imaginative modern, active type. Rows of mute display cases are giving way to exhibitions of working models or the real thing; interrelated moving picture shows with sound effects, and, in some cases, audience participation to arouse a sense of involvement with reality. The military services often envisage their museums as **propaganda** for enhancement of their public relations and as a spur to **recruiting**. Pacifists sometimes point to them as centres for peace studies. Whatever the assumed purpose, military museums everywhere are visited by a great many people, who often pay an entrance fee and belong to societies of friends of the museum.

Musket, *see* **Rifle**.

Muslim wars of expansion. Victory at the Battle of Badr, in 624, consolidated the Prophet Muhammad's new, bellicose religion and gave power to his fanatical Arab followers. After his death in 632, Islamic influence spread rapidly, mainly by force throughout Arabia and into Palestine (where five major battles were won prior to 640) before the conquest of Egypt began. Then the Muslim bowmen and light, highly mobile armies swept northwards into Mesopotamia and Anatolia and westwards along the north African shore.

In 717, with a fleet of 1,880 ships and 80,000 men, they reached Constantinople, but were beaten back after a year's siege and heavy fighting. Meanwhile they entered Carthage in 698; and in 711, a mere 12,000 strong, defeated 90,000 Visigoths at Guadalete in Spain to prepare the way for the conquest of that country and the invasion of France by an army of up to 50,000 in 732. But at this crucial moment, when west European Christian forces were disunited, Charles Martel gathered an army to rout the invaders at **Tours** – a decisive success which, following the victory at Constantinople, threw Arab power into decline; gave breathing space to Christianity; and led to a change in the way the religion would be spread (more by teachers and by traders than the sword) through northern Africa, India and into south-east Asia.

But Muslim belligerence remained a constant threat to Europe and led to a series of wars of **religion**. Beginning with the **Crusades** (1096–1226), they were followed by the wars of the **Ottoman Empire**, after the fall of Constantinople in 1453; reflected in various campaigns of **World War I** and **II**; and extended, more recently, in the **Arab–Israeli Wars**, the Afghanistan War of 1979–89, and the **UN–Iraq War** of 1991.

Currently, the threat of further Muslim expansion by force of arms in, for example, southern Russia remains by no means unlikely.

Mussolini, Benito (1883–1945). A demagogic Italian socialist who, as a soldier, was wounded in a grenade-training accident in

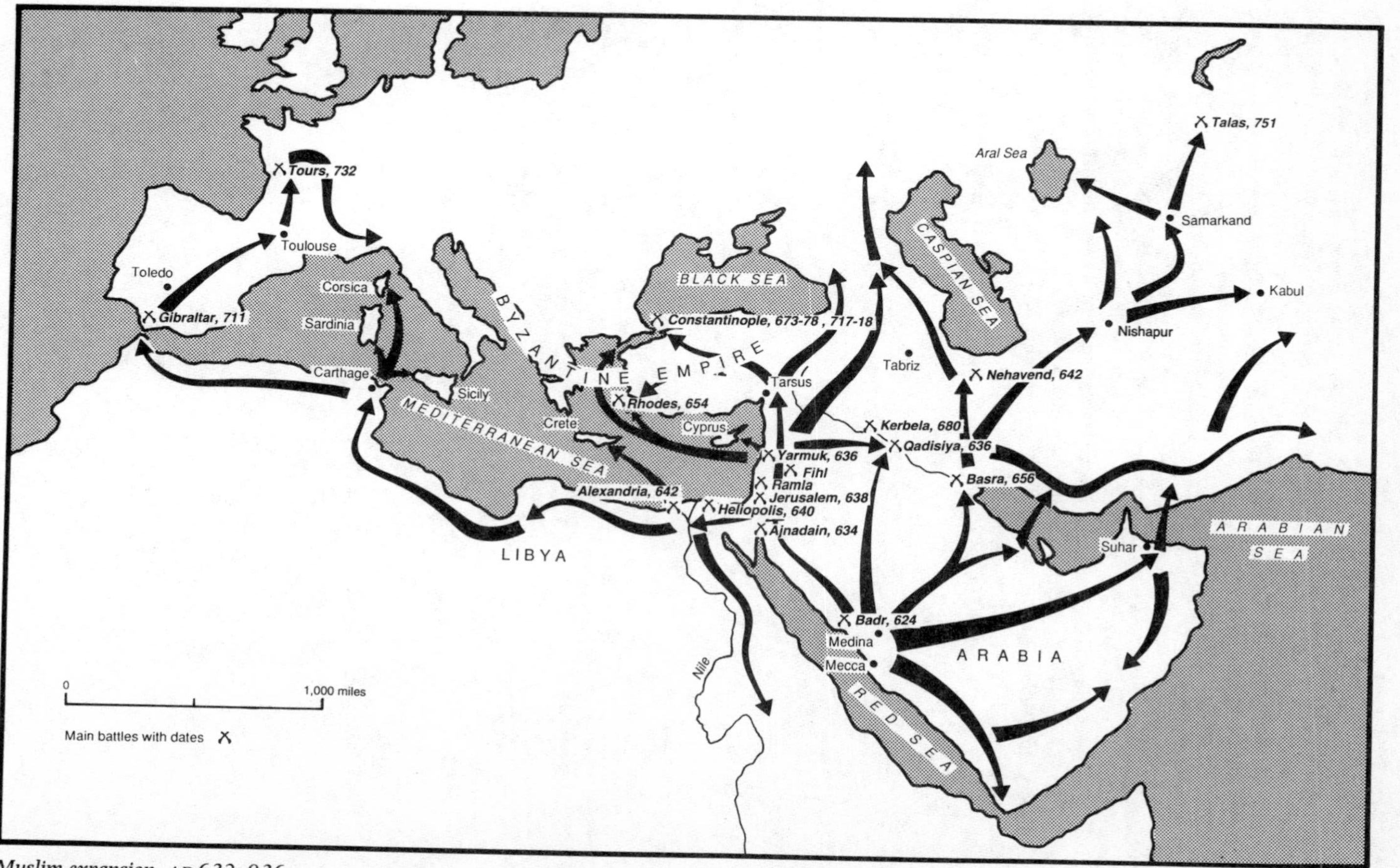

Muslim expansion, AD 632–936

1917. Involvement with Fascism after **World War I** and the use of violence to advance his struggle for power made him Italy's Chief Minister in 1922 and, in 1925, Dictator. His bullying foreign policy led to the Italo-Ethiopian War of 1935, strong involvement on the Nationalist side in the **Spanish Civil War** in 1936 and the occupation of Albania in 1939.

A close association with **Hitler** after 1936 led to the founding of the Axis in May 1939 and entry into **World War II** on Germany's side on 10 June 1940, when he felt little fighting was needed for rich pickings. He found his error in successive defeats throughout the **Mediterranean**, in Greece and in East Africa. It was of no avail when German forces came to his rescue in 1941, despite a glimpse of victory in Egypt in 1942. On 25 July 1943, in the aftermath of the loss of **North Africa** and the impending loss of Sicily, he was deposed and imprisoned. Rescued by a daring German *coup de main*, he spent the rest of the war as Hitler's lackey until caught and killed by Italian partisans on 28 April 1945.

N

Napalm is a mixture of aluminium soap of **naph**thelmic and **palm**itic acids which, when mixed with petrol, forms a sticky, flammable substance used in **flame warfare**. Invented in America in 1942, it had the advantage over previous flame-thrower fuels (FTF) of a slower rate of burning and of coalescing in a 'rod', thus significantly increasing range, accuracy and target effect. Allied forces employed it extensively during **World War II**, including a filler for incendiary **bombs** dropped on Japanese cities. Beginning with the **Korean War**, it was air-dropped in containers on **armoured fighting vehicles** and troops, often with devastating effect.

Napoleon Bonaparte (1769–1821) was an artillery officer whose rise to Emperor of France during the **French revolutionary wars** began with his expert concentration of **artillery** fire during the successful siege of the British in Toulon in 1793. Soon involved in high government politics, as well as in battle, in 1796 he was appointed to command of the army in Italy where he triumphed in a brilliant campaign, culminating in the victory at Rivoli in January 1797. In 1798 an invasion of Egypt, at his suggestion and under his command, was undertaken, but checked when Admiral **Nelson** destroyed the French squadron at the Battle of the Nile.

Leaving his army to its fate, Napoleon escaped to France where he played a decisive role in the *coup d'état* of 9 November 1799 which made him dictator of France at the head of the Consulate. Thereafter, driven by insatiable ambition, he exploited the residual French revolutionary spirit and desire for glory to prosecute a succession of successful gambles and conquests – which eventually were reversed as the result of his inability, after **Trafalgar** in 1805, to overcome British sea power and Spanish resistance, the disaster at **Moscow** in 1812 and defeat at Waterloo in 1815.

Napoleon benefited initially from organizational improvements of the French Army previously carried out by **Gribeauval**, **Carnot** and de Broglie, and the signals **communication** system provided by Chappé. His charismatic leadership, courage, strategic and tactical genius were combined, after victories over outmoded opponents, with brilliant political solutions to turn enemies into temporary allies. But his inability to overcome British sea power (*see* **Naval warfare**) and a tendency to be dismissive of technical innovations (including rejection of **balloons** and **Fulton**'s **submarine**) were fatal defects. His callous indifference to his soldiers' welfare, which permitted the inadequate **logistic** support of the invasion of Russia in 1812, ultimately brought about the fatal massed desertions by Frenchmen in addition to the antagonized allies he had won by the sword to France's detested side.

Napoleon III (1808–73) was the nephew of **Napoleon Bonaparte** who took advantage of the French revolution of 1848 to acquire the presidency and, in 1853, be elected Emperor

– a virtual dictator. As a shrewd politician, he concentrated on the technological improvement of the armed forces and restoring French power. A major arms race began as he modernized the navy with **armoured**, steam-propelled warships and re-equipped the army with the latest weapons, including the Reffye **machine-gun**, but to the neglect of the more important **artillery**. In 1854 he became involved in the **Crimean War**. This was followed by the first of a series of costly adventures, including the Italian wars of independence in 1859, the Mexican War in 1862, and the disastrous **Franco-Prussian War** of 1870, which ended for him in captivity at the Battle of **Sedan** on 2 September.

Napoleonic wars. *See* **French revolutionary wars.**

NASA (National Aeronautics and Space Administration) is an American government agency formed in 1958, based on the National Advisory Council for Aeronautics (NACA), for the **research and development** of **space vehicles** in competition with the Russian space exploration programme. Its pronounced military and **cold war, propaganda** aims were amplified when President John Kennedy proposed putting a man on the moon before the Russians, in the 1960s. Its immense contributions to **technology** in connection with **aircraft, communications, computers, electronics, lasers,** means of **propulsion, navigation, radar,** medicine (*see* **Medical services**), **rockets and guided missiles, space vehicles, surveillance** (including survey), environmental matters, new materials, among many other things, therefore had a revolutionary impact on everyday life.

Nato (North Atlantic Treaty Organization). The organization was brought into being on 4 April 1949 by the North Atlantic Treaty, designed for mutual defence against the Russian threat. The signatories were Belgium, Britain, Canada, Denmark, France, Iceland, Italy, Luxembourg, the Netherlands, Norway, Portugal and the USA. Greece and Turkey acceded on 18 February 1952, West Germany on 9 May 1955, Spain in 1982. It is a permanent military alliance with an international council, secretariat and various political and social committees. The military command structure is international and works under a military committee of the Chiefs of Staff of the member nations. It comprised the Supreme Allied Commander Europe (SACEUR), commanding forces on the Continent and in the Mediterranean, the Supreme Allied Commander Atlantic (SACLANT) and C-in-C English Channel Area (CINCHAN), whose headquarters were responsible for defence planning and supervision and exercising of forces, leaving operational functions to regional headquarters.

The organization has had many political and military vicissitudes, including France's withdrawal from the integrated staffs in 1966 and her insistence that Nato troops moved out – damage gradually repaired after the death of General **de Gaulle**. Turkey's similar withdrawal in 1974 took place at a time of crisis with Greece. Several nations expressed concern at domination by the USA at the same time as the USA pressed for their increased contributions. Nato's solidarity was related to the measure of the Russian and Warsaw Pact (*see* **Warsaw Treaty Organization**) threat. But the collapse of communism and the USSR in 1991 led to a reappraisal of its role, as a security and unifying organization at the service of a community, later to be enlarged by inclusion of ex-Warsaw Pact nations.

Naval warfare. Because the sea is not man's natural habitat he was relatively slow, owing to technological ignorance, in finding ways of moving across, let alone beneath, its surface. The first warlike naval operations were probably **amphibious** with minimal combat between man-paddled vessels, which were incapable of sailing against the wind. Not until 2900 BC is there a record of armed vessels, when the Pharaoh Snafru sent 40 from Egypt to Byblos to obtain cedar (probably for **shipbuilding**), and it is 2700 BC before we hear of a naval expedition to Phoenicia by the Pharaoh Sahu-re to take prisoners as slaves (possibly as oarsmen). Both voyages, be it noted, epitomized a lasting trend of using commercial vessels also for combat.

It was the Phoenicians, between 1100 and 800 BC, who invented the first specialized fighting ship, the bireme **galley**, which could be rowed at sufficient speed to hole an opponent below the waterline with its **ram**. It was this kind of ship, gradually modified and improved and sheltered in storage on slipways, which equipped Mediterranean navies (and others) during the **Greek–Persian wars** of the 4th century BC, throughout the **Punic–Roman** and **Viking wars** until the Battle of **Lepanto** in AD 1571. These were conflicts in which massed galleys charged in close ranks to ram, grapple and board (using the Roman-invented corvus, a plank with a spike, on the end, as a bridge) to engage in hand-to-hand combat. Naval engagements were, therefore, similar to land warfare, including fire support from on-board **siege engines** and **bows and arrows**; the galleys were usually commanded by a master mariner, crewed by a few trained sailors and rowed by slaves, but manned by soldiers for the fighting.

A turning-point occurred in 1340 when, at the Battle of **Sluys**, a British fleet composed of sailing Hanseatic cogs and barges plus galleys (and possibly armed with primitive **artillery**, even if it was not actually fired in anger) defeated the French. For in the next century the investigative thrust of the Renaissance and the outwards urge of **Henry the Navigator** developed the Portuguese trading caravel into the larger, more combatworthy, square-rigged **carrack**, with its high fore-castle and stern. And this produced, in the 16th century, John Hawkins's more seaworthy **galleon** (a carrack without fore-castle), which made galleys obsolete. These were all technical revolutions that produced significant **strategic** and **tactical** results.

Naval warfare was always, of course, to a large extent commerce-related. **Convoys** were used not only for the safe passage of amphibious forces (such as at the Battle of Ecnomus in 256 BC) but also to sustain trade against piracy and raiding. In 1588 the fateful entry into the **English Channel** of the Spanish Armada had as its aim an invasion of Britain, but also was intended to put a stop to British interference with Spanish trade with the Americas. However, a tactical doctrine for fleets of galleons had not been formulated, because ships' masters were not amenable to military discipline and often steered much as they pleased, regardless of the demands of admirals. Not until Admiral Robert **Blake** became General-at-Sea in 1649, and as Admiral in 1653 issued his *Articles of War* and *Fighting Instructions*, was the Royal Navy properly disciplined and its captains compelled to fight in formation, instead of as an uncontrollable gaggle (*see* **Navies, Organization of**).

The *Fighting Instructions* were tried out during the Anglo-Dutch Wars and gradually improved upon over the next 200 years by the principal naval powers – notably the French and British. Fundamentally, they created the tactics of broadsides from lines of battleships pounding each other as they closed to board. But gradually artillery, which merchantmen mounted for self-protection but which fighting ships employed only for supporting fire prior to boarding, assumed a crucial offensive role – even though **gunnery** was often poor and but few ships ever were sunk by gunfire alone. Moreover, the *Fighting Instructions*, which laid down strict tactical manoeuvres, tended to restrict ships' captains' initiative. And lack of quick signal **communications** between ships stultified flexibility – until the introduction by the French in 1763 of a code book of flag signals which, in subsequent years, were copied and greatly improved upon by the British to the benefit of their command, control and tactics.

The **wars of the 17th and 18th centuries** impelled steady development of naval warfare in keeping with ships' improved seakeeping qualities, better **dockyards** maintenance and increases in fire-power, which ensured British and Dutch maritime dominance to the great strategic disadvantage of France and the other continental powers. The politico-strategy of **blockade**, allied to commercial expansion, practised by the Anglo-Dutch during the War of the Spanish Succession, not only enriched those nations and had a decisive **logistic** function which weakened the French economy, but also initiated the more than two centuries of worldwide naval superiority which enormously enhanced Britain's technology and manufacturing industry.

Anglo-French naval rivalry, the **Seven Years War**, the **American War of Independence** and the struggle for survival drove Britain to concentrate more on her naval than on her land forces. Like her opponents, she rejected dubious means of underwater attack by David **Bushnell**'s and Robert **Fulton**'s **submarines** but worked instead on improved ways of smashing timber with artillery such as the carronade. But the key to Britain's ultimate victory over the combined French and Spanish fleet at the Battle of **Trafalgar** in 1805 was the development by Admiral Lord **Nelson** of tactics enabling the concentration of the fire-power of several ships against single enemy ships in succession, in preference to the usual one-to-one broadside slogging matches. Without fighting another major battle, the British fleet-in-being secured the nation's sovereignty, trade and freedom to land and withdraw armies wherever she chose, leaving smaller vessels, such as **frigates** and **corvettes**, to impose blockade and gather vital **intelligence**, among a myriad of other tasks.

But the days of the wooden walls were numbered by the development of iron-built steamships and the enormous increase in the power of artillery. All came to a head during and after the **Crimean War** with the construction of iron-clad warships such as *La Gloire*, *Warrior*, ***Merrimack*** and *Monitor*. The mere threat they posed, let alone demonstration of their influence during the **American Civil War**, made urgently inevitable the scrapping of all existing warships, the design of totally new ships and building techniques, and ultimately a complete revision of gunnery and tactical doctrine. Reluctant sailors were forced to abandon the broadside and introduce rotating turrets and barbettes to house great, breech-loading guns. Despite a strange aberration and rebirth of the **ram** as the result of the Battle of **Lissa**, it was realized that being able to shoot ahead at long, rather than point-blank, range was most effective. Hence sprang the demand for revolutionary **rangefinding** and gunnery techniques to engage and hit targets on the horizon – with the corollary that stronger **steel armour** protection was also an essential and feasible improvement.

The gun-versus-armour race which ensued coincided with the scrapping of sails in the 1870s when steam engines (*see* **Propulsion, Means of**) became much more reliable and masts obstructed turret fields of fire. As if that was not enough, it was realized in the closing years of the 19th century that hits by artillery were not only difficult to obtain but less deadly than explosions below the waterline caused by **mines** and locomotive **torpedoes**, the latter delivered by totally new classes of deadly vessels. From the first 18-knot, steam-driven **torpedo-boats** of the 1870s evolved the larger torpedo-boat **destroyers** of the 1880s and much more reliable **turbine** boats with speeds of 48 knots in 1897. Even more menacing yet than the heavily armed and fast surface ships were modern **submarines** which, because of the invention of electric motors and **oil**-fuelled engines in the 1880s, became viable weapon systems at the start of the 20th century. The moment had arrived when two new, modern naval powers – the Japanese and German – were dramatically to expose their rising potential and intentions.

The shape of future naval warfare was clearly revealed during the **Russo-Japanese War** of 1904–5 when, with the exception of submarines, all the latest naval weapon systems, enhanced by **searchlights** for night fighting and the newly invented **radio** communications, were employed with devastating impact. **Battleships**, **cruisers** and torpedo-boats destroyed each other with gun and torpedo on the surface and were sunk by mines from below. The bottling up of one Russian fleet, and its subsequent destruction by army artillery in besieged **Port Arthur**, demonstrated the vulnerability of **dockyards** to land attack. And the elimination of another fleet at the Battle of **Tsushima** seemed to validate the overriding power of battleships in open water, albeit when firing at ranges up to 7,000 yards and employing the latest tactic of bringing to bear concentrated fire from a line of ships by 'crossing the T' of an approaching enemy in line ahead.

But even as the Japanese were triumphing in the Far East, the British, spurred on by Admiral Sir John **Fisher**, were engaged in a naval race with the Germans, spurred on by Admiral Alfred von **Tirpitz**, and on the eve of building HMS ***Dreadnought*** and the subsequent battleships and **battle-cruisers** which were to make obsolete all existing capital ships. This was the class of vessel with which the major powers failed to win **World War I** because the Germans were reluctant to risk them in major engagements; and the British unable to annihilate them at the battles of **Dogger Bank** and **Jutland**.

Indeed, by 1916 naval strategy, as well as tactics, had been overturned again by technological change. **Aircraft** were playing a small role in **reconnaissance** and attacks on shipping, and **aircraft-carriers** were being improvised. Largely, German submarines and mines were dictating strategy because they seriously threatened commerce with a partial siege of the British Isles. As a result the remaining major operations centred upon the defence of convoys, which proved to be the most effective **anti-submarine** measure.

It scarcely seemed to occur to many sailors that the intricate, massed fleets, deployed for World War I to bring maximum artillery fire to bear, were outmoded before **World War II**. **Jellicoe**'s resistance to the convoy system for merchant shipping in 1917, allied to his reluctance to detach destroyers from the Fleet for escort duties, was symbolic of an unwillingness to recognize change. Change was imposed accidentally by the Washington Naval Treaty, which restricted battleship and cruiser construction but permitted the building of the next generation's capital ship, the aircraft-carrier, along with the dive- and torpedo-**bombers** which would send battleships to their graves.

Naval warfare in World War II was a global business, conditioned by underwater craft and aircraft, but governed by the principles stated by Admiral **Mahan** that commerce and logistics were key issues and sea power decisive. Unfortunately, however, his notion that Anglo-American sea power would generate peace proved false, because Germany paid little attention to it in 1939 and the Japanese, with a high-grade Fleet, gambled on a victory beyond their powers against the world's largest navies. Indeed, although submarines and aircraft did bring the Axis powers immense successes, it was these weapons which also helped crush them in the battles of the **Atlantic**, the **Mediterranean** and the **Pacific**. In this war only one battleship was sunk by gunfire alone, but the rest of its class was hunted and refused domination of the seas. In the struggle visual encounters between lines of major surface vessels were unusual, but combat between groups of destroyers, frigates and submarines was a regular occurrence. Aircraft-carriers dominated the seas but, as at **Taranto**, **Pearl Harbor**, the **Coral Sea**, **Midway**, the Philippine Sea and **Leyte Gulf**, rarely saw the targets they were attacking. Yet, in this naval struggle all contenders were aiming, basically, to maintain or stop commerce in battles, fighting to keep convoys moving or impose blockade, and launching amphibious operations to seize or deny strategic bases.

In these battles the underlying struggle was **technological**: in the races to produce, for example, superior **radar** and **sonar** (or countermeasures to them) in submarine warfare; and to develop tactical doctrines, enhanced by greatly improved radio, to outmanoeuvre the enemy while baffling him with false information and subterfuge. In World War I knowledge of enemy **codes and ciphers** was of great value. In World War II it was probably more crucial to the result at sea than on land and in the air; the Axis powers were totally outwitted in this department.

Since 1945 sea power has retained its crucial importance since only about 9 per cent of trans-ocean commercial cargoes go by air. Though aircraft-carriers have played (and continue to play) important roles in **cold** and **limited wars**, and are prime targets of hunters, it is the submarine – above all the **nuclear**-powered type with **nuclear** warhead **missiles** – which has assumed the role of capital ship along with that of deterrent (*see* **Deterrence**) and peace-keeper. It too is tracked and hunted by air and surface forces in both peace and war; in peacetime because, in the event of an outbreak of hostilities, it will be of paramount importance to all contenders to sink enemy boats, most no doubt already at sea, as quickly as possible. In the meantime merchant shipping is the prime target and in need of constant protection to ensure food supply and, of immediate importance, the movement of reinforcements and materials for major forces across the oceans. It is at the same time involved in the struggles for possession of vital terrain (such as Northern Norway, Iceland and Greenland) controlling access to open waters. It is no exaggeration to say that the opening bouts of any future war are constantly in progress in peacetime as **surveillance** forces study maritime movements, track potential predators and watch for threatening signs.

Navarro, Count Pedro (*c.* 1460–1528). A Spanish sailor who became the foremost military engineer of his time, besides being a **guerrilla** leader in the Italo-French Wars (1495–1515). He won fame in **siege warfare**, including the mining of Turkish forts, the defence of Canosa and Taranto and, above all, his reduction of French forts at Naples (1503). Later siege successes included Vélez de la Gomera (1508) (where he improvised floating batteries), Oran (1509) and Tripoli (1510). He mounted the heavy arquebus à croc (**musket**) not only for the defence of fortress walls but also on **wagons** for a mobile role at the Battle of Ravenna (1512), where he was taken prisoner by the French, with whom he served until his capture in Italy and death as a prisoner of war.

Navies, Organization of. In its modern sense the organization of navies for defensive and offensive operations was first established in the **Greek–Persian Wars**, when the need for special fighting **galleys** (as opposed to adaptation of merchant vessels) was deemed unavoidable, despite the cost. Henceforward the demands of **naval warfare** dictated the evolution of different classes of vessel, the design of fighting ships, the grouping of vessels by classes for administrative, as well as training and operational reasons, and the establishment of an admiralty for political and financial regulation of ships, stores, **dockyards**, communications centres and, in due course, **aircraft** and airfields.

Even in the 5th century BC a form of Admiralty economic control was practised by the laying up on slips of galleys in peacetime and their manning in time of war with reserve crews composed of sailors, soldiers and slaves called up for the duration. Such measures grew ever less acceptable with each advance in technology, especially when the introduction of **artillery** created a demand for more specialists who, in turn, invented techniques which called for dedicated branches of the service in support. Naturally, as naval operations spread worldwide after the 15th century, Admiralties had to decentralize, usually into regional fleets such as Home, Atlantic, Mediterranean and Pacific. Also, with the proliferation of revolutionary technologies and classes of ship which coincided with the passing of sail, still more subdivisions of fleets became necessary into shore bases, task forces, groups, flotillas and squadrons.

By **World War II** a Fleet, by number or name, might comprise a squadron of **battleships**, supported by a **cruiser** squadron, one or more separate **destroyer** and **submarine** flotillas and attached units of auxiliary supply vessels. At the same time, under control of its C-in-C it might, as for example was common in the battles of the **Pacific**, have self-contained **aircraft-carriers** and an **amphibious** fleet attached.

Like the other services, navies in peacetime can afford only what limited funds are available as they cope with current tasks and plan for war. Reserve forces and their utilization in emergencies and on **mobilization** are essential – but can make rather larger use of civil assets than armies and air forces. The Royal Navy, for example, maintains a fleet of auxiliary logistic vessels manned by civilians and armed as necessary. These were the mainstay for **ammunition** and fuel supply, supplemented by vessels called up from trade, of the Task Force sent to the **Falkland Islands** in 1982. As in the past, when merchant ships were permanently armed for self-protection against pirates (among other aggressors) and as was done in both World Wars when ocean liners were turned into armed cruisers to assist with **convoy** escort and **blockade**, these ships fought in action. It is a relatively simple matter to mount guns and **missiles** on deck. In **World War II** some merchant ships carried **fighter** aircraft to be catapulted against enemy **bombers**, while many more were converted to escort aircraft-carriers. The carriage of vertical take-off and landing aircraft (VTOL) simply provides a variation on the theme.

Like all cost-effective organizations, fleets are variable in composition, flexible in operation and subject to changing doctrine under the influence of **technology** and techniques. The organization, for example, of the Federal Fleet in the **American Civil War** operating in the Mississippi River was arranged to satisfy close support of land forces by small, steam-powered vessels using fire-power and amphibious techniques. The Austrian Fleet at the Battle of **Lissa** went into battle in a formation intended to make use of the **ram**, in addition to modern fire-power. Japanese organization during the **Russo-Japanese War** of 1904, mirroring that of the Royal Navy, made best combined use of battleships, cruisers and destroyers to tackle Russian vessels in port, in coastal waters and on the high seas, demonstrating an operational flexibility in grouping greatly assisted by **radio** to implement regrouping at speed. It was the introduction of the **mine**, the submarine and aircraft which demanded the greatest re-organizations. Soon the technology and techniques that had made it necessary for maintenance and training to form specialized **destroyer** and **torpedo-boat** flotillas called for similar organizations for minelaying, minesweeping, submarine and anti-submarine forces, air squadrons and anti-aircraft ships and their weapons. These, in turn, made necessary depot support ships and a proliferation of shore establishments to develop the new weapons and train sailors to use them, a trend escalated vastly throughout both World Wars to cater for advances in signalling techniques and the calls of amphibious warfare. In addition, there was the vital need to integrate the new technology in individual warships and train leaders to make best tactical use of their units in, for example, the highly complex battles between aircraft-carriers and in the escort of convoys. It has all amounted to a never-ending process of evolution since 1945.

Navigation. Man's means to fix his position anywhere on, and later, above, the earth's surface had to await the production of **charts and maps**. Until then he was able only to follow a route by reference to known objects, such as topographical features, while only sensing his way across the sea when out of sight of land. How the ancients managed to navigate at sea is largely conjectural, but reference to the sun and other astral bodies undoubtedly were the most important for keeping direction in accordance with sailing directions, or logs, which mariners compiled from observation of features, currents, sea bottoms, etc., and handed down from father to son. The first pilotage book, the Periplous of Scyclax, dates from *c.* 350 BC, but was related to even earlier, rudimentary charts.

The discovery in 1180 by Alexander Neckam of how to use a magnetic needle as a direction finding compass was a big step forward. But further progress awaited the invention of ways to measure velocity in addition to comprehensive and reliable charts. By stages the former had fully evolved by 1637 through recording, by the use of an hourglass, how long it took to sail 47ft 3in, measured by observation of knots on a trailed line – hence 'knots' is a measurement of speed for a nautical mile (6,080 ft/min) per hour, a method which remained highly imprecise until the introduction of serviceable **clocks and watches**. But it was not until the 19th century that reasonably reliable charts became available.

A basic navigational method is dead reckoning – the calculation of position by measurement of direction and velocity from a known point of departure. Far more reliable, however, is the plotting of known latitude and longitude.

Latitude was first ascertained *c.* 1470 by the Portuguese using a Seaman's Quadrant, an instrument for measuring 'altitude' – the angle of elevation between an astral body and the horizon. It was superior to an adaptation of the astronomical astrolabe but precise only in a flat calm. But used with tables of the sun's declinations and improved on, via the long-staff, until the invention of the efficient sextant in 1757, it helped fix latitude and *longitude* to within acceptable limits – always provided clouds did not prevent the taking of sights. But the fixing of *longitude* also demands the exact time at noon when east or west of the meridian through Greenwich, which was only made possible after the invention of the chronometer in 1761.

There followed a succession of refinements, such as the liquid magnetic compass in 1862 to damp out the major oscillations of the dry-card compass. **Aircraft** required further modifications to the compass and the invention of the pitot tube to measure velocity, along with ways of measuring drift and true ground speed. On the ground there were far fewer problems until the 20th century, although in unmapped territory reliance at times still had to be placed on guides, and it was not unknown for entire armies to lose direction. However, the **motor vehicle** gave the soldier a much greater radius of action, free of the limitations imposed by the **horse**'s need for fodder and water supplies. Long-distance operations in featureless terrain became possible and a requirement for more skilled navigation arose. Armies fighting in the **desert** in **World War II**, and especially such special forces as the Long Range Desert Group, the SAS and Popski's Private Army, soon became adept desert navigators; it is no coincidence that the latter adopted the astrolabe as their cap badge. They mastered the theodolite so as to fix their position by the stars and used the sun compass, originally invented by the Light Car Patrols operating against the North African Senussi in 1915 and perfected by Major R. A. Bagnold for his explorations in the Western Desert in the 1930s. The device, mounted on the vehicle where the driver could see it, relied on the shadow cast by a vertical post on to a horizontal plate graduated in 360 degrees.

At sea and in the air the time-honoured methods, though refined, were used throughout World War II, as well as the electronic aids which began to appear in the 1940s. The Germans introduced their Knickebein directional radio beams to guide **bombers** to their targets, while in the air and at sea Decca developed **Gee**, a radio navigation system around the coasts of Britain, first used for bombers and then to control shipping during the invasion of Europe in 1944. The Americans devised a similar system for their offshore shipping known as Loran C. A further development since 1945 has been the appearance of inertial navigation systems using measurement of the direction of the force of gravity with the aid of **gyroscopes**. Linked with a **computer**, information on heading, velocity, position and corrections for earth rotation can be resolved to provide an accurate fix. This approach has particular attractions in long-range **rocket** missile navigation and has been used extensively for **submarines**.

On land it was the predicted effects of full-scale **nuclear** war, expected to remove known landmarks, which in the 1950s led to work on automatic navigation aids for land vehicles. Various ideas were tried, generally using **computers** fed by compasses and signals from the road wheels to resolve direction and distance data into grid references. The problems of fixing accurately the start point, plus inherent errors from, for example, wheel slip, have precluded a really successful answer.

The future lies with the navigational satellite, already widely used by ships, aircraft and land forces, and, in 1991, as a battle winner for precise desert navigation during the **UN–Iraq War**. Coupled with digital computers giving instantaneous read-outs, these satellites in fixed orbit around the earth can provide positional data to an accuracy of a few metres.

Nelson, Admiral Lord (1758–1805). The son of a parson who won immortality by his achievements in the **American War of Independence** and the **French revolutionary wars**. He became renowned not only for physical bravery (which cost him an arm and an eye in battle) but also for his willingness to risk censure by departing from orders to win **strategic** and **tactical** advantage. Not all the numerous minor actions he took part in were successful, but the failures pale before his key role in presenting Admiral Sir John Jervis with victory at the Battle of St Vincent (1797); his destruction, in a rare night action, of a French squadron in the Battle of the Nile (1798); his decisive disobedience of orders to defeat the Danish fleet at Copenhagen (1801); and, in 1805, his celebrated tracking of the French fleet to the West Indies and back, which eventually culminated in his death at the Battle of **Trafalgar**.

At the root of Nelson's greatness in battle was an insistence, through manoeuvre, upon concentration of maximum fire-power against successive parts of the enemy's line before he could counter-attack. But his genius lay in a vibrant offensive spirit, in humanity and in the devotion he inspired in subordinates by encouraging them to use their initiative with confidence. In sum, these virtues produced what became known as 'the Nelson touch'.

Nightingale, Florence (1820–1910). A child of a rich family whose vocation was nursing and who was made famous when she led a team of nurses to Scutari to care for British wounded from the **Crimea**. She did little actual nursing since her self-assumed role, as a brilliant and extremely forceful organizer, was that of training and disciplining her low-grade nurses and bullying and persuading the Government, the Army and its doctors into a complete reconstruction of **medical services**. She was largely successful owing to strong public support and her own talents as an administrator. Her *Notes on Matters Affecting the Health, Efficiency and Hospital Administration of the British Army* and *Notes on Nursing* became bibles not only for the British forces but also for **health** and nursing services worldwide. After the war she worked to exhaustion establishing nursing training colleges and advising health authorities on the design and running of modern hospitals. She was carried to her grave by six sergeants of the British Army.

Night-vision devices. Until the 1880s when Elmer Sperry and others began experiments with electric–carbon arc lights, night vision was usually, at best, provided by naked flares and fires from burning ships or buildings, enhanced by **optical instruments**. The introduction of **searchlights**, their mounting in **fortifications** and warships by the turn of the century, and their first use on land in the Boer War and at sea in the **Russo-Japanese War**, revolutionized night fighting. Flares, searchlights and **artillery** star-shell were widely used in both World Wars, but it was not until the **electronic** age that devices specifically designed to help the eye see in the dark appeared.

Early equipment depended upon active **infrared** (IR) illumination of the target, often using a searchlight with an IR filter over the lens. The next generation of equipment was passive, relying on the ambient light always present, even on the darkest night, to illuminate the target. The **image intensifier** was developed to assist the eye to see as if in twilight with an enhancement of about 10^5.

Low-light **television** (LLTV) provided another approach to the problem with the advantage of remote viewing and multiple display, but with the disadvantage of bulk, high-power requirement and limited performance against moving targets.

Another technique is the passive use of

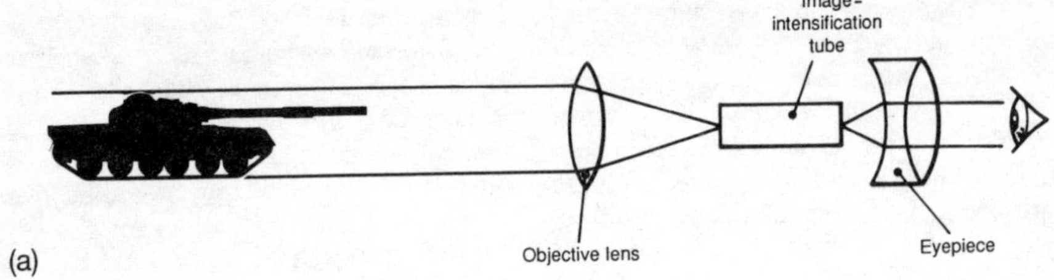

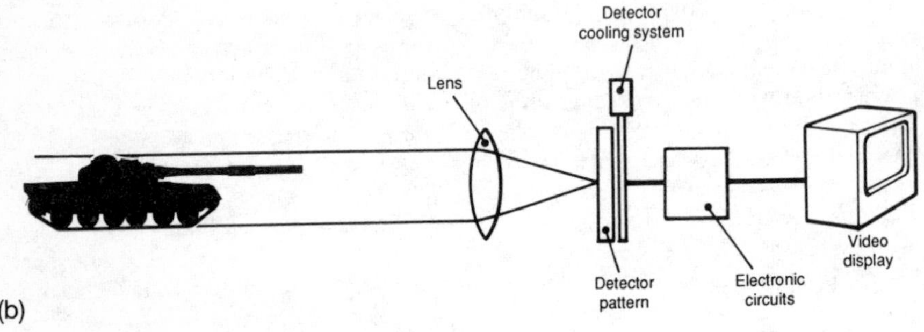

Outline of (a) an image-intensification sight and (b) a thermal-imaging system

the target's contrasting thermal radiations. **Thermal imaging** (TI) devices have been in use since the 1970s, particularly as part of a **tank**'s fire-control and **surveillance** systems to display a picture of the target on a television screen. It was proved vastly superior to other tank night-vision equipments during the **UN–Iraq War** of 1991 and enabled American and British tanks to dominate Russian tanks fitted only with obsolete near IR equipment.

All these devices have made theoretically possible the 24-hour battlefield day. But it is wise to remember that in the Yom Kippur War of 1973, when both sides used night-vision aids, the strain of operating the equipment was such that after 36 hours both sides had to call a halt and go to sleep.

Nimitz, Admiral Chester (1885–1966). Nimitz was Chief of Staff to the commander of the US Atlantic **submarine** force in **World War I**, and thereafter had a fairly diverse career (including being Chief of the Bureau of Navigation in 1939) for one of pronounced character picked for stardom as C-in-C Pacific Fleet on 16 December 1941 after the attack on **Pearl Harbor**. Managing to establish a working relationship with the Chief of Naval Staff, Admiral **King**, he presided over the curbing of Japan's expansion in the **Pacific Ocean War** and the immensely effective 'island hopping' **strategy** aimed at the heart of Japan. In his dealings with such personalities as General **MacArthur** and Admirals **Halsey**, Spruance and **Turner**, he steered a steady course with minimal friction, even with King, in operations he judged correctly when the Japanese shifted to the defensive in 1942. In 1944 he knew when to bow to MacArthur in the controversy over invasion of the Philippines; but he was also very terse with Halsey when that officer left the St Bernardino Strait unguarded during the Battle of **Leyte Gulf**. He was present at the Japanese surrender. From December 1945 to December 1947 he was Chief of Naval Operations in the difficult post-war years of retrenchment. He published no memoirs.

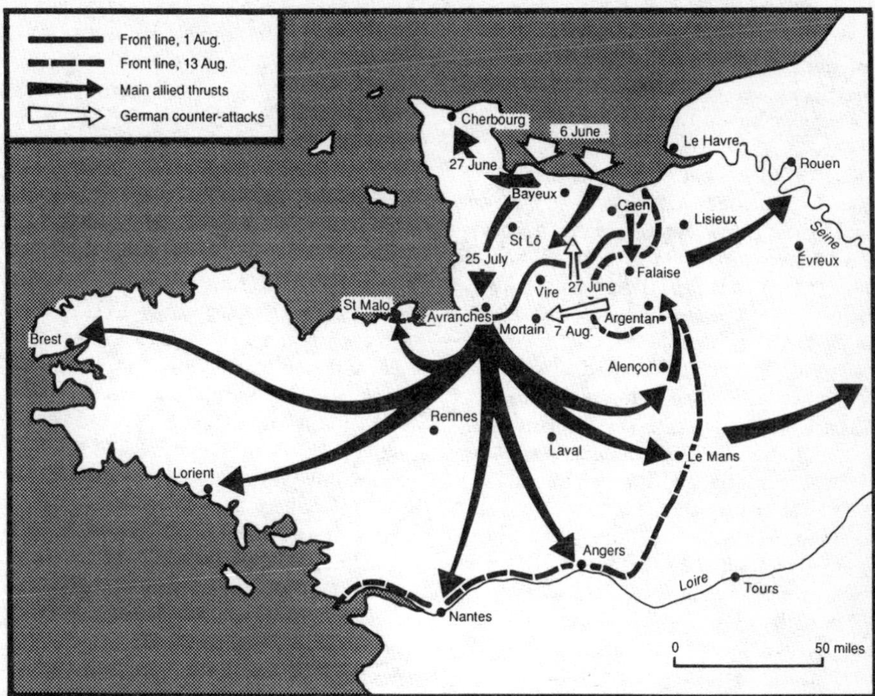

The armoured break-out from Normandy, 1944

Nobel, Alfred Bernhard (1833–96). A brilliant and inventive Swedish engineer and chemist who studied **explosives** and in 1867 devised the stable dynamite after experiments with unstable nitroglycerine. He built up a large manufacturing business and invented a wide range of new explosive compounds and devices, such as the jelly-like blasting gelatine, from gun-cotton, in 1876; the smokeless, powerful ballistite (of crucial importance to **ammunition** and the automatic **machine-gun**, as well as **tactics**) in 1885; and sophisticated detonators. Reviled by many for the development and production of so many war materials, he left most of his fortune to the founding of five prizes open to international competition – of which one was for contributions to peace.

Normandy, Battles of. The Viking dukedom of Normandy in northern France came into being in AD 911 as a prize of the **Viking expansion**. It was the scene of numerous battles and sieges during the **Hundred Years War**, including the major sieges by Henry V of Caen (1417), Rouen and Cherbourg (1418–19). Thereafter it remained fairly free of invasions until the **Franco-Prussian War**, when it was largely occupied; and **World War II** when the Germans overran France in 1940.

For the next four years its people and the German garrison lived fairly comfortably together, only disturbed occasionally at the coast by small hit-and-run raids and from above by increasing air activity, with very heavy bombing before 6 June 1944 when Allied armies landed in great strength.

Preceded by three **airborne** divisions and a stunning naval and air bombardment, the Allied armies under General Eisenhower, commanded in the field by General **Montgomery**, came ashore on a 50-mile front. Against them stood Field Marshal Rommel's

Army Group B under C-in-C West, Field Marshal von Rundstedt, entrenched behind the **Atlantic Wall**. Although resistance to the Americans was very effective on one beach (called Omaha) at the base of the Cotentin peninsula, they won a strong foothold – though it took 12 days to fight through to the other side and cut off Cherbourg. Meanwhile the British and Canadians had captured Bayeux and were in sight of **Caen** on the left flank of D-Day, though unable to go further as the Germans sealed off the bridgehead. Fight as well as the Germans did, they were constantly deprived of sufficient strength, partly owing to interdiction by Allied bombing of the lines of communication and also to the unwillingness of **Hitler** and von Rundstedt to transfer troops from the Pas de Calais where the main blow was still expected. As a result, Rommel's attempts to assemble and deliver a decisive counterstroke were repeatedly frustrated.

Throughout June and July, in close countryside ideal for defensive infantry combat, the Germans managed, by the skin of their teeth, to hold on against an opponent who was superior in fire-power and able, despite stormy seas (which destroyed one of the two artificial **Mulberry harbours** towed from Britain and delayed **logistic** buildup), to more than keep ahead in the reinforcement race. Both sides suffered heavy losses in what was an attritional struggle, one in which Montgomery deliberately used the British and Canadians on the left to attract and smash the German **armoured** forces, leaving the Americans on the right, under General Bradley, to take Cherbourg (27 June) and assemble a massive force near Saint-Lô to smash through in the direction of Avranches. On 20 July the Germans remained pinned to the south of the ruined city of Caen by a strong British armoured attack.

On 25 July Bradley opened his attack against thinly held enemy positions and drove south at ever-increasing speed, compelling Hitler to demand a massed armoured counterstroke against the extending American flank. The Americans fended off the blow with relative ease without delaying their advance into Brittany or eastwards towards Le Mans and the German rear (the Germans still held fast at Falaise). Now Hitler's refusal to permit withdrawal or an end to the beaten counterstroke played into Montgomery's hands. As the American threat to Paris began to develop, he also caught the mass of the German Army in a pocket formed by Americans in the south and British and Canadians in the north. The trap finally closed on the 19th with heavy loss to the Germans. After a great many demoralized escapers had raced for the River Seine, attacked all the way from the air, they arrived at the river only to find, in many cases, that they were in yet another pocket with all the bridges destroyed.

Normandy was a difficult battlefield, particularly for the **infantry**, because **artillery**, **mortars** and **machine-guns** held sway and **tanks** found the terrain difficult to penetrate. Once mobility was restored in the final phase the Allied casualty rate fell when armoured forces bullied their way through. Normandy was one of the most costly single Allied battlefields of **World War II**, but it ruined the German Army in the West and made it possible to end the war in 1944 had the subsequent pursuit been conducted in a more rational and concentrated manner.

North Africa, Battles of (1940-43). The north African shore had been traversed by Arab armies during the **Muslim wars of expansion** and largely remained part of the **Ottoman Empire** after Spain was cleared of Muslims in 1492. In 1911, Italy provoked a war with Turkey and seized the vital ports, including Tripoli, Benghazi and Tobruk. By 1912, in a campaign which included **air warfare** for the first time (with scant effect), and on the eve of the ejection of the Turks from the Balkans, the whole of Tripolitania and Cyrenaica were in Italian hands.

On 10 June 1940 this largely desert territory again became a battlefield when Italy declared war on Britain and France, at the same time as the battles of the **Mediterranean Sea** and the siege of **Malta** commenced. Greatly outnumbered as were the

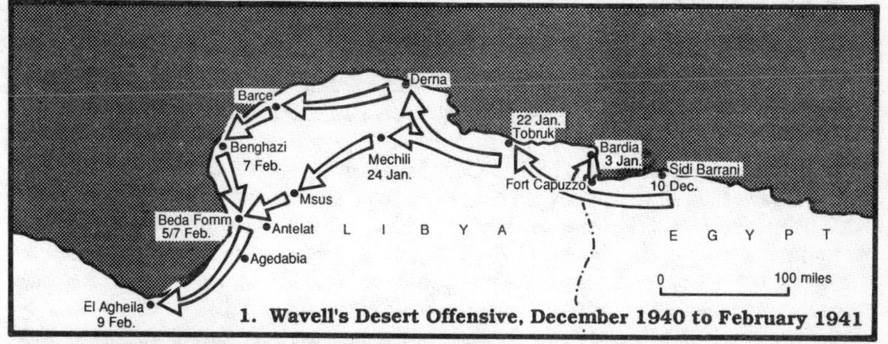

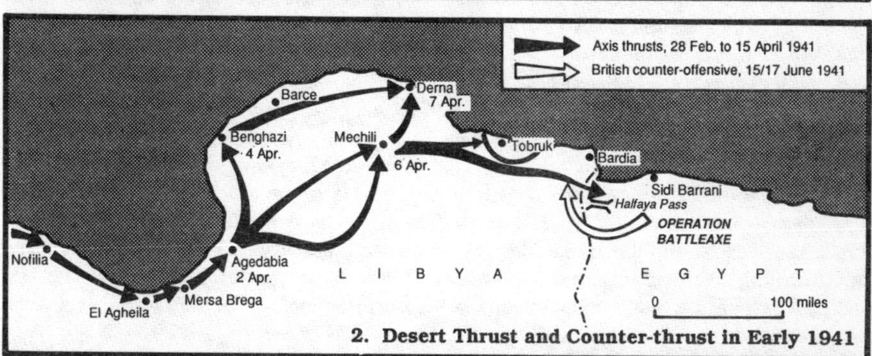

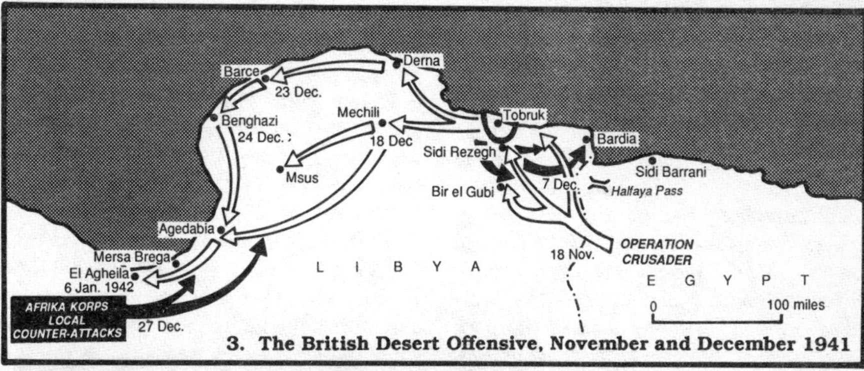

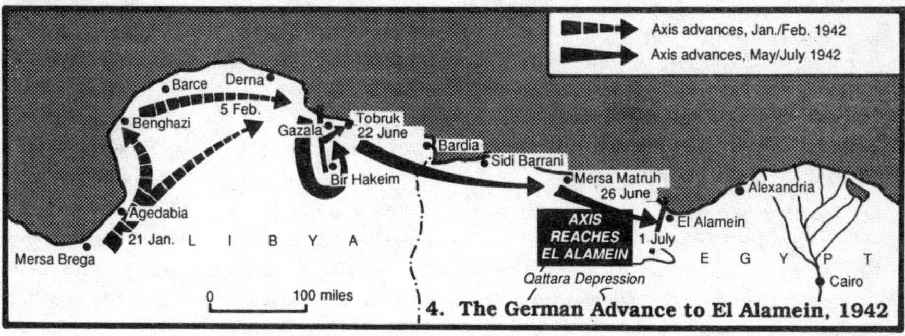

The Battles of North Africa, 1940–42

British in Egypt (under General Wavell), they at once began small-scale desert raids against the Italians in Cyrenaica and established a morale superiority. Activities which were concurrent with the bombardment of the French Fleet at Oran on 4 July and chasing the Italian Fleet into port, but suspended when the Italian Tenth Army, under Marshal Graziani, invaded Egypt on 13 September.

The Italians advanced against light opposition to Sidi-Barrani, where they fortified a number of camps and prepared for the next move. On 9 December, however, the British Western Desert Force of 31,000 men under General Richard O'Connor overwhelmed the camps by surprise and within 48 hours had thrown what was left of Tenth Army out of Egypt, to take refuge in the ports of Bardia and Tobruk. Sidi Barrani fell on 5 January 1941 and Tobruk on the 22nd, the pursuit brushing aside enemy mobile forces and pressing on to annihilate the Italians at Beda Fomm on 5 February, finally raising O'Connor's score in prisoners to 130,000 against his own losses of 500 killed and 1,373 wounded.

Fearing a total Italian collapse, **Hitler** sent a light armoured corps under General Rommel to Tripoli. Contrary to orders he probed towards Benghazi, detected frailty among the British forces and plunged eastwards. Stretching beyond the limits of prudence in **desert warfare logistics**, he reached the frontier on 14 April but was prevented by British and Australian troops from capturing Tobruk. By deciding to hold the port, Wavell denied the Axis army vital supplies, and was able himself to mount a limited offensive at the frontier on 14 May, followed on 15 June by a major thrust (making use of new tanks brought by **convoy** through the Mediterranean) to relieve Tobruk. After three days' fighting this offensive was thrown back by Rommel, whereupon both sides concentrated upon preparing for a race to capture or relieve Tobruk.

On 18 November the British Eighth Army advanced, catching Rommel by surprise and bringing on the Battle of Sidi Rezegh in the approaches to Tobruk. Fortunes waxed and waned; Rommel wasting **strategic** opportunities won by brilliant **tactics**; the British, under General Auchinleck, persevering until Rommel's logistics collapsed and he was forced, at the year's end, to pull back into Tripolitania. From there he sprang eastwards again on 21 January, once more surprising and demoralizing the British, who were driven back to Gazala, where both sides stopped in exhaustion on 4 February.

Another race began to see who could attack first, won by Rommel on 28 May at the start of the Battle of Gazala and the gruelling struggle which led to the fall of Tobruk on 21 June and the pursuit to Mersa Matruh. There Rommel was close to defeat in an attack against a numerically superior British force on 28 June. But the British withdrew in confusion to the shortened, unturnable line at **El Alamein** where, in three battles of attrition in July, August and October, Rommel's dream of reaching the Suez Canal was ended.

Starting on 4 November, the routed Axis army retreated with hardly a stop, retiring into Cyrenaica as news came through on the 8th of extensive Anglo-American landings in French North Africa at Casablanca, Oran and Algiers. A surprise stroke which met with only short-lived Vichy French resistance before the whole of Morocco and Algeria was in Allied hands. A British landing at Bône on the 12th posed a threat to Tunisia. To secure Rommel's rear, Field Marshal **Kesselring** rushed German and Italian forces by air and sea into Bizerta and Tunis, narrowly winning the race to prevent those ports falling into Allied hands, thus bringing on the first of the battles of Tunisia. Meanwhile Rommel had fallen back to the Tripolitanian frontier and Montgomery's Eighth Army was engaged, until 11 December, in replenishing itself by sea and land, prior to striking at El Agheila and resuming the pursuit. The Axis rearguards barely paused before evacuating Tripoli on 23 January 1943 and motoring to the temporary safety of the Mareth Line, which the French had once built to keep the Italians out of Tunisia.

At that moment Axis tenure of North

Africa, despite the Battle of Kasserine Pass and its other spoiling offensives in January and February, was at the mercy of overwhelming Allied naval, land and air power. By 13 May some 275,000 German and Italian prisoners, along with most of their top leaders in Tunisia, were in Allied hands, bringing to a total of about 620,000 the number of Axis troops lost in North Africa since 1940 – approximately three times the Allies' casualties.

North Sea, Battles of. From across the often turbulent waters of the North Sea came the Germanic Angle and Saxon tribes from northern Europe (once they had developed, *c.* AD 250, their seaworthy long boats), followed in the 8th century by the **Vikings**. Interceptions at sea of these raiders rarely took place. Indeed, not until 1588, when the **carracks** and **galleons** of the Spanish Armada were being pursued northwards, was there a major naval engagement in the North Sea.

The Anglo-Dutch Wars (1652–74) featured, however, the first major example of **naval warfare** strategically related to trade and the passage of **convoys**. A fierce struggle developed in which the British lost six out of the 13 major battles, of which 12 were in the North Sea. This laid the foundations of modern **blockade**; of concentrated gunfire by disciplined formations; and of British sea power. This power, exercised in conjunction with allies, was to endure for the next 200 years with only one more major battle in the North Sea – that of Camperdown in 1797 – this time a British victory over the Dutch when, typically, both sides severely damaged each other and none of nine Dutch ships captured were reusable.

Things were different in both World Wars when the North Sea was strategically vital to both Germany and her enemies in the prosecution of blockade, securing the safety of important coastal targets and, for Britain, the movement of military material to and from the Continent. The threat of a German invasion in **World War I** tied down strong British home-defence forces; similarly German troops were diverted to guard the Belgian coast – though only in 1917 did the British seriously plan an amphibious operation.

At the start of World War I both sides initiated offensive surface operations, often in conjunction with minelaying, but the Germans were strongly deterred by the Battle of Heligoland Bight on 28 August 1914. Thereafter, with the exception of a few major forays into the Sea's central basin, they restricted surface action to night raiding by light forces along the Flanders coast and into the **English Channel**. With only 30 **submarines** in commission (against 78 British of which 65 were in home waters), the Germans in 1914 were severely limited in underwater operations, though U9's sinking of three **cruisers** off the River Maas on 22 September plus another torpedoed on 15 October, and a raid on Scapa Flow on 18 October (which made the Grand Fleet shift to Rosyth while nets were installed) was salutary. Henceforward the **torpedo** and **mine** threats loomed large, requiring vast counter-efforts to safeguard ports and clear the way for commerce and warships when they put to sea.

This was the background to the damaging hit-and-run raids by German **battle-cruisers** and cruisers against British east coast ports on 3 November and 16 December 1914, followed by a third intercepted raid on 24 January 1915 which ended with the Battle of **Dogger Bank** and loss of a German cruiser. The German response to this defeat was to institute attacks on 4 February on merchant ships (including neutrals) which made necessary the protection of shipping to and from Holland, Norway, Denmark and Sweden, as well as the introduction of effective convoys along the east coast. The Germans realized, however, that this campaign, far from crippling the Allies, might only lose them neutral sympathy, particularly that of America. They still held hopes of wearing down the Grand Fleet. It was with this in mind that they carried out a hit-and-run cruiser raid against Yarmouth and Lowestoft on 24/25 April 1916, as a preliminary to the sortie of the High Seas Fleet under Vice-Admiral Scheer, which brought on the Battle of **Jutland** on 31 May.

German failure at Jutland, followed by the fiasco of another High Seas Fleet sortie in August (when Scheer withdrew because of a misleading zeppelin report and Admiral **Jellicoe** did likewise because he feared submarines), persuaded the German High Command that an unlimited submarine campaign against commerce was their main hope of a decision at sea which might win the war. Supplemented by three surface raiders, which broke out in November and December 1916 to cause considerable damage on the high seas, the campaign began on 1 February 1917. Henceforward battles in the North Sea and its approaches were between small forces; the British attempting to prevent U-boats slipping through the English Channel and round the north of Scotland; the Germans trying to hit the submarine hunters at the same time as the British tried by surface raids at Zeebrugge and Ostend in April and May 1918 to block access to the sea. Only occasionally did the High Seas Fleet threaten to put to sea again, but at least every four days the Grand Fleet had to send out a battle squadron to protect convoys to Norway which, in October and December 1917, were ravaged by two fast light cruisers.

In many respects, **World War II**'s North Sea battles started as they had left off in 1918, except that air **bombing** and mine-laying were everyday occurrences and much more effective than in the past. The same pattern of convoys and blockade emerged – until the Battle of **Norway** brought both fleets into action against each other (with heavy losses), and won for Germany an admirable mounting base for the movement of submarines and surface raiders out into the Atlantic and Arctic Oceans. The advantage was vastly augmented when the rest of the West European coastline fell into German hands in 1940, and the Battle of the **Atlantic** became an all-consuming struggle. From June 1940, indeed, the North Sea became a subsidiary starting-point for German raiders setting forth, yet a wide trench against German invasion of Britain and against British hit-and-run **Commando** raids against Norway, Holland and Belgium. In endless coastal fights, **bombing** substituted for naval bombardment and the most persistent and dangerous surface raiders of each side's convoys were motor **torpedo-boats** stalking by night with minimum fear of air attack.

In 1942, when Germany reverted to the defensive, she viewed the North Sea as a very likely approach for an Allied invasion of Norway. It was therefore in Norwegian waters that much skirmishing took place, though without encounters between major units except when **Arctic convoys** fought their way through. After the reconquest of the French and Belgian Channel ports in October 1944, it was in the mine-infested approaches to the Heligoland Bight and off Holland and Norway that submarine hunters concentrated when seeking the latest U-boats as they emerged from German ports.

Norway, Battles of. From the start of **World War II** the Norwegian government gave tacit permission for German ships to use her territorial waters. The Allies reacted to this by occasionally intercepting German ships in Norwegian waters, notably the *Altmark* on 16 February 1940. **Hitler** then correctly reasoned the British were about to enter Norway and, feeling also that possession of Norwegian bases would assist air and sea attacks against Britain, pre-empted the move. As the Royal Navy began laying **mines** in Norwegian waters on 8 April they encountered ships of the German invasion force, sank a troopship and damaged others. Next day Germany entered Denmark without resistance and in Norway put ashore troops under General Nikolaus von Falkenhorst, at Olso, Kristiansand, Stavanger, Bergen, Trondheim and Narvik. Mostly they seized their objectives without much trouble and often with the connivance of the Norwegian **fifth column** under Vidkun Quisling. Indeed, if an officer, in disobedience of a spurious order, had not won time by sinking a German **cruiser** in Oslo fiord, the King and the Government would have been captured.

Norwegian resistance was patchy and mainly concentrated north of Bergen. At

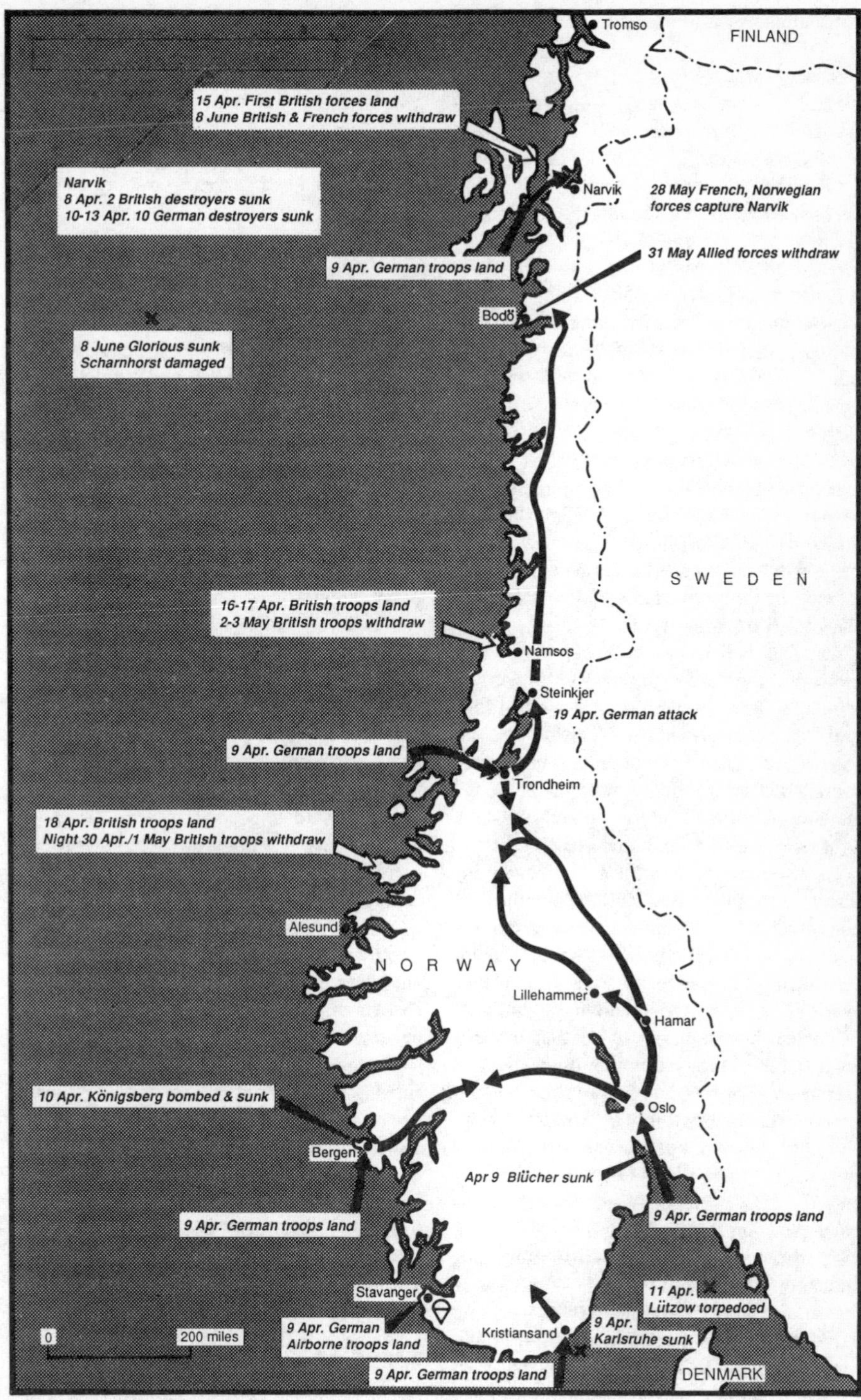

The Battle of Norway, 1940

Narvik, however, British naval actions on 10 and 13 April sank all 10 German **destroyers** and the transports prior to landing a small force on the 15th. But it was not until 16 April that the British Army under General Bernard Paget began landing in strength at Namsos and, on the 18th, put a weaker force ashore at Åndalsnes, while reinforcing the detachment at Narvik. Meanwhile the Germans had lost two more cruisers but had consolidated in the south and were beginning to advance up-country to link with the garrisons at Trondheim and Narvik. Always they benefited from complete air superiority, which seriously hampered the British, whose **logistic** arrangements were all along in chaos without any extra pressure from the Germans. By 30 April the British at Åndalsnes were in dire straits and had to be withdrawn. It was the same at Namsos on 2 May. At Narvik, on the other hand, Norwegian, British and French troops had driven the Germans into the mountains by 28 May, but by then the Germans were besieging **Dunkirk** and a mere foothold in Norway was an irrelevance. Allied evacuation was completed successfully on 8 June, though for the loss of an **aircraft-carrier** and two destroyers in a fight with two German **battle-cruisers**.

For the remainder of the war Norway lay under German occupation, its populace at first divided in its loyalties but gradually swinging against the invaders. As a base for attacks into the Atlantic and, from 1941 onwards, against **Arctic convoys** it was invaluable to the Germans. It was also a target for the first of many **Commando** raids on 4 March 1941 – this one against the Lofoten Islands which caused much damage to shipping and industry but also alerted the Norwegians to the perils of German retribution after the raiders had departed. In fact, it was naval and air raids which, throughout the war, did most damage with numerous attacks on coastal convoys and warships lurking in the fiords – of which that on the **battleship** *Tirpitz*, often damaged and finally sunk on 12 November 1944 by two 12,000lb **bombs**, was the most notorious. All along it was Hitler's conviction, fed by British hints, that the Allies most certainly would invade Norway which made him retain a very large garrison there instead of in action on other over-stretched fronts. They surrendered peaceably in May 1945.

Nuclear energy and weapons. In 1938 Professor Dr Lise Meitner, in collaboration with Dr Otto Hahn and Dr Fredrich Strassman, coined the word 'fission' to describe their theory that the splitting of **uranium** into nearly equal fragments would cause a reaction which would allow uranium to 'burn' and produce 10 to 100 times the energy of other, known nuclear disintegrations. In so doing they carried a stage farther the research into radiation by the Curies, Albert Einstein, Ernest **Rutherford** and others, inducing a realization of the feasibility of an **atom bomb**.

When the first **atom bomb** was exploded on 16 July 1945 it was revolutionary not only in its destructive power (including that of **radiation**), its **strategic** and **tactical** potency, but also as a source of power for means of **propulsion** and **generators**. Therefore, against the background of the **Cold War** and the likelihood of a future world energy crisis, scientific **research and development** was divided between *military* and *peaceful* applications.

Military investigations were concentrated upon the development of strategic and tactical weapon systems delivered by **aircraft**, **rockets and guided missiles** and by **artillery**; or laid in **mines**.

Strategic, fission warheads were usually rated at the nominal 20 kilotonnes (KT), and above, of the bomb dropped at Hiroshima. Anything less was considered tactical, which by the mid-1950s could be 2KT or even less, for battlefield use in relatively close proximity to friendly troops. But the strategic family of weapons were increased enormously by the testing by America on 1 November 1952 of a fusion, or thermonuclear, **hydrogen bomb**; a warhead with a yield said to be 7 megatonnes – the equivalent of 7 million tonnes of conventional **explosives** and the production of extremely lethal radioactive fall out from gamma rays. This test was matched by the Russians in August 1953, thus launching the major powers into the competition for nuclear equality, if not superiority, which became the fulcrum in the balance of **deterrence**.

Peaceful nuclear applications centred upon electricity generators and ship propulsion. By 1951 an experimental fission breeder reactor was working, and in 1956 a full-scale power station was operating in Britain – the first of many of various types to be built worldwide. Because of the need massively to shield against radiation from the core of a nuclear reactor and its products, sheer bulk and weight, along with elaborate safety precautions, were essential in construction, with the result that their use in aircraft and **motor vehicles** was soon ruled out. But ships also had to be most carefully designed to minimize danger in the event of an accident. The vast majority of nuclear-powered vessels are thus **submarines**, plus a few **aircraft-carriers** based on special naval **dockyards** – and very few commercial ships owing to almost prohibitive civil safety precautions.

However, nuclear-powered **space vehicles** with minimal shielding have been launched – and at least one (Russian) has already crashed to earth in Canada, scattering radioactive debris.

Nuclear energy and weapons are here to stay, no matter what the objectors say and however much **disarmament** is implemented. It will remain beneficial but also extremely menacing, which is why limitation and control treaties must be enforced, and why it is vital that the technology and weapons should not be allowed to fall into the hands of irresponsible and predatory people or nations.

Nuclear strategy. *See* **Deterrence.**

Nuclear winter. *See* **Deterrence.**

O

Oil fuel, as a source of light, heat and for **flame warfare**, and in bitumen for ships' caulking and adhesive bonding of weapons, has been used since earliest recorded time, wherever found near the earth's surface. Its large-scale application derives, however, from James Young's discovery in 1850 that oil could be distilled from shale in Scotland. Followed by the drilling by Edwin Drake, in 1859 in the USA, of the first commercial well.

The growth of the oil industry to its present size stemmed from commercial demand for lubricants and a 50 per cent cheaper, cleaner, more easily handled source of energy than **coal**. It flourished as the fuel and lubricant of the practical internal combustion and diesel engines, without which there would have been no **mechanization** or **aircraft**. Its uses as a feedstock for synthetic materials (*see* **Plastics**) came later.

Military uses for oil were at once apparent. In consequence, by **World War I**, its relatively few known sources (mainly in the USA, Russia and parts of the Middle East) had assumed vital strategic and **logistic** importance, along with the routes followed by ships, **railways** and pipelines. Inevitably oil supply has been involved with political crises and at the core of modern wars, as well as being a limiting factor owing to **blockade**. Nevertheless, sophisticated exploration and production methods have greatly increased the locations and recovery of deposits, thus reducing earlier dependence on only a few fields. But the availability of supplies remains a crucial logistic factor and a challenge to inventors and technologists against the day when the wells run dry.

Oppenheimer, Robert (1904–67) was a physicist who studied in America and Europe and was a leading teacher in the USA of quantum and **nuclear** theory. In 1943 he was put in charge of a team of prominent scientists at the secret Los Alamos laboratory, tasked to make an **atom bomb** ahead of Germany. A successful fission device was exploded in July 1945 and the first bomb dropped on Hiroshima on 6 August. In 1954, as a result of his publicly pronounced concern about the consequences of nuclear developments, he was declared a security risk and removed from office.

Optical instruments. In military applications the purpose of optical instruments is generally twofold: to improve eye performance in the sighting of weapons; and to improve the eye's **surveillance** capability, especially in low-light conditions. Increased magnification helps to make a target more easily identifiable, but the field of view is reduced in proportion and aiming errors are increased; a compromise to meet the requirement is usually necessary.

In 1589 G. B. della Porta described the grinding and polishing of magnifying lenses, and in 1608 Hans Lippershey and Zacharias Janssen combined a convex and a concave lens to invent the telescope as a way to make distant objects look nearer. Not until the mid 17th century, however, were telescopes in general naval use to help read bunting and recognize ships. But it was the increased range of **artillery** which made obsolete simple open sights, and magnifying instruments essential.

Towards the end of the 19th century armies were obtaining the first binoculars and navies were on the eve of adopting **rangefinders**, clinometers and fire directors to improve **gunnery** control. Navies also began installing **periscopes** in **submarines** in 1896 – a device which was used also in **fortifications**.

World War I created enormous demands for optical instruments, including simple periscopes for trench warfare and **tanks**, and telescopic sights for **fighter aircraft**. These demands posed the Germans no problems, since they were leaders in optical glass manufacture, but were serious for the British since they were largely dependent on imports from Germany. This problem was solved by ordering 32,000 German binoculars, through the Swiss, in return for rubber to replenish stocks which had been depleted by the Allied **blockade**.

By 1939 the manufacture of optical glass in the UK was well established and predicted fire from artillery of all kinds was entirely dependent on optical instruments, especially the rangefinder; however, the monostatic instruments then in use were not entirely satisfactory, errors increasing with the square of the range. The gravest disadvantage of all optical instruments was, of course, their comparative uselessness in poor visibility. The particular long-range demands of naval gunnery and of air defence saw rangefinding and surveillance by **radar** gradually supplanting optical fire-control systems.

In the field, however, and for aircraft gun and **bomb**-sights, the optical instrument still seemed to provide the best answer. A number of British firms working with the Royal Aircraft Establishment combined to produce the equipment needed for British aircraft, while in America Sperry's produced sights for use in their aircraft – simple and robust but, since they assumed steady flight, not of such practical value as the more sophisticated British sights; the latter were subsequently adopted by both countries.

For tanks and direct-fire artillery weapons the simple straight-through telescope was in general use throughout **World War II**, usually fitted with a graticule (US: reticle), consisting of a series of accurate marks which enable the gunner to apply deflection and range without removing his eye from the instrument. The disadvantage for the tank was that the telescope had to be in generally the same horizontal plane as the gun barrel, which meant that much of the turret had to be exposed above cover to take aim. Later designs adopted the more complicated periscopic sight set in the turret roof, thus reducing exposure. Tactically this was sound, but it resulted in some complex linkages to connect gun to sight. Such sights rely on a combination of prisms and tilting mirrors to provide the required optical path, with modern lens coatings permitting the use of more optical elements than would previously have been possible.

Modern small arms have also called for more elaborate optical sights, especially for night work, and the British Trilux, of prismatic construction similar to half a binocular, is now in service as a typical example.

Though rangefinding and long-range surveillance has been largely overtaken by **electronics**, the need for optical instruments for sighting and surveillance in the combat zone remains.

Ordnance services. No matter how good the **artillery**, ammunition or any of the myriad items needed to equip military forces, they are useless unless received by users in sufficient, serviceable condition. The ordnance organizations therefore have always been crucial to **logistics**, although simple to operate and maintain before the advent of **explosives** stimulated new **technology**, above all artillery in the 14th century. Prior to that **engineers** improvised on the spot from local materials, or the **armaments industry**, such as it was, delivered direct to users who usually arranged their own storage, care and maintenance. Early **dockyards and naval bases** functioned as ordnance depots to maintain and repair ships and store sails, ropes and timber. But it was in the 15th century, in conjunction with the decline of decentralized feudal systems, that governments felt compelled to establish ordnance services for national forces.

For example in 1455, after the **Hundred Years War**, the British centralized the provision and care of arms and military equipment in the Tower of London and, under the Tudor monarchy, established a Board of Ordnance in 1597, which was associated with the Admiralty, which in 1540 had begun building the first government naval dockyard at Portsmouth.

Procurement and administrative systems evolved with the proliferation of technology, the military commitments of the time and in accordance with national characteristics. Classification of stores and distribution systems were therefore inconsistent to say the least. For example, some nations included food with ordnance stores; others kept separate, or almost overlooked, **clothing** and **medical** supplies. The European **wars of the 16th, 17th and 18th centuries** impelled the growth of arsenals in company with sporadic attempts to introduce standardized weapons, distribution and accounting procedures. As chronic aggressors, the Swedes or French often tended to take the lead, notably during the **Thirty Years War** and **French revolutionary wars**.

Nevertheless, the often risky French logistic arrangements during the Napoleonic Wars marked the beginning of the end of an epoch. The logistic disasters of the **Crimean War** and subsequent 19th-century wars; and the revolution of mobility and fire-power caused by **railways**, the **telegraph** and breech-loading weapons (including the **machine-gun**) made almost obsolete existing ordnance services. From the 1860s onwards, under political pressure and the control of General Staffs on the German model (*see* **Staffs, Military**), more rationalized systems of procurement, nomenclature, storage, maintenance and distribution evolved for both navies and armies. Generally speaking these were established by the end of the century – ready for testing in the Boer War and the **Russo-Japanese War**. Yet this was the eve of the introduction of new weapon systems such as the **submarine**, quick-firing artillery, **radio**, **motor vehicles** and **aircraft** which had to be catered for in **World War I**.

The need for regulation and identification of innumerable items in depots and along lines of communication created, inevitably, vast bureaucracies which fanned out from ministries to unit quartermasters and, eventually, the fighting man. As the traditional systems of magazines and foraging gradually gave way to comprehensive provisioning, firm principles of indenting, of priorities and, wherever possible, automatic supply to sustain holdings at standardized scales at all levels were created in order to save units in action from having 'to look over their shoulders' for vital ammunition and equipment.

Mechanization complicated matters enormously. Numerous were the failures in both World Wars to meet demands for equipment in the right place and time, and in working order. Notably for **armoured fighting vehicles** (**AFVs**) and aircraft, special delivery organizations were created to deliver combat-ready machines to front-line units. In the 1990s the increasing complexity of ordnance services is simply an extension of the intricacies observed in the 1950s. So complicated are they, indeed, that their effective functioning depends almost entirely upon **electronics**, **computers** and **mechanized** handling. Without that assistance the fast British reaction in the **Falklands** War might have been fatally impaired. And despite the latest methods there were many occasions when units of the United Nations in the **UN–Iraq War** of 1991 were hunting among the vast depots for individual items – in much the same way as they had hunted from ship to ship off the beaches of **Normandy** in 1944.

Ottoman Empire, Wars of the. The Muslim Turkish Ottoman (Osmanli) Empire was founded by Osman I in the early 14th century in Asia Minor. From here it began the

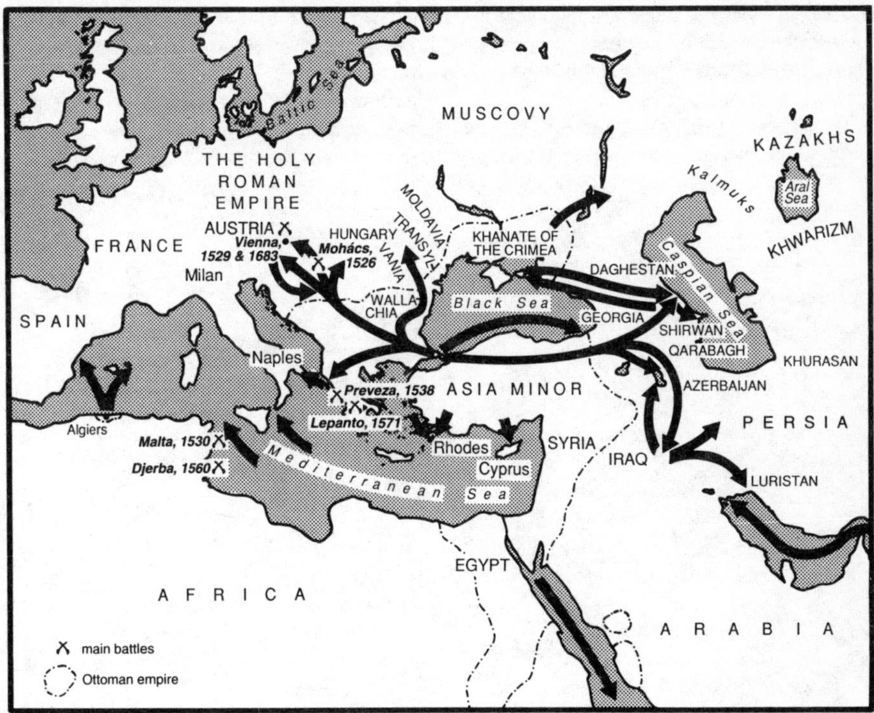

Ottoman expansion to 1683

final erosion of the Byzantine Empire by expanding into the Balkans. And, despite a serious set-back at the hands of the **Mongols** under Tamerlane in 1402, it captured Constantinople in 1453 after a desperate three months' **siege** in which heavy **artillery**, **mines**, **amphibious** attacks and sulphurous **smoke** were profusely used.

For unlike Huns, **Vikings**, Magyars and **Mongols**, the 15th-century Turkish navy and armies were not only well equipped and supported by good engineers, but also ahead of some European armies in tactics and technology which were ruthlessly applied by a formidable élite corps of **armoured cavalry**, **infantry** and bowmen – the Janissaries who were recruited from enslaved Christians and ingrained with a fanatical discipline. Zealously the Turks expanded, via Egypt, along the north African shore; into Hungary in 1526 where, at Mohacs, they outnumbered, outmanoeuvred and routed the Habsburg army; then to the siege of Vienna in 1529, when they withdrew in **logistic** chaos before the Austrians; and eastwards during the 16th and 17th centuries into the Caucasus to the Caspian Sea, through Persia and Iraq and thence by sea to the Indian subcontinent and round Arabia. But these brutal campaigns provoked steadily improving resistance along with effective alliances among the Persian and European powers.

As the Turks became overstretched and her enemies more technically sophisticated, the tide turned gradually against the Ottoman Empire. Thrusts like that to Oran in 1563, the disastrous siege of **Malta** in 1565 and general depredations in the western Mediterranean were countered after the Turkish Fleet was destroyed at **Lepanto** in 1571. Nevertheless it was 1683 before contraction set in when the Vizier Kara Mustapha's lunge at Vienna was crushed by a politically and logistically remarkable combination of central European forces under John Sobieski, a victory which opened the way to the Habsburg conquest of Hungary, and the Ottoman Empire's accelerating surrender of territory in the 18th and 19th centuries from successive wars against the Balkan nations, Russia and Italy, culminating in the aftermath of **World War I** when the entire politically, economically and militarily weakened structure collapsed.

P

Pacific Ocean War, 1941–5. As a basis of the long-intended Japanese war of expansion, domination of the Pacific was essential, something well understood by the Americans and British when they noted the progressive **fortification** of islands in Japanese possession and studied the infiltration of China and South-East Asia in the 1930s. It was Japan's tragedy that she fell into the hands of a bellicose military clique, who had the support of their Emperor Hirohito, and that in 1940 the weakening of Britain at the hands of the Germans tempted them to gamble upon winning a decisive naval victory against the Americans. This would have helped them to seize a strategic perimeter defence from the Aleutian Islands through Marcus and Wake Islands to New Guinea and the Dutch East Indies, extending also into **Malaya** and **Burma**. The gamble was forced upon the Japanese when, in September 1940, President **Roosevelt** tried to deter Japan by placing an embargo on shipments of **oil**, scrap **iron** and **steel** from America, a measure which posed Japan the choice of curbing her ambitions or going to war. On the admission by Admiral **Yamamoto** that he could only guarantee 'playing Hell for 18 months', they opted for an undeclared war.

On 7 December 1941 they made the surprise attack on **Pearl Harbor** and began their rampage through the Philippines, at Hong Kong, and into Malaya, the Dutch East Indies and Burma, via Indo-China and Thailand. Destruction of the Allied fleet off Malaya and in the Battle of the Java Sea opened the way to completion of their initial plan by April 1942, during a remarkable period of successes in which the Americans lost their bases on Guam and Wake Islands. However, with the **aircraft-carriers** that had survived Pearl Harbor the Americans in this period raided Japanese bases in the Gilbert and the Marshall Islands in February, as well as Marcus and Wake Islands, but were repulsed when attempting a raid on Rabaul. In March they struck at Lae and Salamaua before, in April, bombing Tokyo. At this point the Japanese decided to extend their perimeter by capturing the Aleutians, Midway Island and Port Moresby in New Guinea, thus also posing a further threat to Australia, Port Darwin having already been bombed in February. But surprise was lost. This scheme, like most other Japanese plans, was fully revealed to the Allies by decryption of **radio** signals.

At the battles of the **Coral Sea** and **Midway** the Japanese aircraft-carrier fleet was ruined, thus removing the vital, mobile force that alone could secure the static island bases of the perimeter defensive system, and permitting the first tenuous Allied **amphibious** counterstroke at Guadalcanal in August. Henceforward the initiative rested with General **MacArthur** in the South-West Pacific area and Admiral **Nimitz** in the Central and South Pacific areas. MacArthur developed the counter-offensive in New Guinea which, in conjunction with Nimitz's advance through the **Solomon Islands** in 1943, led to the isolation of Rabaul. Nimitz opened his drive

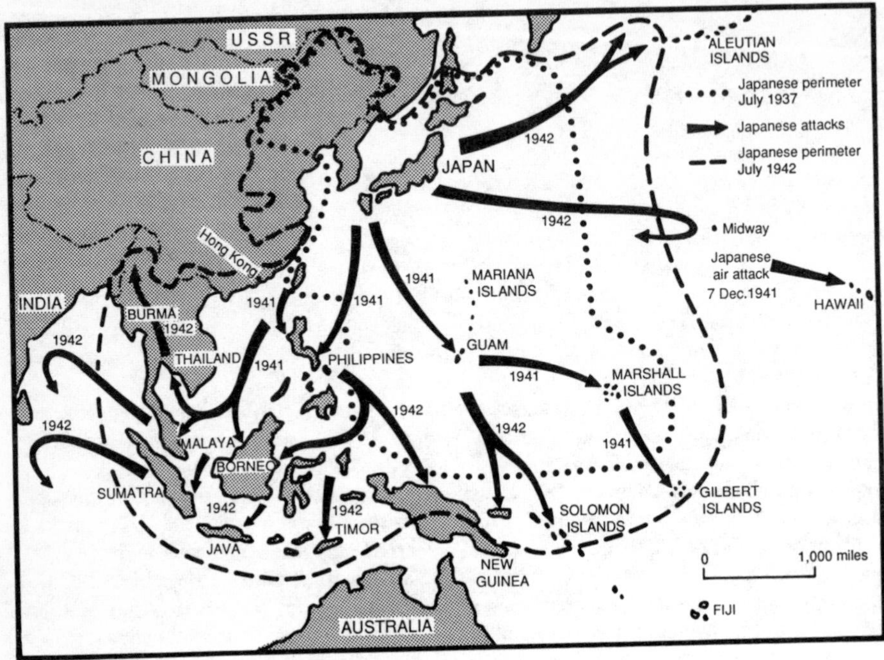

The Pacific theatre of war and the Japanese expansion, 1941–2

towards Japan in the **Gilbert Islands** in November, followed in February 1944 by the taking of Kwajalein Island and Eniwetok in the Marshalls – the latter operations remarkable for the wiping out of some 10,000 Japanese at a cost of only 771 American dead. Next came the assault on the Mariana Islands and Guam and the related naval Battle of the Philippine Sea in June, leading to President Roosevelt's decision, for political as much as military reasons, to support MacArthur's desire to invade the Philippines instead of, as Nimitz wished, going straight to Formosa or China, before invading Japan. As it was, the **blockade** of Japan by **submarines** (mainly American), which sank 57 per cent of the 8.5 million tonnes of merchant shipping lost in addition to 201 out of 686 warships put down, plus Nimitz's neutralization of Palau Island and the victory of his aircraft in the Battle of Formosa (Taiwan), in October, went far to achieving his original aim. For in conjunction with the assault on the Philippines (which brought on the conclusive Battle of **Leyte Gulf**) in October, began the **Battle of Japan**, made possible by American possession of air bases in the Marianas from which long-range B29 bombers could operate. The aerial bombardment multiplied in destruction after the capture of Iwo Jima in March 1945 and, in June, the conclusion of the bloody battle for Okinawa. At this point Japan was prostrate. As her forces in the Philippines were broken, Australian troops landed in Borneo to recapture the oilfields she had gone to war for in 1941. The dropping of the **atom bombs** on Hiroshima and Nagasaki in August merely provided a convenient excuse to call off a war she never had much hope of winning.

Panama Canal. First mooted in the 16th century but seriously projected by Ferdinand de Lesseps in 1879, work on this 50-mile-long combination of canals, locks and lakes across the Panama isthmus finally commenced in 1904. By then the USA had acquired possession in perpetuity of the 10-

mile-wide Canal Zone and it was the US Engineer Corps which increasingly ran the project, which was viewed as of rather more strategic than commercial importance. For, with a width of 1,000ft and a depth of over 41ft, the biggest **battleships** and liners could pass, giving the USA a short cut while forcing enemies to go round Cape Horn from the Pacific to the Atlantic. Its construction cost many lives from disease (thus intensifying anti-malarial and yellow-fever measures).

There have been occasional anti-American movements, the latest put down in 1989.

Parachute. The first parachute descent from a **balloon** was made by André Garnerin in 1797. But, until **World War I**, most descents were more for entertainment than life-saving. **Air warfare** and the need to preserve the lives (and **morale**) of tethered balloon crews, as well as dropping agents from aircraft behind the enemy lines, changed that. Yet, except in the German air force later in the war, air crews mostly flew without parachutes – partly to save weight and partly to deter premature 'bail outs'.

Between the World Wars the issue of parachutes (usually made of silk) to combat crews became normal. Then the formation of **airborne forces** created enormous demand, with the result that synthetic materials, such as nylon, were increasingly used not only by personnel but also for cargoes. Initially parachutes had been packed in bags attached to aircraft and pulled out as the parachutist fell. Then came manpack-parachutes opened by pulling a rip-cord or by a static line attached to the aircraft. Later barometric devices, set to work at the desired altitude, were developed.

The faster aircraft flew, the more difficult it became for aircrew to escape, let alone open the parachute once clear. As a result ejector seats were developed to throw individuals clear and automatically open the parachute, even on take-off or at supersonic speeds.

Parachutes have also been arranged for deployment as an aircraft's air-brake during landing and for delaying the impact of **bombs**. The latest kinds, useful for **special forces**, are steerable to help gliding and accurate landing on the objective.

Paris, Siege of (1870–71). After the initial German victories of the **Franco-Prussian War**, including **Sedan**, part of the French Army withdrew into Paris, which was besieged on 19 September 1870. Behind two formidable fortress lines, a garrison of 220,000 (of whom only about 50,000 were trained) plus 300,000 untrained *gardes nationales* of revolutionary outlook, came under command of the President, General Louis Trochu. Few preparations had been made for **siege**. Supplies were short. But General von **Moltke** had neither the desire nor the means to assault when so many of his resources were employed elsewhere and his lines of communication harried by **guerrillas**. Indeed, not until 5 January was it possible to assemble the siege train with sufficient ammunition for a sustained bombardment. Meanwhile a revolt by the *gardes nationales* had been put down by Trochu on 31 October and two ineptly managed sorties by the garrison had collapsed in November and December as starvation and disease loomed.

Von Moltke was under severe political pressure to finish the job quickly, but wisely depended upon the bombardment and time to do so. Failures by the French armies in the field to defeat the Prussians, and a third disastrous attempt at sortie on 19 January 1871 (when the *gardes nationales* fired on their own side), finally compelled Trochu to conclude an armistice on the 26th. Civil war ensued in the spring. The lessons from this siege intensified the current debate about **fortifications**, with great influence on future defence policies.

Parsons, Sir Charles (1854–1931), served his engineering apprenticeship with the **Armstrong** Company, but in 1884 became a partner in the firm of Clarke, Chapman and invented the first steam-**turbine dynamo** for ships' lighting. In 1889 he founded his own firm to produce power-station machinery and ships' **propulsion** systems. In 1897 he built *Turbinia*, the first of many turbine-propelled vessels of all types, including those with high-pressure-geared turbine engines. Acclaimed as the most original engineer since James **Watt**, he also took control of a firm which developed and manufactured **optical** glass.

Passchendaele, Battle of (also known as the Third Battle of Ypres). It was conceived in 1917 by General Sir Douglas **Haig**, ostensibly with the aim of seizing the Flanders ports at the request of the Royal Navy, when he undertook a campaign of attrition to maintain an Allied offensive strategy while the French Army recovered in the aftermath of its May mutiny. By assaulting wetlands whose drainage system had been wrecked by shellfire, Haig made it impossible to use **tanks** to pave the way for infantry against a well-forewarned enemy who effectively used the new mustard gas. Starting in rain on 31 July 1917, the attackers were mired in a morass of the **artillery**'s making. Sheltering in **concrete** pillboxes and shattered villages commanding the vital Passchendaele Ridge, the Germans held the whip hand with their system of 'elastic defence' (rebounding with counter-attacks after giving ground under pressure). Yet their casualties piled up to some 260,000 against about 300,000 British and 8,500 French, before it was called off on 10 November after a mere five miles' advance.

Artillery had failed on its own but **mechanization** was vital. For **logistically** the supply of sufficient **ammunition** by the Allies (107,000 tonnes in 35,666 truck-loads for the 14 days' preliminary bombardment alone) would have been impossible without **motor vehicles** and **railways**.

Patton, General George (1885–1945). An American cavalryman with charisma and flair who ably led a unit of the US Army's Tank Corps in **World War** I but, between the wars, played a rather negative part in the development of **armoured** forces. In North Africa in **World War II** he commanded the assault on Casablanca on 8 November 1942, and took over and revived II Corps in the aftermath of its defeat at Kasserine. He led Seventh Army in the Battle of Sicily in July 1943, but was afterwards dismissed for the much-publicized slapping of a soldier suffering from **combat fatigue**. Given command of Third Army in 1944, he led it in the break-out from **Normandy** and the pursuit to Germany, in which his intriguing and his selfish cornering of supplies had a controversial effect on Allied strategy. His counterstroke during the Ardennes Offensive in December 1944 was highly effective, as was the dash with which he led Third Army in the Battle of **Germany** in 1945. He was killed in a motor accident in December 1945.

Pearl Harbor Raid. As the main US base for the US Pacific Fleet, the military installations on Hawaii and the **dockyard** at Pearl Harbor were obvious targets for attack by the Japanese. Moreover, the Americans had precise **intelligence** that such a blow was to be expected on 7 December 1941, but failed to take measures, apart from sending the **aircraft-carriers** to sea, to thwart it. Admiral **Yamamoto**'s plan to sink the entire Pacific Fleet in harbour by air attack was based on the concept of a single crushing blow such as Japan had delivered against the Russians at **Port Arthur** and **Tsushima** in 1904 and 1905 respectively. He gambled upon permanently destroying America's capability to restore her strength in the **Pacific Ocean War**.

The well-prepared blows by 200 **aircraft** in two waves, under the command of Vice-Admiral Chuiki Nagumo, sank, crippled or damaged all eight **battleships** present and severely damaged many other vessels, including three **cruisers** and four **destroyers**. Many aircraft were also destroyed and over 2,200 servicemen killed at a cost to the Japanese of 100 men, 29 aircraft and five midget **submarines**, the latter failing in their missions.

The attack gave the Japanese complete freedom of action in the Pacific but was not the decisive victory Yamamoto desired. By failing to locate and sink the aircraft-carriers and, on Nagumo's part, not launching a third strike to wreck the **oil**-storage tanks and dockyard, these assets survived to help speed American recovery and ruin the Japanese striking force within the next six months at **Coral Sea** and **Midway**. But the shock to the United States was the most telling blow and its iniquity embittered the nation, enforcing its determination to destroy Japan. The subsequent controversy over culpability has kept speculation alive ever since.

Penicillin, the first antibiotic, was accidentally produced and noticed in 1929 by Alexander Fleming, and developed urgently at the start of **World War II** by Oxford scientists led by Howard Florey. For, because antibiotics are antibacterial and attack other microorganisms, penicillin's revolutionary, non-toxic curative potential was realized, especially in the treatment of wounds and venereal disease (*see* **Health services**). Many lives were saved by penicillin after its large-scale production was initiated by Florey in America in 1941. Subsequently bacterial resistance began to occur, but different strains of antibiotics have been developed in the treatment of numerous infections.

Periscopes are **optical** instruments to facilitate observation of the surroundings while remaining behind cover or armour or when submerged. In 1854 E. H. Marie-Davy invented a tube with two mirrors angled at 45° and facing in opposite directions for use in **submarines**. In 1872 prisms were substituted for mirrors. Subsequently tilting prisms, weapons control, **night vision** and **photographic** devices have been incorporated in submarines' periscopes along with machinery and lenses to operate the periscope telescopically and provide image magnification.

Periscopes were also included in late 19th-century **fortifications** for **surveillance** and **gunnery** control; and, in **World War I**, improvised for observation and sniping without raising the head above a trench parapet. **Armoured fighting vehicles**' crews also required periscopes and these were developed in **World War II** to incorporate telescopic sights for gunners, and grouped in commanders' cupolas to provide all-round vision. Periscopes have rarely been used by **aircraft**.

Pershing, General John (1860–1948). An American officer who acquired immense experience in the American–Indian War of 1886; in Cuba during the Spanish–American War; in minor Philippines actions; as an observer of the **Russo-Japanese War**; and as Commander US forces in the Mexican War against Pancho Villa in 1916, before being appointed C-in-C US Army in Europe after America entered **World War I** in April 1917. For over a year Pershing had too few troops to play the part of an independent force leader. Unwillingly, he was compelled by circumstances to place individual divisions under French or British command in helping to stem the **Hindenburg** offensive in the spring of 1918. Indeed, his army was largely equipped with Allied heavy weapons, because of America's unreadiness for war, and never became self-sufficient. Only at the battles of Saint-Mihiel in September and Argonne in October was a degree of independence achieved, in battles which revealed the troops' inexperience offset by courage. After the war, from 1921 until retirement in 1924, he was a conservative Army Chief of Staff who was lukewarm to **mechanization**.

Pétain, Marshal Philippe (1856–1951). A French officer who, as a brigade commander at the age of 58 with no active service experience in 1914, became a corps commander at **Arras** in May 1915 and in June, an army commander. Sent to **Verdun** on 26 February 1916 to save the situation, he became a hero of France in doing so. Likewise, when made C-in-C in May 1917 at the nadir of the French Army's fortunes, he was able to quell a mutiny and sufficiently restore fighting spirit to withstand the **Hindenburg** offensive in 1918 and bring the war to a successful conclusion. Regarded as having greater military wisdom than perhaps he merited, he tended in various appointments after 1918 to ossify the French Army's development by failing, like many another, to understand the latest effects of **technology**.

When French resistance was broken by the Germans in June 1940, he was called upon to form a government and make peace.

By retaining control over the fleet, a small army and the colonies, he was able to salvage some slight French self-respect and pursue an unrealistic policy aimed at isolating the nation from the war. Senility condemned him to the role of a mere figurehead. Tried after the war for his complicity in post-1940 events, the sentence of death was commuted to life imprisonment.

Petersburg, Siege of. In the aftermath of the costly Wilderness battles of the **American Civil War**, General Grant crossed the James River on 13 June 1864 and marched on the fortified, but only thinly defended **railway** junction of Petersburg in an endeavour to encircle Richmond. This most skilful operation caught General **Lee** off balance, but the Federal Army fumbled its attempt to rush the city, giving Lee time to reinforce and win a three days battle on 18 June. From then on Grant's side-steps round Petersburg to the south and west were repeatedly countered by the Confederates. On 30 July the explosion of a four-tonne **mine** under a redoubt opened a gap which the Federals casually failed to exploit. Subsequently, as fast as the Federals advanced, admirably supplied from their base on the James River, the Confederates dug new trenches or counter-attacked in a grim battle of attrition which ground to a halt in cold and wet on 28 October. It left the citizens of Petersburg and Richmond to suffer a winter short of supplies, on the verge of defeat. Yet despite Grant's renewed attempts in 1865 to take the town, it held out under severe pressure until 2 April when Lee withdrew on the eve of the war's end.

Philosophers, Military. There have been many military philosophers but very few who have made a profound and lasting impression on the art of war. This entry concentrates on that few, beginning with **Sun Tzu** who lived *c.* 500 BC and ever since has strongly influenced the Chinese and Japanese – but not people in the west even though his brief *The Art of War* contains the distilled essence of the military art. Sun Tzu's ideas were governed by the **technology** of the day. And therefore, like all military philosophers to come, subject to imagination, guesswork and chance when indulging in futurology.

No doubt **Vegetius**' *The Military Institutions of the Romans* (AD 390) retained their authority until the 16th century because neither weapons nor technology had advanced much since the great days of the Roman Empire; and **strategy** and **tactics** stayed moribund. Moreover, his book was only a synthesis of Roman military customs in an attempt to check the Army's further decline. Furthermore, not until the advent of mechanical **printing** in 1450 and the subsequent spread of literacy, could more than a very few privileged persons have access to or then read existing manuscripts.

Machiavelli was to some extent spared these handicaps. Printed copies of *The Prince* (1514) and *On the Art of War* (1520) had a remarkable circulation of seven editions in the 16th century alone, plus translation into some European languages. Yet even he, well supplied with the lessons of recent wars, seriously underestimated the potential of **artillery** and, fearing that they provoked hostilities, ingenuously denigrated **fortifications**. Yet, situated at the heart of the Renaissance in Italy and at the start of the accompanying revolutions, his influence was vital, though of shorter duration than his predecessors'. For in the 17th and 18th centuries the brilliant engineer **Vauban** and the successful generals Prince **Eugene** and the Duke of **Marlborough** created new philosophies of war, more by deeds than words.

These three, pre-eminent among many great soldiers and with immense experience of the latest technological effects at war, made obsolete many of Machiavelli's concepts in the same period that Admiral **Blake**'s *Fighting Instructions* were, in contests with Admirals Maarten Tromp and Michael de Ruyter, in the Anglo-Dutch wars, and against the French in the War of the Spanish Succession, revolutionizing **naval warfare**.

Yet when, in 13 fevered nights in 1732, the outstandingly imaginative Marshal de **Saxe** wrote *My Reveries on the Art of War* (published 1757), he drew heavily upon

Machiavelli's *On the Art of War*, in addition to his own experiences. His intention was to 'establish a new method of the art of war', and included visions of technological and tactical improvements to breech-loading artillery and small arms, precise aiming of weapons instead of volley firing, more practical **clothing**, cadence marching, looser formations and redoubts instead of entrenchments. Ideas which took nearly 200 years to fulfil and which neither **Frederick the Great** nor **Napoleon Bonaparte** paid much heed to because Saxe was, technologically, too far ahead of his time; and later justifiably taken to task by **Jomini** in the suggestion that war was 'a science covered in darkness', and not an art.

In any case, King Frederick's celebrated *Military Instructions for the Generals* (first published 1747) was, in modern parlance, more like Standard Operational Procedures (SOP). Napoleon's *Maxims* (1827), like Vegetius' work, did not visualize future wars. Later 18th-century futurology, indeed, was more in the province of the fashionble Comte de **Guibert** in his celebrated *Essai général de tactique* (1772), which had a profound, long-term effect on military development by advocating the *levée en masse*.

But it was the Napoleonic Wars which produced those rival giants, generals von **Clausewitz** and **Jomini**; highly experienced senior staff officers and teachers whose involvement with the decisive campaigns helped them to analyse war in detail. Perhaps in too much detail, it can be argued apropos Clausewitz (who strongly influenced the burgeoning European General **Staffs**); but far more succinctly by Jomini who swayed the Americans and who, unlike Clausewitz, studied naval strategy. Yet the gathering pace of technological progress and the effects of the **Franco-Prussian** and **Russo-Japanese wars** and **World War I** thrust into prominence the next generation of historians and seers: Admirals **Mahan** and **Fisher**; the banker Ivan Bloch (who, in his *Future War* (1898) none too shrewdly heading a team of military writers, foretold trench warfare but overlooked air power); and Captain Murray Sueter who, in 1910, envisaged the decisive importance of air supremacy.

After **World War I** they were followed by Generals **Douhet** and **Mitchell** as the apostles of air power; and **Fuller** as a major philosopher and enthusiast for **mechanization**, **armoured** and **guerrilla warfare**. Yet these, who influenced admirals such as **Yamamoto** and generals such as **Trenchard**, **Guderian** and **Student**, were among the last of the great know-alls. For such was the complexity and diversity of 20th-century technology, strategy and tactics (including **nuclear weapons** and **deterrence**), even the best operational **analysis** and **research and development** groups were hard pressed to cope.

Military philosophy remains in the realms of intellectual conjecture; seeing no farther ahead than five years by guesswork. Like gunners, military gurus tend to fire many shots, forget their numerous misses but count their hits.

Photography (*see also* **Cameras**). The impact of photography on war may be said to have started with the work of Roger Fenton in the **Crimea** in 1855 and with the remarkable pictures obtained by Brady, Gardner and O'Sullivan during the **American Civil War**. However, these pioneers were merely recording events, not assisting commanders in making combat decisions. Nevertheless, the implications were clear, especially when in 1858 Gaston Tournachon was able to take aerial photographs of Paris from a **balloon**. The first use of aerial photography in war was during the Italian War of Independence in 1859.

Reconnaissance was, by 1914, clearly seen as a role for **aircraft** and the development of the portable roll-film camera by Kodak in 1888 made photography from the air a practical possibility. It was in 1914 that the British had, in one day, photographed from an aircraft the whole of the Isle of Wight and the Solent fortifications, developing the negatives in the air ready for printing on landing.

No army could afford to ignore this new

spy in the sky. As **World War I** progressed techniques improved so that it was not long before a complete mosaic of the defences along the Western Front had been produced for the commanders to study. An elaborate programme of **reconnaissance** flights ensured that the picture was regularly updated and changes noted.

The 1930s saw the military emphasis concentrating on aerial photography, as it has continued to do. It was during this period that development or adaptation of aircraft specifically for the photoreconnaissance task began, a typical example being the German Dornier Do17, fitted with the latest high-definition cameras and capable of speeds over 250mph. It was during this interwar period too that photoreconnaissance developed beyond merely observing and recording the enemy's defences. In the conquest of Abyssinia in 1935 the Italians successfully used aerial survey to plan their advances into otherwise unmapped terrain.

It was, however, during **World War II** that photoreconnaissance really came into its own. By the 1940s the hand-held cameras of 1914 had long been supplanted by properly fitted equipment, usually mounted in the floor of the aircraft and pointing straight down. It was important to ensure that each exposure overlapped the next so that a complete mosaic of the area could be built up. This required straight and accurate flying, often coupled with a shutterless technique whereby the film ran continuously past a slit at a rate matched exactly to the image movement in the camera's focal plane as the aircraft flew along (known as image-motion compensation).

Improvements in equipment meant that the menace of the **reconnaissance aircraft** was even more evident, together with the need to destroy such aircraft before they could regain their own lines. The only defence was speed and so, to reduce weight, reconnaissance aircraft were stripped of all non-essentials, including armament. Flying reconnaissance missions required no little personal courage and, as the bombing offensive over Europe developed, the need for ever deeper photographic penetration into enemy territory arose; to be met by the British with the development of the De Havilland Mosquito, specifically designed for this role (though subsequently used for many others).

The task was not, of course, confined to taking pictures. An essential phase, having got the film safely back to base, was the timely interpretation of what the camera had revealed. This was highly skilled, sometimes intuitive, work and special photo-interpretation units were formed. Their job was often made easier by stereoscopic photographs produced by twin cameras mounted with lenses 2·5in apart to simulate the view seen by the left and right eye respectively. The picture had to be looked at through special stereoscopic viewers to get the three-dimensional effect, but otherwise flat pictures could be seen in sharp relief in this way, revealing much more information.

Developments since 1945 have revealed the need for two levels of photoreconnaissance. In the combat zone low-level photoreconnaissance is provided by drones or **remotely piloted vehicles** (RPV), flown over enemy lines. They are most useful for such spot tasks as discovering whether a bridge is intact or not; strip searches on a straight flight path can also be carried out, but the reaction time will be governed by the time taken to interpret the film when it has been processed. **Cold War** requirements were for high-level deep penetration of the opposition's territory, and for this the Americans produced such aircraft as the U2, flying so high as to be virtually out of sight from the ground. However, it is the reconnaissance satellite which, with the latest camera techniques, has provided the real breakthrough. It is now possible to obtain detailed imagery of almost any part of the earth's surface from suitably positioned satellites. The pictures, though taken from several hundred miles up, reveal intimate details of ground activity and are an essential tool in providing accurate and timely information of a potential opponent's intentions.

Finally, while the main emphasis has been on aerial reconnaissance, it should not be forgotten that photography has had a considerable indirect influence on the development of military equipment; in particular, high-speed cameras used extensively in weapon trials in order to observe the behaviour of projectiles in flight. There are numerous other examples in weapons development.

Pike. The pike was an **infantry edged weapon** developed by the Assyrians and Greeks from the throwing **spear** by lengthening the shaft. In Roman times it largely fell into disuse, but was revived in the 16th and 17th centuries to reinforce riflemen and musketeers whose slow rate of fire, due to reloading difficulties, made them vulnerable. The invention of the **bayonet** again made pikes obsolete, although there was another desperate revival in 1940 when the British issued a few to deal with the threat of German invasion.

Pistols are small hand-held firearms dating, it appears, from the late 14th century in Sweden and which, by the end of the 15th century, had acquired a wooden butt. Initially they were for self-defence at close quarters (say, 20ft); as they continue to be to the present day at only slightly greater range since they are inherently difficult to aim. By 1525 **cavalry** had used the crude petronel at the Battle of Pavia in support of **infantry**. But at the Battle of Cerisoles, in 1544, **cavalry** fired petronels against infantry in lieu of charging to close quarters – tactics which were developed until, in 1631, **Gustavus Adolphus**'s cavalry peppered enemy infantry before charging.

The technical development of the pistol (wheel-lock, flintlock and so on) ran in parallel with other firearms (*see* **Machine-guns**, **Rifles** and **Muskets**). Important as was its simplicity and reliability, Samuel **Colt**'s revolver of 1835 was at least 100 years behind less practical predecessors. And, in the opinion of many, more reliable than the semi-automatic, magazine-fed pistols which entered military service in the German Army in 1908 with the Luger 08. Despite disadvantages, pistols remain in service for close combat and moments of sheer desperation.

Plastics are synthetic materials made from chemicals and from natural substances such as **oil** and **coal**. Generally malleable and light in weight, they can incorporate vital properties such as toughness, strength, flexibility or heat resistance, etc.

In 1833 Jon Berzelius, a founder of modern chemistry, coined the word polymer to describe the synthesis of large and small molecules, a formulation which, within ten years, led to the preparation of nitrocellulose, vinyl chloride, styrene and acrylic acid. The invention of celluloid (or xylonite) by the Hyatt brothers followed in 1870 and bakelite by Leo Baekland in 1909; materials which were sold as alternatives for **wood**, rubber, leather, glass and metals. The process of substitution was accelerated by shortages caused by **blockade** during the World Wars, inaugurating a revolution in lifestyles along with a host of chemical and technical applications which proved many plastics were superior to and cheaper than traditional materials.

For military purposes, plastics have been invaluable for **photographic** film, **ammunition**, **explosives**, **armour**, **clothing**, **missiles**, **aircraft**, **motor vehicle** and **space vehicle** components, electrical and thermal insulators and a host of other uses.

Port Arthur, Sieges of. This ice-free port fell into Japanese hands on 19 November 1894 during the Chinese–Japanese War when 10,000 Chinese put up timid resistance to the Japanese assault under General Maresuke Nogi. But the fruits of victory, to Japanese rage, were snatched away by Russia in a diplomatic manoeuvre.

Thus Port Arthur was a prime Japanese objective at the start of the **Russo-Japanese War** on 8 February 1904 when the fleet within was attacked without warning by **torpedo-boats** prior to a very damaging bombardment of ships and forts by the Japanese fleet. By no means prepared for war, let alone the siege General Nogi had again been ordered to impose, the garrison of 40,000 under General Anatoli Stössel prepared for battle as the Russian fleet was defeated at sea and driven back into port. Stössel's modern forts (*see* **Fortifications**), which were put in order before Nogi launched his first attack on 25 May, were well protected by **barbed wire**, **searchlights**, **artillery** and **machine-guns** with ample **ammunition** and supplies enough for six months for the garrison plus 18,000 civilians.

Anxious to seize the port before the fleet put to sea, Nogi attacked without the help of a siege train and was bloodily repulsed. It took two months to bring in heavy guns, yet progress was still slow and casualties enormous after they had arrived. Moreover, gun barrels began to wear out as ammunition supply was depleted, and as the Japanese were weakened by disease. Not until 5 December, after a prolonged battle, did they capture the key 203 Metre Hill dominating the harbour, enabling the emplacement of artillery to destroy the ships at anchor and apply irresistible pressure on Stössel. Still well supplied by ammunition, his casualties amounted to 31,000, leaving few survivors fit for duty who were not ravaged by scurvy and starvation. He surrendered on 2 January 1905 to an opponent who had suffered 59,000 battle casualties and whose 34,000 sick included 21,000 with beriberi.

Poulsen, Valdemar (1869–1942). A Dane who in 1898 invented a machine which, **electromagnetically**, could record and store speech through a microphone on piano wire (the first forerunner of the tape recorder). His USA company to market his telegraphone failed for lack of public interest. In 1903, however, he modified William Duddell's so-called 'singing arc' to generate continuous **radio** waves; and incorporated a 'tikker' to chop them up, thus creating radio audio notes which, in 1908, were demonstrated over a distance of 150 miles as the precursor of voice radio, public broadcasting and a host of military and civil uses.

Pound, Admiral Sir Dudley (1877–1943). An officer who won a reputation on Admiral Lord **Fisher**'s staff before **World War I**, he commanded a battleship at **Jutland** and was Director of Naval Operations 1917–18. He was C-in-C Mediterranean Fleet at the time of the **Spanish Civil War** and the Italian invasion of Albania in April 1939, prior to becoming First Sea Lord and Chief of Naval Staff. He therefore bore the brunt of the Royal Navy's transition to **World War II** and guided it through adversity until fortunes improved in 1942 and the Battle of the **Atlantic** was won in May 1943. In this period he suffered strong criticism for over-centralization and errors during the Convoy PQ17 affair. As a member of the Joint and Combined Chiefs of Staff Committees he knew no respite. He died shortly after resigning following a stroke.

Power-generation equipment. *See* **Batteries**, **Dynamo**, **Propulsion**, Means of and **Turbines**.

Printing. Mechanical printing has always been related to the availability of suitable paper, ink, **metallurgy** and **power** supply. Thus Chinese impressions of the 5th century with ink on paper may well be the first examples of a craft which, by the 8th century in Korea, was capable of printing one million copies by manual labour from a few handset blocks. But usually John Gutenberg is credited with inventing the modern printer, *c.* 1450 (although there were competitors), followed by numerous engraved-letter presses throughout Europe by the end of the century, significantly stimulating the Renaissance and, for example, the dissemination of such books as **Machiavelli**'s *Art of War*.

It was 1814, however, before Friedrich König's steam-powered, flat-bed, cylinder press (1,100 sheets per hour) was built to print the London *Times*, and from that moment the restrictive practices of the printing-craft guilds and unions were gradually eroded in response to the demand for mass-produced, long-run editions with lower unit costs. The process accelerated when *The Times* introduced John Walter's rotary press in 1866 and when, in the 1880s, William Church's print casting of 1822 was superseded by Ottmar Mergenthaler's Linotype print-setting machine using a Christopher Sholes' type **typewriter** keyboard of 1867.

The 20th century has witnessed an accelerating revolution through lithography, offset web, gravure and many other processes in company with photocopying, thermofax transmissions by line and **radio**, **laser** printing and **computer** technology.

Propaganda is the organized use of publicity material to spread information, doctrine or practices. As an instrument of **psychological warfare** it is as old as history. For example, Julius **Caesar**'s *Conquest of Gaul* was political and personal propaganda, as have been many military personal reminiscences. The word, however, only came into use in 1662 by derivation from the missionary work of the Roman Catholic Church's Congregation for the Propagation of the Faith. Most strikingly, in Britain during the **wars of the 18th century**, with the growth of the **printing** industry, pamphlets and newspapers became powerful instruments of political propaganda that influenced the military art and campaigns.

In modern parlance it deals in white (truth) and black (lies) information and embraces every possible civil and military activity – offensive and defensive – before, during and after a war. Its efficacy depends upon the means of dissemination and the literacy and insight of the people addressed, plus the skill of the propagandists in judging how, when and where to direct their messages. For example, the educated British electorate was readily influenced by William Russell's impassioned reports of the **Crimean War**, which brought down the Government, just as peoples on both sides of the **American Civil War** were fairly easily and quickly reached and influenced via the latest **communications** through leaflets and journalism. Modern propaganda, indeed, emerged with the technological effects (*see* **Technology, Effects of**) of **telegraph**, **telephone** and **radio**, allied to the introduction of the **typewriter** and rotary **printing** machines, and of mass-circulation newspapers after the 1860s.

World War I first indicated how decisive and more easily distributed propaganda had become. On the Allied side in 1914 a Campaign of Hate against Germany was generated by the newspapers and supported by official Belgian and British committees investigating alleged atrocities, very few of which were fully substantiated as breaches of the existing **laws of war**. Along with the propaganda generated in 1915 from public indignation at the execution of Nurse Edith Cavell for organizing the escape of English soldiers from Belgium, public opinion was steered into supporting a total, 'just' war, which stimulated **recruiting** and the introduction of conscription in Britain and elsewhere. Germany's rebuttals of these accusations were not always well conceived, especially in neutral countries and, above all, the USA (where both sides conducted a war of words to win public support for their cause), where it induced resentment and was often countered.

More difficult was the introduction of propaganda among the enemy, by techniques which the British and French developed most successfully from 1915 onwards in association with **blockade**. Through interception of enemy mail, for example, sensitive subjects were detected and attacked by special propaganda organizations. Leaflets and other printed material which improved in sophistication with experience based on evidence of impact, were introduced through neutral countries, dropped from the air among the Central Powers, or fired by **mortars** into the front line. They refuted anti-Allied propaganda and spread rumours which eventually convinced the people on the home front, as well as the fighting men, that the war was lost. The process was also adroitly used by the Germans in helping start the Russian Revolution in 1917.

Between the World Wars, propaganda techniques were mightily enhanced by the introduction of public **radio** broadcasting systems which could reach even the illiterate in every part of the world, a facility grasped most ruthlessly by the Communists when indoctrinating the masses to spread the Revolution, and by the Fascist powers as they also planned expansion. The German Propaganda Ministry directed by Josef Göbbels won enormous successes by the manner in which **Hitler**'s chosen victims and their allies were undermined and sometimes taken over without a shot being fired. Göbbels's principle, that an oft-repeated lie eventually would be believed, became standard practice in dictatorships. He employed every vestige of the propagandist's art to mislead the German people into an unshakeable, apathetic resistance against hopeless odds – as well as to assist the armed forces win victories over opponents baffled by a barrage of misleading information.

In **guerrilla** and **cold wars** it has been commonplace, in the struggles for public support of causes, to emphasize indoctrination and terror by propaganda. Every political and diplomatic initiative is supported and every advantage taken of opposition weaknesses. Many military attacks are designed and executed specifically for propaganda purposes. The separation of truth from fiction has become so difficult that, quite frequently, there is rejection of both in an atmosphere of confused cynicism. The suggestion is that, as the people's insights are improved by education, they find it easier to recognize the real issues through the subterfuge sometimes called 'disinformation'.

Propulsion, Means of. One of man's earliest impulses was the urge to overcome his muscular and stamina limitations or find a substitute for them. *Animal* power above all came from the **horse**, and camels, oxen and **elephants** were all pressed into service. But they suffered from physical limitations and fear, and therefore required so much training, feeding and attention that their utility in war was always uncertain. So, at an early stage, man looked to the elements for help – to water and wind power to drive inanimate boats and machines.

River and tidal currents would have provided the first means of propulsion of dugout logs in *water*, assisted and, to some extent, steered by paddles and oars. No doubt sails soon were invented to catch the *wind* and certainly the earliest evidence of oar- and sail-powered war vessels comes from Egypt *c.* 2900 BC. On land, however, man was restricted to muscular power until Thomas Savery invented a practical, reciprocating-piston, *steam engine* in 1698, although in the 14th and 15th centuries there were a few fantastic proposals for windmill-driven combat vehicles. In the meantime, the **shipbuilding** industry progressed with ever larger sail spreads to cater for **naval warfare** and commerce (*see* **Shipping, Commercial**).

Not until the cylinder-boring machine invented by John **Wilkinson** made feasible the technical improvements to steam engines by James **Watt** in 1763 and 1769, could Nicolas Cugnot's steam-engined vehicle of 1769, with its endurance of 20 minutes at 2·25mph, show its paces; or in 1775 could John Rumsey's boat be propelled by a small steam-pump which sucked in water at the bow and squirted it astern. These project studies proved the acceptable economic feasibility of steam engines to drive ships, road vehicles, **railway** locomotives and **airships**. Thermally inefficient as they were, they possessed most of the things animals had not, including fearlessness, fair reliability and a potential for immense improvement. This was shown in France, in 1783, by the Marquis Claude d'Abbans's 182-tonne *Pyroscaphe* with paddle wheels turned by a double-ratchet mechanism connected to the engine; though it was 1837 before John **Ericsson**'s *Surveyor* was driven by a screw propeller, only to have the idea rejected by sailors who said it was impracticable for warships.

At an early stage in the development of steam engines, requirements were as clearly stated for higher power-to-weight ratio, greater fuel economy, miniaturization and reliability, as for improved performance – specifications which, ever since, have conditioned all kinds of power plant. The huge beam engines which drove industrial plant and pumping engines were too big for ships and useless in railway locomotives, cross-country tractors and road vehicles. Moreover the problems of handling and stowing **wood** and **coal** fuel, and then feeding it to a boiler remote from the engine, gave impetus to the hunt for self-contained, liquid-fuelled power plants of greater efficiency – gunpowder and hydrogen having failed early project studies.

Nevertheless, as early as the 12th/13th centuries the principle of the **rocket** powered by an **explosive** had been known and practised by the Chinese and Mongols. In 1792 it had been fired by Indian troops against the British, whereupon William **Congreve** adapted the idea for use against the French in 1806. Rocket motors posed a fire-risk, however, and, for the time being, were rejected in favour of safer means of propulsion.

In the meantime the invention in 1834 of a storage-battery-powered *electric motor* by a blacksmith, Thomas Davenport, gave promise of a future means of propulsion – until it was realized that the size, weight and need to charge heavy and bulky **batteries** were serious handicaps. Even so, as a starter motor, the electric motor would one day become an ancillary to the next vital invention, the *internal combustion engine*.

Numerous were the projects to build a practical safe and self-contained internal combustion (with electric ignition) or *compression–ignition engine* along the lines of the theory described by Sadi Carnot in 1824. After the drilling of the first successful **oil** well by Edward Drake in Pennsylvania in 1859 and the vision of unlimited petrol supply, these projects culminated in a four-stroke engine by the firm of Otto & Langen in 1867. In this motor a mixture of petrol and air was drawn into a closed cylinder by a downwards-moving piston; the mixture compressed by the piston moving upwards and then electrically ignited; the subsequent explosion thrust the piston downwards; the cycle was completed by the piston moving upwards, thus expelling the exhaust gas. In 1885 the principle was applied by Nikolaus Otto and Gottlieb **Daimler** to the building of a practical, vaporized-petrol-fuelled, four-stroke engine propelling a small vehicle.

The prospects and many uses for the Daimler-Benz engine were seized upon immediately; initially by **airship** constructors for whom its safe, compact, high power-to-weight characteristics and easily handled and clean fuel were fundamental attractions. Most important of all was the use by the **Wright brothers** in 1903 of their own 12hp engine, weighing 170lb, to power their *Flyer* in the first flight of a man-lifting heavier-than-air machine. Henceforward, although the majority of petrol reciprocating engines, with carburettors to control fuel vaporization and sparking-plugs for ignition, would be made for land vehicles, it was aircraft engines which led the way in performance, compactness and power-to-weight ratio. Weight and cooling problems were tackled by two kinds of multi-cylinder engine, using the air-cooled radial kind, with propeller attached to the finned cylinders mounted around the crankshaft, or attached to the rotating crankshaft; or the more complicated but more powerful in-line water-cooled kind, with a radiator for cooling.

But already, in 1827, Benoît Fourneyron had demonstrated that the water-powered **turbine** – a wheel with curved blades attached to its periphery, rotated by reaction to a water jet – had shown promise as something more than a stationary power generator (*see* **Generators, Power**). Then, in 1831, William Avery built one as a steam engine, and abandoned it because of defects. Many others experimented, but it was Sir Charles **Parsons** in the 1880s whose multi-stage *steam turbine* proved really practical and efficient. Using a large number of staged turbine rotors to increase power and economy,

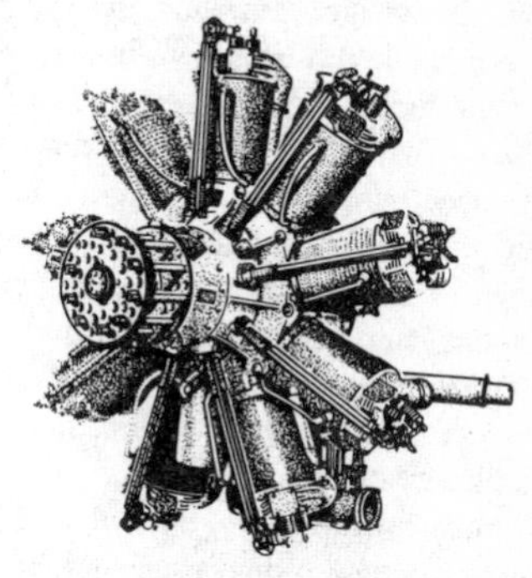

Rotary engine

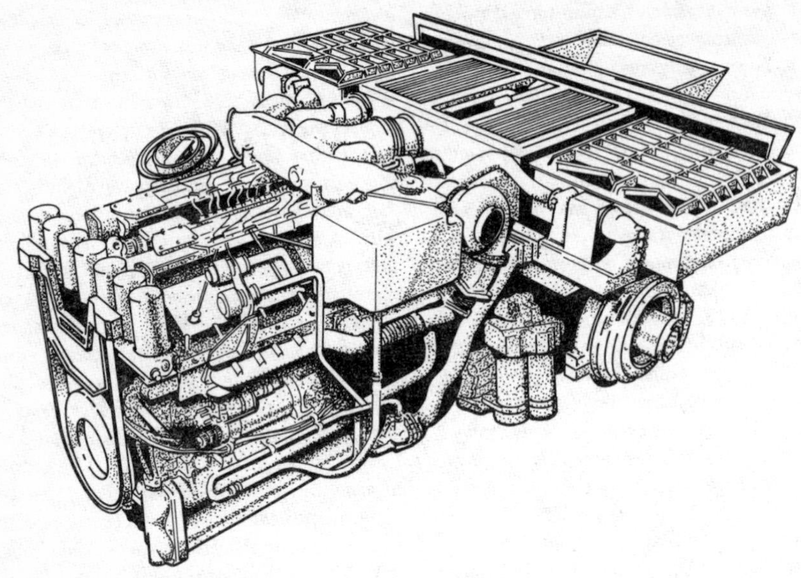

In-line diesel tank engine, c. *1990*

Parsons also, incidentally, eliminated the vibrations inherent in reciprocating engines. In 1897 a 2,000hp Parsons turbine engine drove the 44-tonne boat *Turbinia* at 34 knots, indicating that the marine reciprocating engine's days were numbered.

Fundamental to all engines were advances in **metallurgy** to solve the taxing problems of stress, strain, heat and so on created by high performance and to save weight. **Steel** and steel alloys were among the most important materials, but as time went by metals, such as **aluminium** alloys and **titanium** were employed, as well as **plastics**. In 1919 aero engines weighed about 2lb/hp; ten years later that figure was halved and that was not the end of the story. Moreover, collaboration between chemists and engine designers produced petrol which could be used with modified engines of much higher compression ratio. It culminated, in the mid-1930s, with Eugene Houdry's discovery of catalysts to produce higher yields from basic oil feedstock, as well as better spirit.

In the meantime, in 1892, Rudolf Diesel had invented the *compression–ignition engine* in which low-grade, though more energy-efficient, oil fuel was injected into the cylinder, where it ignited spontaneously once the air was sufficiently compressed by the piston, thus dispensing with electric ignition and portending the replacement of coal fuel for ships. It introduced too, for **submarines**, a far safer engine than the petrol kind used to drive them on the surface and to charge the batteries for the electric motor used when submerged.

The approach to and outbreak of **World War I** stimulated the immense growth of all means of propulsion. In 1900 **battleships'** reciprocating engines developed power in the order of 13,000–18,000hp for 18 knots; in 1905 HMS ***Dreadnought*** was powered to 21 knots by much more reliable 23,000hp Parsons turbines; and come 1919, battleships boasted engines generating 60,000hp and some **battle-cruisers** 160,000hp. While many aircraft in 1914 were powered by engines of about 70hp, those of 1919 gave as much as 450hp. Moreover, aircraft engines, such as the in-line American Liberty of 338hp, increasingly were installed in **tanks** as, in due course, would be radial aircraft engines.

Prior to **World War II** research and

development, stimulated by air and land record-breaking and racing, as well as calls for **motor vehicles** with low fuel consumption, tended to concentrate on internal-combustion engines. A large variety of in-line engine configurations appeared, such as horizontally opposed and H- and V-cylinder-bank types designed to pack more power into a small space. At the same time supercharging by forcing air into the cylinder to increase performance, particularly at high, rarefied atmospheres, became essential and common. In parallel, fuel injectors, which substituted for carburettors on compression–ignition engines, were developed for petrol engines (of over 2000hp by 1945), with considerable, if costly, improvements to performance.

Meanwhile, the Germans were experimenting with rocket motors for **space vehicles** and aircraft, *closed-cycle engines* for submarines and, concurrently with the British and Italians, turbine **jet** engines for aircraft. All three types would be in service by 1945 and, with the exception of the closed-cycle engine, revolutionize means of propulsion.

The closed-cycle engine was the invention of Dr Helmuth Walter in his efforts to dispense with **batteries** for submarines, whose discharge rates restricted speeds and limited submerged running to 400 miles at best. Walter's engine used hydrogen peroxide broken down by a catalyst to generate gas which drove a turbine. The gas was then put through a condenser to separate the carbon monoxide, which was discharged into the sea. It was a dangerously unstable system which suffered from numerous difficulties and, within a few years, was made obsolete by *nuclear* engines.

The enormous promises of greater speed and endurance from nuclear propulsion systems have been largely unfulfilled because of emotional and political restraints related to their much-publicized threat of pollution of the environment, as well as economic reservations. They have been installed in some Russian space vehicles but never in aircraft and only in a very few warships and **aircraft-carriers**. In submarines, however, they have been in large-scale and significant use since being fitted in the USS *Nautilus* in 1955, without causing serious pollution, despite several accidents, mainly in Russian boats whose safety procedures are notoriously lax.

Fuel cells – which generate power through the reaction of a gas or liquid to electricity and air or oxygen – have been known of since 1839. But only since Francis Bacon developed the idea in the 1930s have they become a practical means of propulsion in specialized underwater, space and some land vehicles. They have the advantage of compactness and an efficiency as high as 80 per cent. But they require refuelling, have slow discharge rates and, therefore, cannot be used for high-speed working. Even if electricity superconductor materials become practical, it is unlikely that fuel cells will compete with maritime nuclear engines, though they might well be viable in land vehicles.

Prussian wars. The Kingdom of Prussia began to assume a dominant role in Europe in 1688, soon reinforced by its part within the anti-French alliance during the War of the Spanish Succession. But it was King Frederick William I (1713–40) who turned it into the bureaucracy which merged a large semi-militia into the state's structure. And **Frederick II (the Great)** (1740–86) whose belligerent foreign policy led to an invasion of Silesia and involvement in the Austrian War of Succession (1740–48), followed by the **Seven Years War** (1756–63) and a **guerrilla war** in Poland after engineering her partition and acquiring much territory in 1772.

The Prussian Army emerged as a strictly disciplined, well-trained and flexible force – a model for other armies. Until, that is, its involvement in the **French revolutionary wars**, defeat by **Napoleon** and the collapse of the state in 1807 led to a period of reform and the alliance with France which lasted until 1813, following defeat in Russia; and Prussia's joining the anti-French alliance to play a major part in the campaigns of 1814 and 1815.

After 1815, Prussia not only maintained a strong army but also began to design and make her own weapons as part of the industrialization led by firms such as **Krupp** and **Siemens**. The army, prompted by Otto von **Bismarck** and led by General Helmuth von **Moltke**, won the Danish War of 1864, the Austro-Prussian (Seven Weeks) War of 1866 and the **Franco-Prussian War** of 1870–71, equipped, as never before, by German technology and influenced by the military philosophy of Karl von **Clausewitz**. The ensuing Reich, created from the German states but dominated by Prussia, absorbed all their armies (with the exception of Bavaria's) and then set about building a navy to challenge Britain's, in moves which contributed to **World War I** and the destruction in 1918 of the Prussian Kingdom.

Psychological warfare. Attack upon and defence of the minds and behaviour of leaders and followers of armed forces and the populace always have been natural and essential activities in time of peace and war. Although in some people's minds synonymous with **propaganda** warfare, its ramifications and execution go well beyond that single though important indoctrinating function. In the assault upon **morale** and resolution, intimidation by demonstrations of overwhelming power through displays of strength and deafening noise has frequently decided an issue by crucially undermining or cowing resistance. The well-timed and coordinated application of terror against unprepared forces through the use of concentrated **artillery** and **rocket fire**, **bombing**, **chemicals**, **flame**, **tanks** and, indeed, any kind of unusual weapons system, have had strong psychological effects by encouragement and discouragement. An element of psychological warfare is also found in the propagation of threats of secret weapons and stimulated dread of the unknown.

Wise commanders seek psychological advantages and guard against their opponents' ploys to obtain surprise by subtle, indirect methods. Examples abound throughout history – such as Joshua's noisy and mystic demonstrations during the siege of Jericho; **Hannibal**'s use of **elephants** to frighten the Romans; **Genghis Khan**'s spreading of false rumours about **Mongol** strength; the exploitation of Jeanne d'Arc's inspired zeal to restore French morale in the **Hundred Years War**; the screaming feathers of charging 17th-century Polish **cavalry** and chilling silence of a line of British infantry awaiting attack during the **wars of the 18th century**.

These were ploys many European leaders lost sight of during **World War I** as they indulged in head-on brute force without effective attacks on minds and resolve. It was too late when defeated Germans subsequently conceded that **blockade**, propaganda, tanks and superior equipment had fatally undermined their morale. But realistic to try hard to apply the lessons in **World War II**.

A feature of both Fascist and Communist Staff organizations, post-World War I, was the appointment of specialist officers in headquarters to advise on and implement psychological measures. In fact they were usually involved mainly with propaganda in its various defensive and offensive aspects, but their mere presence was significant, as many charismatic leaders, such as Admiral **Halsey** and Generals **Guderian**, **MacArthur**, **Montgomery** and Rommel, realized when they copied politicians by employing propagandists to publicize their talents and, in Rommel's case, actually win him the admiration of the enemy.

Psychological warfare is of the essence of **guerrilla** and **cold wars**. In such special circumstances related to emotive causes, the vital necessity to attack the morale of outnumbered but dedicated partisans has long been recognized, just as the best partisan leaders, such as Giuseppe Garibaldi and **Mao Tse-tung**, have held their followers together by persuasive measures and sound propaganda. Few anti-guerrilla campaigns have been waged with such skill as by the British during the Malayan War of Independence. The Communists antagonized their supporters by cruelty, while General Gerald Templer shrewdly won what he termed 'the battle of hearts and minds' by demonstrating the cause of democratic freedom through minimum force, a psychological strategy which, if applied in **Vietnam** by the French, might have saved much bloodshed. Indeed, a determination to minimize casualties, as practised by the United Nations during the **UN–Iraq War**, significantly boosted **morale** both in the forces engaged and on their home fronts; while Iraqi promises of inflicting heavy losses tended to weaken UN resolve.

Punic wars. The three wars between Rome and Carthage which began in 265 BC were for strategic control of the central **Mediterranean** and freedom to trade – and therefore a **naval** and **amphibious war** for which the Carthaginians, with many **galleys**, held initial superiority. But in 260 BC the Romans built 120 copies of a Carthaginian galley in 60 days, and at Mylae in 260 BC and Ecnomus in 256 BC won decisive victories. Yet they were to suffer the loss of 284 ships out of 364, with 100,000 men, in a storm in the Sicilian Straits in 255 BC and a crushing defeat, with the loss of 93 ships and 28,000 men, at the Battle of Drepanum in 249 BC, when Carthage lost no ships at all. But off the Aegatus Islands in 241 BC, in connection with a Roman invasion of Sicily, the loss of 120 out of 200 galleys forced Carthage to sue for peace and evacuate the island.

The Second Punic War began in 219 BC when **Hannibal** established a base at Saguntum in Spain, with a view to avoiding Roman **blockade** by taking the land route through southern France and across the Alps to strike at Rome via northern Italy. Hannibal's ensuing 16 years' campaign, with 13 victories out of 16 battles, was a **strategic**

The Punic wars

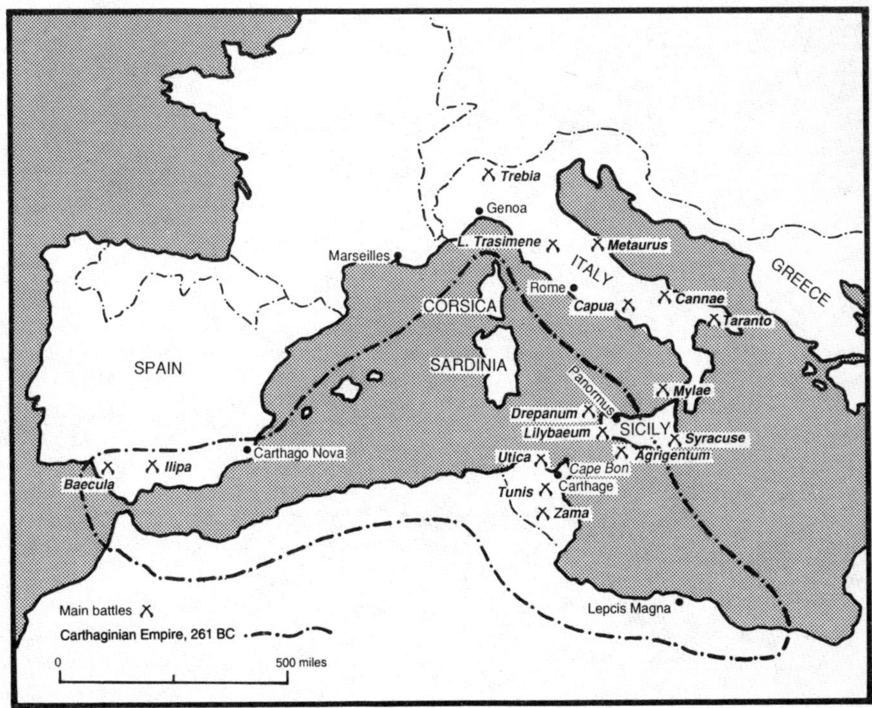

and **logistic** masterpiece, of which the rout of the Romans at Lake Trasimene in 217 BC and the classic envelopment at **Cannae** the following year were the **tactical** high spots. Yet the Romans (commanded by Scipio Africanus from 210 BC) made astute use of sea power to supply Rome when it was threatened in 211 BC; instituted **guerrilla warfare**; and gradually built up their strength and wore down Carthage's, until defeat at Metaurus in 207 BC forced Hannibal southwards. Scipio thereupon felt free to invade Africa in 204 BC and, at Zama in 202 BC, inflicted the defeat which led to Hannibal's suicide and the end of the war.

Carthage survived, although it was no longer a serious threat. But Rome still feared her and, in 149 BC, reopened hostilities, although it took two years and the reduction of the population of 250,000 to 50,000 (who were sold into slavery) before the city fell and was razed to the ground.

Q

Q-ships. As an anti-**submarine** measure in **World War I**, the British in 1915 began employing merchant ships with concealed armament as decoys. They sailed alone to lure the **submarine** into surfacing to use its gun in order to save **torpedoes**. Some crew members would simulate a panic abandonment of ship while the others waited until their attacker closed the range before disclosing their guns and opening fire. Increased caution by U-boat commanders gradually reduced their effectiveness, which was never immense. By the war's end only 11 U-boats had been sunk for the loss of 27 Q-ships, most of which went down after introduction of the **convoy** system in May 1917. In **World War II** the Japanese also used them in desperation, but to little effect.

R

Race to the Sea. When the German retreat from the **Marne** came to an entrenched halt at the Battle of the Aisne on 18 September 1914, both Germans and the Allies continued in their attempts to outflank each other by a succession of northwards sidesteps, manoeuvres which were commensurate with the ability to transport by road, **railway** and, on the British part, by sea sufficient men and resources for offensive action at selected places. From the outset possession of the Channel ports was acknowledged by both sides as crucial, though inevitably each shuffle ended in yet another collision and an extension of the wired-in trench barrier. On 23 September a French thrust from Roye threatened the vital German railhead at Saint-Quentin but was brought to a halt and thrown on the defensive next day as German pressure there and also north of the Somme was asserted – encounters which tested the French severely between Albert and Bapaume on the 29th. Yet it was a drawn battle of similar consequence to the struggle brought on by the well-prepared assault by Sixth German Army at **Arras** on 2 October, which so nearly encircled that strategic centre but also ended in stalemate.

Meanwhile the Belgian Army, under siege in Antwerp (which fell on 9 October), was with British help clinging to Flanders and its ports as two German **cavalry** corps thrust towards Hazebrouk, a breakthrough which was checked only by the timely arrival in strength of the British Expeditionary Force (BEF) and thrown back east of Ypres, thus ending on 15 October the so-called Race to the Sea. For that day the Belgians secured Nieuport on the left by the coast, the French covered the Yser Canal and the BEF had filled the gap by joining hands with the French on the La Bassée Canal on the eve of the First Battle of Ypres.

Radar. A system which measures the following **electronically**: *range*, by timing the passage of **electromagnetic** waves to and from a target; *direction*, by using a suitable antenna to project the waves in a narrow-beam, antenna direction indicating target direction; *velocity* or *target movement*, by the Doppler effect, that is, the difference between the transmitted frequency and the reflected frequency from a moving target.

The German scientist Heinrich **Hertz** in 1885 was the first to demonstrate in his laboratory the reflection of **radio** waves from a metallic object. It was experiments in ways to improve long-range radio signals which led to the discovery in the 1920s of the ionosphere. This in turn led to studies of the application of very-high-frequency microwaves and, with the development of the cathode-ray oscilloscope, provided the essentials for a practical radar system.

The Germans were still in the lead when, in 1933 and secretly, their Navy's signal research division was able using radar to detect a ship in Kiel harbour; soon afterwards the French installed radar in the liner *Normandie* as an aid to iceberg detection.

In 1934 the British took a major step into

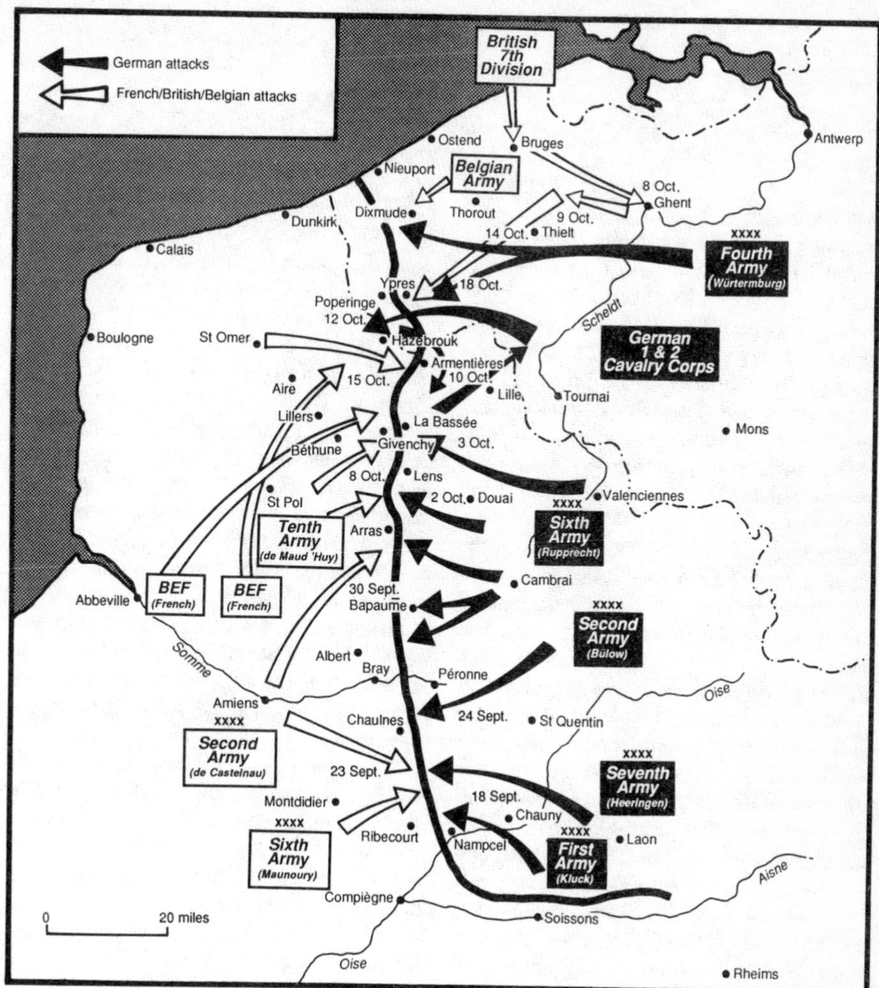

The Race to the Sea, 1914

the field when Robert **Watson-Watt** of the National Physical Laboratory showed that an aircraft could be detected by 'floodlighting' using a 50m-wavelength beam. The aircraft could be indicated on a cathode-ray oscilloscope in such a way that position, altitude and course could be plotted. By fitting friendly aircraft with a pulse repeater the ground operator could **identify friend from foe (IFF)**; intercepting **aircraft** could thus be directed to a tactically advantageous position from which to attack the enemy **bombers**. A historic experiment was carried out at Daventry in 1935 which confirmed Watson-Watt's findings and led to the Chain Home system of radar stations which, by the time war came in 1939, stretched from Netherbutton in the Orkneys to Ventnor in the Isle of Wight.

Effective though the Chain Home system was (60 per cent reliability out to 70 miles at 20,000ft) it was but a first step; further developments soon opened up the possibility of dramatic improvements in the world of

surveillance, **rangefinding** and target acquisition. To start with, however, both German and British radar followed broadly similar lines. Chain Home operated on 8–13m wavelength while early British 'beam' sets (Chain Home Low) used 1.5m. The German Freya worked on 2·4m, usually in conjuction with Würzburg providing direction for **anti-aircraft** guns and **searchlights**. The German Seetakt ship-watching and ranging set operated on a wavelength of 80cm.

By 1937 the British had developed a first airborne radar which enabled a **fighter** aircraft to detect and fly to within visual distance of a bomber at night. A 6m GL radar for anti-aircraft guns was capable, by 1941, of reducing the kill rate from 20,000 rounds per aircraft in 1940 to 4,000 rounds per aircraft.

However, a further major step was in the offing, described by Watson-Watt as 'the centimetric revolution'. It had long been realized that a narrow-beam radar of, say, 10cm wavelength, would be a fundamental advance, providing an accurate measurement of bearing angles and able to avoid both the clutter of random echoes and most enemy jamming. The problem lay in generating adequate power; the valves then available were only really satisfactory for wavelengths of about 1.5m. The breakthrough came with the resonant-cavity magnetron valve developed by Professor Oliphant and his team at Birmingham University. In 1940 a trial device produced hundreds of watts and a production version later that year gave 10kW at 10cm, thus providing the essential element for centimetric radar.

A number of developments flowed from this during the remainder of **World War II**: improved fire control and rangefinding for air defence; H2S, an airborne centimetric radar which, combined with the Plan Position Indicator developed from the cathode-ray oscilloscope, created a radar map helping a bomber crew to find their objective; and the Type 271 ship-borne radar, developed for spotting surfaced U-boats, but equally successful in detecting other surface objects.

Radar was initially seen as an aid to detection of aircraft and ships. The clutter from objects on the ground at first made it an unlikely tool for the soldier. However, since 1945 great strides have been made in refining radar capabilities. Small radars for close-range ground surveillance, for rangefinding and for missile tracking are in general use in many armies.

Radar is of course an active system and is thus susceptible to jamming. However, jamming is not always successful and other ways to avoid detection have been sought. One such is the development of the '**stealth**' aircraft such as the recently introduced American B2 bomber, which has the radar signature of a much smaller aircraft. Radar reflectivity is best from flat metal objects, worst from non-conducting materials with rounded surfaces. B2 is therefore constructed with as few flat surfaces as possible and maximizes the use of plastics and other non-conducting materials; the engines, for instance, are buried within the non-conducting wings. Similar 'stealth' materials have been used to coat the outer surfaces of combat vehicles in an effort to reduce their radar signature.

Radar can be used on land, sea and in the air, for the detection and location of aircraft, ships, moving targets on the battlefield and of rockets and artillery shells. In addition radar has a good rangefinding capability. However, the equipment can be a massive static installation with large rotating antennas; on the other hand, radars may be mounted on ships or aircraft, on vehicles or even be hand-held. While it is little affected by climatic conditions and is effective at night, it is often bulky, the antenna may be prominent – even the most modern phased-array versions with no rotating parts – and it is an active system open to countermeasures.

Radiation. Devastating though the blast and thermal effects of **nuclear weapons** are, it is the unseen radiations resulting from a **nuclear** explosion which terrify many people most. These radiation effects are not necessarily the most lethal; it is immediate radiation, occurring within 60 seconds of a burst, which releases alpha, beta, gamma and neutron particles, which is. The last two can cause serious casualties to exposed troops, since they travel several thousand metres. Residual radiation, consisting of beta and gamma particles, which can kill, is found in the ground after a fireball has touched it and in the fallout from the dust sucked up into the atmosphere – a hazard to vehicles trying to cross the area and to personnel downwind of the target.

Radio. The principle of radio **communications** had been postulated by the mid 19th century by Michael **Faraday** and James Maxwell: that **electromagnetic** energy could be transmitted through space at the speed of light without a connecting cable. It was Sir William Preece who, in 1892, managed to transmit a radio signal over a distance of 364 metres. But it was Captain Henry **Jackson** and Guglielmo **Marconi** who, independently in 1895, first demonstrated radio's real potential.

Their transmitters and receivers were founded upon an induction coil, an untuned spark gap and a simple antenna, the signal being activated by **telegraph** key. Marconi's equipment was able to transmit over a distance of 1·6km and only two years later he was transmitting Morse code from the Isle of Wight to a tug at sea. Progress was swift and by 1901 Marconi was able to send messages routinely to ships at sea and even across the Atlantic – some 4,800km.

As in so much to do with **electronics**, technological improvements followed hot on the heels of each other and in 1904 the heterodyne principle was first used, a significant refinement; the vacuum diode valve was introduced by the Briton John **Fleming**, who coined the word electronics; there were rapid improvements in the reliability and miniaturization (though still bulky by modern standards) of equipment; and trials with radio direction-finding had taken place, with military thinkers already appreciating that the location of a radio transmitter could provide valuable intelligence of enemy deployment.

Despite the traditional military resistance to innovation, the possibilities that radio opened up both for land and sea operations were rapidly appreciated, the Germans in particular being quick to grasp the potential. By the start of **World War I** horse-drawn mobile radio detachments were in service with the German Army, operating over about 200km; static sets at GHQ having a range of about 1,100km. However, the Chief of Signals had not been informed of the operational plan in 1914 and this lack of coordination meant that his equipment was not deployed to best advantage when the German Army advanced into enemy territory.

Right from the start of active operations the problem of security arose, in that radio transmissions could be picked up by friend and foe alike. Both sides in World War I set up monitoring stations to intercept the enemy's radio traffic and to break the **codes** which each used to conceal their intentions. Thus it was seen that radio could be a two-edged weapon.

At sea, communications over long distances had been well established by 1914. Since the **Russo-Japanese War** of 1904 Admiralties ashore were able to control navies afloat in a way which foreshadowed the direct governmental influence over far-away military operations. Evidence, perhaps paradoxical, of the power of radio in naval operations at that time is that during the Battle of **Jutland** the radio silence imposed on ships of the Royal Navy was a significant contribution to Admiral **Jellicoe**'s success.

On the German side there was particular interest in radio control of their **submarine** fleet, limited by the fact that the boats could only communicate when on the surface. Radio was used to guide U-boats to their quarry, though with little success as the British were able to intercept the signals and take appropriate action.

The feasibility of radio communications from **balloons** was demonstrated in 1905; in **airships** in 1909; and in **aircraft** in 1910, although the weight and bulk of equipment were handicaps throughout **World War I**. But the use in the 1920s of ever higher frequencies in smaller, lighter sets, and the introduction of voice radio (first demonstrated by Valdemar **Poulsen** in 1908 over 150 miles), enabled aircraft to communicate fluently with ground controllers and between each other by **World War II**. At the same time smaller and more robust sets could be fitted in land vehicles, giving armies true mobile communications.

It was soon appreciated that voice radio (radio telephone – RT) was the only realistically rapid way to control large numbers of **tanks**; 1931 saw the first demonstration of a complete brigade (in which all vehicles were fitted with crystal-controlled radio sets) deploying under direct orders from their commander, Brigadier Charles Broad. Navies, however, were rather slower to adopt RT, despite it being a far better command and control system (*see* **C³**) when hunting submarines than telegraphic and visual signals.

Throughout World War II most radio communications were in the low frequency (LF 30–300KHz), high frequency (HF 3–30MHz) and very high frequency (VHF 30–300MHz) bands. The latter was vital, for example, for RT control of **fighter** aircraft during the Battle of **Britain** and for German tank units throughout the war.

The military environment continued to call for ever smaller, simpler to operate, more robust, interference-free and longer-range equipment. The unceasing electronic developments of the late 20th century have made possible improvements to military radio that were inconceivable in 1945. No single system provides all the user's needs and on the battlefield one can expect to find radios operating in the HF and VHF bands for communications between units and back to brigades. The equipment may be vehicle-borne, manpack or hand-held; the different types may have varied capabilities, but all will be compatible with each other in use.

For communication between higher formations, brigade-division-corps, radio relay (first used in World War II by the Germans and later between General **Montgomery**'s headquarters in Normandy and London) has found British favour, using VHF, ultra high or super high frequencies. The system is generally line of sight, which can impose location problems, but it does provide secure telephone-type conversations. An arrangement of interlocking nodal stations ensures that if one station is destroyed the network is not disrupted. The nodes of such a system, static when operating but able to close down and move without affecting the rest of the network, could be distributed, for example, over a corps area of operations, forming a lattice into which mobile users can connect at any point. A system of **computerized** switching routes messages automatically through the most convenient nodes.

Finally, **space vehicles** increasingly play their part also, via communications satellites in geostationary orbits at about 36,000km above the earth. Within its 'footprint' the satellite is continuously available to the user and the lack of screening means that the higher radio bands can be used, providing high-capacity good-quality communications. Initially such links could only be provided between static ground stations and ships at sea, chiefly on account of the large antenna dishes required to capture the signal. However, the increasing power of satellites means that antennas small enough for vehicle-mounted, manpack or even hand-held stations are now available. Tactical satellite communications such as these are the likely ways forward in the future, aided by the revolution in microelectronics – as demonstrated most effectively by the news media during the **UN–Iraq War**; and for accurate long-distance moves, often in poor visibility, by the Allied ground forces.

Raeder, Admiral Erich (1876–1960). A man of diminutive stature who was Admiral von Hipper's Chief of Staff at the Battle of **Jutland** and became professional Chief of the German Navy in 1928. He thus was responsible for its recovery from the cuts imposed by the Versailles Treaty and the building programme prior to **World War II**. He gave priority to **battleships** and other surface vessels over **submarines** and **aircraft-carriers**. As a brilliant strategist, Raeder disapproved of **Hitler**'s premature entry into war, but once embarked on that course aggressively promoted the invasion of **Norway** and the Battle of the **Atlantic** in 1940. At the same time he did little to encourage invasions of Britain or Russia, arguing instead in favour of strangling Britain through a **blockade** imposed by occupation of **North Africa**, domination of the Atlantic and closure of the **Mediterranean Sea**. Hitler only half-embraced this strategy, largely because he failed fully to grasp the meaning and importance of maritime warfare. Yet Raeder did not put all his weight behind the vital submarine campaign and resigned as Navy C-in-C in January 1943 when Hitler ordered demobilization of surface ships as surface raiders. In 1946 he was sentenced at Nürnberg to life imprisonment for war crimes, but released in 1955.

Railways. Although there is evidence of wagons with flanged **wheels** being pushed or towed along wooden rails in the mid 16th century, the introduction of commercially viable railways had to await the development in Britain, in 1825, of George **Stephenson**'s reliable, steam-powered *Locomotion* running on iron rails. But it was the Prussians, placed in mid-Europe, who took the lead with military development by strongly influencing the layout of track to satisfy **strategic** as well as commercial purposes; and who, in 1848, moved an army corps of 12,000 men with their **horses**, **artillery** and supplies by rail to Cracow and thereafter built trunk routes to serve **mobilization** and swift redeployment of forces along the east/west axis. This far-sighted policy was copied by every principal power, so that from the **Crimean War** onwards railways were to play vital **strategic**, **logistic** and **tactical** roles.

The **American Civil War** was the first great railway war because of the long distances and vast numbers involved and the existence of networks, fragmented as they might be in 1861, to create a new mobility. Rail centres, such as Atlanta in the Confederacy, became important objectives as part of a vital strategic dimension. But with enhanced mobility came limitations. Track was vulnerable to **guerrilla warfare**, tying down large forces to guard it (*see* **March to the Sea**), and there remained the problem of transferring stores from railheads to front-line troops, for which manual labour and horses and carts were still needed until replaced by **mechanization** and **motor vehicles** in the 20th century.

World War I and trench warfare in Europe brought railways into their own. The strategic advantage of the German system was partially matched by the French and to a lesser extent by the Russians. The French system was an essential link in the Allied logistic chain, with convenient spurs laid from the main lines to assembly areas and stores depots immediately behind the front. For instance, the problem of moving **tanks** from factories in the UK to the combat zone, in the total absence of the road transporter, was really no problem at all: almost every factory had its private railway spur and, apart from the Channel crossing, the vehicles were delivered to the troops in the assembly areas entirely by rail.

For the transport of supplies forward from the railheads, tramways were built, often running right into the front line, sometimes on wooden rails which could be relaid in a new direction with comparative ease. The motive power for these light railways came from small motor tractors.

Railways were also of significance elsewhere; one was laid to accompany General **Allenby**'s troops advancing across the Sinai from Egypt into Palestine; and the railway from Damascus to Medina in the Hejaz – a Turkish lifeline – was a frequent target for T. E. **Lawrence**'s Arab raiders. A 50-mile section of that same railway was torn up by Australian engineers in **World War II** and relaid from Ma'an to Ras al Neqb to provide a railhead for stores flooding in through the port of Aqaba in southern Jordan, on their way to Russia via the Iran supply route.

By 1939 the flexibility of motor transport meant that railways, except for the strategic movement of stores and men, did not have so great an impact on logistics as in previous wars. They did, however, have one curious, perhaps unexpected, impact. It had been decreed that all British tanks should be able to travel anywhere on the British railway network. This requirement automatically limited a tank's width, thereby limiting the turret-ring diameter, which in turn limited the size of gun that could be mounted in the turret, a factor which dogged British tank design throughout World War II.

Since 1945 railways, owing to the growth of civil road and air transport, have been severely pruned. The result is that military usage, except with ex-**Warsaw Treaty** powers, has shrunk and in Nato logistics no longer assume a significant priority.

Ram. The ram, a sharp spike or strengthened projection attached beneath a ship's bows, appears to have been invented by the Phoenicians *c.* 1100 BC and incorporated in bireme **galleys** for underwater attack by ramming at full speed. It became obsolete with the galley's passing but, illogically, was revived when steam power appeared, despite the powerful **artillery** threat which made close-quarter fighting improbable. Nevertheless it scored a notable, if misleading, success at the Battle of **Lissa** in 1866 and therefore often was fitted until *c.* 1910.

Rangefinders. In the mid 19th century the only way to determine the range to a target was by *visual estimation*, a highly inaccurate method still in use for tank guns till well after **World War II**. However, as effective ranges and the intrinsic accuracy of direct-fire weapons improved, the need for an accurate rangefinder became overwhelming; and nowhere more so than in the naval sphere with the introduction of capital ships like ***Dreadnought*** and the new **battle-cruisers**, operating at high speed and at much greater distances from the enemy.

Optical solutions were the *coincidence rangefinder*, first produced by Barr and Stroud in Britain in 1880, and a more accurate device, the stereoscopic rangefinder, being pioneered by German firms at about the same time. These instruments were incorporated in the new fire-control equipment being developed by both countries for their respective warships and were in service in time for **World War I**. The weakness of the British equipment, which normally required the mean of several readings to produce a reasonably accurate range, was revealed all too clearly at **Jutland** in 1916 where, as the official history says, there was 'totally insufficient time . . . in some cases for a single range to be obtained' as fleeting targets appeared briefly through the haze. The German *stereoscopic rangefinders*, inherently more accurate and quicker to use, coupled with their electrical fire-control equipment, enabled the guns to be brought to bear more rapidly. For a much smaller expenditure of **ammunition** the Germans destroyed three British battle-cruisers in short order.

Both types of optical rangefinder depend for their accuracy on the base length between the lenses, anything less than one metre being of very little use; indeed some of those developed for anti-aircraft guns in World War II had a base length of up to 5·5m. Since 1945 optical **rangefinders** have appeared on **tanks** also, limited in effectiveness by cross-turret dimensions. The latest development for tanks is the **laser** *rangefinder*, accurate to within 5m at 10,000m range. The technique relies on the time taken for light pulses to be reflected from a target to give a measure of range. Measurement by **radar** works on a broadly similar principle using radio wave pulses, such systems being more suitable for air defence and an aircraft environment.

Reconnaissance, Means of. The axiom that time spent on reconnaissance is seldom (or never) wasted is eternally true. Foolhardy is the commander who ignores it. Yet there are numerous examples in history of its neglect, including inertia in taking full advantage of the latest technology to enhance **intelligence** gathering – that vital process in the conduct of war.

Reconnaissance may be performed by fighting for information, or by **stealth**, which is the more usual way. Until the invention of practical **optical instruments** and the man-lifting **balloon** in the 18th century and the subsequent revolutions in technology, the acquisition and identification of targets was dependent on the human senses: the naked eye, hearing, smell and, occasionally, touch. Whatever the method employed, the people involved require considerable specialized skills, insight and, quite often, daring when observing or penetrating enemy deployments, be they at sea, on the land or in the air. But wherever or in whatever circumstances the search is conducted, or the means employed, the need to make use of dominating positions (notably from a concealed overview at altitude) is often paramount, and usually benefited by **surveillance instruments** of some sort or another. Spies have also always been regarded as reconnaissance agents and as such are mentioned by **Sun Tzu**, the first military **philosopher**.

Further reference should be made to **aircraft**, **airships**, **armoured fighting vehicles**, **AWACS**, **balloons**, **cameras**, **helicopters**, **night-vision devices**, **optical instruments**, **photography**, **radar**, **radio**, **robots**, **sensors**, **sonar**, **space vehicles**, **stealth**, **surveillance instruments**, **television** and **thermal imaging**.

Reconnaissance aircraft. Virtually all **aircraft** have a potential reconnaissance capability in peace and war, even if not specifically designed or equipped for the purpose. From 26 June 1794, when a French captain spied out the Austrian deployment from a tethered **balloon** at the Battle of Fleurus, commanders on land and sea have usually rated this way of gathering **intelligence** the most important aspect of **air warfare**. The task appeared all the more promising when Gaspard Tournachon **photographed** Paris from a balloon in 1858 and, in one day in 1914, British army officers photographed the entire defences of the Isle of Wight and Solent from 5,600ft, developing the negatives in mid-air, ready for printing on landing.

It was the benefits of aerial photography which, in **World War I**, overcame the instinctive reluctance of commanders to believe what airmen reported and encouraged the establishment of organizations to task the airmen and rapidly process and disseminate their results. Inevitably this led to demands for specialized aircraft manned by experts. Observers who went aloft in tethered balloons to direct **artillery** fire were trained gunners. **Airships** were crewed by experienced naval or army men. But slow and vulnerable, lighter-than-air machines were soon superseded by heavier-than-air types, preferably two-seaters carrying a trained observer who also operated the **camera**.

Up to **World War II** the prospective contestants, in addition to covertly taking photographs of military installations from civil aircraft, also engaged in **electronic** intelligence-gathering. For example, German civil airships when flying over Britain and France were equipped to detect **radar** transmissions. Already there was a recognition of the difference between short-range **tactical** and long-range **strategic** missions. But soon, as a result of heavy losses among converted **bombers**, it was realized that the aircraft most likely to reach its target and return safely, at low or high level, was a fast **fighter**. As a result, long-range fighters, stripped of armament and fitted with cameras for horizontal and oblique photography, were developed for very fast, high-altitude work – perhaps the most famous being the British single-engine Spitfire and twin-engine Mosquito with pressurized cabins. These, with operational ceilings above 40,000ft and ranges in excess of 1,600 miles, could rove out of range of **anti-aircraft artillery** and with little danger of interception by fighters.

This immunity was made all the greater by the **jet engine** with its inherently superior performance at the highest altitudes. In fact, the principal handicap to visual air reconnaissance post-1940 was the camera (including the **television** kind) whose versatility, including penetration of cloud cover, was steadily improved. At the same time, however, the amount of intelligence which could be acquired by monitoring vast quantities of electronic emissions extended the scope of air reconnaissance enormously and created a demand for sophisticated, in-flight detectors and recording devices.

To carry a variety of new equipment far

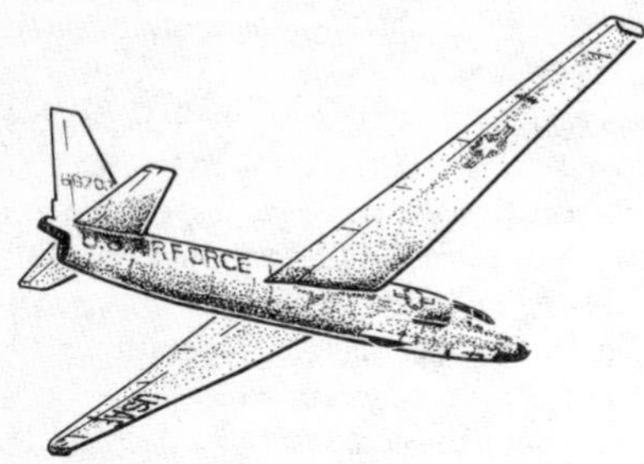

High-altitude reconnaissance aircraft, c. 1955 *(US Lockheed U2)*

over foreign and potentially hostile airspace, machines such as the American Lockheed U2, jet-powered sailplane with its ceiling of 90,000ft, and the Lockheed SR71 with a speed of Mach 3·5 and ceiling above 90,000ft, were developed. They filled the gap between slow **helicopters**, searching for tactical targets, flying nap of the earth, and **remotely piloted vehicles** and reconnaissance satellites (orbiting **space vehicles**). All obtain vital operational information in peace and war.

Recruiting. The recruitment of suitable officers and men prepared to face the dangers and hardships of service in armed forces is among the most difficult of fundamental military activities. There are very rarely enough naturally dedicated candidates with the required qualifications, even when called upon to serve patriotically in a popular, just war. Through history inducements of high rewards and glory have been essential, particularly to obtain leaders and technologists in short supply. **Conscription** in one form or another was found unavoidable and became common from the 18th century to fill the ranks of nations in arms, but a cadre of long-service volunteers remained essential. Also, as frequently demonstrated, small groups of well-trained volunteers were often superior in battle to an impressed mass.

The rules and inducements of recruiting are mostly established at national level and implemented either centrally or regionally by the armed services involved. Territorial and community groups, colleges, universities and technical institutes have usually been recruiting grounds, with particular emphasis on the latter as the quality of person demanded by **technology**'s effects rose. Moreover, as higher educational standards improved sophisticated insights in the 20th century, no longer was the right person attracted without reasonable offers of good standards of living, care and training, along with pay comparable with civil rates. As a general rule, successful recruiting has been the result of carefully considered campaigns, with a strong public relations bias, aimed at people who are looking for a special lifestyle of adventure and comradeship, and with sound **training** which civil life does not always supply.

Red Cross Society. After the Battle of Solferino in 1859, Henri Dunant wrote *A Souvenir of Solferino* which aroused public indignation at treatment of the wounded. As a result, what became the International Committee of the Red Cross was formed in 1863 in Geneva by 14 nations; the Committee wrote the first Red Cross Convention. Over the years it also undertook responsibilities for prisoners of war and grew, along with the Muslim Red Crescent Society and the Iranian Red Lion and Sun Society, into a formally recognized, international organization dedicated to the welfare of both the military and civilians in peacetime as well as war. It is related to the **Geneva Conventions**, at the heart of the **laws of war** and usually allowed to function effectively, even by the most inhumane nations and combatants, because the odium of not doing so has so often prompted punitive diplomatic, criminal and economic proceedings.

Regimental system. A *regiment* is an **army** (and sometimes **air force**) **organization** of both operational and administrative purpose. The word *regiment* itself has several different meanings, dating from the 16th century. Usually it means a single unit of **cavalry** and sometimes of **infantry**, too; though more often than not a regiment of the latter has consisted, on both sides of the Atlantic, of two or more units (battalions). In the British Army, and some armies derived from it, the Corps of **Artillery** is still known as the Royal Regiment of Artillery, though it comprises, like the Royal Tank Regiment, a varying quantity of numbered units.

Many regiments trace their ancestry to founding commanders and/or territorial connections, and cherish their historic traditions and battle honours as the basis of efficiency and **morale**. They also draw immense strength from the sense of family in them, which reflects the quality and common sense of commanding officers and the senior warrant and non-commissioned officers. For administrative purposes, regiments require a home depot where its affairs are ordered, **recruiting** arranged, veterans' welfare cared for and, with some, **training** carried out. Occasionally the system is criticized by cost-conscious accountants and politicians as an expensive luxury. But the American Army, which once dispensed with it, has repented and revived it as a vital element of combat spirit.

Religion, Wars of. From the moment mankind adopted forms of mystical worship there came into being variations of doctrine and practice which led to the growth of rival sects and factions (often within sects), which provoked armed strife. It is a regrettable fact that religious wars have been among the most ferocious, exceeded in ruthlessness only when associated with racial strife.

Arguably the earliest wars with religious substance were the **Greek–Persian** contests in which militant Zoroastrianism, practised by **Darius I** and some Persians, came into conflict with Grecian beliefs in capricious and often belligerent gods and goddesses. These were wars in which the eventual Hellenic victory consolidated their faith and also that of their rivals, the Romans. But although most so-called pagan people worshipped and turned to numerous gods for help and comfort in time of conflict, the world had to await the coming of the militant and missionary Christian and Islamic faiths before wars became tyrannized by religion. Yet, in the meantime, the religions of the Far East – Hinduism, Buddhism and Confucianism – had caused less violence than the inherently aggressive ideologies of the Middle East and Europe.

If the spread of Christianity was anything but placid, it did avoid the warlike dogmas of Islam. Thus it is to AD 632 and the outbreak of the **Muslim wars of expansion** that the escalation of modern wars of religion may be dated. The Christianized West and East Roman Empires resisted the European barbarians and the Arab onrush, and the **Crusades** were launched, after the Battle of Manzikert in 1071, to save Byzantium and restore access to Palestine (the Holy Land). It was this involvement of Western feudal armies with the advanced military technology and techniques (notably **fortifications** and **metallurgy**) of Byzantium which provided the impulse for a revolution in **armour**, **edged weapons**, **bows and arrows**, **tactics** and castles. They helped stem the Islamic expansion as the Arabs, weakened by internal schism, overreached themselves – and as the invention of gunpowder and firearms in the 14th century gave Christians the capacity to wage their internecine religious wars as well as eventually checking the **Ottoman Empire** at Vienna in 1683.

If the schism within Islamic sects was brutal, it was well matched in ferocity and technological progress by Christian divisions caused by the sundering of the Roman Catholic Church; by the dissident Hussites of John **Zizka** in the 15th century; and by the Protestant uprisings in the 16th century. Thus the intellectual Renaissance and religious extremism inspired the intensified military activity which spurred technology to new endeavour. Advances in an era of strife inevitably made the **wars of the 16th and 17th centuries** (above all the religious **Thirty Years War**) prime movers of the Industrial Revolution and the **wars of the 18th century**, in which the art and technology of war began to advance by leaps and bounds.

Despite improved religious tolerance in many civilized countries in the 19th and

20th centuries, many wars continue to be related to religion, if not as the direct cause, as a convenient rallying factor; indicating that tolerance is but a veneer. Christianity and Islam retain their inherent belligerence, both externally and internally; evidence of this can be found by reference to **guerrilla warfare** in Northern Ireland, the **Afghanistan War**, the **Iran–Iraq War**, the **UN–Iraq War**, the struggles for power in Yugoslavia, and many more.

Remotely Piloted Vehicles (RPV). *See* **Robots, Military.**

Research and development (R&D). Beyond much doubt the origins of military R&D were founded by the trial and errors of Stone Age weapons makers and boat-builders. Naturally, technical effects (*see* **Technology, Effects of**) advanced in proportion to the expenditure, effort and resources applied to R&D. They were historically reflected in the pace, for example, of the **shipbuilding** industry in moving from design of the first fighting **galley** of *c.* 3000 BC, by evolution and trial to the caravel of *c.* AD 1470, the **carrack** of *c.* 1560, the **galleon** of 1570, the steam **frigate** of 1859, the dreadnought **battleship** of 1905 and the **nuclear submarine** of 1957; or the appearance of the short bow in *c.* 4000 BC, the crossbow *c.* AD 1100, the longbow *c.* 1160, muzzle-loading firearms in the 1330s, breech-loading, rifled firearms and **machine-guns** in the mid 19th century as precursors of the 20th century's heavy, automated and **electronically** controlled weapons firing **nuclear** projectiles.

Rudimentary as R&D was until the growth of reasoning was stimulated in universities during the Renaissance, it remained expensive in relative economic and intellectual terms – and above all required the motivation of necessity and inspired innovators and inventors. For example, the wars of **religion** gave spur to necessity for improved weapons, **metallurgy** and **fortifications**; the intellectual and entrepreneurial drive of **Henry the Navigator** and the Portuguese revolutionized **navigation** and shipbuilding in the 15th century; and the genius of James **Watt** in the 1760s and Michael **Faraday** opened the doors to, respectively, the steam-power and **electromagnetic** age, discoveries which were exploited by a host of prolific inventors such as Thomas **Edison** and Sir Robert **Hadfield**.

Weapons R&D in its modern sense may be said to have started during the Industrial Revolution as a means to reduce over-engineering, escalating costs and waste. To begin with it was mainly privately funded, although there are records in Britain, for example, of formal weapons and **gunnery** trials by Admiralty and the Ordnance Board dating back to the 18th century – a tendency to increased government funding caused by the growth of the Industrial Revolution and the arms race post-1850, and the far greater complexity and cost of weapons systems such as battleships, **submarines** and **aircraft**.

Aircraft, indeed, introduced a new dimension without historical precedent. Notably, fundamental research into the principles of flight with a wind-tunnel was carried out by the **Wright brothers** and successive aviators, but France soon involved both scientists and the military. It was the British, however, who took the state-controlled lead in 1908 with an aeronautics advisory committee, under Lord Rayleigh, which included the Director of Artillery, representing the Balloon Factory at Farnborough, which grew into the Royal Aircraft Establishment (after **World War I**) and the present-day Defence Research Agency (Military and Aerospace) dedicated to R&D.

When war came in 1914, R&D was nevertheless still in its infancy in relation to the wide requirements of modern war, and World War I revealed to both warring sides the inadequacies of their research programmes. Committees and establishments to meet various needs proliferated in France and Britain. Coming into the war late, the United States was able to benefit to some extent from others' mistakes, with the formation in 1916 of a National Research Council. Russia, though possessing talented scientists such as Vladimir Ipatieff, was heavily dependent on Western technological aid and did little research of her own. Italy had a Department for Invention and Research under Vito Volterra, but was also dependent on others for technological support. In Germany the situation was rather different, with industrial research already very well established. There seemed little need for the independent groups of scientific advisers found necessary amongst the Allies. Nevertheless individual scientists were recruited to help with particular projects such as new weapons for trench warfare and **radio** for aircraft and submarine communication. One establishment that did play a part was the Kaiser Wilhelm Institute for Physics and Electrochemistry, which became involved in **chemical-warfare** research.

Following the Armistice in 1918 the attitude to military R&D varied from country to country. In Germany, industrially based research departments, including the Kaiser Wilhelm Institute, reverted to peaceful projects without interference from the victorious Allies, thus remaining available to the military when the need arose. The exception was the German Navy's signal research division, which remained in being and enabled Germany to enter the **radar** field in 1933. In France the scientists soon lost interest in defence matters and in America the dream of an independent body to coordinate military scientific research was never realized.

Only in Britain – where admittedly many scientists were equally disenchanted with military work – were a significant number of R&D establishments kept alive: for example the Admiralty Research Laboratory at Teddington, the Explosives Research Department at Woolwich (looking at armour-piercing ammunition and flashless propellant), the nearby Wireless Experimental Establishment (considering vehicle-borne radio) and, of course, RAE Farnborough.

Thus the experience of World War I set the scene for what was to come, the increasingly complex equipment requirements of **World War II** calling for equally elaborate R&D 'programmes', not only to ensure that the equipment reaching the soldier was fit for its task, but also that it should be cost-effective, though the military and Treasury view of what constituted cost-effectiveness frequently differed.

There can be no doubt that the various R&D establishments that were set up in Britain during World War II played a crucial role in eventual victory. The two most prominent to emerge for land-service equipment were the Fighting Vehicles and Armament Research and Development Establishments, achieving an international reputation in their field in the post-1945 military world in which the universities and operational **analysis** played an enlarging part.

In the USA, however, they were matched to an increasing extent for large-scale projects during the **Cold War** by the very large industrial companies and corporations, such as Rand, under contract to the Defense Department and **NASA**. In the USSR, where science and technology were an integral part of the social and centralized bureaucratic system, priorities were allocated with a bias to the **space**, **nuclear** and three armed services demands, guided by the Academy of Sciences with its underlying research institutes.

Post Cold War, however, a tendency to switch from public to privately funded programmes has come into vogue. Be that as it may, the need for sophisticated R&D is greater than ever for both peaceful and military projects. The process has been made infinitely more cost-effective since **electronic computers** were introduced with their vast and rapid capabilities to store and analyse data, solve complex mathematical problems and produce models prior to building prototypes.

Rickover, Admiral Hyman (1900–1986) was a US Navy officer who, after seagoing experience, graduated in electrical and **nuclear** engineering. In 1947 he took charge of the Navy's nuclear-propulsion programme and became responsible for *Nautilus*, the world's first nuclear-powered **submarine**, which went to sea in 1955. Thereafter he served as a driving, and controversial, leader of American nuclear R&D for ships and electrical power plant; as well as a trenchant critic of the American educational system.

Rifle. This entry deals only with shoulder-fired small arms. But reference to **artillery**, **ammunition** and **pistols** is suggested.

The rifling of **iron** barrels with spiral grooves, to improve range and accuracy by spinning the projectile, dates from the 15th century. Until the 19th century, however, it found little favour owing to manufacturing problems and, prior to the introduction of breech-loading weapons, the slowness and difficulties of loading.

For, prior to the 19th century, almost all firearms were handmade, smooth-bore muzzle-loaders in which the gunpowder charge was poured in by hand, followed, in succession, by a cloth wad and the projectile (a ball) rammed down with a rod. Priming and firing of the earliest pieces was by spark ignition of a pinch of gunpowder through a tiny hole to detonate the main charge, characteristics retained for nearly 450 years, although steadily changed by numerous kinds of lighter, more efficient and reliable weapons. For example:

The very long and heavy German *Hakenbusche* or *Arquebus*, which required a support during firing and, *c.* 1390, had a slow-burning cord (or match), dipped in saltpetre, for ignition. This method developed by 1425 into the lighter, matchlock arquebus with a trigger which lowered the match into the touch-hole – thus enabling the firer to concentrate on aiming. The system was improved upon, *c.* 1515, by the *wheel-lock* – a *matchlock* actuated by spark from a chip of iron pyrites scraped against a rough-edged wheel, rotated by the trigger, thus dispensing with the unreliable but much cheaper match. Then came:

The *flintlock*, invented *c.* 1556 and fired when the trigger was squeezed to strike sparks from a flint located above the primed flash-pan. The numerous versions of this device eventually made the matchlock and wheel-lock obsolete, when the word **musket** (derived from the Italian for hawk – *moschetto*) came into use to describe what became, in the 17th century, an almost standardized *flintlock* weapon of between 40 and 64 inches in length, weighing 15–20lb, which, none too accurately, fired a 2oz lead ball 150–200 yards. With various modifications, including the fitting of a **bayonet**, the musket remained in service until made obsolete in the mid 19th century by the *breech-loading* rifle with cartridge and conical projectile ammunition.

Rifles, some of them breech-loaders, began to enter military service in Europe and America in the mid 18th century. They were accurate up to 300 yards, played a significant part in the **American War of Independence** and were often issued to élite **infantry** units. But they were expensive and did not come into general service until the 1830s, when one-piece rounds with percussion-cap cartridges made feasible reliable breech-loading; and as industry (including state firms such as Enfield) reduced prices by mass-production methods in lieu of traditionally crafted pieces.

First among breech-loading rifles in gen-

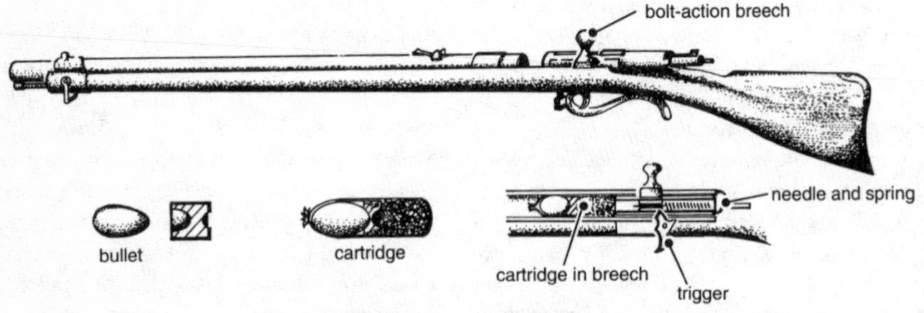

Needle-gun, 1848 (Prussian)

eral (Prussian) service (1848), with *bolt action* (to facilitate rapid loading and ejection of the spent cartridge), was Johann von Dreyse's needle gun, invented in 1828. Simultaneously, magazine-fed repeater rifles were being developed, of which the Winchester, Tyler Henry's and Christopher Spencer's models won fame during the **American Civil War**. These were later improved by the Springfield Company and European firms, such as Enfield, Mauser and **Schneider**, to equip the armies of both **World Wars** with five- to ten-round magazine, bolt-action rifles of about 0·300in calibre which could fire ten aimed shots per minute. Weapons which, with telescopic sights, were deadly beyond 1,000 yards in the hands of expert snipers.

Meanwhile the **machine-gun** had entered service, followed in World War I by a few unsuccessful semi-automatic, *self-loading* rifles. These were so strongly objected to by pundits, for many bigoted reasons, that all armies, except one, were armed mainly with outmoded, bolt-action rifles at the outbreak of World War II. In 1936, however, the Americans had adopted John Garand's gas-operated, eight-round model with a 24in barrel on the eve of a revolution in infantry tactics.

For, with the help of operational **analysis**, it came to be realized that infantry's firepower was suppressive and mostly from ranges up to 400 yards. After 1945, therefore, lightweight, self-loading, short-range rifles, such as the Russian Kalashnikov AK47, were adopted along with much improved ammunition and higher muzzle velocities. They were followed in the 1980s by even lighter, general-purpose weapons, such as the British 5·56mm SA80, with simple sights and mechanism, which can be used as a rifle, a sub-machine-gun and a **grenade** launcher.

Riot control. Before modern police forces (armed in most instances though unarmed in Britain) were formed in the mid 19th century, it was customary for outbreaks of rioting to be quelled by military forces. This implied the use of arms rather than minimum strength and persuasion, such as civil police controlled by popular governments are trained to employ. Yet generally the use of deadly force to deal with rioting, particularly of the armed political kind, has been deemed a military matter once the police admit, owing to the size of a problem or the inadequacies of their equipment and training, that matters have gone beyond their control.

Popular governments in the 20th century have learnt gradually the hard way how to contain riots by subtle methods and have come to appreciate the political value of moderation. It was discovered that suppressing riots without care for life did not necessarily stamp out resistance. Probably the merciless quelling of unpopular, anarchic riots during the Siege of **Paris** in 1871 was unavoidable under the threat of civil war, but the pitiless

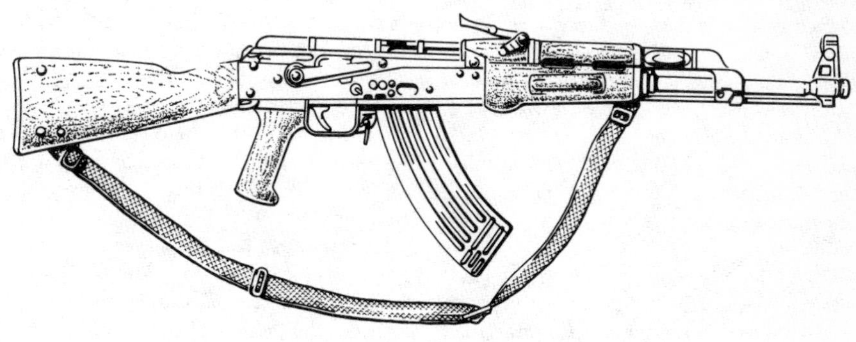

Automatic rifle (Russian AK47)

crushing of political gatherings in Russia, which admittedly bordered on insurrection from 1906 to 1914, indicated panic on the part of rulers. Their insensitive use of brute force in the manner accustomed exacerbated a resentment and mistrust which in 1917 proved fatal as revolution broke out. Similarly, as the British discovered in the aftermath of political protest in India in 1919, the use of excessive force of arms by soldiers can be counter-productive if exploited by shrewd **propagandists**. It was one of the ironies of the Amritsar Massacre that General Dyer's military overreaction to moderate Indian protests against aspects of British rule probably shortened the British regime by many years.

Yet Amritsar produced another ironic and beneficial effect since it forced the British to recognize the political ineptitude of uninhibited force for riot control and made them adopt the restrictive rules for applying minimum force which have stood the test of time and much practice. Henceforward the military would be used in support of the civil power only when formally requested. Thereupon the military commander was not only in complete charge but also had to complete the task of restoring order by the use of prescribed minimum force. Firearms were not to be used except in self-defence, or by single shots aimed to kill identified trouble-makers after due warning. The system proved reasonably effective and was copied by most nations which had proper regard for humanity and public opinion. Furthermore, the rules could include use of such incapacitating weapons as high-pressure water hoses, non-toxic gas and baton rounds (rubber or plastic bullets) to help disperse unruly crowds. That was better, it was discovered, than the bullying use of **tanks**, which were not only the antithesis of efficient minimum force but sometimes highly provocative and a useful **propaganda** weapon in hostile hands.

But nations (mostly the dictatorial kind) with scant regard for and a deep fear of public opinion have many times flagrantly used maximum force to cause excessive loss of life. Fascist Italian and German forces of law and order had little regard for minimum force as they came to power in the 1920s and 1930s. Communist Russia in East Germany, Hungary, Czechoslovakia and her own provinces since 1945 used everything from tanks downwards to crush riots, methods copied by the Chinese in 1989 in their own country.

Robots, Military. The beginnings of automatic mechanical devices are closely linked to knowledge of **electromagnetic effects** and with remotely controlled weapon systems during **World War I** when the British designed a guided, explosive-laden, Unmanned Air Vehicle (UAV) to glide into German **airships**. By 1919 the Americans had built an unmanned, **explosive** UAV controlled by **gyroscope** and aneroid barometer; the British had a wireguided explosive **tank** (Unmanned Ground Vehicle (UGV)); and the Germans (without much success) actually used unmanned motor-**torpedo-boats** to ram and explode alongside enemy vessels.

In essence these military robots were intended as inexpensive ways to remove the human factor when carrying out especially dangerous or difficult tasks. Between the World Wars development of **radio**-controlled UAVs as flying **bombs** and also as targets for **anti-aircraft** practice continued. But it was the German-led development of **rockets and guided missiles** which, with the need for control, in particular, of the semi-automatic **V1 cruise missile**, the **V2** rocket and various anti-aircraft and wire-guided missiles and small, wire-guided demolition tanks, induced a revolution in UAVs and UGVs. These laid the foundations of modern UAVs (once called Remotely Piloted Vehicles (RPV)) and clever (or 'smart') munitions.

Modern UAVs may be divided into those designed for **reconnaissance** and **electronic** warfare (EW); and those for demolition missions.

Reconnaissance and EW systems have been either jet-propelled or driven by an airscrew; fly at supersonic, but usually at subsonic, speeds; weigh between 200lb and 10,000lb; and, because of their relatively small size, are difficult targets to detect and hit. Most are of the winged variety, although **helicopter** types and hovering platforms have been made. They have the advantages of being cheap and easily handled from small, mobile ground stations which service, launch and recover them. To accomplish their tasks, which include visual and electronic surveillance of tactical activity, target acquisition, radio relay, electronic jamming and the suppression or decoying of missiles, they carry **photographic** or **television** cameras and/or various kinds of electronic **sensors** which, in order to save time and possible loss of information, can transmit their discoveries in flight. Guidance can be manual/**optical** by radio from a ground or air control post, through on-board television or by pre-flight programming.

Considerable use was made of UAVs, in **Vietnam**, where the Americans flew 3,435 missions between 1964 and 1975. They were used extensively by the Israelis in the Yom Kippur War of 1973 and in Lebanon in the 1980s to jam radar. They provided vital information during the **UN–Iraq War**.

Demolition systems tackle the problem of destroying pinpoint targets – a business which was usually very cost-ineffective until the introduction of the so-called smart bomb. An early example was the 498kg Walleye gliding bomb, deployed in the Vietnam War in 1967 by the United States. Controlled from the parent aircraft via a television camera in the nose of the bomb, which had movable fins, the device could be steered to its target with some accuracy.

Far more deadly was the next generation, which included the 908kg *electro-optical guided bomb* (EOGB) and the 1,362kg *laser-guided bomb* (LGB). The first was a development of Walleye, enabling the controller to locate the target on his television screen and designate it to the bomb; then leaving the bomb, after release, to find its own way to the target. For the LGB the target is illuminated from the launch aircraft with a **laser**, a **sensor** in the bomb detecting the illumination and using the reflection from it to guide the bomb on its way to a strike. The success rate against isolated targets in Vietnam was remarkable, with no fewer than 106 bridges being destroyed during a single three-month period in 1972. A virtue of such 'fire and forget' weapons is the ability to attack with precision military targets which might otherwise have to be left alone because of their vicinity to innocent civilians.

The concept of smart munitions has an artillery application also, in the need to attack **armour** at long range with indirect fire, with the particular aim of breaking up a concentration before an attack can develop. Accuracy can be achieved if some form of guidance can be provided in the shell itself, reacting to laser target marking operated by a forward observation officer. Such a weapon is known as a *cannon-launched guided projectile*. Future developments using micro-electronics and millimetric wave radar may well allow the projectile to seek out the target for itself.

Finally there are *Unmanned Underwater Vehicles* (UUVs), which since 1958 have been developed, chiefly by the Americans, French, British and Russians, to assist with recovery of sunken vessels, missiles and broken cables; and also for **mine** detection and clearance. Their potential as **anti-submarine** vehicles is also obvious, particularly since robots are as yet only on the threshold of achieving full automation.

Rockets and guided missiles. Even if, as is possible, some sort of rocket was used by the **Mongols** in the 13th century, by the 15th they had fallen out of favour and remained forgotten for about 300 years. After the British Army had been subjected to Indian rocket fire during the siege of Seringapatam in 1799, Colonel William **Congreve** was asked to develop a system for use by the Royal Artillery. His successful design of a warhead attached to a long stick for stability in flight had a range of several thousand metres.

Rocket systems continued to be developed during the 19th century, with William Hale producing a spin-stabilized version which was 100 years ahead of its time. Both

Congreve and Hale rockets were used during the **American Civil War** with a variety of payloads – solid shot, **grenades**, musket balls, incendiaries – but without marked success. Rocketry then languished until **World War I** when there was a somewhat abortive British attempt to develop rockets for the attack of **balloons** and **airships**. However, interest revived again in the 1930s during the general fear of massed air attack, with an idea for rocket barrages against aircraft. The project got to the trial stage, with firings taking place in Jamaica to preserve secrecy.

Congreve's dictum of a century before – 'the facility of firing a great number of rounds in a short time, or even instantaneously, with small means' – was still attractive, and in Germany and Russia, and later in America, interest in multiple-launched rocket systems reawakened during **World War II**. Rockets were not accurate, but the ability to saturate a whole area with high explosive at short notice could provide very effective support for hard-pressed infantry. Specially equipped **landing-craft**, for example, supported the **Normandy** landings in 1944 in this way; but large-scale firings were mainly confined to the fighting in Russia and the lead was in German hands under Doctor Werner von **Braun**.

Aircraft rockets were also under development in Britain as early as 1940, with the first airborne trials conducted over Chichester Harbour in 1941. The aircraft rocket was seen primarily as an **anti-tank** weapon, though armour-piercing warheads as originally fitted were found to be ineffective; surprisingly, high explosive proved to be better, the steep attack angle tending to dislodge the tank turret or even blow it off altogether.

All this activity, culminating in the German **V2**, led to the development of the long-range *free-flight rocket* (FFR) of the post-war era. FFR accuracy has improved markedly over the years, but it still cannot compete with the gun for close-support tasks, mainly because of its long minimum range; the French 145mm Rafale, for instance, even with the use of air brakes, cannot engage a target closer than 10km. In addition, though many rounds can be fired simultaneously or very quickly from a multiple launcher, reload time compares unfavourably with that of a gun. However, large rockets can carry a variety of payloads, including a **nuclear** warhead (*see* **Atom bomb** and **Hydrogen bomb**), for attacking an area target. They are perhaps best suited as carriers for terminally guided sub-munitions, designed for top attack of armoured vehicle concentrations, typically at ranges of about 30km.

In response to the wide range of Soviet FFR systems, America has introduced the Multiple Launch Rocket System (MLRS), built by the Vought Corporation. MLRS has two pods each of six rockets mounted on a tracked launcher. The 12 227mm rockets, each weighing about 270kg and containing 644 bomblets of various capabilities (including **mines**), have a 'footprint' of 300m × 300m at 30km; and can be ripple-fired in about one minute. A pre-surveyed site is not needed, so redeployment is rapid. This weapon system played a very significant role in the **UN–Iraq War** in 1991 and has made heavy artillery virtually obsolete.

The *guided missile* (*see also* **Robots**) is a rocket which can be guided all the way to the target. Guided missiles are usually classified as: *surface-to-surface* (SSM), of which the *anti-tank guided weapon* (ATGW) is a special case; surface-to-air (SAM) for air defence; *air-to-surface* (ASM) for aircraft ground attack and *air-to-air* (AAM) for aerial combat.

Before the end of World War II Germany was already working on an ATGW system called X7 which, though it never came into service, foreshadowed most types in use today, all steered to the target by signals transmitted by the operator via a fine wire paid out by the missile as it flies. Later developments have allowed for semi-automatic guidance, relieving the operator of the worst aspects of tracking; the future for ATGW probably lies in some form of heat-seeking 'fire and forget' system where the missile homes automatically on to the target after launch.

SSMs, excluding ATGW, vary from tactical systems in the 200–2,000km bracket, where the FFR would be hopelessly inaccurate, to the very-long-range *intercontinental ballistic missile* (ICBM). Most such systems use some form of inertial guidance, the **cruise missile** coupling this with a terrain-following system providing continuous course monitoring. SSMs have proved particularly attractive for ship-versus-ship engagements and have largely supplanted the long-range naval gun. Strategic SSMs are vital to **deterrence** and assumed a new importance with the introduction in 1957 of the American nuclear-**submarine**-launched, two-stage Polaris with a range of 2,000km; a class of weapon system which has been considerably developed with much more powerful, multiple guided missile warheads, such as Trident, and which is a potent threat since it is difficult to detect. At the same time American and Russian land-based ICBMs, which also were the launching vehicle for **space vehicles**, were, for protection's sake, either made very mobile or located in hardened, underground silos; and also, as with the American MX, fitted with multiple missile warheads, given an orbiting capability and often submarine-launched to protect the launch platform.

SAM systems range from low-level missiles such as the American Stinger, to the very advanced Safeguard system designed to protect the USA from ICBM attack. In the medium range Britain has Rapier, effective out to 7km at 3,000m, the nearest Russian equivalent being SA8 (12km at 5,000m).

ASMs are comparatively short-range, though they can attack over-the-horizon targets; systems are usually radar or, in some cases, **television**-guided. AAMs, in forms such as the American Sidewinder, first used successfully in **Korea** in the 1950s, and now the more recent Firestreak, have supplanted the **machine-gun** and cannon for aerial combat.

Roman wars. Rome's acquisition of power in Italy by 264 BC was as much the result of sound organization and astute political and commercial manoeuvre as of military action. Starting *c.* 500 BC, the various squabbling tribes were brought under her hegemony; invaders were ejected and a start made with the building of a road network. At the same time an **army organization**, based on the heavy **infantry** legion, with light **cavalry** in support, evolved into a highly trained and motivated force. At the outbreak of the **Punic wars** in 265 BC, it was compelled to build a **navy** in order, successfully, to gain control of the central **Mediterranean**, forcing **Hannibal** to invade Italy via the Alps in 218 in what, eventually, proved an attritional land war that culminated in the elimination by Scipio Africanus of Carthaginian power at Zama in 202.

This triumph stimulated further forceful expansion so that, by AD 180, Rome had achieved the total dominance of the Mediterranean, as well as conquering Spain, Gaul (France), Britain, southern Europe, Asia Minor and Mesopotamia (thus, incidentally, eliminating Greek influence).

Yet, until the appearance of Scipio Africanus, there were no military commanders of genius, nor significant innovations in technology or tactics on land or sea. The legions were simply more flexible Greek phalanxes. Short **swords** and **armour** were made of **bronze** and, only after AD 400, ferrous metals; while **chariots** and **bows and arrows** were not rated highly. At sea the **galleys** were copies of earlier Greek and Carthaginian biremes and quinquiremes, though with the addition of improved grappling devices and the corvus for boarding, along with 'castles' built fore and aft. Indeed the only major Roman technical improvements were their **siege engines** and as, latterly, **morale** declined and the Empire was thrown on the defensive, high-walled **stone fortifications**.

Rarely was the Empire at peace. Gradually it became overstretched through fighting either to extend its conquests, under ambitious captains such as Julius **Caesar**, or to put down serious revolts, as did Octavian at Actium in AD 31; and losing territory to barbarians who, in 410, had captured Rome and by 476 had virtually eliminated the Western Empire. The latter had come into being in 285 when, for administrative necessity, the unwieldy Empire was divided into two;

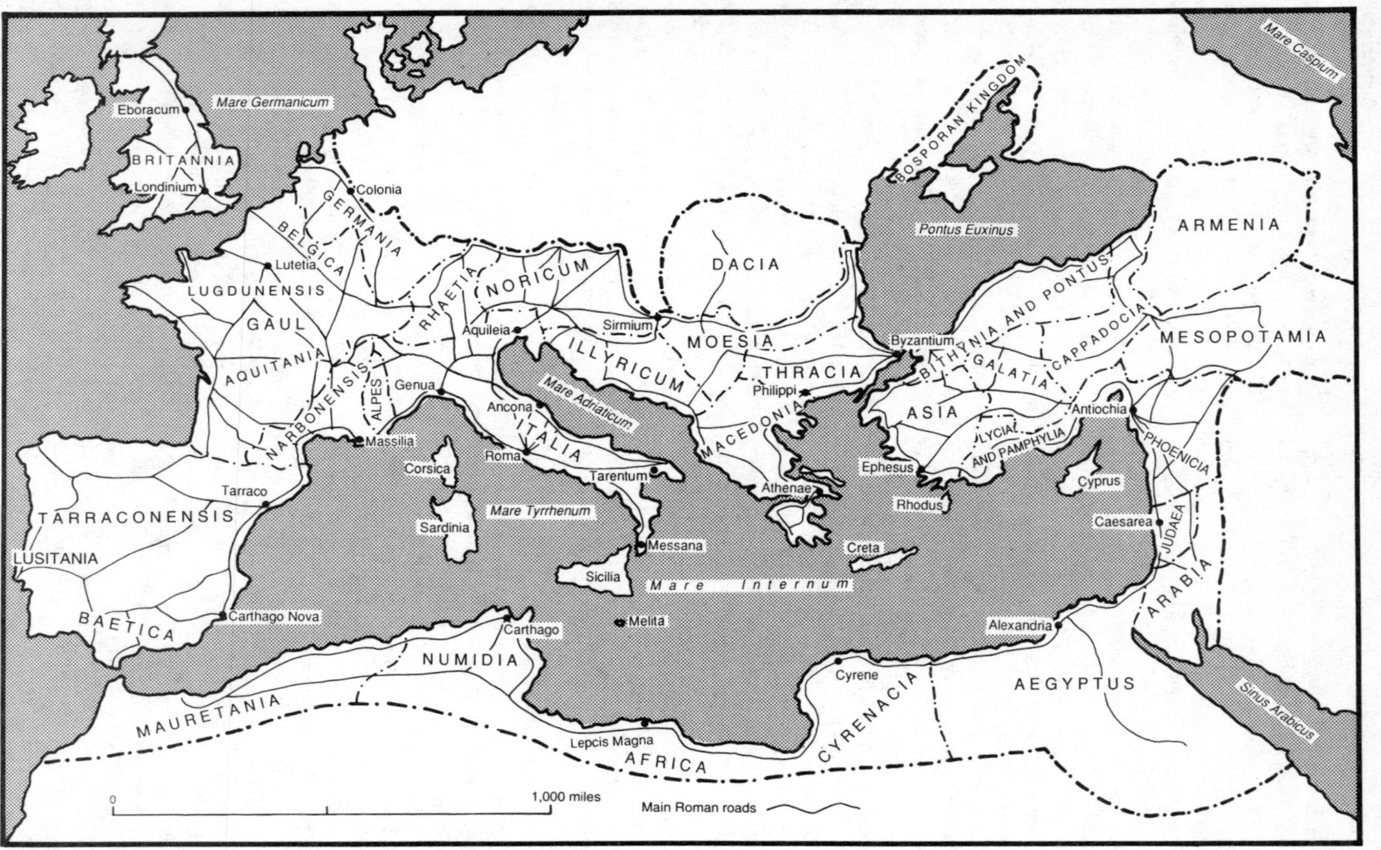

The Roman Empire, with major provinces and road system, AD 280

the Eastern Empire located its capital at Byzantium (later called Constantinople and now Istanbul).

After 476, Byzantium became the hub of Roman resistance, not only to the spread of the barbaric Huns and other Asiatic tribes into Europe, but also to the Arab **Muslim expansion** from out of the Middle East. The reconquest of Italy and part of the Western Empire by the Emperor Justinian and his great general **Belisarius** at the start of the 6th century were but the first of a succession of wars of race and **religion** which included the Christian **Crusades**. They constituted a remarkable 1,200-year-long rearguard action, which terminated with the fall of Constantinople to the Turks in 1453 as **explosives**, firearms and **artillery** made obsolescent the decayed Roman military technology and techniques; and opened the way for a renewed Muslim thrust into Europe as part of the wars of the **Ottoman Empire**.

Roosevelt, Franklin D. (1882–1945). A Democrat of immense political charisma and astuteness who became Assistant Secretary of the US Navy in 1913 and a keen student of naval affairs. As President of the USA in 1933, when faced with the task of reviving the nation's economy, a programme of re-armament to stimulate industry was among his measures. As C-in-C the armed forces he was compelled at the outbreak of **World War II** in 1937 to prepare for hostilities with Japan and, in 1939, the danger posed by war in Europe. After the fall of France in 1940 he accelerated the expansion of the armed forces, placed an embargo on the export of **oil** and **iron** products to Japan and, in line with a belief that war with Germany was likely, steered the nation to support those powers, Britain first, who fought the Axis.

After **Pearl Harbor** he concentrated upon collaboration with Britain, Russia and China in the formulation of a strategy which gave priority to the defeat of Germany before Japan. With Winston **Churchill** he had sound relations but found Josef **Stalin** and Chiang Kai-shek more difficult to deal with. In his Army Chief of Staff, General **Marshall**, he found a kindred spirit, but the prickly, less than brilliant Chief of Naval Staff, Admiral **King**, often caused him problems. His **strategic** decisions, influenced by the Allied Combined Chiefs of Staff organization set up in 1942, usually reflected political as much as military factors. He had an innovative mind and readily supported **guerrilla warfare** and the making of the **atom bomb** with secret funding, but in his shaping of the post-war world a belief that he could 'handle' Stalin proved illusory – as he may have come to realize, shortly before his death on 12 April 1945, when Stalin's intransigence and expansionist activities already were producing what became the **Cold War**.

Rossbach, Battle of (*see map overleaf*). This decisive engagement of the **Seven Years War** was between an Austro-French army of 64,000 men, under Duke Charles de Soubise and Prince Joseph, and King **Frederick**'s 24,000 Prussians. The disparity in numbers was well compensated for by French indiscipline and Soubise's ineptitude. On 5 November 1757, Soubise sensed that Prussian retreat from high ground was imminent and that a flank attack would threaten Frederick's base. Advancing without **reconnaissance** in three columns, he did not realize that Frederick had observed the move and redeployed his **infantry** and **artillery** to dominating ground and his 4,000 **cavalry** to a flank. A cavalry charge caught the unwary columns by surprise on one flank and, a few minutes later, they were raked by cannon from the other as seven Prussian infantry battalions doubled to the support of the cavalry. The Prussian artillery was decisive in this copybook all-arms battle. The Allies broke and ran, losing 8,000 men, 8 generals and 67 guns; Prussian casualties were only 541.

Russo-Japanese wars. After being diplomatically humiliated when compelled to yield **Port Arthur** to Russia after capturing it in the Chinese–Japanese War of 1894, the Japanese were bent on vengeance. Signs of Russian political confusion, their failure to reinforce

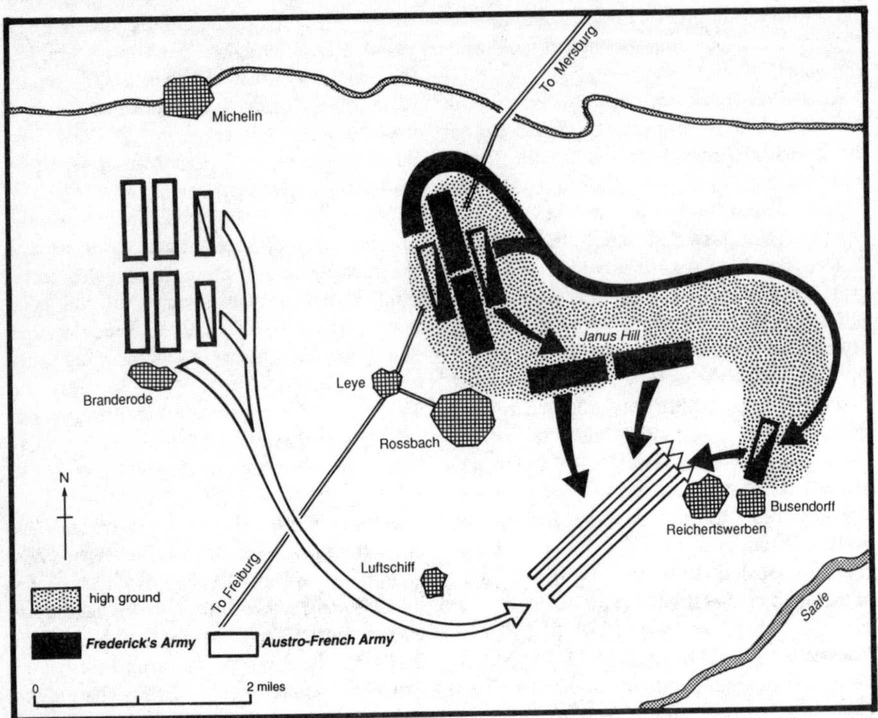

The Battle of Rossbach, 1757

Manchurian forces and the near completion of the Trans-Siberian **railway** persuaded the Japanese to strike without declaration of war on 8 February 1904. That night **torpedo-boats** entered Port Arthur harbour, sank several ships and withdrew under cover of a bombardment by Admiral **Togo**'s fleet. Meanwhile a Japanese invasion fleet on its way to Chemulpo (where the army landed on the 17th) fought off Russian warships to complete domination at sea as the main Russian fleet remained in Port Arthur.

Outnumbered 10 to 1 on land, and in **logistic** chaos, the Russians went on the defensive while they began transferring forces from the west, the gap in the railway at Lake Baikal making this difficult. Meanwhile Admiral Stepan Makarov took command of the fleet and began harassing Togo, a strategy brought to an end when Makarov was drowned on 13 April after his flagship struck a **mine**. By then the Japanese Army was approaching Port Arthur, defeating the Russians under General Zasulich at the Yalu River on 1 May, and landing on the Laiotung Peninsula on the 5th before commencing the siege on the 25th. By this time the Japanese Army under Marshal Iwao Oyama was starting to concentrate for the invasion of Manchuria. On 14 June a Russian attempt to relieve Port Arthur was broken at the Battle of Telissu and on the 23rd a sortie by the fleet, now under Admiral Vilgelm Vitgeft, came to nothing.

In mid-July the Japanese began to advance into Manchuria, winning the Battle of Moteinlung on the 31st and compelling the Russians under General Alexei Kuropatkin to withdraw, provoking the Russian warships at Port Arthur to break out to join the rest of the fleet at Vladivostok. But at the Battle of the Yellow Sea on 10 August, Togo won a crushing victory – and thereby set in motion the Russian plan to send their Baltic

Fleet to the Far East where it too met its doom at the Battle of **Tsushima** on 27 May 1905.

Meanwhile Oyama defeated the well-entrenched and reinforced Russians at Liaoyang on 3 September, at a cost to his army of 23,000 men and to the Russians of 19,000. Following up the Russian retreat with 170,000 men, he again defeated Kuropatkin's 200,000 at Sha-Ho on 17 October with respective losses of 20,000 and 40,000. These were engagements which showed only too impressively the immense killing power of modern weapons, how very difficult it was to exploit local successes and the problems of keeping a modern army's **logistics** working adequately. At Sandepu on 27 January 1905, Oyama was fortunate in a snowstorm to avoid defeat by Kuropatkin when the Russians failed to recognize their opportunities in a counterstroke. The stalemate persuaded the Russians to withdraw to Mukden and fight the culminating land battle of a war which ended in mutual, utter exhaustion, but with Japan's accomplishment of her war aims.

Twice more the two nations fought in Manchuria as further instalments of the continuing Japanese wars of expansion. These were the series of incidents on the Mongolian frontier in 1938, culminating in the Japanese rebuff at the Battle of Khalkin Gol in 1939, and the Russian invasion of 9 August 1945 when their experienced mechanized armies overran a weak Kwantung Army of about 600,000 in six days, and swept on into North Korea at the close of **World War II**.

Rutherford, Ernest (Lord) (1871–1937). A physicist whose work on **electromagnetic effects** led him to the study of **radiation** as emitted by radium and **uranium**. In 1899 his vital discovery of alpha and beta rays led, in 1910, to his crucial announcement of a **nuclear** structural theory. His discoveries, in due course, would open the way for others to make the revolutionary **atom bomb** and production of nuclear power. He was awarded a Nobel Prize in 1908; and in **World War I** worked on **submarine** detection. With colleagues in 1930, he published *Radiations from Radioactive Substances*.

S

Salamis, Battle of. When King Xerxes invaded Greece in 480 BC in the second phase of the **Greek–Persian Wars** and the Greeks retired from Athens into Corinth, the Greek fleet of about 325 **triremes**, under Themistocles, concentrated in three rows in the western narrows north of Salamis. The Persian fleet, reduced by battle and storms to about 800 triremes, advanced to the attack while infantry were landed on the flanks and Xerxes watched from the land. The heavier Greek triremes did considerable damage with their **rams** to the Persian vessels which yawed under a stern wind to expose their sides. In the boarding which followed, the Persian archers were outmatched by the 6,000 better-armed and armoured Greek **infantry** crews. As the Persian front row fell back to mingle with the following rows, they were prevented from making use of their numbers by mounting confusion. Many boats went ashore or were sunk to a total of about 200, against only 40 Greek. Xerxes fled as the Greeks mopped up the Persians who had landed. The Persian invasion was fatally blunted by a combination of superior technology and tactics.

Salerno, Battle of. On 9 September 1943, six days after the Italian campaign began in Calabria and timed to coincide with the announcement of an armistice between Italy and the Allies, Fifth US Army under General Clark began landing in the Bay of Salerno. Neither Field Marshal **Kesselring** nor the German Tenth Army Commander, General Heinrich von Vietinghoff, were entirely surprised, but their concentration of troops at Salerno was conditioned by the major distraction of disarming the Italian Army. The invaders were met by alert defenders and by nightfall had established only four small, unconnected bridgeheads. Meanwhile German air attacks that included guided **bombs**, against shipping (scoring 85 hits and sinking four transports, a **cruiser** and seven **landing-craft**), were part of a race to mount a decisive counter-attack to break the Allies before they were strong enough to break out. The Germans nearly won the race on the 12th and 13th when their six divisions made headway, but were finally checked by the dropping of an American parachute brigade, intense naval and army **artillery** fire and air attacks.

By the 15th a stalemate had been reached, to be broken when General **Montgomery**'s Eighth Army, approaching from the south, made contact on the 16th. Only then did Kesselring order a withdrawal.

Saxe, Marshal Count Maurice de (1696–1750). A profligate German infantryman who fought under **Eugene** at the Battle of **Malplaquet** (1709) and in 13 days in 1732 wrote his highly innovative *Reveries upon the Art of War* (*see* **Philosophers, Military**). A stern disciplinarian, he went on, in the French Army, to demonstrate his military genius during the War of the Polish Succession (1733–8) and the War of the Austrian Succession (1740–48). To his credit was the capture of Prague (1741) and his brilliant conquest of Flanders through victory at the decisive Battle of Fontenoy (1745) and the subsequent capture of Tournai, Ghent, Bruges, Oudenarde, Ostend and Brussels; and victories at Raucoux (1746) and Lauffeld (1747), followed by the capture of Maastricht in 1748.

Schlieffen, General Count Alfred von (1833–1913). A particularly dedicated member of the Great German General Staff who was at the Battle of Königgrätz and took part in the **Franco-Prussian War**. He is best known as the Chief of the General Staff who, between 1891 and 1905 in the belief that a European war was inevitable, drew up the famous Plan which gave priority to knocking out France before crushing Russia. The Plan, with modifications, was put into operation at the beginning of **World War I**, and was revealed as defective because it underestimated the **logistic** problems as well as French and British determination and combat skill. Instead of winning a short war, it brought defeat at the Battle of the **Marne**, to ensure a losing long war of disastrous long-term political effects such as he was incapable of visualizing.

Schneider–Creusot Company (*see also* **Le Creusot foundries**). In 1837, when the brothers Adolphe and Eugène Schneider took over the Forges et Fonderies du Creusot at Le Creusot, France (which William Wilkinson had started in 1782 as an **iron**-making munition factory and had supplied the **French revolutionary wars**), they turned more to making **railway** lines, locomotives and small steamships. After the **Franco-Prussian War** the firm expanded rapidly to produce **steel**, **armour** and **artillery**, including the quick-firing 75mm gun and, in **World War I** the first French **tank**, the Chars d'Assault of 1916. During **World War II** the great works were severely damaged by bombing and **guerrilla** attacks. The company remains an important supplier of armaments, including tanks and self-propelled artillery.

Scott, Admiral Sir Percy (1853–1924) was a naval **gunnery** specialist who, spurred on by Admiral **Fisher**, built **training** aids for gunlayers, developed **rangefinders** and electric circuitry and invented the **Director Fire Control System** in time to incorporate them in the dreadnought **battleships**. Despite lack of tact and contempt for officers of 'the old school', he managed to push Royal Navy gunnery into a technical lead prior to **World War I**; and at the war's outbreak established the **anti-aircraft artillery** defences of London against **airship** attack. No great tactician, he was, however, a visionary who, in 1912, correctly spoke in favour of building only small warships because **aircraft** and **submarines** soon would make battleships obsolete.

Searchlights. In 1846 W. E. Staite invented automatic-feed carbon arc lamps, which gradually were developed for factory and street lighting and for a searchlight of 11,000 candlepower in 1874. With the invention of **dynamos**, ships' searchlights (some with Swan Company incandescent filament lamps) became feasible in 1881. Henceforward, with reflectors and lenses to enhance and concentrate the beam, they were prime **night-vision devices** which were used on land during the Boer War and extensively in **naval** operations during the **Russo-Japanese War**. Additionally in **World War I**, they were given an air-defence role to illuminate the target; later they incorporated **sound locators** for pointing, until the latter were superseded in mid-**World War II** by **radar** – which soon also made air-defence searchlights obsolete.

During World War II searchlights were frequently used at sea and as part of an airborne, radar-directed **anti-submarine** system (the Leigh Light). On land they were used for battlefield illumination; and in Britain the CDL (Canal Defence Light), a 13-million-candlepower carbon-arc **tank** searchlight, with a flicker device to blind the enemy, was produced, though never much used in action. In the 1950s, however, the Russians mounted **infrared** (**IR**) tank-mounted searchlights and were copied by **Nato** armies, with the British Chieftain incorporating a three-million-candlepower, Xenon white or IR light into its **gunnery** system.

Electronic passive night-vision devices, however, reduced the military role of searchlights to only a minor one; but they continue to have tasks, including a **helicopter**-borne one in counter-insurgency, **riot control** and rescue operations.

Secret services. Secretive organizations to provide **intelligence** of enemy or rival activities and to run counter-espionage as part of a nation's defences are as ancient as conflict. **Sun Tzu** in his *The Art of War* (*c.* 500 BC) laid emphasis on spies; **Genghis Khan** and his **Mongol** armies employed agents to spy upon his intended victims; Queen Elizabeth I had an elaborate organization to keep watch and ward against internal and external foes; and the Duke of **Marlborough** employed a spy in the highest level of King Louis XIV's entourage to give vital warning of French intentions during the War of the Spanish Succession – to name but a few typical instances.

Without prior knowledge of hostile intentions and lacking defences against enemy prying, the best-laid military plans can be ruined. Stories about Secret Service work are numerous because of their mystery and the insights they offer about the inner workings of government and war-making. But they are rarely historically complete and generally inaccurate, if only because records usually remain, of necessity, eternally secret. A remarkable exception, though itself leaving much unsaid, including the vital budgeting, is the five-volume British official history of intelligence (*see* Bibliography) in World War II which, among many revelations, categorically demolished a host of legends and journalistic myths.

Secret Service work is not restricted to purely military matters. For example, the American Secret Service was set up in 1865 to combat counterfeiting of the new paper currency and only subsequently acquired other duties, including protection of the President in 1901, investigation of fraud and, in **World War I**, of violations of neutrality. Tsarist Russia, with its traditional passion for concealment and conspiracy, ran numerous agencies, many of them tasked to quell subversion, such as the Cheka, the GPU and the OGPU, which changed names and colour but greatly expanded their activities during and after the Revolution to become the NKVD, NKGB and KGB with tentacles of espionage and subversion reaching throughout the world. Germany, too, had its Prussian police force for internal security and, in due course, the agencies of Nazism, including the RSHA and the Gestapo. Britain formed its Secret Service in 1909 under the Foreign Office, in response to the deterioration of international relationships, to 'be a screen between the Service departments and foreign spies; to act as an intermediary between the Service departments and British agents abroad; to take charge of counter-espionage'. Then, military attachés in foreign embassies played an important role. Today, cultural attachés and trade missions are added to the list of agencies involved, especially with industrial espionage.

The ramifications during World War I of Secret Service work linked to diplomacy and **propaganda**, military operations and subversion, security **radio** intercept and **code and cipher** breaking, taught vital lessons for prolific use in **World War II** and the subsequent **Cold War**. Never in a tranquil environment, they performed their duties in anonymity, furtiveness and an atmosphere of intrigue, mistrust and interdepartmental feuding which frequently made people wonder whose side they were on. There were traditional rivalries between the different Service intelligence branches and problems of coordination. Ever more vital **technological effects** required closer monitoring as well as providing improved methods of **electronic surveillance**. The creation in Britain in 1940 of the overlapping Special Operations Executive (SOE) and in the USA in 1942 of the Office of Strategic Services (OSS) produced much friction. Later, in the USA, there were overlappings and jealousies among the Federal Bureau of Investigation, the Secret Service and the **CIA** (Central Intelligence Agency).

The presentation of Secret Service matters to the public is usually either guarded or sensational, and often deliberately misleading or provocative as part of the struggle between agencies. Not infrequently one adversary denigrates or tricks another into an indiscretion. The pressures applied by politicians of one persuasion to undermine the opposition by disclosures and scandals concerning secret activities are simply part of the game. Bland requests, on democratic grounds, for increased freedom of access to Secret Service work are not always what they seem – and certainly the antithesis of the essence of what intelligence-gathering and security work is about. But woe betide any organization, civil or military, which drops the guard that secret information from reliable sources provides.

Sedan, Battles of. The ancient town of Metz, guarding the River Meuse at one of the gateways to France, has experienced many wars but few battles more crucial to the history of France and Europe than those of 1870 and 1940.

1870. During the Franco-Prussian War, after the French Army of Marshal Bazaine had retired into Metz, the Army of Châlons under Marshal MacMahon with **Napoleon III** in company, moved to its relief. But on its approach, on 29 August, the French collided with the Germans at Beaumont and retreated in disorder on Sedan where MacMahon intended to reorganize. Seizing the opportunity, General Helmuth von **Moltke** ordered the Third and Meuse Armies to encircle the city from two directions, a brilliant manoeuvre completed on the 31st before the French could escape. Next day at first light, when short of **ammunition** and supplies, the French (with MacMahon among the wounded) awaited their fate in positions largely overlooked by the Germans. Watched by the Kaiser, Chancellor **von Bismarck** and von Moltke, the Germans held the line of the Meuse and attacked the French flank and rear where the defences were frailest. Fight with desperation as the French did, their case was hopeless. At 1100 hours, 2 September, Napoleon surrendered with MacMahon's entire army amounting to 104,000 men and 419 guns. German casualties were about 9,000 in a complete victory which decided the war.

1940. At the start of the Battle of France on 12 May, the advanced guard of General **Guderian**'s XIX Corps reached the outskirts of Sedan in three days after brushing aside French covering forces in the Ardennes. Next afternoon, in preparation for a pre-planned assault across the River Meuse on either side of the town, the French **artillery** positions were neutralized by prolonged bombing while high-velocity **anti-aircraft guns** fired directly into the embrasures of French pillboxes on the south bank. Under cover of this fire, German infantry crossed to secure bridgeheads with relative ease and expand them during the night while rafts and bridges were constructed to carry tanks and other heavy equipment. French counter-attacks did not materialize owing to command inertia and a corrosive panic which seized their soldiers. On the evening of the 14th Guderian appreciated that the 12-mile gap which had opened in the French defences allowed his Corps to pass through to the west in the drive towards the English Channel. That day, too, French counter-attacks began to develop at Stonne on his southern flank. These, along with air attacks on the bridges (which incurred extremely heavy losses to the **bombers**), caused some concern and continued for several days to come – but never in sufficient strength to restore the situation.

Seeckt, General Hans von (1866–1936). An officer who served on the Staff throughout **World War I** and won acclaim as General August von Mackensen's Chief of Staff in the Gorlice and Serbian offensives of 1915 and the Romanian campaign of 1916–17; and as Chief of Staff to the Turkish Army in 1917 and 1918. After the war he commanded Frontier Force North in the approaches to East Prussia in a difficult political/military situation which he managed with consummate skill.

In 1919 he became head of the **Truppenamt**

which, following the Versailles Treaty, did duty in place of the proscribed General Staff. In 1920 he was made Commander of the Army (*Reichswehr*). In these appointments he saved Germany from successive uprisings (*putsch*), disbanded the swashbuckling anti-Communist *Freikorps* and took the Army out of politics while having it support the government in power. At the same time he remodelled the Army in the light of the lessons of World War I and the constraints of Versailles; established secret industrial arrangements to develop the latest **technology** and weapons with countries such as Sweden, Argentina and, above all, Russia; and encouraged the brilliant members of the *Truppenamt* such as von Bock, von Rundstedt, Walter Wever, **Guderian**, **Kesselring** and Kurt **Student** to update doctrine, introduce the latest organizations and methods and create, in secret, an **air force**, **tank** force and **airborne force**. Political indiscretions in 1926 lost him the job. But he entered politics as a member of the *Reichstag*, from 1930 until 1932, and in 1934 and 1935 headed a mission to help Chiang Kai-shek reorganize the Chinese Army and fight a **guerrilla war**.

Sensors. Since surprise is often achieved by **stealth**, the need for devices to enhance **night vision**, hearing, smelling and feeling is essential. Telescopes in the mid 17th century were the first naval **optical instruments**. But it was not until **World War I** that electrical listening devices, such as microphones to detect enemy movement, geophones to sense tunnelling, **sound locators** to give warning of **aircraft** and **hydrophones** and **sonar** for underwater detection came into use.

Prior to **World War II electronic** aids, in the form of **infrared** photoelectric cells, **radar** and **television** led to the development of the highly sophisticated electro-optical sensors of the late 20th century, which can give 24-hour all-weather **surveillance** in addition to a capability to sniff out some **explosives** and **chemicals**. Some sensors can be located in aircraft or **space vehicles** or on the ocean bed; those with remote acoustic or seismic characteristics are the most common, and can be hand-emplaced, fired by **artillery** or air-dropped. Sensors do not think and can be deceived randomly or by deliberate countermeasures; yet some can differentiate, for example, between types of vehicle or men and livestock. Their information can be fed by cable or **radio** into **computers** and made to cause automatic and discrete reactions.

Modern sensors, indeed, are most potent **reconnaissance** systems and part of **C^{3}I** which give commanders ears, eyes *and* striking power behind the enemy lines without much risk to lives.

Sevastopol, Sieges of. The naval base of Sevastopol was the prime Anglo-French objective of the misconceived **Crimean War** in 1854 – and by no means ready for the siege which began on 8 October. Indeed, it might well have fallen but for the incompetence of the Allies and the efforts of Colonel Frants Todleben in rapidly putting right the **fortifications** and stocking the magazines. In fact, although the Allied Fleet **blockaded** the place, which came under bombardment on the 17th, it was never entirely isolated. And such were the effects of Russian counter-battery fire, the explosion of the main French magazine and a storm at sea, which wrecked ships, that an assault was postponed until 7 June 1855, when the outer defences were at last seized. But attacks on the key Malakoff and Redan fortresses 10 days later were abject failures, with 4–5,000 men lost to both sides. Not until 7 September, in a most carefully prepared assault, timed for surprise by synchronized watches, did the French capture the Malakoff, from where they dominated the Redan, which the British had again failed to take. That night the Russians completed demolitions and withdrew from the city and port.

In October 1941 the German Army invested the city, although the siege was not complete until General von Manstein's Eleventh Army had cleared the Crimea in May 1942. His four-week-long assault, with a pulverizing **artillery** and **air** bombardment of the strong fortifications and meticulous preparation, was notable for relatively low

German casualties, nothing like the some 120,000 Russian losses, of whom 90,000 were prisoners along with vast booty.

On 9 October 1943, after a brilliantly conducted withdrawal by General Ewald von Kleist from the Kuban across the wide Kerch Strait, the Russians were poised to recover the Crimea and Sevastopol. But von Kleist had punished the Russians severely and managed to remove 256,000 men, 73,000 **horses** with their 28,000 wagons, 21,000 **motor vehicles** and 1,815 guns, and was not followed up. When in April 1944 the Russians re-entered the Crimea, **Hitler** for once permitted a yielding defence of this prestigious bastion. The final Russian assault on Sevastopol on 7 May was merely checked for two days to allow most of the garrison to be evacuated by sea.

Seven Days Battle, The. The first prolonged struggle of the **American Civil War**, which decided the fate of the Federal invasion of the Yorktown Peninsula in 1862. After the landing in March, General George McClellan, with 90,000 men, advanced without undue haste towards Richmond against 15,000 Confederates, coming within sight of the city on 25 May – by which time the Confederates numbered 60,000. On the 1st the drawn battle of Fair Oaks was fought, stopping dead the Federal advance. The same day General **Lee** took command of the Confederates and on the 25th began a series of manoeuvres which pushed McClellan back at Oak Grove, Mechanicsville, Gaines Mill, Savage's Station, Frayser's Farm and Malvern Hill. Only at Gaines Mill did Lee win the fight while, at Malvern Hill on 1 July, the Federals won a resounding victory, which McClellan failed to recognize when persuaded to abandon offensive operations and retire to a beachhead at Harrison's Landing, whence the Army was evacuated. The Federals suffered some 10,000 casualties, the Confederates 21,000, figures which spoke for themselves in terms of generalship.

Seven Years War (*see map overleaf*). In 1756, when Russia, France, Austria, Sweden and Saxony formed an alliance to curb the growing power of **Frederick the Great**'s Prussia, Frederick occupied Saxony and began a war in which Britain, already at war with France, was his only ally. Greatly outnumbered, Prussia survived because she had a well-equipped and trained army, the advantage of operating on interior **strategic** lines; and the financial and indirect military support of Britain against the French at sea, in North America and in India.

Prussia fought 17 battles (11 won, 4 lost and 2 drawn) in seven campaigns. The important battles were Prague (1756); Kolin, Gross-Jagersdorf, **Rossbach** and Leuthen (1757); Zorndorf (1758); Minden (with vital British participation) and Kunersdorf (1759); Liegnitz and Torgau (1760); and Wilhelmstal (with British participation) (1762). Although Berlin was seriously threatened by the Allies in 1757 and raided by Austrians and Russians in 1760, Frederick managed always to avoid final defeat, by the narrowest of margins. He did so notably in December 1761, when his army was down to 60,000 men and the British were contemplating desertion. But, instead, the Alliance collapsed when Russia's desertion, at the request of Tsar Peter III, made peace negotiable in 1763.

Meanwhile, following the British victory at Plassey in 1756, the French had been driven from India; had lost naval parity when their fleets were crushed off Pondicherry, at Lagos Bay and at Quiberon Bay in 1759; and in 1759 had been defeated at Quebec, ultimately to lose control of Canada. The Prussian and British victories created a new balance of world power. But they also prompted the French urgently to reconstruct their armed forces on the eve of the technologically vital Industrial Revolution (*see* **Wars of the 18th century**).

Sherman, General William (1820–91). A West Point graduate who fought in the Mexican War in 1846 but left the Army in 1853. He rejoined on the Federal side for the **American Civil War** and served as a colonel at First Bull Run. At the Battle of Shiloh on

The Seven Years War in Europe, 1756–63

6/7 April 1862 he won the lasting confidence of General Grant and thereafter played leading parts in the Mississippi River campaigns, the Battle of **Vicksburg** and, as Commander of the Army of the Tennessee, the battles of Chattanooga in 1863. But it is for the thrust to Atlanta in 1864 and the **March to the Sea** across Georgia for which he is most renowned (and by some Confederates reviled) as he struck northwards to wreck the South's war potential.

After the war he was sent west to deal with the Indians and settle the country while the transcontinental **railways** were under construction. In 1869 President Grant recalled him to be Commanding General of the Army, a post he held until 1883 with as little political involvement as possible and the distinction of forming a main Army Training Centre at Fort Leavenworth.

Shipbuilding, Naval. Until Xerxes built special fighting **galleys** for his invasion of Greece in 480 BC and the Greeks built **triremes** to combat him, ships were mostly designed for trade and as transports. For centuries to come, nearly all vessels were for commerce (*see* **Shipping, Commercial**); made of wooden frames with planks fastened by pegs or nails; propelled by oarsmen and a single sail; and modified for war by increasing the oarsmen and by adding, as did the Romans, 'castles' fore and aft. Indeed, naval architects evolved design very slowly until **artillery** made **siege engines** obsolete and the Portuguese, pushed by **Henry the Navigator**, built the faster, more efficient caravels and **carracks** which sailed closer to the wind and dispensed with oarsmen. They led to John Hawkins's **galleon** of 1570, which could be strengthened for broadside-mounted cannon and provided with **ammunition** magazines below decks.

Yet, until the mid 17th century, merchant ships, for self-defence, favoured artillery while navies preferred boarding and close combat. The fierce artillery duels of the Anglo-Dutch wars changed that tactical philosophy, leading to concentration on firepower incorporated in the multi-deck ships of the line, built on slipways or in dry docks in naval **dockyards**. This class lasted until the appearance of **iron**, **coal**-fuelled steamships in the 19th century as suitable timber became scarcer.

Commercial lines were ahead of navies in ordering the latest technology. Copied later by naval architects, their shipbuilders replaced wood frames and nailed planks by iron frames and riveted plates. But the entry into service of Dupuy **du Lôme's** wooden, ironclad steam/sail *La Gloire* in 1859, John **Ericsson**'s turreted, all-iron, steam-powered *Monitor* in 1862 and the ensuing development of heavy, breech-loading artillery, dictated that the sails and broadside configuration of the wooden walls were obsolete; that commercial vessels had only a minor role in **naval warfare**; and dockyards required many new skills and techniques. But the new **steel battleships**, **cruisers** and **destroyers** with rotating turrets of the **Russo-Japanese War** would be as much the products of private industry as of the official yards. Complex **gunnery** systems and weapons such as the **torpedo** would be incorporated; and hulls divided into watertight compartments to minimize the effects of underwater damage.

By the outbreak of **World War I**, coal fuel was being replaced by **oil**; **submarines** and **aircraft-carriers** were stretching ingenuity to revolutionary limits. Navies, therefore, demanded far stronger, more powerful and costly vessels than ever before. This led to **research and development** in efforts to compromise between combat-effectiveness and affordability. For example, the Germans introduced cheaper, weight-saving electro-welding in place of riveting for the so-called pocket battleship *Deutschland* in 1929. And the Kaiser Corporation employed mass-production of cargo ships by prefabrication, to replace Allied losses, to a significant extent during **World War II**.

The threat from **aircraft**, the **atom bomb**, submarines and **mines** compelled the mounting of **anti-aircraft** guns as well as **radar**, **sonar** and a whole host of new **electronic** devices for which space and power supply had to be found. But at the end of World War II it appeared that battleships were obsolete. So naval architects were told to develop **nuclear**-powered submarines and aircraft-carriers; and smaller, lightly armoured vessels with **helicopters** and **rockets and guided missiles**. Minesweepers, faced by nigh unsweepable **mines**, required **plastics** and fibreglass hulls, **electronics** and **robots**. Naval shipbuilding, more than ever, involves almost every sort of technology – at vast expense.

Shipping, Commercial. As Admiral **Mahan** pointed out, commerce and **logistics** are key issues and sea-power decisive. Maritime trade, **amphibious** and **naval warfare** have always been strategically and, because warships (*see* **Shipbuilding, Naval**) were often adapted merchant ships, structurally linked. Therefore it was inevitable that in so many great wars, such as the **Punic**, Anglo-Spanish, Spanish Succession, **Seven Years**, **American Civil**, **French revolutionary**, and **World War I**, the commercial trading factor was weightier than many others.

For economic reasons, it once was feasible to restrict expenditure on specialized warships and adapt cargo or passenger vessels for combat; or to encourage so-called privateers (sometimes alluded to as pirates) to prey on enemy ships for profit, under Letters of Marque (first issued by Spain and England in the 16th century). The latter practice led to the gradual development of international Maritime Laws concerning **blockade**, prize rules, belligerents' rights of search, etc., control of seaways and the arming of merchant vessels.

Rarely, however, could merchant ships fend for themselves against warships. Usually they required escort or **convoy**. With the introduction in the 19th century of steam **propulsion**, the substitution of broadside **artillery** batteries by rotating turrets and the plethora of complex devices which attended these revolutions, merchant ships had slighter chances of survival on their own. And even less against **submarines** and **aircraft** in the 20th century.

Nevertheless, in **World Wars I** and **II** they usually were armed for self-defence and often adapted for subsidiary roles. Such as the equipping of fishing boats for minesweeping, the use of cargo and passenger ships as seaplane carriers, the arming of liners for convoy escort and **Q-ships** against surfaced submarines, and the addition of an aircraft catapult or a flight deck to give convoys air defence; the latter concept was taken a stage further with the introduction of **helicopters** and VTOL Harrier **fighters**, as demonstrated during the **Falklands** War of 1982.

Despite the vast expansion of air transport (which has, except in exceptional circumstances such as the Falklands War, made troopships obsolete), over 95 per cent of transoceanic cargo still moves in ships, many of them huge bulk carriers or container ships. So Mahan's pronouncement remains valid and is likely to be so far into the future.

Trebuchet

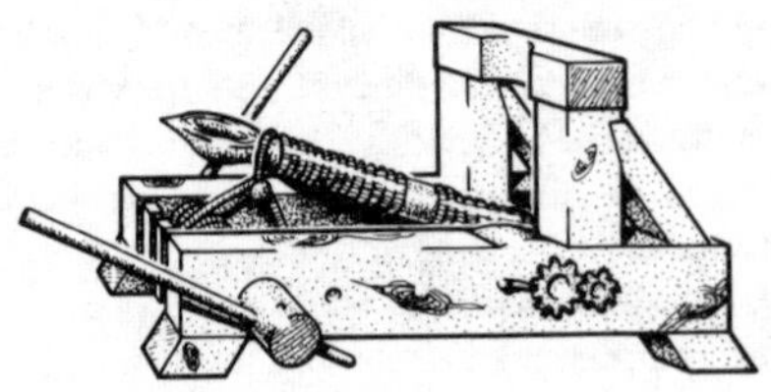

Catapult

Shrapnel, Colonel Henry (1761–1842), was a British **artillery** officer who, in 1786 at his own expense, invented hollow shells with a gunpowder filling, fused to explode in the air at selected ranges and shower the target with bullets. This first air-burst **ammunition** was successfully tried in action in 1804 in Surinam; and decisively at the battles of Vittorio (1813) and Waterloo (1815), as the Duke of Wellington acknowledged. But the 'shrapnel' shell was so secret that its inventor was denied both public recognition and substantial reward. Nor did he benefit greatly from his other improved ammunition, gun carriages and artillery-range tables.

Siege engines. No sooner did **wood** and **stone fortifications** become difficult to breach than requirements evolved for demolition and assault machines which were either mobile or built on site and manned by élite technicians. It is unlikely that the Assyrian army was the first to be equipped with battering-rams and fighting-towers from which archers could dominate high battlements and fire flaming arrows to spread **flame** and panic among the besieged. But from *c.* 700 BC onwards bigger and more effective machines, including tension- or torsion-powered catapults and ballistae, were developed by the Greeks and Romans, respectively, to hurl projectiles, rocks or grappling-irons either to smash weapons and breach walls and gates or to help gain a purchase prior to scaling walls from fighting-towers equipped with a drawbridge. These engines were either dragged from place to place, or dismantled and carried by pack animals. Sometimes the lighter types were used in open battle; or mounted in **galleys** for **naval warfare**.

A variation of the catapult was the *c.* 11th-century trebuchet, a very inaccurate, missile engine which could adjust its range by changing a counterweight of up to 10 tonnes. It remained in service until outmatched by **artillery** in efficiency, range and accuracy in the 16th century. In **World War I, armoured fighting vehicles** were added developed as the basis of siege engines of the future.

Siege warfare was initiated by the first primitive man who defended his abode from encircling enemies; and developed in response to improvements to **fortifications**, **siege engines**, weapons technology, **tactics**, **communications** (surface and **signal**) and **logistics**.

It is generally supposed that the Sumerians and Assyrians (*c.* 1000 BC) were forerunners in creating organized ways and means to overcome well-defended places by encirclement, starvation and, if strategically necessary, breaching and assault of the fortifications. Their successors – notably the Greeks and Romans – merely made improvements. However, until the Emperor Charlemagne created a rudimentary logistic system (*c.* AD 800), sieges were never feasible in winter; and often close-run episodes depended upon who, besieged or besiegers, was starved out first. This phenomenon, which still occurs in the 20th century, underlines the crucial logistic factor and shows that a large stationary force tends to exhaust local supplies.

Laying siege is a committal to the most deliberate and costly of military operations. Preparations by defenders and assailants must be carefully planned. The former needs to strengthen defences, lay in adequate stocks and arrange for timely outside relief. The latter must reconnoitre the enemy positions; make logistic provisions; establish a defended base and cut off the objective (perhaps by lines of circumvallation); acquire **engineer** stores; assemble siege engines and begin the breaching operation, which often is undertaken against time.

Wall-and-moat fortresses and towns, until the 15th century when heavy **artillery** made them obsolete, usually were tightly surrounded and the fighting at close range. If walls were deemed unassailable, breaching was sometimes accomplished by **engineers** undermining and collapsing the walls. For economic and chivalric reasons, the more civilized nations gradually introduced codes of conduct to allow garrisons to surrender honourably once a breach was rated assailable; but on penalty of barbaric treatment if a reasonable summons was refused (*see* sieges of Acre, Cherbourg, Constantinople, Harfleur, Jerusalem, **Malta**, Pontoise, Syracuse). **Mongols**, however, were among many peoples who rarely gave quarter.

When heavy artillery began to outrange other pieces, and high walls were replaced by lower-profile, thicker emplacements with bastions and underground shelters, the revolution in siege warfare grew in pace and ingenuity,

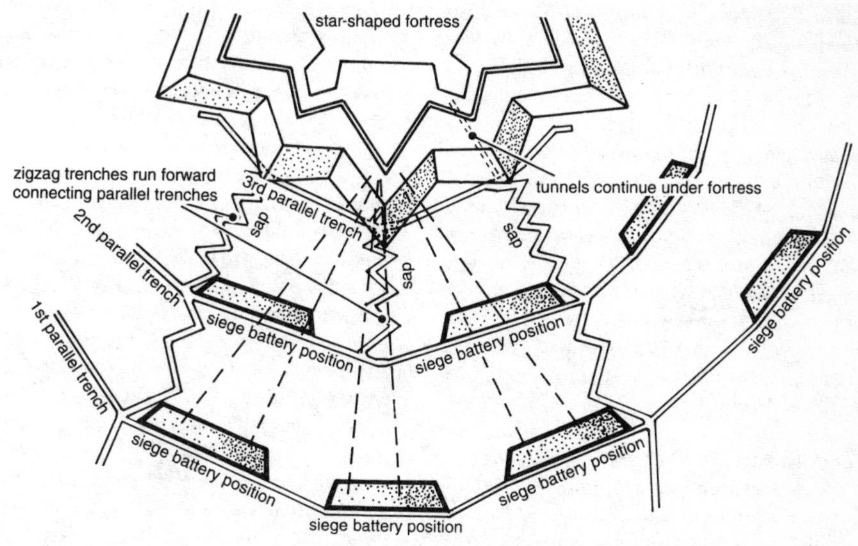

Siege warfare technique against Vauban-type fortress

culminating in Marshal **Vauban**'s star-shaped forts of the 17th century and his siege methods. Moreover, since the **wars of the 17th** and early **18th centuries** included far fewer major battles, sieges often decided the outcome of campaigns. Yet, while the principles and conduct established during the Middle Ages remained the same, the techniques and scale of siege warfare vastly changed.

Perimeter lines of circumvallation and counter-circumvallation (to defend the besiegers against external threats) grew longer, because of the extended range of gun batteries. Batteries had to be protected by earthworks and tasked to give fire-support as engineers and **infantry** dug zigzag saps towards the enemy walls and lateral parallels to connect them. Then the guns would concentrate on suppressing the enemy artillery and opening a selected breach. Meanwhile the defenders would try to delay the process by gunfire and sorties. Sometimes sieges continued for weeks on end through enormous logistic outlay and, as at Lille in 1708, into the winter. (See also the sieges of Toulon and Vienna.)

After the **American Civil War** and **Franco-Prussian War**, star forts were deemed obsolete and replaced by **steel**-and-**concrete** forts which were located at a distance from strategic places; and often connected by entrenched lines which were so well defended by **barbed wire** and **machine-guns** that formal siege operations were needed to overcome them. For examples note **Port Arthur**, **Verdun** and, indeed, examples throughout most of **World War I**. The process caused stalemate until **tanks** acted as siege engines; and, in **World War II**, air transport (*see* sieges of **Kut-al-Amara**, Imphal and Kohima) began to make resupply and reinforcement practical. (For other 20th-century sieges *see* Bataan, Corregidor, Khe Sanh, Leningrad, Malta and Tobruk.)

Siegfried lines. Two of the strongest and most extensive **fortifications** ever built by the Germans were named after the legendary and usually triumphant Teutonic hero Siegfried.

In **World War I**, in 1916/17, they constructed a triple-lined entrenched position, with dense **barbed wire** and a few pillboxes, extending from Arras to Soissons; and withdrew into it in March 1917. Known to the Allies as the Hindenburg Line, it was partially turned by the British at **Arras** in April, and breached in two hours on 20 November during the **tank** Battle of **Cambrai**. Following the collapse of the **Hindenburg** offensive and the Battle of **Amiens** in August 1918, the Germans withdrew once more into the Line. Again, at the end of August, it was penetrated at Arras during a general Allied advance which finally broke through the Line near Cambrai in mid-October as the Germans continued their well-conducted withdrawal.

In 1937, in readiness for **World War II**, **Hitler** ordered construction of another Siegfried Line of concrete-and-steel emplacements from the Swiss frontier to Luxembourg, with lesser works extending along the Belgian and Dutch frontiers. This was not tested until September 1944 when Allied troops reached the vicinity of Aachen. There, from mid-November to mid-December attempts in adverse weather to break into Germany made only marginal progress and were brought to a halt when the Germans launched an offensive in the Ardennes. Between 8 February and 10 March, however, the Line was outflanked by the Canadians and British in the north, and then penetrated with relative ease between Roermond and Trier by American thrusts which outflanked the strong southern sector.

Siemens Company. In 1847 Werner von Siemens (an artillery and telegraphy officer) formed a company to make **telegraph** apparatus and insulated cable. As the firm grew it developed underwater cables, armatures for electric motors, **dynamos** for electric-light enterprises; and, under the influence of Werner's brothers, Karl (Sir William) and Friedrich, electroplating, regenerative condensers for steam engines, and the open-hearth furnace for **steel**-making. In 1866 the firm branched into Britain, where it became British under Sir William and his nephew, Alexander Siemens. It grew into, and remains, one of the world's most important manufacturing and **electronic** concerns. In **World Wars I** and **II** it supplied both sides with vital equipment.

Simulators are **training** aids which replicate and save wear and tear of expensive operational equipment. *Circa* 1900, Captain Percy **Scott** began to make mock-up, classroom models to practise naval gun drills, plus a so-called 'Dotter' and a 'deflection' teacher for gunlaying in a rolling ship. In the 1920s the British Tank Corps fitted mock-up turrets with air rifles, to train gunners to engage targets on an indoor pellet range while rolling, yawing and pitching. And in the 1930s the Miles Link Trainer (an **aircraft** cockpit with similar motion) was developed to teach instrumented 'blind' flying.

The expense of training air crew in the 1950s compelled air forces and airlines to call for flight simulators to reduce in-flight training. Using **electronics**, cathode-ray tubes and **television**, **computers** and **lasers**, machines which realistically simulated most in-flight characteristics, including roll, yaw, pitch, acceleration and deceleration, were devised. They enabled more efficient and safer initial training; followed by subsequent routine checks of aircrew competence without necessarily going aloft. They also contributed significantly to **research and development** (R&D) into new machines.

Similarly, in the mid-1960s, the British developed **armoured fighting vehicle** driving and gunnery simulators (to minimize the firing of costly **ammunition** and to save fuel and track wear). A Solatron, laser-activated **gunnery** and battlefield training system not only made feasible 'dry', unumpired field exercises, down to 'killing' individual **infantrymen**, but also provided a valuable tool in **tactical** R&D (*see* **Tactics**).

Navies also now benefit from on-shore or on-board training in, for example, damage control, **anti-aircraft** and **anti-submarine weapons** and tactical simulations before exercising, expensively, at sea.

Simulators are great time and cost savers but have limitations. It is still essential to practise the real thing.

Slessor, Air Marshal Sir John (1897–1979). An airman with much operational experience in **World War I**, the 1920s and 1930s who, as a Staff officer, contributed significantly to Royal Air Force policy in **World War II**. In 1943 he commanded Coastal Command at the height of the Battle of the **Atlantic** and in 1944 was Deputy Air C-in-C Allied Forces in the Mediterranean. He retired in 1952 when Chief of the Air Staff and was responsible for formulating and publicizing a rational strategy of ultimate **deterrence** based on **nuclear** weapons.

Slim, Field Marshal (Viscount) William (1891–1970). An infantryman who saw service at Gallipoli and in Mesopotamia, France and Belgium in **World War I** and subsequently in India. In **World War II** he commanded a brigade in Eritrea and a division in Iraq and Iran before taking command of I **Burma** Corps in 1942 during the retreat to India. In October 1943 he took command of Fourteenth Army in time to cope with the Japanese offensives in the Arakan, at Imphal and Kohima. He exploited those victories brilliantly with resources that were rarely more than adequate – though he did benefit from substantial tactical air and transport support in the drive past Mandalay to Rangoon in May 1945.

For a short period in 1948 he left the Army to become Deputy Chairman of British Railways, but was recalled as Chief of the Imperial General Staff in November when he found himself poised for instant resignation owing to a clash with the Government. None the less, he survived in the post until 1952 when, as a Field Marshal, he was appointed Governor-General of Australia.

Sluys, Battle of. When, at the start of the **Hundred Years War**, French ships raided the English coast, King **Edward III** accumulated a polyglot fleet of 180 **galleys**, cogs and barges and brought the French fleet, of equal numbers, to battle in Sluys harbour on 24 June 1340. Probably English **artillery** was present but there is no evidence of use. Instead their 12,000 archers slaughtered on deck the packed French, including crossbowmen; and their 4,000 **armoured** men-at-arms boarded for combat as if on land. The French lost all but 24 of their ships, with a very high number of killed and drowned. That day, with immense portent, the British won a naval superiority over the French they were never to lose.

Smoke in warfare. Firearms introduced a special dimension to the battlefield when the simultaneous discharge of large numbers of cannon and muskets blanketed the area in dense clouds of smoke, to the extent that troops were often firing blind into the murk; smoke also exposed otherwise concealed positions. Hence the search, in the latter part of the 19th century, for smokeless powder to be used as a propellant.

Once smokeless propellants were available it became obvious, if it had not been appreciated before, that while 'the fog of war' was a hindrance to good shooting, it could be the soldier's friend in concealing his own tactical movement. Paradoxically, therefore, the need arose during **World War I** for screening smoke. The task was not as easy as it might appear. Though there are a number of chemical substances which, when ignited or exposed to the atmosphere, will give off clouds of smoke, the problem lies in building up a screen of sufficient density quickly enough. There are two main substances used: white phosphorus (WP) and hexachlorethane (HCE).

WP has the advantage that it produces a cloud of very dense white smoke almost instantaneously; ideal for a short-term local screen. However, it dissipates quickly, pillaring into the air; furthermore WP has unpleasant antipersonnel burning effects which can affect friendly forces. HCE, which is not harmful, tends to cling to the ground, can be built up into a long-lasting screen, but takes time to develop. It is usually contained in several small canisters carried in artillery shell, bursting on impact.

One of the original ideas for smoke shell was a filling of 'noxious gases' (*see* **Chemical warfare**), intended to make an enemy position untenable. The need to deal with the armoured personnel carrier (*see* **Armoured fighting vehicles**) has resurrected this idea, since conventional anti-armour rounds may well immobilize the vehicle but leave the infantry section within comparatively unscathed. A round has been designed which will penetrate the crew compartment and, once within, release a smoke agent, forcing the crew to dismount where they can be neutralized by small-arms fire.

In war at sea the introduction of the **coal**-fired steam engine immediately produced concealment problems for ships otherwise hidden below the horizon. Careful trimming of the furnaces helped to conceal the tell-tale smoke and some ships were provided with 'smoke boxes', to which exhaust smoke could be diverted for short periods, then released to atmosphere later. At other times exhaust smoke from the funnels could be deliberately increased to provide tactical concealment. In a modern diesel-engined ship the order 'make smoke' can be met by passing unburnt diesel fuel over the hot exhaust, a similar device for armoured vehicles having been pioneered by Russia and more recently copied in the West – wasteful of fuel, but producing rapid, effective concealment.

Smoke can thus be a friend or enemy on the battlefield, but in either case always subject to the vagaries of wind and weather.

Smuts, Field Marshal Jan (1870–1950). A great South African statesman who worked for Anglo-Boer understanding but during the Boer War led a Commando against the British. Before **World War I** he was Minister of Defence and helped General Louis Botha eject the Germans from South-West Africa in 1915. In 1916 he was C-in-C in East Africa, but in March 1917 joined the War Cabinet in London to play a strong role in creating the Royal Air Force, among other valuable services. From 1919 to 1924 he was Prime Minister of South Africa and, as Deputy Prime Minister in 1939, defeated the anti-British Prime Minister, J. B. Hertzog, over South Africa's neutrality. Then, as Prime Minister and C-in-C the Armed Forces, he carried the divided nation into **World War**

II, in which capacities he travelled the war zones and contributed notably to British and Allied War Councils.

Solomon Sea, Battles of the. The Japanese presence on Guadalcanal island and throughout the Solomons in May 1942, along with the Battle of the **Coral Sea**, denoted a turning-point in the **Pacific Ocean War**. Within two months the Allied counter-offensive had opened on Guadalcanal, forcing the Japanese to commit major forces to support their expanded empire's perimeter. In and around the 'Slot' – the central channel among the islands – intense naval and air battles took place to sustain the fight for Guadalcanal. Admiral **Fletcher**'s decision on 7 August to withdraw the **aircraft-carriers**, for fear of loss, gave the Japanese an advantage which was enhanced on the 9th at the Battle of Savo island when four out of eight Allied **cruisers** (one of them Australian) and a **destroyer** were sunk; only one Japanese cruiser went down. This forced the **amphibious** ships to retire from Guadalcanal. Yet the Japanese missed this opportunity to reinforce Guadalcanal until 22 August, by which time Fletcher's three carriers had returned to engage three Japanese carriers under Admiral Nobutake Kondo. At the subsequent Battle of the Eastern Solomons, one American carrier was damaged. But although the Japanese lost a light carrier, a destroyer and 90 aircraft, and had a seaplane carrier damaged, they again failed to press home their attack with **battleships** and cruisers.

On 31 August the Americans had another carrier damaged by a submarine and, on 15 September, lost the carrier *Wasp*, depriving them again of carrier support off Guadalcanal and also that of the battleship *North Carolina*, which was badly damaged by a

The Solomon Islands and southern approaches

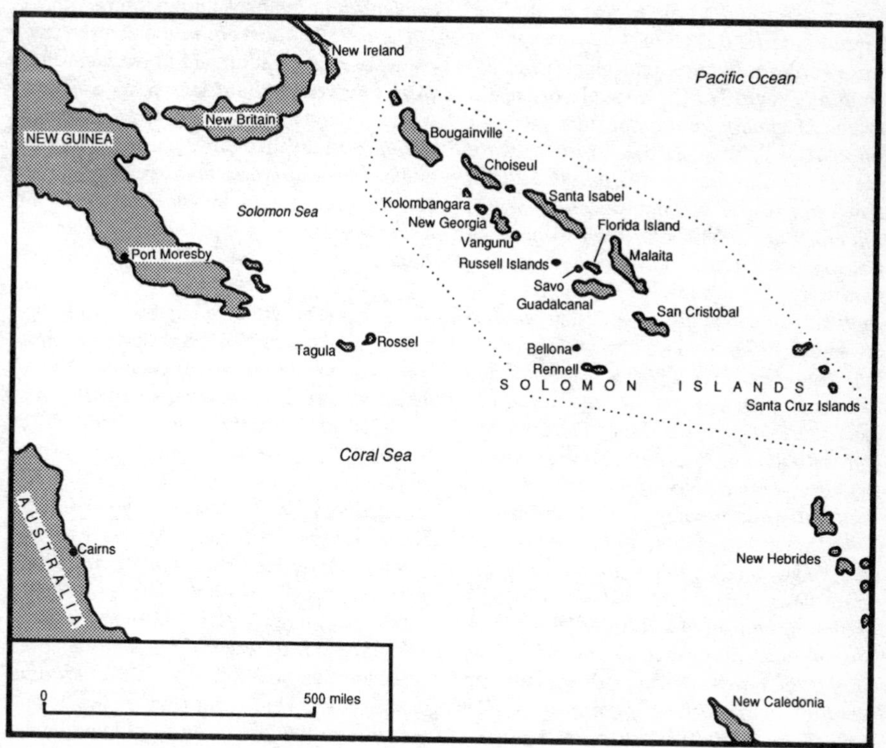

submarine's **torpedoes**. Yet from 11 to 13 October at the Battle of Cape Esperanto, both sides simultaneously reinforced Guadalcanal in a remarkable series of events in which the Japanese lost or had damaged, in a night action, three cruisers and three destroyers. They still managed, under Admiral Raizo Tanaka, to take control of the Slot on the 13th and 15th and, with two battleships supported by cruisers and destroyers, bombard Henderson Field on Guadalcanal.

On 18 October Admiral **Halsey** took command in the area and Admiral Thomas Kincaid, now with two carriers, relieved Fletcher. This change almost coincided with an attempt by Admiral **Yamamoto** to have Admiral Kondo win a decisive victory against the weakened Americans. At the Battle of Santa Cruz Islands on 26/27 October Kondo nearly succeeded, sinking one and damaging the other American carrier, though suffering damage to two of his four carriers from simultaneous attacks by the Americans. But Kondo also lost 100 aircraft and withdrew instead of finishing off the remaining American carrier, as well he might have.

On 12 November the struggle reached a climax of almost continuous encounters in three days' fighting as the Japanese strove desperately to reinforce Guadalcanal. With only five cruisers and eight destroyers, Rear-Admiral Daniel Callaghan tried to head off Kondo's two battleships, two cruisers and 14 destroyers in a tumultuous night battle at short range. Kondo lost a battleship and two cruisers (with all other ships damaged) and Callaghan (who was killed) two cruisers and four destroyers, plus all but one of the rest damaged. Day and night the slaughter continued as Kondo forced through his **convoy** with the loss of seven transports. But when Kincaid sent in two battleships (one of which was instantly put out of action), the *Washington*, using **radar**, fought 14 ships, left one battleship and a destroyer sinking – and conclusively secured American dominance of the Slot. For although the Japanese charged again on 30 November (when they sank an American cruiser off Tassafaronga), on 29 January they declined battle off Rennell's Island, 10 days before successfully evacuating Guadalcanal.

Somme, Battle of the. To take pressure off the French Army at **Verdun**, the British Army, with French support, launched an offensive against the Germans on 24 June 1916 on either side of the River Somme. Like so many attacks in **World War I**, this one by General **Haig** was bent on attrition with no strategic objective. After seven days' bombardment by **artillery**, which lacked sufficient medium and heavy guns and whose **ammunition** was 25 per cent defective, the assault went in on 1 July on a 20-mile front. It made little progress and cost the British alone 60,000 casualties. It continued until 13 November, crawling forward only eight miles and costing the British 420,000 casualties, the French 195,000 and the Germans 650,000. But though, once more, artillery and **infantry** failed to open a gap for the **cavalry**, the surprise use of 32 **tanks** on 15 September offered hope for the future.

The battle did succeed in making the Germans desist at Verdun, and it also compelled them to revert to the defensive on all fronts when Generals **Hindenburg** and **Ludendorff** recognized that their Army was demoralized and had lost the cream of its combat leaders. At the same time the British painfully learnt lessons which they applied in subsequent battles.

Sonar. An acronym derived from *so*und *na*vigation and *r*anging, a piezoelectric system using ultrasonic, **electromagnetic** waves to detect underwater objects, particularly **submarines**. Invented by the British and French in 1918, it directs short pulses towards an object (and, of course, the seabed) which are reflected and then detected, the lapse of time being measured to give distance to the object. It was not ready for use in **World War I**, however.

Developed during the 1920s and 1930s and called ASDIC (after the *A*nti-*S*ubmarine *D*etection *I*nvestigation *C*ommittee) by the British, it was assumed (in insufficiently exhaustive trials which, even so, produced only

50 per cent detections) to provide the answer to the submarine menace. This complacency deflected attention from submarine warfare and **anti-submarine weapon** systems, until experience by British **destroyers** hunting submarines during the **Spanish Civil War** discovered that all was not well. The troubles were multifarious. Sometimes the pulse would reflect from debris or a shoal of fish, placing demands upon the interpretative skill of the operator – or a submarine might escape detection from the pulse altogether by diving below different temperature levels. This meant that, even after a firm contact had been made, an astute submarine commander could evade the enemy by skilful manoeuvres. On the other hand, the click heard by submarine crews when the pulse made contact on the hull could have a demoralizing and deterrent effect, especially when followed by the crash of nearby **depth charges**.

As **World War II** progressed, improved sonar equipment was produced and integrated with more effective weapon systems and improved tracking techniques to deal with deeper-diving submarines; in April 1943 the British developed an Asdic which could, in some circumstances, maintain contact with a deeply submerged boat; and Asdic Type 147B could measure a submarine's depth – a capability which made it worth fitting sonar in submarines. Then in February 1945 a British boat, using passive sonar, detected and plotted an enemy boat and struck it with a straight-running torpedo (noting that when a submarine uses active sonar to detect its prey, it also advertises its own presence).

Since World War II intensive efforts, in company with **hydrophones** to hear sounds, have been made to improve sonar performance against deep-diving, quiet-running submarines – and also, incidentally, to detect **mines**. The Doppler effect's change of note has helped track moving submarines – and also whales and shoals of fish. But more powerful and sensitive sets are not enough.

It has been found expedient to use **aircraft** far more; for example, by dropping patterns of sonar buoys to detect, fix and transmit to the aircraft or a ship sufficient data to give the exact location of a boat which happens to be beyond the range of ship-borne instruments; or by employing hovering **helicopters** with a 'dipping sonar', a device which is winched below the temperature layers to improve the chances of a contact and send back sufficient information, via **computers**, upon which to base an accurate attack. Sonar is not infallible; just another important aid among the many weapon systems of anti-submarine forces in a never-ending contest of technologies and techniques.

Sopwith, Sir Thomas (1888–1990). A self-taught pilot in 1910 who, in 1912, founded the Sopwith Aviation Company and in 1914 designed a seaplane which won the Schneider Trophy. During **World War I** the company specialized in biplane **fighter** aircraft, including the Pup and Camel, as well as the Triplane fighter and Cuckoo **torpedo bomber**. Between the wars he took control of the Hawker-Siddeley Group which produced, among many machines, the biplane Fury and the monoplane Hurricane and Typhoon fighters, designed by Sir Sydney Camm. After **World War II** the firm made the Hunter and the Vertical Take-Off and Landing (VTOL) Harrier fighters.

Sound locators. Rather in the manner of a cupped ear or an ear trumpet, the sound locators which were developed by the British in 1915 to give warning of the approach of enemy **airships** and **aircraft** were means to amplify sound. But the binaural effect of double horn locators, some with microphones incorporated, also gave better indication than the human ear of the direction in which an object was travelling, thus making it possible for **searchlights** to find and illuminate targets for anti-aircraft guns to engage.

During **World War I** sound locators often found targets, although they could as easily be friend as foe. But their effectiveness diminished in the 1930s owing to greater aircraft speeds and altitudes. It was indeed fortunate for air defences when, by 1942, **radar** had made them obsolete.

Two or more microphones of low-frequency resonance, if surveyed into a base-line, can locate hostile **artillery**, as a prelude to counter-battery fire, by differentiating between the sequential noises of a gun firing and a shell in flight and when bursting. The results of detection, when transmitted by cable or radio to a control centre, make ascertainment of the gun's position possible with reasonable accuracy, dependent upon wind, temperature and humidity; and provided activity is not intense. This passive method, pioneered in **World War I**, is still in use, though being superseded by active radar which is more accurate and much quicker to obtain results. (*See also* **Hydrophone**.)

Space vehicles. The German development, led by Dr Werner von **Braun**, of **rockets and guided missiles**, ensured during **World War II** the development of space travel in special vehicles. When, in 1957, the Russians, under the urge of Sergei **Korolev**, put a 184lb sphere called Sputnik I into earth orbit of between 143 and 584 miles, they initiated fierce competition between the superpowers and a new sort of military contest. Indeed, the vast majority of the innumerable vehicles sent into space since have been of military use. They fall into three categories: those orbiting the earth at altitudes below 22,300 miles; those which are in geostationary orbit at that height; and those which escape the earth's gravity and fly into deep space. It is the first two with which this entry is mainly concerned.

Military space vehicles have many functions: **communications**, **surveillance**, weather forecasting, **navigation** and use as missile platforms. From the start **radio** signals linked vehicles to earth for monitoring and control. It was only a matter of time, therefore, before messages and, in 1961, **television** pictures, would be relayed from one part of the earth to another via a satellite vehicle. The technique was rapidly enhanced by **electronic** discoveries and **computers** which enabled solar-powered, geostationary vehicles to provide continuous, multi-channel communications worldwide, or to eavesdrop on

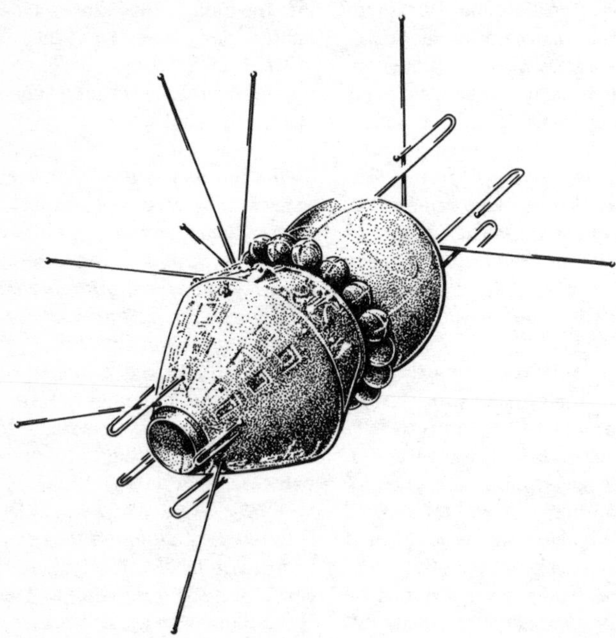

Space vehicle (Soviet Vostok 1*)*

electronic communications not normally receivable by any other method. **Cameras** (*see* **Photography**) were also set up at an early stage to bring back or transmit pictures of the earth's surface – with significant **reconnaissance** and **charts and maps** potential. Not only was it possible for orbiting satellites to obtain detailed pictures for **intelligence** purposes, far more accurate charts and maps also could be drawn which made possible precise navigation as well as pinpoint delivery of missiles to their targets. The process is amplified by the use of **sensors** to detect **electronic** radiations from an almost unlimited range of sources. This facility made concealment and *electronic countermeasures* (ECM) more vital and difficult than ever.

American space vehicles began to be used in combat situations for weather forecasting in 1965 during the **Vietnam War**. In the 1973 Yom Kippur War (**Arab–Israeli wars**) Russian surveillance satellites kept the Egyptians informed of the extent of the Israeli penetration across the Suez Canal. In 1982, America diverted orbiting satellites to assist the British during the **Falklands** War. There have been other occasions when these vehicles have been of immense use in providing intelligence to a client, but they are of far greater **deterrence** value since they help reduce the possibility of military surprise by such secretive nations as Russia, which has traditionally forbidden foreign access to its lands – not just sensitive localities. Equipped with cameras and sensors satellites are constantly on watch for signs of unusual and threatening activity.

To help achieve maximum effect from missiles it might be desirable to destroy space vehicles that can detect their launch. Anti-satellite vehicles and weapon systems thus have a part to play and are being developed – and might also have to deal with weapon systems aboard orbiting space vehicles. It would also be desirable to eliminate or disrupt enemy communications and surveillance systems. Such attacks could not be performed in isolation, they would break treaties and constitute acts of war, with the same consequences as violations in international waters or land invasion. (*See also* **Strategic Arms Limitation Treaties, Strategic Defence Initiative** (SDI) and **Technology, Effects of.**)

In the aftermath of the **Cold War** the reduction of tension marginally retarded development of space vehicles. The Russians still man the Salyut space stations which were initiated in the late 1960s after their Moon programme faltered owing to rocket failures. The Americans still operate space shuttles for commercial as well as military missions – such as those which played such an important role during the **UN–Iraq War**.

Spanish Civil War. The fall of the Spanish Monarchy in 1930 and the divisions between the Left and Right political groups, plus various provincial demands for autonomy which frequently lapsed into violence, by the summer of 1936 had made Spain almost ungovernable. On 13 July an army mutiny, under General Franco in Morocco, provided the spark for a civil war throughout the mainland. The contest rapidly resolved itself into a fight between Franco's Nationalist Fascist forces (supported by Germany and Italy) and the Communist-inspired government forces (supported by Russia and Mexico and left-wing enthusiasts).

From its initial power bases in the south and north-west (less the Basque province in the north), the Nationalists developed a strategy aimed at seizing **Madrid** and clearing the country from west to east. With the cream of the Army on his side and the so-called **fifth column** working underground in **guerrilla warfare** Franco was able in August to make rapid progress. The Nationalists seized Badajoz (to link up their western holdings) and advanced to within striking distance of Madrid, which was besieged in November. Soon, too, the Basques were driven into Bilbao, which was besieged on 1 April 1937, and chased out of Guernica after a heavy tactical **bombing** attack on the 25th in support of ground forces, which Russian **propaganda** brilliantly exploited as an example of typical Fascist bestiality. Attempts by the League of Nations to contain the war went on against a background of manpower and

logistic assistance by the Fascist and Communist powers. Both sides attempted **blockade** to cut off the import of equipment and supplies, but neither had naval or air forces capable of the task. Even so, there were many incidents involving British, French and American warships (of the International Patrol) which were trying to protect neutral merchant ships (many of which were sunk or damaged) against Spanish, German and Italian **aircraft** and warships, including **submarines.**

By 1937 the Germans and Italians had substantial ground and air forces involved (including the **Condor Legion**). By then too the Government had also been reinforced by modern Russian equipment and sympathetic international volunteers for the fight against Fascism, and that year Nationalist expansion was largely contained. Two Italian divisions made a very poor showing in the Battle of Guadalajara in March. Madrid held out, although Bilbao fell in June and the Government offensives with Russian **tanks** and **aircraft** near Madrid in July, and in Aragón in August and September, merely prolonged a stalemate. The situation partially loosened after the Nationalists checked a Government attack at Teruel early in 1938 and, starting in February, drove the enemy back towards Tarragona and Barcelona. The victory was denied conclusive rewards when the Government counter-attacked along the River Ebro at the end of July to stabilize the front.

But the Government forces were suffering from the same innumerable differences of opinion within their ranks that had made them vulnerable in 1936. When the Nationalists attacked again towards Tarragona and Barcelona on 23 December, this vital front began to disintegrate along with Government **morale** everywhere. Barcelona fell on 26 January 1939. When Madrid fell on 28 March it was the signal for the Government to lay down its arms next day.

The Spanish Civil War, 1936–9

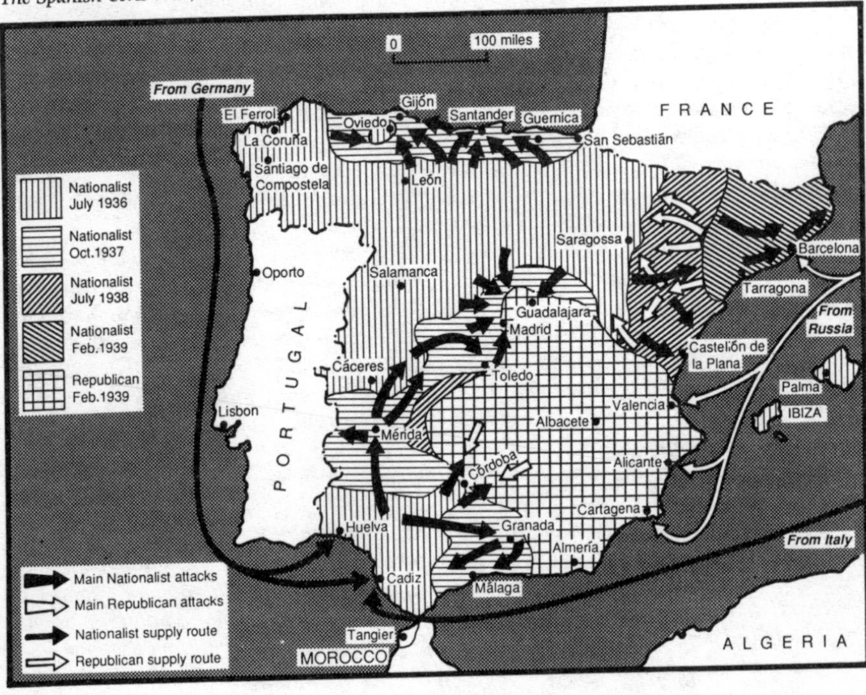

Various estimates put total losses from battle, air raids and political executions at close on one million. The country was ruined in what, for the military of the major Fascist and Communist powers, was treated as a trial for some of their latest equipments, though by no means all their techniques. For example **armoured** forces were not properly employed in the deep penetration role, and **strategic bombing** was not deliberately practised. In the main, these 'new' arms used embellished **World War I** tactics, without revealing their true potential or flaws.

Spanish wars, 1495–1603. Once the last Muslim strongholds in Spain were conquered and Christopher Columbus began his first voyage of exploration of the Americas in a 100-tonne caravel in 1492, Spanish expansion and acquisition of wealth gathered great momentum. The result was that she became embroiled with the French, British and Dutch in dynastic and trade wars in the Mediterranean and the Netherlands.

Employing European **tactics**, **horses**, weapons and **armour** (*see* **Wars of the 16th century**) against the far less sophisticated American 'Indian' peoples, the Spaniards' technical superiority enabled commanders such as Hernando Cortés and Francisco Pizarro to conquer and pillage as they pleased. Come 1585 they had, through sea-power, won the lion's share of the Americas' treasure trade, wealth which was used for aggrandisement, further expansion in Europe and to contest the resurgence of **Muslim expansion** which began in 1520 and at times threatened the Iberian peninsula. In the 1570s, after the Battle of **Lepanto**, when 81 of her obsolescent **galleys** fought hard for Christianity, Spain's influence reached its peak.

The Spanish expansion, 1492–1588

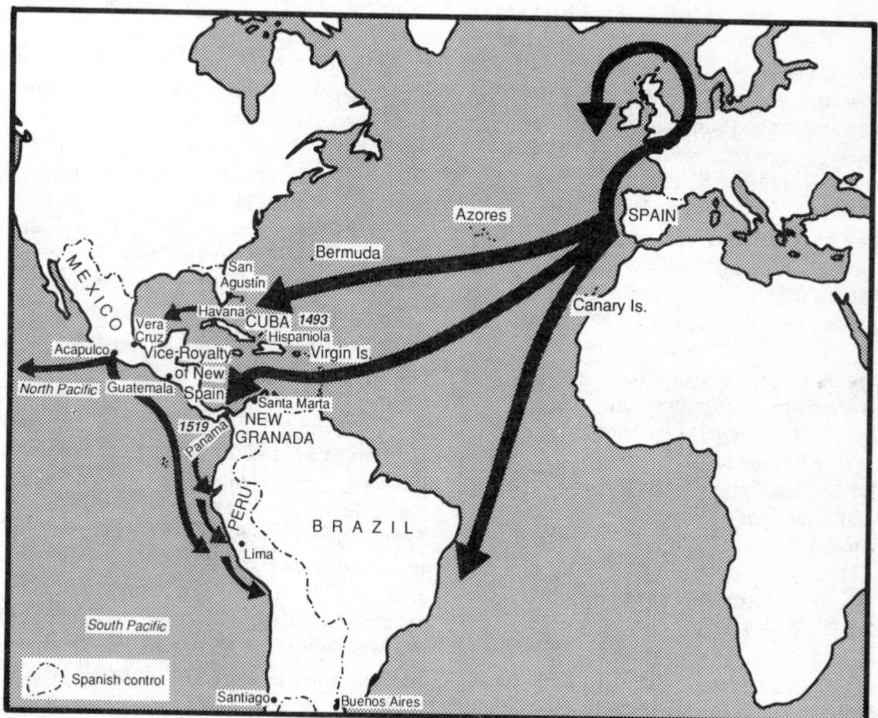

Then ever-increasing military commitments, including the Netherlands' rebellion, which festered from 1567 to 1609, and the expense of modernizing the fleet to match the nimble British **galleons** (which in 1577, under Sir Francis Drake, began preying on the Spanish colonies and treasure ships), placed Spain under intolerable strain. The year 1585 was a turning-point, when Protestant Britain began supporting the Dutch in the Netherlands and King Philip II decided to launch an invasion of England, escorted from the Netherlands by a great Armada.

This plan was thwarted in 1587 after Drake destroyed 33 Spanish ships in a raid on Cádiz, but revived in 1588 when 20 great galleons (of between 700 and 1,000 tonnes), 4 galleys, 4 galleasses and 44 armed merchantmen, plus 58 other vessels, sailed from Spain and entered the English Channel on 19 July (English calendar). Engaged by upwards of 100 British ships of between 200 and 1,100 tonnes, it fought its way to Calais, where it anchored on the 26th to learn that the Army was not ready. Driven to sea in panic on the 28th by fireships and harried by the British until they ran out of **ammunition**, the Armada scattered before the wind, its ships buffeted and wrecked by storms on passage round the British Isles until only 45 survivors reached home.

As Spain entered slow decline, a **naval warfare** revolution founded on more seaworthy, heavily armed 'wooden walls', enabled Britain to prosper, build an Empire and take part in the European wars of **Religion**.

Spears are two-edged, thrusting or throwing polearms of the greatest antiquity and used mainly, like the **pike**, by **infantry** to keep an enemy beyond arms' reach in close combat. *Javelins* are simply lighter, throwing versions, while the longer *lance* is usually carried by **cavalry**. In modern armies they all, belatedly, were obsolete in **World War I**. But primitive infantry sometimes resort to them for lack of anything better.

Special forces. The history of war is littered with élite units formed to carry out specialist roles requiring outstanding prowess and a high standard of training. More often than not they have been recruited from volunteers of good intelligence and strong physique. In this entry only uniformed units are considered, although even these have been known to operate in plain clothes on the fringes of the **laws of war**. The specially trained, green-uniformed **infantry** riflemen, who in the later **wars of the 18th century** were employed as marksmen and skirmishers, may be said to have entered this category, bordering on the role of guerrillas.

The action by Major John Norton-Griffiths and his batman in organizing the destruction, with civilian assistance, of the Romanian oilfields in November 1916 may be considered a special forces epic – just. Such acts were rare in **World War I**, when **guerilla warfare** was unusual, but not infrequent between the wars and, for example, during the **Spanish Civil War** by members of the **fifth column** and Communist groups.

In its modern sense the term may be said to originate with the use by the German military intelligence service (*Abwehr*) of the Brandenburg Regiment for raiding and counter-guerrilla tasks. But the Germans were never enthusiastic raiders, leaving it to the British with their formation in July 1940 of the **Commandos** to develop hit-and-run attacks against the enemy coast. From their ranks emerged, within a few weeks, **airborne forces** and a Special Boat Section to operate from folding canoes. The latter would expand into Special Boat Squadrons (each known as SBS) which, like Commandos, still are part of the Royal **Marines**; Combined Operations Pilotage Parties (COPP), whose role in canoes or other small craft was beach **reconnaissance** and the guidance of **amphibious** forces to their landing-places; and such canoe demolition parties as the Royal Marine Boom Patrol Detachment (RMBPD) and 14 (Arctic) Commando which, like the SBS, also were experts at attaching limpet mines to enemy ships. Also derived from the Commandos was the Special Air Service (SAS), along with units like Popski's Private Army, which began life in **North Africa** in 1941 for desert raiding alongside the Long Range Desert Group (LRDG), whose primary role was reconnaissance behind the enemy lines in the **desert** but who readily adapted to other environments. The latter were both disbanded after **World War II**, along with Army Commandos, but the SAS was re-formed as a very special, raiding élite in **cold** and limited **war** operations, as well as for stay-behind parties and, in many celebrated incidents, armed aid to the civil power when a terrorist situation got beyond police capability.

There have been many imitators. In World War II the Americans formed Marine Raider and Army Ranger units which copied and, to begin with, were trained by the British Commandos. Within the Office of Strategic Services (OSS) they also formed uniformed Operational Groups which helped organize guerrilla parties behind the enemy lines in Europe under British Special Operations Executive (SOE) directions. Many governments have raised SAS-type units for counter-terrorist actions that require extremely sophisticated weapons and techniques.

In 1952 the Americans formed Army Special Forces on Operational Group lines to work with guerrillas behind the enemy lines. In 1957 they were introduced into **Vietnam** to organize civilian groups against North Vietnamese infiltration into the South. They developed and expanded this role throughout the war until, at peak, they were controlling 42,000 members of Civilian Irregular Defence Groups (CIDG), besides forming South Vietnam's special forces. In 1961 President John Kennedy authorized them to wear a green beret, an honour they were to justify in many other areas of conflict besides Vietnam.

Russia and her allies also field special forces, generally known as Spetsnaz, whose role is raiding and the support of subversive elements in enemy territory. They were used for counter-insurgency work in **Afghanistan** and are fully expected to be in the lead of any major Russian operation of the future.

In the **Falklands** War of 1982 and the **UN–Iraq War** special forces performed important reconnaissance and sabotage tasks.

Speer, Dr Albert (1905–81), was **Hitler**'s personal architect with responsibilities for prestigious German building projects. In February 1942 he was made head of the state Todt Construction Organization and soon afterwards also Minister for Armaments and War Production. Remarkably he rapidly rationalized much of the cumbersome German industrial effort and, despite strategic **bombing**, raised production enormously, besides investing heavily in **research and development** of **aircraft**, **armoured fighting vehicles**, **rockets and guided missiles**, **submarines** and a Pandora's Box of other weapons which, during the **Cold War**, were developed by the Allies. Found guilty of war crimes, he served 19 of a 20-year sentence.

Staffs, Military. Naval and military commanders have always, in peace and war and in one form or another, required staff officers to help them organize, control, train and arrange **logistics**. Normally the number and quality of officers naturally are governed by the size and complexity of the units and formations under their command. As a general rule, therefore, numbers were small in proportion to force size until the growth of technology (*see* **Technology, Effects of**) and the **recruitment** of masses created highly complex organizations and operational environments. Commanders such as Xerxes, Julius **Caesar**, **Edward III**, **Genghis Khan**, **Blake** and **Marlborough** made do with a handful of picked and experienced deputies and liaison officers because their subordinate organizations were few, frontages usually in full view, weapons basic and logistics, as often as not, decentralized and hand-to-mouth.

A demand for more formal arrangements and an élite of specially trained staff officers emerged during the **wars of the 17th century**. War academies, like the one set up in Holland in 1697 by John of Nassau, became essential and in the 19th century, as a result of the War of the **French Revolution**, had to be developed into specialized staff colleges (*see* **Education, Military**). These were followed by the setting-up by General Helmuth von **Moltke** of the Great Prussian General

Staff, composed of most carefully selected staff officers of outstanding character, intellect, initiative and high moral tone. Its officers were trained to express themselves fearlessly and control subordinate formations and units through the issue of general directives rather than detailed orders. The system which called for precise procedures linked to carefully studied doctrine, became the model for all staffs to follow. It applied not only to armies but also to navies (the US Naval War College was the first, in 1884) and, in due course, to air forces.

The increasing complexity of technology and logistics, after 1850, naturally imposed far heavier demands on staffs. To begin with they tended to be slow in adapting to advances in science, which required them either to become themselves strongly aware technically, or to welcome specialist technical staff officers within their cult. For example, the German General Staff, for all its reputation, did not come fully to grips with the characteristics of the latest **communications** systems and **mechanization**. As a result, Germany lost the Battle of the **Marne** in 1914. Nor did the opening of Technical Staff Colleges prior to **World War II** close the divide between General Staff officers and the technical officers whom they disdained, although commanders and staff officers such as Field Marshal **Kesselring**. General **Guderian** and Air Marshal **Dowding** were notable exceptions.

World War II also underlined the need, already realized, for joint and combined staffs. The more it became necessary for sea, land and air forces, as well as allies, to work together under one commander, the more he required harmonized staffs, regardless of service, arm or nationality. The supreme example of this was the Allied Expeditionary Force which, under General Eisenhower, invaded Europe in 1944, controlled by an integrated, international staff. It was this organization which began to develop the standard operating procedures that **Nato**'s international staff system employs today, largely using English for its highly diverse communication system.

A modern formation staff is usually composed of five branches which often are titled by the letters G or S and a number. They are coordinated by the most senior staff officer or a Chief of Staff. In Nato, for example, responsibilities are as follows: (G1) personnel, discipline, awards and ceremonial; (G2) intelligence; (G3) coordination, establishments and operations; (G4) logistics; and (G5) civil affairs.

At all levels branches, which have access to arms, technical and logistic advisers, most of whose heads have access to the Commander, are staffed by graded officers. The 3rd grade is often a captain who deals with routine, day-to-day matters; the 2nd grade a major who deals with policy, planning and future events; the 1st grade a lieutenant-colonel who coordinates in consultation with the Chief of Staff or Commander while thinking well ahead.

Stalin, Josef (1879–1953). A Georgian who came to note as agitator, **propagandist** and **guerrilla** leader in the Russian Revolution of 1905 and as a prominent politico-military leader during the Russian revolutionary wars. In 1917 Vladimir Lenin made him People's Commissar with a seat on the Military Council under Leon Trotsky, who used him as an Inspector of the Army in a trouble-shooting role which included organizing the defence of Tsaritsyn (Stalingrad) in 1918, Petrograd (Leningrad) in 1919, and in 1920 the Polish Campaign and the defeat of General Petr Wrangel's White Army. In 1924 he succeeded Lenin to become dictator and generalissimo of Soviet Russia, a position he used to reorganize and modernize the armed forces, with emphasis on the Army and Air arm.

An utterly ruthless person, he purged the nation and armed forces so thoroughly of their best leaders in the 1930s that, when the threat of war with Japan and Germany became unmistakable, all were in a deplorable state of fear, inefficiency and demoralization. The fumbled Finland Campaign of 1939/40 exposed deficiencies which could not be rectified in time for the German invasion of June 1941. This was one reason for Stalin's desperate attempts to appease **Hitler**, his refusal to adopt defence measures until the last moment, and thus a cause of the Russian débâcle. During the Russo-German War he overcentralized command, although gradually growing to trust and delegate more authority to selected generals, above all **Zhukov**.

Politically, Stalin won immense post-war gains for Russia in the 'liberated' territories, a traditionally Russian expansionist policy he pursued with vigour to bring about the **Cold War** and such incidents as the **Berlin** siege in 1948–9 that stiffened Western resistance.

Stealth. In combat it is often fundamentally desirable to avoid being heard or seen. Until the introduction of engines (*see* **Propulsion, Means of**) in the 19th century this was achieved by, for example, the muffling of oars, hooves and footwear and by camouflage when in the vicinity of the enemy. Engines could be muffled by silencers but never enough to prevent detection by **sensor**, especially those of powered **aircraft**, by **sound locators**. Indeed, with the arrival of **radar** in the 1930s, aircraft needed ways of avoiding detection or presenting a delusive image on visual display units (VDU); the dispensing from aircraft in 1943 of metallic chaff went part way to meet this requirement.

Research showed that, at a price, low observability could be built into the configuration of **reconnaissance aircraft** such as the Lockheed U2 and SR71 of the 1950s and 1960s. This method evolved into very costly projects in the 1970s and 1980s to shield heat sources (such as engines) and reduce **electronic** signatures, as well as to design special shapes to disguise size. The first combat examples of this were the Lockheed F-117A stealth fighters which, during the **UN–Iraq War**, flew 1,300 sorties without loss.

Stealth is also being built into surface warships, **submarines**, **rockets and guided missiles** and **armoured fighting vehicles**, especially by the shielding of heat sources, although with only marginal chance of success with absorbent materials against **thermal imagers**. Countermeasures to stealth have yet, indeed, publicly to disclose their full potential. It is not inconceivable that a combination of multi-**sensors**, **lasers** and **computers** could cancel out the highly expensive outlay on a technology which is, after all, only a tactical and not absolutely protective aid. At present, however, stealth does provide a distinct and, in some respects, decisive advantage; and has the potential to be retrofitted into existing weapon systems.

Steel. It is a fairly well-established metallurgical fact that steel was discovered *c.* 1200 BC through the process of refining **iron** and that the earlier heating and hammering ways were too difficult and expensive to produce an affordable metal in quantity. Furthermore, even when the first blast furnace was invented in AD 1380, in response to the demand for improved armour and weapons, steel – above all hardened steel – remained a comparatively rare metal reserved for the rich.

Ways to improve refinement, including oxidizing, made slow progress until Henry **Bessemer** invented the converter in 1855. By blasting air through melted pig-iron the metal was rapidly purified to create mild steel ready for forging or milling. By comparison with existing methods, this raised production and reduced costs by enormous margins – and stimulated competition. A year later the **Siemens Company** was first among many producers to devise production of cheap steel without infringing Bessemer's patents. Now the demand for improved weapons, **armour**, machinery and buildings encouraged metallurgical **research and development** with steel alloys, of which Robert **Hadfield**'s manganese-armoured steel of 1882 was among the most important. Today tungsten, nickel, molybdenum, boron, chromium (to resist rust), silicon, carbon, **aluminium** and **titanium** feature among the principal materials employed to satisfy the ceaseless demands of technologists and manufacturers, who also have introduced many new methods of heat, chemical and electrolytic treatment to cut costs, raise production and meet specifications related to extremes of temperature, oxidation, high stresses and so on.

Stephenson, George (1781–1848) and **Robert** (1803–59). George was a watchmaker who, in 1812, made the first practical miner's safety lamp. Work on pit-pumping engines led to his production, *c.* 1813, of a steam locomotive to pull **railway** wagons; and, in 1825, *Locomotion* to pull the first passenger train. Henceforward he was the foremost locomotive maker and the chief engineer of many railway companies during the period of 'railway mania' in the 1840s.

His son, Robert, was a trained engineer whose many inventions improved the efficiency and speed of locomotives; but who specialized in the construction of tunnels and cuttings and, above all, **bridges** of the tubular-girder type.

Stone and clay always have provided essential materials for **fortifications** and roads. For economic reasons stone usually has been obtained for local use; but if unavailable, clay *bricks*, since at least 3000 BC, have been baked as substitutes. *Cement*, first made from clay and later from lime, gypsum and water, was developed for bonding stone by the Assyrians, the Egyptians and Greeks. But it was the Romans who, by mixing slaked lime with volcanic ash and water, produced 'cement rock' *concrete*. In fact, the word concrete was not coined until 1635, nor patented by Joseph Aspdin, as finely ground Portland Cement, until 1824 – 30 years after others had invented similar cements. But from then on cement, mixed with gravel, sand and crushed stone, evolved as structural concrete. And, in the 1890s, it was *reinforced* by **steel** bars to create a strong material for innumerable construction purposes.

Strategic Arms Limitation Treaties (SALT). In 1968 the nuclear powers, less France and China, signed the Non-Proliferation Treaty to which more than 100 other nations added their signatures. However, Israel, India, Pakistan and South Africa, which either had aspirations to possess or were on the verge of possessing nuclear weapons, abstained – a familiar pattern in **disarmament** matters. After this damp squib came talks from 1969 to 1979 between the nuclear powers, but chiefly the USA and the USSR, which produced the SALT 1 and 2 agreements designed to limit and control nuclear weapons. But by then a fresh arms race was in progress involving deployment of the latest Russian missiles, which led to fresh suspicions of insincerity on the part of both sides and refusal by the Americans to ratify SALT 2. Since then Britain and France have refused to include their nuclear weapons in any agreement between the USA and Russia and negotiations have taken a different aspect under pressure of Russia's President Mikhail Gorbachev's political initiatives. Both Russia and America have withdrawn or destroyed most of their newest missiles from Europe and have agreed to the witnessing of those removals as well as limited inspections of various other military weapon systems, including the **chemical** kind. But though, in political terms, progress seems to have been made (along with troop withdrawals), the number of weapons remaining is almost as large as ever and rather more daunting in an atmosphere of dubious bonhomie.

Strategic Defence Initiative (**SDI**). Sometimes known as Star Wars, this $30 billion project, announced by President Ronald Reagan in March 1983, was said to be aimed at a very sophisticated, complex ground and space **surveillance** system, designed to provide protection for the USA, probably by missiles, X-rays and **lasers**, against missile attack. To those sceptics who doubted its technical feasibility and feared the cost, had to be added people, the Russians to the fore, who claimed it destabilized the current **nuclear** balance by acquiring a new technology for the West which Russia could neither afford nor, perhaps, acquire – even though Reagan offered to share the **technology**. SDI remains a very costly weight in the politico-strategic balance and a factor in **Strategic Arms Limitation**

talks. Meanwhile research continues by both sides into the feasibility of all manner of devices which are likely to add significantly to knowledge and the **deterrence** equation.

Strategy, Military. In simple terms of the art of war, strategy is the technique of planning campaigns by selecting the aims and solving the **logistics** problems connected with moving men and resources to their battle positions, where, if a decision has not been reached through diplomacy, they are used tactically (**Tactics**) in battle.

Sun Tzu, *c.* 500 BC, was the first to record the prime importance of strategy, along with **logistics**. Nevertheless, until the 17th century the number of nations and commanders who formulated long-sighted strategies were few and far between – **Alexander III** and **Edward III** were among the few outstanding strategists prior to **Marlborough**, **Eugene** and **Frederick the Great** (*see also* **Wars of the 16th, 17th and 18th centuries**).

Since then, impelled by the effects of **technology**, strategy has progressed beyond being seen as a collection of stratagems, cunning and trickery into a more intellectual process connected with evolving political systems, **communications**, society and the influence of philosophy (*see* **Philosophers, Military**). The principles of **war** continue to rule every progression but the ramifications multiply; from British naval operations and the continental campaigns of **Napoleon** during the **French revolutionary wars**; through the guileless naïvety of the **Crimean War**; the far more sophisticated conduct of the **American Civil War**, with the Federal side's application of **blockade** through sea-power, mobility through **railways**, and the land attack upon the South's economy and logistics; to the distraught, misguided strategies of **World War I** (with their initial belief in short wars) and the evolution, prior to **World War II**, of grand strategy. This is the discipline which embraces the entire apparatus of state politics, economics, logistics and **propaganda** alongside the threat or application of military force. It appeared piecemeal during the Russian revolutionary wars and the associated long wars to spread Communism worldwide.

Without perhaps fully realizing it, Adolf **Hitler** developed grand strategy in the 1930s and, by so doing, baffled his opponents by bluff before they came to understand and create an adequate deterrent to his attacks. And yet his military strategy, such as it was once the bluff was called, fell far short of being grand, arguably, indeed, being opportunist and thus lacking the foresighted thread of anticipation which is of the essence of sound military strategy. For no matter the age or the circumstances, a workable strategy is not to be conjured up at a moment's notice; aims have to be identified and selected with a view to their desirability; then exposed to critical examination of feasibility and the likely enemy strategy and reactions. Once a course or courses of action have been decided upon, sufficient time is required to plan and then position men and resources – which in global war can be a long-drawn-out process not easily amenable to the sudden switches of aim or execution Hitler often indulged in. Like **mobilization**, which is closely related to strategy, large-scale projects are not always reversible at short notice.

With the coming of **nuclear weapons** and of **aircraft** and **rockets and guided missiles** to deliver them from long range, the associated strategy of the **deterrent** has appeared, thus broadening the scope of strategy itself. For, as was soon discerned, there is a considerable difference between the somewhat ponderous preparation of a knock-out blow by World War II means and the feasibility of achieving that aim by pressing a few buttons to deliver that blow. The realization concentrated minds upon the, at times, abstract calculations and postulations concerned with striking so-called credible nuclear balances to ensure that one side could not steal a telling advantage over the other. This balancing of power created, by default as well as deceit, the vast stocks of superfluous warheads neither side wanted to use but which were expanded in the name of insurance, plus vast and intricate preparations to fight a major war without weapons of mass destruction in case nobody risked using them.

But while as an essential, integral part of the implementation of deterrence, profound public intellectual debate proceeded, conventional **limited wars** of long duration were fought under the shield of the nuclear deterrent. The **Arab–Israeli**, **Korean**, **Vietnam**, **Indo–Pakistani**, **Iran–Iraq** and **Afghanistan wars** were simply the most long-standing among many smaller ones which rarely reached a conclusion because the strategy of prolonged resistance by dedicated minorities using modern techniques of armed resistance and subversion is so difficult to counter. In the days of formal wars, strategic thought was based, for the most part, upon achieving results quickly and at the least cost. Since the Russian Revolution, the world has been afflicted by philosophers and leaders whose strategies are messianic in their fervent beliefs and perseverance in reaching for objectives, regardless of how long it takes or how much it costs. There is no saying, in a setting of almost perpetual war in which rival superpowers are inextricably involved, where that sort of Third World War will lead.

Student, General Kurt (1890–1978), was an infantryman who in 1913 joined the budding German Air Force in time for combat in **fighters** in **World War I**. After the war he helped create the secret Luftwaffe, with many responsibilities for technical development, including **parachutes** and **gliders**. He was made Commander of the newly formed 7th Air Division in 1938, tasked to develop **airborne forces**. These he led in the invasion of Belgium and the Netherlands (*see* **West European Campaigns in World World II**), when he was wounded. As Commander XI Air Corps he was involved in the invasion of Greece in April 1941 and of **Crete** in May. But after **Hitler** turned against airborne forces, he was mainly engaged in conventional land operations and planning special airborne missions such as the abduction of **Mussolini** in September 1943.

Submarines. The first attempt to use a submersible boat at war was by Sergeant Ezra Lee in David **Bushnell**'s screw-propelled *Turtle* against a British warship on 6 September 1776. It failed because Lee could not attach the explosive charge to the ship's bottom. Numerous projects went on being experimented with, but not until the **American Civil War**, on 5 October 1863, was there a successful attack by a submarine on shipping, when the semi-submersible, steam-propelled *David* severely damaged a Federal iron-clad with a spar **torpedo** – shortly before practical locomotive torpedoes became available.

Not until electric and **oil**-fuel motors were available in the 1880s did a practical mechanically driven, fully submersible boat become feasible. In 1886 Lieutenant Isaac Peral of Spain built a **battery**-powered electrically propelled boat. The following year a Russian electric boat with four torpedoes appeared at the same time as the periscope was introduced elsewhere. The next major step forward came in 1895 with *Plunger*, the streamlined boat of John **Holland** of America, propelled by electricity when submerged and by a steam engine on the surface, which also charged the batteries. He vastly improved upon the system in his boat of 1900, which substituted a petrol engine for steam; and it was further developed in Maxim Laubeuf's *Aigret* in 1904 with its safer, oil-fuelled diesel engine. By 1910 surface speeds of 10 knots and of eight when submerged were normal; Holland's earlier methods of depth control by variable water ballasting and tilting elevators (hydroplanes) further advanced; and double-skinned hulls provided other advantages, including deeper diving capability.

By **World War I** in 1914, indeed, there were submarines in service with surface speeds up to 16 knots and 10 submerged, displacements of 700 tonnes, and crews of 35, armed with a gun as well as four torpedo-tubes that could be reloaded from reserves when submerged. They made the submarine into what Admiral Lord **Fisher** had recognized in 1910 as a revolutionary weapon system for offensive effect. How right he was became apparent during the **English Channel** and **North Sea battles** when sub-

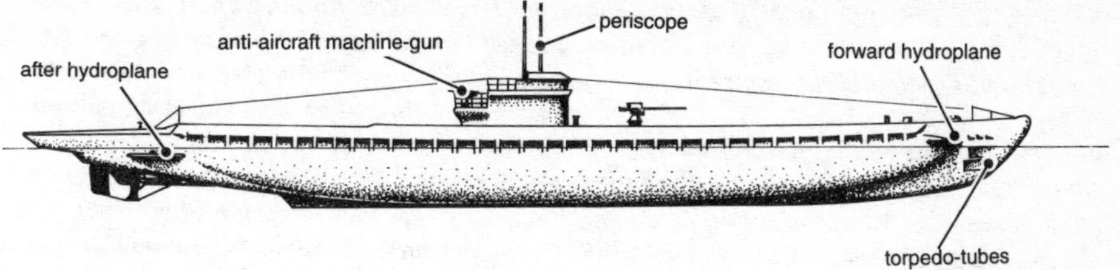

World War II submarine

marines, used for **reconnaissance** and ambush, had a major **tactical** impact on fleet operations, quite apart from their **strategic** effect through their reviled employment by the Germans, from 1915 onwards, as an instrument of **blockade**, with unrestricted attacks against merchant shipping from February 1917. The introduction of **convoys** and, in due course, **sonar** tended to inhibit submarine design developments between the wars. But in Germany Karl **Dönitz** was demonstrating by trials the feasibility of controlling U-boat 'packs' by **radio** to detect and then concentrate them against convoys; and also the tactical advantage of attacking by night on the surface. But if at first neither his boats nor those of the French and British made much impact in **World War II**, it was apparent by 1940 that sonar did not sound the submarine's death knell.

Dönitz's submarines and methods were offered a wonderful chance after June 1940 in the Battle of the **Atlantic**. Given greater resources and more assistance at the outset he might have won. As it was, in the interests of quantity production, he condoned delay in the introduction of significantly faster boats and could not keep pace with the cumulative effects of Allied countermeasures which forced changes of tactics upon him.

Deliveries of supply boats to keep operational U-boats longer at sea were slow and too late; improved counter-**electronic** devices were retarded; faster and deeper diving boats, designed by Helmutt Walter, with improved batteries and powered by a closed system fuelled by liquid hydrogen peroxide; along with the schnorkel tube which enabled a boat to breathe while running its diesel engine and charging its batteries submerged (thus saving it from being exposed to surprise air attack when surfaced), were not ready until 1944. As it was, the introduction of faster boats produced a crisis that might have been catastrophic for the Allies if the war had not ended in May 1945. The catastrophe would have equalled that suffered by the Japanese in the **Pacific Ocean War** where, because they had neglected **anti-submarine weapons**, their warships and merchant vessels suffered such enormous losses that, in 1945, the blockade of Japan was complete, with disastrous consequences.

Since World War II the submarine has become a capital ship of dominating strategic importance. The introduction by the American Admiral Hyman **Rickover** in 1955 of a **nuclear**-powered boat with an environmental regeneration plant and high performance **communications** and **navigation** systems made feasible underwater voyages restrained in time only by the endurance of their crews and the food stored aboard. The true submarines that followed were developed by the Americans, the Russians, the British and the French into boats displacing 29,000 tonnes with a reputed speed of 40 knots and a depth-holding capability of between 2,000 and 3,000ft. These characteristics, when enhanced by **stealth** technology to minimize sound, electronic and heat emissions, and counter sonar, made them extremely difficult to detect, track and attack. Such vessels could carry, in addition to torpedoes, guided missiles (see **Rockets and guided missiles**) such as Polaris, with a nuclear warhead that could be launched from beneath the surface to strike targets that, in the present day, can be 3,000 miles inland. Making the boat armed with nuclear weapons a most effective, elusive arm of the **deterrent**.

Sun Tzu (*c.* 500 BC) seems to have been a successful Chinese commander and military **philosopher** who lived in a period of almost perpetual war. His *Art of War* is not only the oldest known military treatise but also a masterpiece of succinct wisdom which established the Principles of War (*see* **War, Principles of**). Especially by his dictums that 'All men can see the **tactics** whereby I conquer, but what none can see is the **strategy** out of which victory is evolved'; and 'All warfare is based on deception' backed up by sound **intelligence**. Nor did he forget to emphasize **morale** and **logistics**.

Surveillance instruments. At sea, on land and in the air the observation of enemy and friendly forces is vital to provide sufficient and timely **intelligence** of activity and deployments. It makes use of seeing, hearing and smelling, human sensory procedures which, until the appearance of **optical instruments**, were unassisted and depended upon practice and ingrained abilities. Until the invention of the man-lifting **balloon** in the 18th century and **aircraft** in the 20th, scouting techniques were specially practised to obtain close observation, making use of **photography** as soon as it was available. But the employment of small craft at sea and of patrols on land were and are vital.

With the arrival of **electronics** the range of surveillance techniques was at once widened and exploited. **Sonar** had a surveillance role. Intercepted **radio** transmissions made possible the location through *direction finding* (DF) of a transmitter's position, helping draw useful deductions. Over the years the interpretation of patterns of electronic emissions, including **navigational** beams, made it possible to discern enemy deployments and intentions without necessarily reading a word of what was being transmitted. In the 1930s, **radar** began to supersede **sound locators** to add enormously to the ability to detect aircraft (and also ships) from longer and longer ranges, a system of surveillance with a remoteness that permitted radical changes in scouting and patrolling methods – but yet another source of emissions for intelligence.

When poison gas (*see* **Chemical warfare**) was first used in 1915, only the human sense of smell and respiratory discomfort gave clear warning. But in due course **sensors** were used for detection and subsequently developed extensively to monitor, through heat, sound and 'sniffing' the presence of targets of interest to the special organizations created to control and co-ordinate a mass of information, gleaned from the meanest person with binoculars to the loftiest **space vehicle** with **sensors** and cameras, which is central to **C^{3}I**.

The mass of information now produced by surveillance requires very sophisticated sifting of fact from fictional deception, followed by rapid, helpful presentation of results. For this purpose **computers** play a vital analytical, display and disseminatory role.

Swords. These, the most handy of cutting and thrusting **edged weapons**, appeared in the **Bronze** Age and became obsolete in the early 20th century when close combat went out of fashion in face of modern firearms. Varying in length from the short stabbing weapon of the Greek **infantryman** to some of 42 inches in the 16th century, they either were pointed and double-edged or, like scimitars and cutlasses, single-edged. Bronze blades began to give way to harder **iron** *c.* AD 200, which in turn were replaced by **steel**, as available, and because iron blades easily lost their cutting edge or bent when striking something hard. Most early swords had a simple protected hilt, but later examples benefited from elaborate, complete hand guards. (*See also* **Bayonet**).

T

Tactics are the art of fighting a battle at sea, on land and in the air but subordinate to **strategy**. For example, no matter how skilfully the German commanders in **World War II** won numerous tactical victories, they could not compensate for **Hitler**'s inept strategic insights. Tactical applications in wars and battles are to be found in many entries in this Encyclopedia and also under **Air**, **Amphibious**, **Chemical**, **Desert** and **Flame warfare**, **Fortifications**, **Guerrilla**, **Jungle**, **Mountain**, **Naval** and **Psychological warfare**. It will be seen that the best tacticians searched for the enemy's weak spots in flank, rear or by overhead approach, while at the same time guarding against surprise and enemy countermeasures. They made good use of the principles of war (*see* **War, Principles of**) to achieve mobility, firepower, protection and cunning in pursuit of their aims in attack, defence and withdrawal, and often were those who understood best how to harness **technology's effects**. The most important weapon systems have been warships (from **galleys** to **aircraft-carriers**), **aircraft**, **armoured fighting vehicles**, **artillery**, **bows and arrows**, **chemicals**, **machine-guns**, **mines**, **rifles**, **rockets and guided missiles**, **smoke**, **submarines**, **swords** and **torpedoes**.

War's multiplicity demands the formulation of doctrines to guide tacticians in the heat of battle. Tactical doctrine is a touchstone for leaders under pressure and is central to **training**, **drills** and Standard Operating Procedures (SOP). But doctrine is neither immutable nor sacrosanct.

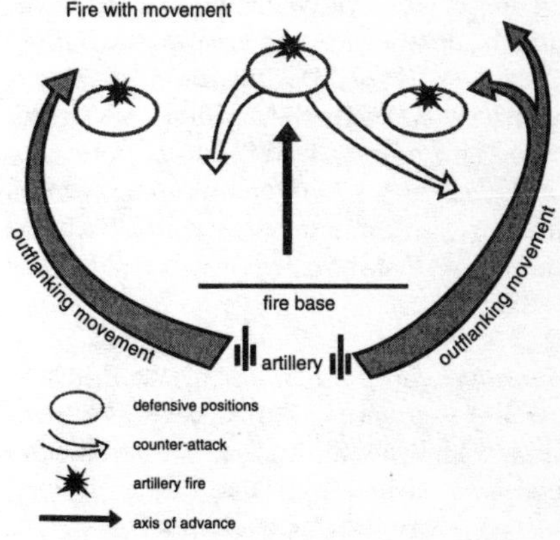

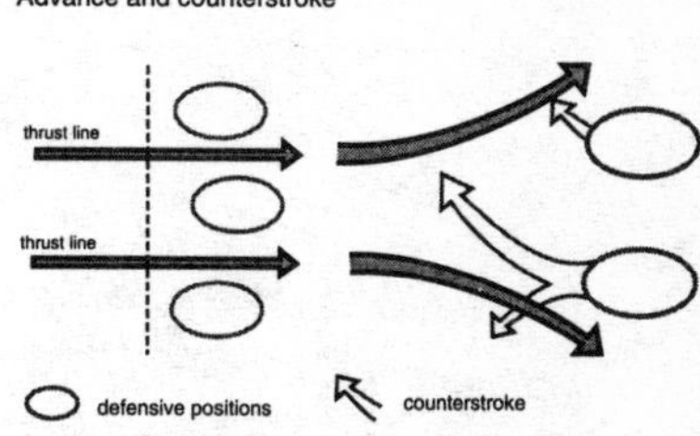

Land-battle basic tactical movements

Among the greatest tacticians are to be found innovators such as Admirals **Blake**, **Dönitz**, **Nelson**, **Togo**, **Yamamoto** and **Yi Sung Sin**; Generals such as **Alexander the Great**, Julius **Caesar**, **Frederick the Great**, **Fuller**, **Guderian**, **Hobart**, **Marlborough**, von **Moltke** (the Elder), **Napoleon**, **de Saxe**,

Sherman and **Zhukov**; airmen such as **Douhet, Mitchell** and **Trenchard**; and guerrilla leaders such as **Lawrence** and **Mao Tse-tung**.

Tal, General Israel (b. 1924). An NCO in the British Army in **World War II** who fought in Italy (Italian campaign, 1943–5) and then joined in the Israeli fight for independence. As commander of Israel's Armoured Corps he procured modern British and American **tanks** and, as he trained this armoured force, concentrated most on the need for excellence in **gunnery**. In June 1967 he commanded an armoured division in the Six Days War and made the decisive breakthrough from Gaza to El Arish. During the Yom Kippur War he was Deputy Chief of Staff. Afterwards he became adviser to the Minister of Defence on development and organization, with a leading part in designing and developing the innovative, now battle-proven Merkava tank.

Tank was the word chosen by the British in 1916 to provide security cover for their rhomboidal-shaped, tracked **armoured fighting vehicles** (AFV). Some were labelled 'Water tanks for Mesopotamia' or 'Leningrad'. The word caught on so that, ever since, almost any AFV is liable to be called a tank, including **armoured cars** and armoured personnel carriers. In modern parlance, however, the word usually refers only to tracked AFVs with a fully rotating gun turret, with **radio** and, as often as not, **night-vision** equipment, manned by a crew of up to five people.

Tannenberg, Battles of. *1410*. Called Grunwald by the Poles, this day-long struggle between 32,000 Germans, with 100 cannon and 21,000 chain-mail **armoured cavalry** of the Teutonic Order (under Ulrich von Jungingen), and Władysław II Jagiello's 46,000 Lithuanians, Poles, Bohemians (under John **Zizka**) and Russians, with only 16 cannon, was decided when the 18,000 plate-armoured knights of the Alliance outflanked and encircled the over-confident and poorly generalled Teutons. Von Jungingen and most of his 205 knights were killed; the Order's **morale** collapsed and nearly all their castles fell without significant resistance, giving Jagiello a triumph of sound **tactics** over superior **technology**.

1914. At the start of **World War I** the Germans, while they attacked France, intended their Eighth Army (under General Max von Prittwitz) to stand on the defensive in East Prussia against the anticipated Russian offensive. The Russian First Army (General Pavel Rennenkampf) advanced

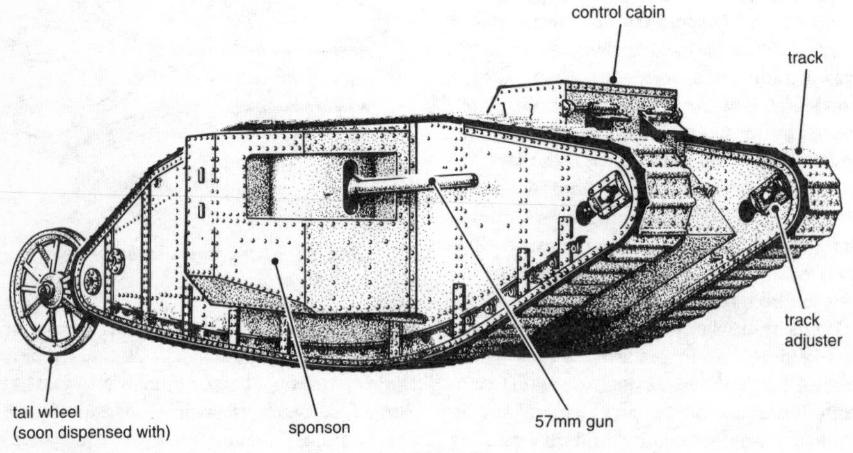

The first tank, 1916 (*British Mother*)

towards Königsberg, fought the inconclusive battles of Stallupönen (17 August) and Gumbinnen (20 August) but failed to follow up when the Germans withdrew. Meanwhile Second Army (General Alexander Samsonov) advanced northwards to cut the **railway** between Allenstein and Deutsch-Eylau, and thus take Prittwitz in flank and rear. Prittwitz lost his nerve, ordered the abandonment of East Prussia – and was at once replaced by Generals **Hindenburg** and **Ludendorff**. When they arrived it was to learn, through **radio** intercept, that Rennenkampf was almost stationary. They were presented with a plan to screen Rennenkampf with a single **cavalry** division while safely concentrating the entire Eighth Army against Samsonov. This was achieved by the 25th – and almost undetected by the Russians, who dispensed with cavalry reconnaissance.

Already on the 24th, Second (Russian) Army's centre had been checked at Orlau, but this minor set-back did not stop the wings advancing on a 50-mile front between Soldau and Ortelsburg to Bischofsburg. On the 26th the Germans closed in upon both Russian flanks finally to complete an encirclement on the 30th which netted 125,000 men and 500 guns – a defeat that forced First Russian Army to withdraw eastwards, pursued by the Germans. The Russians never fully recovered from this defeat. But the victory reinforced German self-confidence besides giving them for better or worse the redoubtable Hindenburg–Ludendorff duo.

Taranto Raid. On 11 November 1940 21 single-engine Swordfish biplanes took off from the **aircraft-carrier** HMS *Illustrious* to attack the Italian fleet at anchor in Taranto harbour. While 10 Swordfish dropped flares and dive-bombed oil-storage tanks, the other 11 scored six hits from 10 functioning **torpedoes**, sank one **battleship**, beached another and put a third out of action for six months. Two Swordfish were lost with the loss of two lives. At a stroke the balance of naval power in the **Mediterranean Sea** was changed by an exploit which satisfied Admiral **Yamamoto** that battleships were obsolescent, a lesson the Japanese rubbed in 13 months later at **Pearl Harbor**.

Technology, Effects of. Man's destiny has been shaped by technology even in its most primitive, **Stone** Age form, although it is arguable whether weapon technology or **tactics** came first to mind. As likely as not it was the discovery of **bronze** and/or **iron** which inspired tactical reform, rather than a requirement for tactical innovation which suggested a feasibility study (as in the 20th century) into, for example, improved **edged weapons** or **bows**. Beyond much doubt, however, **research** by trial and error of **metallurgy** lay at the root of basic technical growth by indicating cheaper ferrous metals and alloys which, along with the (presumably) accidental invention of **explosives** and the more efficient use of fuels for **power generation** and **propulsion**, promoted man's urge to discover and exploit new knowledge.

War is a perennial activity which has intensified significantly since the growth of technological effect became associated with the more sophisticated economically driven industrial systems, linked to the improved **health** and **medical services** and population growth of the late 18th century – the so-called Industrial Revolution. The latter really started in Europe during the **Hundred Years War**, the **Ottoman Empire**'s wars of the 15th century and the invigorating Renaissance, because of the increasing demand for **iron** and **steel** to make better **armour**, **artillery** and **ammunition**, a demand stimulated by radically improved **shipbuilding**, **fortifications** and many kinds of weapons.

Nevertheless, the Industrial Revolution and the **French revolutionary wars** greatly accelerated the inventive process, notably from the impulse of **electromagnetic** discoveries in the first half of the 19th century, most of which were seen to have military applications. From numerous entries in this Encyclopedia can be gleaned the impact of: improved **communications** (those of signals, **railways**, roads and air); new materials, such as **aluminium** and **titanium** and various kinds of **explosive** (including **nuclear**). These inventions and developments had a sharp effect on thinking, organizations and methods. This effect ushered in a period of such rapid and all-embracing change that leaders in government, administration, commerce and the military services (the vast majority of whom had not enjoyed a scientific and technical education) were unable to assimilate (let alone benefit from) what was on offer.

The underlying theme of war since 1850 is that of the struggle between old and new concepts, interwoven with social, educational and institutional changes that have created wide divisions between classically and technically orientated groups. In addition to these changes, which continue to be resisted strongly by those who fear for their occupations and peace of mind, it can be seen that although weight of numbers in war still has a significant bearing, it can be overborne by more skilful use of new technologies, like the original **mechanized** forces on sea, land and in the air which demonstrated that, even when few in number, they could rout larger, technically inferior numbers. On the other hand, the sheer weight of technological effect sometimes produces a self-defeating result. The unbridled search for new knowledge through extravagant **research and development** (R&D) can produce confusion and also obscure the identification of essentials, in addition to wasting energy, time and money. This aspect of technology has led to much heart-searching and numerous endeavours at control which are often related to the cry for **disarmament**. Setting aside the possibility of mutual destruction through loss of control of technology, it is not to be expected that the progressive urge will be abandoned. This makes it all the more essential that military people should at least be technically aware and never dismissive of new ideas.

Telegraph and teleprinters (*see also* **Communications, Signal**). Mechanical, visual telegraph systems were in use during the **French revolutionary wars** – one such system invented by the Reverend John Gamble was in service for ship-to-shore signalling until 1940. But it was experiments, starting *c.* 1747 with electrical impulses transmitted down wire, which, when powered by **batteries**, led to the first practical telegraph systems. Samuel Morse's in 1837 (inspired by a meeting with Michael **Faraday**) was probably the most significant. In 1850 his code and 'sounding' key enabled skilled operators to write down messages as received, thus increasing traffic volume considerably.

The first wartime use of the electric telegraph was during the **Crimean War** (1854). A submarine cable 547km long was laid from the British base at Varna in Bulgaria to the Crimean peninsula. This link, connected by land cable to Paris and London, enabled Napoleon III to harass his commanders with unhelpful suggestions and the British War Office to do the same with theirs. These problems were not helped by depredations on the cables laid overland, by soldiers removing the insulation for pipe stems and by the local inhabitants, who found the copper wire irresistible. The telegraph did not really have a significant effect on operations in the Crimea, unlike the Indian Mutiny of 1857 where, since the system was controlled by the British, it proved to be a decisive factor.

The telegraph really came into prominence during the **American Civil War**, when troop and supply movements were effectively controlled by the rapid and accurate passing of information. However, this war also exposed the vulnerability of line communications, the cutting and large-scale removal of which was a frequent objective of enemy raiding **cavalry**. It was during this war too that real efforts were made to speed up the laying of line near the combat area. The North equipped special telegraph wagon trains with insulated cable and lightweight poles (an example soon followed by the European armies). By linking such lines with the permanent civilian telegraph system messages could be transmitted from, for example, the Government in Washington to commanders in the field.

Meanwhile, technological improvements were growing apace, with Gintl's duplex circuits appearing in 1853, whereby two messages in opposite directions could be sent simultaneously on the same line. Followed in 1874 by Thomas **Edison**'s quadruplex system allowing four simultaneous messages to be sent. Meanwhile J. M. Baudot's 1872 time-division multiplex system had appeared, still used in some teletype machines.

The invention of the **typewriter** and its development the teleprinter, linked to the telegraph system, added a further dimension to communications. Consisting of a typewriter keyboard and a printer, the key strokes generated electrical impulses which were translated into the printed word at the receiving end, the system being secure as long as land line was used. Radio teleprinter (US: teletype) systems introduced in the 1920s had to use encoded signals. Teleprinter circuits were used extensively throughout **World War II**, notably in the forward areas by the Germans in Russia, using the latest multi-channel-carrier frequency systems, sending along bare wires simultaneous electronically separated modulations in enormous quantity. In the post-World War II period teleprinter facilities linked by **radio** to all formation headquarters were a commonplace.

Although during **World War I telephone** and in later years voice radio to some extent supplanted the telegraph, the use of microwave radio and **space** satellites carrying as many as 1,800 channels on a single circuit, combined with teleprinter terminals capable of printing 1,000 lines per minute, means that telegraph systems still have a significant part to play in military communications.

Telephone (*see also* **Communications, Signal**). Alexander Graham Bell, a speech therapist, invented the telephone in 1876 and by 1880 it was in general use. The basic principles of operation are that air-pressure changes induced by the human voice cause a diaphragm to vibrate, the vibrations in turn creating changes in electrical current through a variable resistor. At the receiver an electromagnet translates the fluctuations in current once more into vibrations on a diaphragm, which the listener hears as human speech. The only fundamental change to Bell's invention, though there have been many refinements of it, has been the modern use of digital transmission in the form of pulses, rather than the earlier analogue system of current changes. This makes for greater clarity and an ability for lines to carry more simultaneous conversations.

The telephone was used by the US Army during the Spanish–American War at the end of the 19th century, during the Boer War and during the **Russo-Japanese War** of 1904; but in none of these did the device play a significant part, the equipment being insufficiently robust for the rigours of field conditions. It was **World War I** which really gave the telephone its chance, especially under the comparatively static conditions of the Western Front. The German, French and British all had well-developed signal services by 1914. It has been said that Russian army communications, on the other hand, were approximately at the level of the Americans' at the time of their Civil War some 50 years before; the Russian defeat at **Tannenberg** was in large measure a result of this.

On the Western Front telephone lines spread everywhere in a vast network, both behind the combat area and right into the front lines, so that it was not long before virtually every platoon had direct telephone contact with its company headquarters and so on up the chain of command. The effect on operations was significant, in that far larger forces than before could be effectively controlled from the rear; the voice and hence the personality of a commander could be heard and felt at a lower level; massed **artillery** fire could be controlled directly by a forward observer in the trenches, or in a tethered **balloon**.

The chief problem lay in the maintenance of the miles of cable, constantly disrupted by shell-fire, and the threat to security from the enemy tapping in to lines and listening to conversations. In an effort to combat phone tapping the British introduced the Fullerphone. By setting up an impedance coil and two condensers, Major A. C. Fuller, the inventor, was able to control the rise and fall of current waves up and down the line. A commutator prevented the current surging back into the line, it being momentarily stored in the condenser. Morse signals could thus be sent securely down the line, though normal voice conversation was not protected. This combination of telephone and telegraphy was not new, a device having been produced by van Rysselburgh in 1902, but the military application had not previously been developed.

The weakness of the telephone emerged once mobile operations became more general – as the Germans had already found during their invasion of France and Belgium in 1914. Nevertheless, the telephone remained an important part of the field army equipment throughout **World War II**, special vehicles being able to lay line at up to 160km per day. The instruments became more robust and the use of multi-core cable and more sophisticated portable exchanges meant that telephone communication could be relied upon under even the most adverse conditions.

It is only now in the late 20th century that, following improvements in **radio**, for forward units the telephone is no longer essential in combat communications.

Television. The conversion of a scene in motion, plus its attendant sound, into an **electronic** signal transmitted to a receiving antenna and converted to a visible picture on a cathode-ray tube. The idea had first been proposed, in essence as it finally evolved, in 1908 by the British inventor A. Campbell Swinton; translated into the prac-practical transmission of pictures in 1926 (though by mechanical means); but set on its true course in 1923 when the Russian–American Vladimir **Zworykin** patented the iconoscope cathode-ray camera tube on which television would ultimately depend. The system relies on the bombardment by electrons of a fluorescent-coated screen in a scanning motion to create the picture. Work to achieve success was going on in a number of countries in the 1930s, the leading British engineer in the field being J. Logie Baird.

There were a number of military applications, once commercial television became a reality, of which **radar** was the first to bear fruit. To use television pictures for direct military **surveillance**, an obvious use, was, however, another matter; the early receivers were both bulky and fragile, and television cameras totally unsuited for use in the field. This application was not seriously pursued during **World War II**, but towards the end of that conflict the Germans had already proved the feasibility of television guidance for missiles, though nothing had been put into service.

The bulk and fragility problems were not really overcome until miniaturization of electronic components became a reality with the development of the transistor in the 1950s, and later the electronic chip. Thereafter developments were rapid and the American television-guided 453kg Walleye *gliding bomb* was used successfully in 1967 during the **Vietnam War**, particularly for attacks on bridges. Walleye had a television camera in its nose, transmitting pictures back to the parent aircraft, standing off at a distance to allow the controller to guide the bomb to its target comparatively unhindered. There are now a number of such guidance systems available. Further developments include the so-called smart bombs, typically the 906kg *electro-optical guided bomb*, which enables its controller to locate the target on his television screen, designate it to the bomb which then, after release, finds it own way to the target. (*See* **Robots.)**

Television for surveillance has had some success and has been used in action by the Israelis. The most likely application is to have a camera mounted on a **remotely piloted vehicle** (RPV), either for general 'over the hill' surveillance or to provide the necessary reference picture for a **laser** target designator. The problem chiefly lies in the wide bandwidth required for television channels, plus the extra communication links needed to guide the RPV and point the television camera in the required direction, with channels always at a premium in the forefront of the battle. An RPV-mounted television surveillance system needs a not inconsiderable control organization to take full advantage of the facility and this must be seen against the added hazard of electronic countermeasures, to which television is always susceptible. Nevertheless, a number of countries have developed, or are developing, television surveillance systems for use in the field.

Television as a news media and **propaganda** instrument made its first major impact during the **Vietnam War** by bringing home to people the horror of an arguably unjustifiable conflict. During the **UN–Iraq War**, its direct transmissions from the front line, via **space** satellites, introduced not only an unprecedented immediacy to reporting but also considerable, uncontrollable security leaks to the enemy.

Tet Offensive. Tet is a national holiday in **Vietnam** and was chosen by General **Giap** as the time for a decisive offensive against South Vietnam by the North Vietnamese Army (NVA) and the Vietcong (VC) guerrillas. As a diversion, on 21 January 1968, the US bastion at Khe Sanh was attacked. Then, on the 30th, Saigon, Hué and most other important centres were struck. Fighting was fierce. The US and the South's forces had been expecting something of the sort and inflicted heavy losses. Hoping to strangle the enemy's **logistics**, Giap had neither allowed for a massive supply operation by **aircraft**, **helicopters** and **armoured fighting vehicles** being able to maintain every base intact, nor for South Vietnamese unwillingness to rise up on his side. After about two weeks he admitted defeat, though the siege of Khe Sanh lasted into April.

Yet Giap and Ho Chi Minh had timed the offensive well, to coincide with the American public's belief that the war was almost won and over. For the extensive **television** coverage and false news-media **propaganda**, which implied that the US and South Vietnam had been surprised and defeated, actually handed a strategic victory to the North, which in fact had been beaten comprehensively. Indeed, so severe were Vietcong losses that never again was it the same. Nor was the NVA fit for a major offensive for two or more years. But the USA suffered from a self-inflicted wound when its government, lacking public support as well as faith in victory, sought ways of escape from a long war which still continued.

Teutoburg Forest, Battle of the. In AD 9 a Roman army of 20,000, with 10,000 auxiliaries and camp followers under Quinctilius Varus, advanced into the dense Teutoburg Forest to quell German **guerrilla** forces. Deserted and ambushed by Germans under Arminius, and hearing that his base at Minden was surrounded by hostile Germans, Varus marched northwards. But in close country his three well-equipped and trained legions and camp followers were harassed and almost annihilated. This battle did more than deter future Roman occupation of Germany. It demonstrated the potential of irregular forces fighting tenaciously on ground of their choosing against poorly generalled regulars.

Teutonic expansion. The Teutonic Order was formed during the **Crusades** by German merchants to nurse the sick. But in 1198 it was militarized and, in 1211, well trained and armed under Hermann von Salza, began a ruthless advance into eastern Europe, building castles as it went. Regarded by its arrogant Knights as a 'crusade' against pagan Prussians and Lithuanians, it eventually threatened the already Christianized Poles and other nations. By 1410 the Order held most of the Baltic coast to Narva. Defeat at the crucial Battle of Grunwald (*see* **Tannenberg**), however, brought about the Order's collapse and its extinction in 1466.

Thermal imaging (TI) devices (*see also* **Infrared devices** and **Night-vision devices**). TI devices operate in the upper end of the so-called Far IR band of the **electromagnetic** spectrum. They enable the observer to see the target by means of the thermal-radiation contrasts between one part and another – for example, between a tank's hot engine deck and its cooler glacis plate. The lens system of a thermal imager functions much as an **optical** lens system, except that the lenses are made of material such as germanium which will transmit Far IR radiation. The output is very similar to that of a standard **television** signal and the display is usually on a television monitor. A typical example is the British tank fire-control system known as TOGS – Thermal Observation and Gunnery Sight – installed in the Challenger tank.

TI was used very successfully on a large scale during the **UN–Iraq War** when, at night and in poor visibility, American and British tanks fitted with it totally dominated Russian types mainly equipped with nothing better than Near IR **searchlights**.

Thirty Years War. Usually portrayed as the last major European war of **religion** (Roman Catholics versus Protestants). Lasting from 1618 to 1648, this very complex political and constitutional struggle actually started in 1609 and continued sporadically until 1661. Concurrently with peripheral wars, such as the Russo-Polish (1609–18) and First Anglo-Dutch War (1652–4), there were 13 interrelated wars: those of the Julich Succession (1609–14); Bohemian and Palatine (1618–23); Graubunden (1620–39); Swedish–Polish (1621–9); Danish (1625–9); Mantuan Succession (1628–31); Swedish (1630–35); Smolensk (1632–4); French and Swedish (1635–8); English Civil (1642–6 and 1648–51); Swedish–Danish (1643–5); Franco-Spanish (1648–59); First Northern (1655–61).

The war has also been described in exaggerated terms as being of exceptional violence and destruction. Undeniably there were many kinds of excesses, invariably associated with religious and racial strife, as exemplified by the holocaust at the end of the siege of Magdeburg in 1631 when ravenous, undisciplined Roman Catholic troops behaved as atrociously as victorious besiegers so often have. But most campaigns were of short duration; many parts of Europe escaped the fighting; the vast majority of deaths were from endemic pestilence; populations, including Germany's, increased overall; cultural and scientific activity and several economies thrived.

Militarily and technologically the Thirty Years War was vital since it brought to prominence commanders of the calibre of the Catholic Count Johann Tilly and the Protestants, Count Albrecht von Wallenstein and King **Gustavus Adolphus**, whose organizational, technical and **tactical** innovations with **artillery**, **cavalry** and **infantry** were revolutionary.

In the closing stages there appeared the French Viscount Henri de Turenne, who preferred **strategic** manoeuvres and **sieges** to costly battles; and the Marquis Sébastien de **Vauban**, who established a new doctrine for siege warfare and construction of **fortifications**. (*See also* **Wars of the 17th and 18th centuries**.)

Thornycroft, Sir John (1843–1928) was a revolutionary naval architect whose **research** into resistance of water to ships' hulls led in 1871 to a 16-knot, **steel** craft. This he followed in 1873 with the first **torpedo-boat** (Norway's *Gitana*), armed with a spar **torpedo** (a mine on a pole); and in 1877 the 18-knot HMS *Lightning*, with a swivelling locomotive torpedo tube, the forerunner of innumerable fast torpedo-boats and **destroyers**, often powered by water-tube boilers and, in due course, **turbine** engines. Next he experimented with hulls which skimmed, instead of cutting through the water, thereby evolving stepped hulls and floats for seaplanes and, in 1916, small twin, petrol-engined torpedo-boats with speeds in excess of 30 knots. In 1904 he had been associated with the design of **Dreadnought battleships** and, later, in addition to **shipbuilding**, with **motor vehicle** construction.

Tirpitz, Admiral Alfred von (1849–1930) was a **torpedo** specialist who became Chief of Staff to the Prussian Navy in 1892, C-in-C Far East squadron in 1896 and, from 1897 to 1916, Secretary of State to the Ministry of Marine. He was a good tactician who promoted night actions and the use of **smoke-**

screens. With Kaiser Wilhelm II's backing to increase German influence, he unwisely challenged British naval supremacy by creating a fleet which, by 1912, was projected to include 41 **battleships** and 60 **cruisers**. This began the **shipbuilding** race, in which Britain outmatched Germany in numbers and quality, and which was one cause of **World War I**, a war in which Tirpitz failed to persuade the High Command to risk the decisive fleet encounter he always had desired. Ironically, Germany's defeat at **Jutland** occurred two months after his resignation; and it was her **airships** and **submarines** that he had neglected which eventually proved strategically more effective than her great surface warships.

Titanium (a very common metal) was first discovered in 1791 by William Gregor and rediscovered by Martin Klaproth in 1795 when researching minerals; but refinement, despite many attempts, was not achieved until 1936 by Wilhelm Kroll. Production, mostly by a chemical process, is expensive, but the product is in much demand because it is easily alloyed with other metals and, among many attributes, resists corrosion, performs well at very high temperatures and is very strong and tough. It thus has numerous civil and military uses, for example in means of **propulsion**, **space vehicles**, **shipbuilding**, **rockets and guided missiles**, and as **armour** plate.

Togo, Admiral (Count) Heihachiro (1847–1934). On 25 July 1894, when commanding the Japanese **cruiser** *Naniwa*, Captain Togo fired the first shots in what, on 18 July, became the declared Chinese–Japanese War. He won glory at the Battle of the Yalu River and later played a leading role in preparing the fleet for the **Russo-Japanese War** of 1904, when, as Commander of the Combined Fleet, he directed the initial naval attack on **Port Arthur** on 8 February. His leadership in the ensuing campaign, including the Battle of the Yellow Sea, was brilliant and decisive in what were the first modern naval battles, a prowess he underlined by his skilled conservation of ships in his conduct of the **blockade** of Port Arthur and the annihilation of the Russian Baltic Fleet at the Battle of **Tsushima**. It was Togo's innovative and disciplined spirit which inspired the Japanese Navy throughout the campaigns of **World War II**.

Torpedo. Named after an electric-ray fish, the torpedo originally was a moored sea-**mine**, invented in 1805 by Robert **Fulton**. This entry, however, deals only with locomotive torpedoes, the combined invention of Robert Whitehead and an Austrian naval officer, Giovanni Luppis. In 1866 Whitehead demonstrated a 14ft-long underwater missile with 18lb of **explosive** in the nose, driven by a compressed-air motor at six knots to a range of 700 yards. It was launched from tubes, controlled in depth by a hydrostatic valve operating elevators called hydroplanes and, to begin with, ran straight. Within 10 years specialized fast **torpedo-boats** were being built to mount it and before 1900 it was recognized as the **submarine**'s natural and principal weapon.

By extending the scope of underwater attack, the torpedo revolutionized **naval warfare**. First used successfully in action with indeterminate results in 1877 during the Russo-Turkish War, it helped win major victories for the Japanese at Wei-Hai-Wei in 1895 and in the surprise attack on **Port Arthur** on 8 February 1904. By this time an Austrian, Ludwig Obry, had adapted the gyroscope to give directional control, which considerably improved tactical flexibility. In addition, warheads were increased in diameter to 18in, along with improved motors driven by burning fuel with compressed air to give speeds up to 30 knots, plus extended range. Torpedoes such as these began to hold sway in **World War I**. The mere suspicion of approaching torpedo tracks had tactical influence by inducing manoeuvres which, by evading hits, affected the outcome of major engagements such as **Jutland**. In the submarine campaigns against commerce they became increasingly important once boats were compelled to attack while submerged,

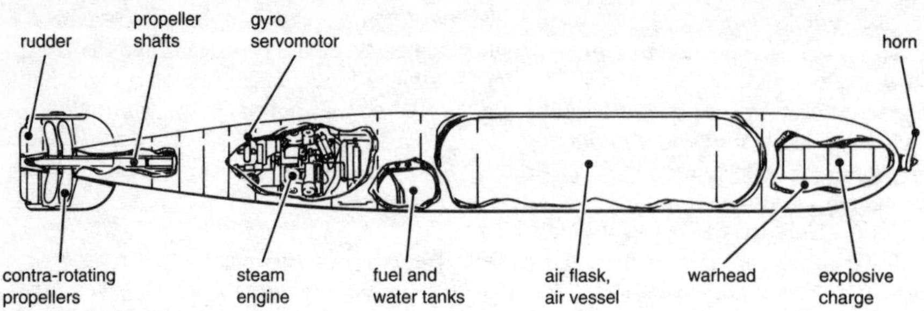

Torpedo

instead of surfacing to engage more economically with gunfire. At the same time the practice of dropping torpedoes from **aircraft**, first done by an Italian, Capitano Guidoni in 1911, announced yet another new era in naval warfare.

Between the World Wars extensive **research and development** raised the performance of torpedoes, although without always improving results. Although the introduction of electric motors enhanced tactical versatility by almost eliminating the tell-tale track, there were numerous malfunctions, notably from the temperamental, magnetic-effect exploders intended to replace contact pistols. Indeed, only Japan produced a fully reliable torpedo for **World War II**: the 24in Long Lance with a speed, powered by a liquid-oxygen motor, of 36 knots for 22 miles, or 49 knots for 11 miles. This was a battle-winning weapon introduced with deadly effect at the Battle of the Java Sea in February 1942. Indeed torpedoes were the vital weapon without which naval victories from the sinking of Italian warships at **Taranto** in November 1940 to the battles of the **Atlantic**, **Pearl Harbor** and **Pacific Ocean**, would have been very different if not impossible.

During World War II further advances in design and performance were made. The Germans took a lead with torpedoes which ran zigzag or figure-of-eight courses, but most promisingly introduced the acoustic kind which 'homed' on to their target's noise; systems which called for equally sophisticated yet quite simple countermeasures. Since then the search has been, in general terms, for torpedoes which can outpace, outdive and follow to impact the fast, deep-diving submarine; or can skim and submerge at higher speeds, **rocket**-assisted, to targets far over the horizon. They are fuelled by ever-improving compounds, incorporate **sonar** and **electronic** technology with miniaturized components and are fitted with all manner of devices to provide immunity from countermeasures. For example, the wire-guided type (based on *anti-tank guided weapon* (ATGW) systems) have security attractions and can be controlled from hovering **helicopters** which, by no means incidentally, are ideal carriers of torpedoes, including special, small-sized versions.

Torpedo-boats (TBs) were the outcome of the invention in 1866 of the locomotive **torpedo** and of John **Thornycroft**'s experimental 16-knot **steel**-hulled boat of 1871. Followed in 1873 by the spar torpedo-boat *Gitana*, the swivelled, locomotive-torpedo-tubed HMS *Lightning* in 1876; and in 1877 a French 14-knot boat with two underwater tubes. There then evolved in the 1880s a tactical philosophy based on the theory that speed, as compared with heavy and costly **armour**, gave 40-tonne TBs adequate protection; and that flotillas attacking in mass, preferably covered by night or **smoke**, would pay off. The Japanese demonstrated this successfully at night at Wei-Hai-Wei in 1895 and **Port Arthur** in 1904, in actions which prompted the construction of torpedo-boat **destroyers** (TBDs).

The concept of the light boat nevertheless

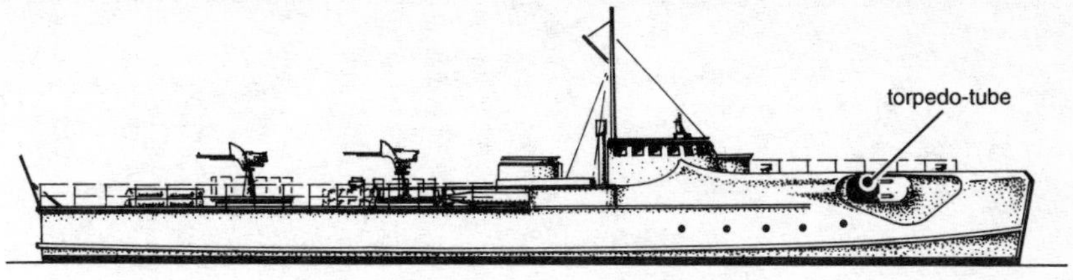

Motor-torpedo-boat, 1940 (German E-boat)

survived, in spite of high losses of such vessels when exposed to fire from destroyers, the secondary armament of larger ships and from **coastal defences**. In **World War I** they were maids of all work, particularly in narrow waters such as the **English Channel**. Elusiveness was vital there, plus shallow draught, as it was for the British 40-tonne Thornycroft *coastal motor-boats* (CMB) or *motor launches* (ML), craft powered by petrol engines with speeds over 35 knots, and the Italian MAS (*motor anti-submarine*) boats. Built in large numbers, they were used against **submarines** and all kinds of surface vessel, as well as for **convoy** escorts and for raids such as that against Zeebrugge. They were, in fact, the forerunners of the hard-chine boats that were developed in the mid-1930s as *motor torpedo-boats* (MTB), with fixed, forward-axial tubes. The British developed these extensively and introduced them to the USA as PT (*patrol torpedo*) boats; unlike the German *Schnellboote*, they were made of wood, not metal. The German 105-tonne S boats (known by the Allies as E (for Enemy) boats) were fine examples of the genre with two tubes, 20mm and 37mm guns, a speed of 42 knots and good sea-keeping ability. First built in 1929 they, like their Italian counterparts, spent **World War II** fighting the British off the shores of Europe in countless engagements, mainly by night.

Similar boats are still in service, armed with **rocket missiles**. Russian- and French-designed types have been used in the **Arab–Israeli wars** with great effect. The Danish and Norwegian navies see a vital role for them among fiords and islands. They are attractive to small maritime nations intent on low-cost defence, or in the fight against smugglers and gun-runners; operations, in other words, where speed is vital and the faster hydrofoil boat comes into its own.

Tours, Battle of. An encounter in 732 between loot-encumbered **Muslim** raiders under Abd al Rahman and a better-disciplined Frankish army under Charles Martel. From the sketchy information available it appears that, in an endeavour to save the plunder, Abd al Rahman felt compelled to launch his outnumbered light **cavalry** against a phalanx of dismounted men-at-arms on commanding ground. The better-armoured phalanx stood firm, Abd al Rahman was killed and the Muslims fled without their loot. This victory, coming after serious earlier Muslim reverses at Constantinople and Adrianople in 718, decisively ended Muslim incursions into Western Europe and also enhanced the importance of **armour**.

Trafalgar, Battle of. When **Napoleon** threatened to invade England between 1803 and 1805, the Royal Navy's principal, unfulfilled aim was to destroy, or at least **blockade**, the combined Franco-Spanish fleet. When Napoleon abandoned the invasion and turned east in August 1805, he ordered Admiral Pierre Villeneuve to enter the Mediterranean in support of the campaign in Italy. But, with 33 ships of the line against Admiral Lord **Nelson**'s 27, he was brought to battle off Cape Trafalgar on 21 October. To a

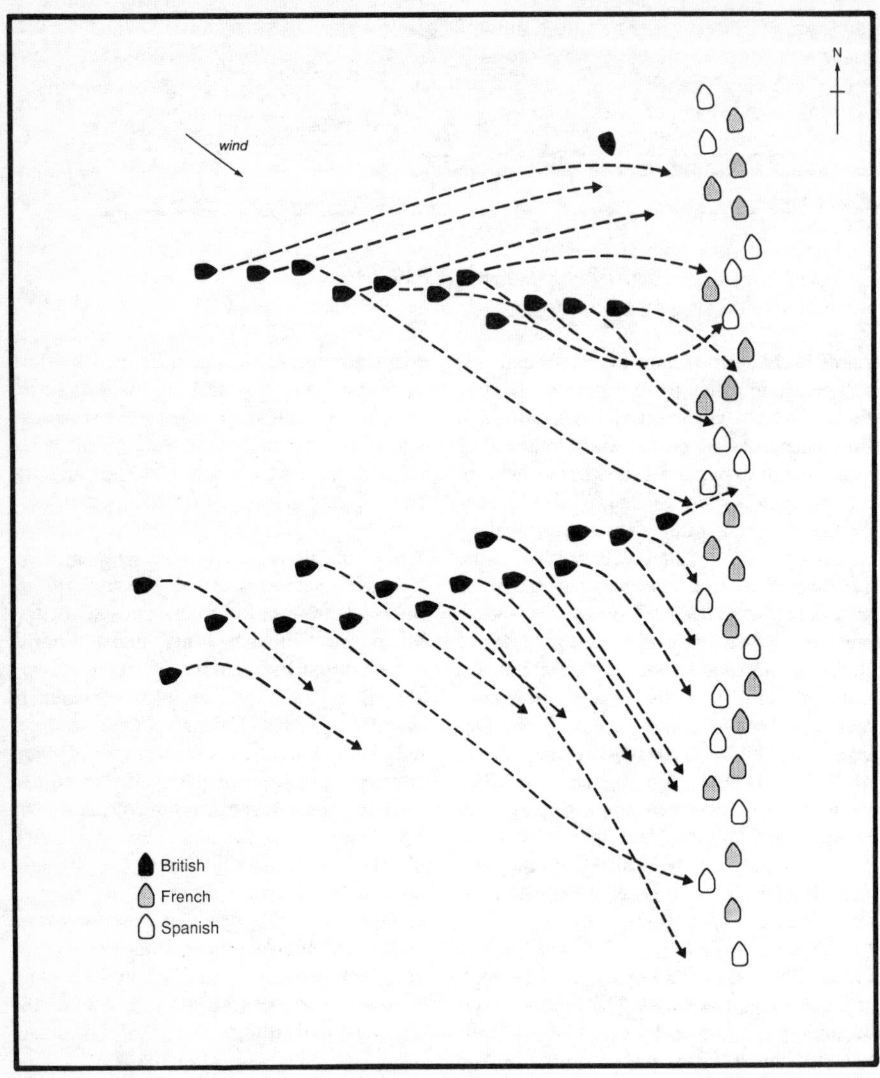

The Battle of Trafalgar, 1805

pre-arranged plan, Nelson attacked in two columns with a view to concentrating the fire of his 2,148 guns on the allied centre and rear before the enemy van could turn and come into action with all their 2,628 guns. A furious four-hour mêlée ensued during which Nelson was killed, Villeneuve captured, 18 allied ships taken and four sunk, without the loss of a British ship. In this, the last major fleet action between sailing ships, Nelson's innovative **tactics** and the superior fighting qualities of his men and ships won a **strategic** victory which outlasted the **French revolutionary war** by more than a century.

Training, Military and technical. Training is at the root of military prowess and **morale**. When reading this entry reference to **Drill**,

Education, Military and **Staffs, Military** might well be useful along with reflection on entries elsewhere giving insight into what effect training, together with **technology**'s effects, has had upon military history over the generations. They underline the vital need for systematic and enlightened teaching practices to prepare and, to some extent, condition all ranks for the performance of their tasks. These duties are in certain respects unique, but in others simply a variation of current civilian practices; but they also indicate how current trends, doctrine and social behaviour impinge upon training policy, its content, methods of instruction and application of technical aids in a comprehensive, cost-effective manner.

Before the invention of mechanical printing and the ready availability of written military instructional manuals, teaching was in the hands of the educated minority, who had access only to a few master works, such as **Sun Tzu**'s *Art of War*. Most information was passed down by word of mouth and much lost in the process. Owing to lack of systematic analysis of recorded experience, therefore, only limited progress was made. The setting-up in the 17th century of military colleges, with their own libraries and archives, lay at the heart of profoundly studied doctrines and the establishment of training directorates. From them the staff disseminated syllabi in writing and implemented them through the training programmes of the colleges, schools and established units, whose commanders were personally responsible for their subordinates' training.

Training reflects weapon characteristics and their effects on **tactics**. Until the invention of gunpowder the only technical specialists were sailors and the **engineers** engaged in the construction of roads, **fortifications** and **siege engines**. Skill at arms, notably that of archers, was taught by repetitive practice and was cheap in expenditure on materials. With the coming of firearms in the 14th century, the conversion of engineers to gunners and the need to train sailors and soldiers in shooting and care of arms and **ammunition** raised overhead expenses. Likewise the founding in the 16th century of naval **dockyards** to care for the more sophisticated **carracks** and **galleons** created a demand for more and diverse craftsmen. From this demand shore training establishments were developed slowly until given a fresh impetus by steam power (*see* **Propulsion, Means of**), the inventive genius of the 18th century Industrial Revolution and the **French revolutionary wars**.

Nevertheless, 1850s instruction was largely repetitive and drilled, without much call on initiative and usually unit-based, with extremely rare dependence upon very few central schools of instruction. Thereafter, the introduction of more complicated weapons, requiring attendant skills and supporting organizations, called for teaching methods beyond commanding officers' capabilities. Experts to operate steam engines, **railways**, breech-loading **artillery**, **machine-guns**, **telegraph**, **telephones**, **radio** and **aircraft** had to be taught in special schools and supervised within the complex systems devised within the training directorates of the Service ministries. Over the years there evolved training programmes for recruits, who would be given basic instruction in military behaviour and weapons before progressing to higher proficiency as part of a group, probably learning a specialist trade on the way and, desirably, how to teach, lead and command others.

Sooner or later the recruit would join a ship's crew, as part of a team trained in some specialized function on board; or become a member of an army or air force unit. He would take part in group exercises which increased in scale and complexity as a ship 'worked up' to full efficiency; or in a year's training programme progressed by stages to major, formation exercises. Training should be a never-ending process, one in which there is always something new to learn as well as old to revise and improve upon. Innovation should evolve alongside repetition to raise standards of performance to an excellence in which the individual learns to do many different tasks and officers to function two or three levels above their present rank.

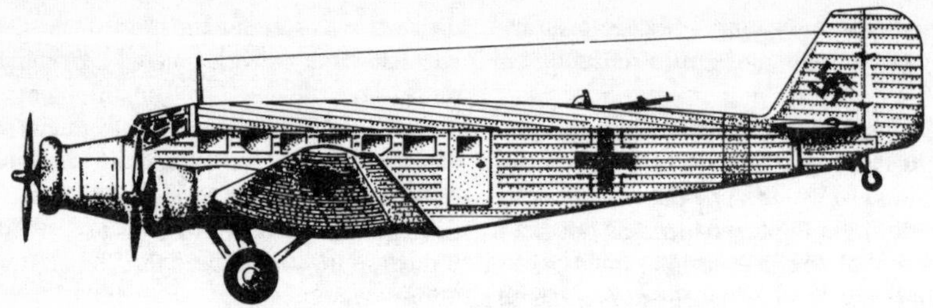

Transport aircraft (German Junkers Ju52)

With the increase in numbers of weapons of high technology and the rising intensification of military activity since the 1920s, realistic training has become extremely difficult to arrange economically. In a more crowded and environmentally sensitive world, space for exercises is harder to come by. At the same time practice with operational ships, army equipment, aircraft and their weapons is extremely expensive in wear and tear as well as fuel and **ammunition** expenditure. Since the 1920s increasing use has been made of classroom **simulators**, to teach **tank** crews, for example, how to drive and aim their weapons before working on and, perhaps, damaging the real thing. In the 1960s economics and the **electronic** revolution, apart from common-sense teaching practice, demanded extensive use of realistic simulators. A strong lead came from **air forces** and civil airlines in the training of air crew. Soon the armies were acquiring improved driving and **gunnery** simulators, while the navies developed systems for teaching tactics without going to sea. **Computers** and **television** contributed vitally to the development of complex **war-gaming** facilities. They were not inexpensive, but cheaper and almost as realistic as the real thing, besides having the advantage of 'playback' to help rub in lessons prior to repetition of the exercise.

Transport aircraft. Before **World War I**, **airships** were used to carry passengers and in 1917 the Zeppelin L59 nearly succeeded in taking 15 tonnes of supplies from Bulgaria to East Africa for German **guerrillas** in East Africa. But despite the symbolic attempt in 1916, with totally inadequate aircraft, to supply the besieged British garrison in **Kut-al-Amara**, the design of transport aircraft lagged behind combat types until after the war.

Initially in the 1920s, the aircraft used by nascent airlines and by the British for air trooping were converted **bombers**. Gradually specially built multi-engine machines, including flying-boats, were produced, of which the German all-metal, tri-engine Junkers Ju52 monoplane of 1932 was the most significant. As multi-engined **aircraft** with greater capacity, longer range, higher speeds and improved reliability came into service, the military seized upon the advantages of mobility offered. The movement by air of General Franco's Moroccan troops at the start of the **Spanish Civil War** was as significant as the adaptation of airliners for **parachute** and other **airborne** operations throughout **World War II**. In parallel with air transport's demonstrations of flexibility and capacity for **logistic** and operational purposes, its mobility was augmented by the essential construction, worldwide, of the airfields and maintenance facilities its own logistics demanded.

In the course of the air supply of Tunisia and Stalingrad in 1942, and in support of the **Pacific Ocean War** and the **Burma** battles, military air transport laid the foundations of present-day civil air networks. The process was reinforced by the **Berlin Airlift** in 1948–9 and the reinforcement of Korea (*see* **Korean wars**) in 1950, and given its most significant boost by the introduction into service in 1950 and 1952 of, respectively, the British Viscount turbo-jet and the Comet jet airliners.

The vast power of the **jet** engine made feasible more reliable transports of such economy that, by the 1960s, air transports had outmoded troop-ships by a substantial margin of cost-effectiveness and had spurred the construction of cargo aircraft, such as the Lockheed Galaxy with its 117 tonnes lift, which could solve emergency logistic problems as by no other means. Probably the most important transport built is the rugged Lockheed C130 which, initially powered by four piston engines and subsequently turbojets, fulfils a host of different roles including tactical assault on rough terrain, tanker for air-to-air refuelling, gunship, **electronic surveillance**, search and rescue, aerial survey and polar supply fitted with skis.

By the 1980s vast air-transport fleets revolutionized **strategic** mobility to an undreamt-of extent, with immense impact on national military policies and budgets, including the reinforcement of theatres of war with troops and heavy equipment, yet without in any way supplanting sea transport as the main bulk carrier.

Trenchard, Air Marshal Sir Hugh (Viscount) (1873–1956). A soldier of dominating personality who fought in the Boer War but learnt to fly in 1913, ready to take a leading part in **World War I**. In August 1915 he took command of the Royal Flying Corps in France as it was fast expanding. He developed a tactical doctrine based on offensive action over the enemy lines and thus greatly helped the ground forces. But it was very expensive in air crews. In January 1918 he became Britain's first Chief of the Air Staff, to play a vital role in the creation of the Royal Air Force (RAF) on 1 April. In June he took command of the **Independent Air Force**, tasked (at his suggestion) to carry out **strategic bombing** of industrial targets in Germany. This proved a costly and by no means very effective **strategy**, yet it convinced Trenchard that air forces were capable of winning by independent action, an article of faith similar to those of General **Douhet**, and one he imposed on the RAF when again made Chief of the Air Staff and began a fight, from 1919 until 1929, to save the RAF from extinction.

Trenchard was a modern administrator with technological awareness. To his credit stand the RAF's cadet, staff and apprentice colleges upon which its excellence and spirit were founded, a system he tried, controversially between 1931 and 1935, to impose when reforming the Metropolitan Police. But when offered the job of Minister of Defence by Winston **Churchill** at the nadir of Britain's fortunes in May 1940, he declined in doubt of the task's practicability.

Trireme. A powerful Grecian-designed **galley** with three banks of oars and, occasionally, sails, and fitted with a **ram**. For short bursts

A Roman trireme

it was capable of a speed of eight knots and was used in **Mediterranean Sea battles**, notably by the Romans, until *c.* AD 1200.

Truman, President Harry (1884–1972). A bank clerk who became Vice-President of the USA and succeeded President **Roosevelt** when the latter died on 12 April 1945. A man of strong resolution, it was his destiny to take military decisions of world-shaking importance as **World War II** drew to a close. He gave the order to drop the **atom bomb** on Japan at the same time as the Potsdam conference was indicating the brewing troubles of the war's aftermath. Recognizing the Russian Communist threat to Europe, he pronounced in 1947 the so-called Truman Doctrine, pledging the USA's support for free people resisting armed minorities, and leading, at the start of the **Cold War**, to the **Marshall** Plan to restore European prosperity; the setting-up of **Nato** in the shadow of the **Berlin** siege; and taking the lead in United Nations intervention in the **Korean War**.

Lacking military training, Truman nevertheless had a Commander-in-Chief's attribute in his ability to analyse factors and convert them into positive decisions. His sacking of General **MacArthur** in 1951 was by no means the least courageous act of this great President, whose domestic politics were set in a low key.

Truppenamt (Troops Bureau). A unit formed by General Hans **von Seeckt** within the remodelled Reichswehr after the German General Staff was outlawed by the **Versailles Treaty** in 1919. This small élite would include technically aware staff officers of the calibre of **Guderian**, **Kesselring**, Franz Halder, Eric von Manstein and **Student** who, in secret, applied the lessons of **World War I** to develop German military doctrine and equipment against the day in 1935 when Adolf **Hitler** repudiated the Versailles Treaty, announced rearmament and re-established the General Staff; and who in **World War II** would be innovative leaders of **mechanized**, airborne and **air forces**.

Tsushima, Battle of. After the elimination of the Russian Far East Fleet at the Battle of the Yellow Sea on 10 August 1904, and with **Port Arthur** tightly besieged, the Russians decided in desperation to send their Baltic Fleet, under Admiral Zinovy Rozhdestvensky, to Vladivostok in hope of restoring the situation. Neither its 12 **battleships**, nine **cruisers** and nine **destroyers**, nor their crews were fit for battle. This was especially so at the end of the 20,000 miles voyage, during which mechanical defects were chronic and **logistics**, particularly coaling arrangements by all manner of devious means, a nightmare. A few went through the **Suez Canal**, the main body round the Cape of Good Hope. **Morale** declined from the day of sailing on 10 October to the evening of 26 May 1905 when, with the **Russo-Japanese War** already won by Japan, they neared Tsushima Straits.

Here cruised Admiral **Togo** with 10 battleships, 18 cruisers, 21 destroyers and 60 **torpedo-boats.** Fully repaired, faster by two to three knots, battle hardened and well informed of the Russian strength and course by **radio** intercepts and scouting cruisers, the Combined Fleet was superior in all respects. From first contact on the 27th, Togo and his captains outmanoeuvred Rozhdestvensky, whose formation lapsed into confusion under cruiser fire before Togo's battleships got within range at 14.05 hrs. With precision Togo repeated the classic 'crossing of the T' to bring concentrated fire against the outclassed Russian battleships. One by one they were sunk or chased into the gloom where they scattered and were hunted through the night and into the following day. Statistics speak for themselves. The Japanese lost only three torpedo-boats and 117 killed. Only three Russian ships reached Vladivostok; all their battleships were sunk, captured or interned in neutral ports; and 4,830 men were killed and 5,917 captured. Of 100 Japanese **torpedoes** fired only seven hit, but they accounted for two battleships and two cruisers.

Tukhachevsky, Marshal Mikhail (1893–1937), was an **infantry** officer of the Russian

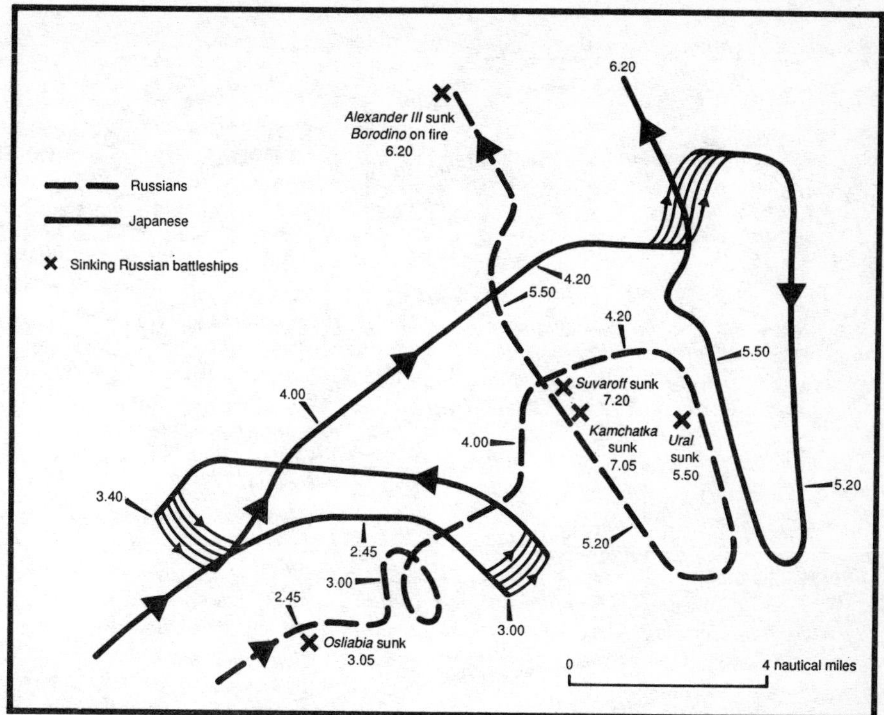

The Battle of Tsushima, 27 May 1905

Army during **World War I**, a protégé of Josef **Stalin** who, though lacking staff training, rose to command the Russian West Front during the Russo-Polish War of 1920. Though defeated at the Battle of Warsaw, he continued in favour to command the Red Army and, after the Russian Revolutionary War, began its modernization as part of Stalin's **armaments** industrialization policy. He gave priority to the development of **mechanized, armoured** and **air forces** modelled on British, French and German lines and had made considerable progress prior to his liquidation during Stalin's purge of the officer corps.

Turbines. Any of various devices that convert the energy in a stream of fluid (water, steam, gas) into mechanical energy by passing the stream through a system of fixed and moving fan blades, causing the latter to rotate.

The first water-wheel may have been invented in the 1st century BC; Hero of Alexandria's tube, rotated by steam from attached nozzles, was, in the 1st century AD, the first reaction motor; and Giovanni Branca's use in 1629 of a steam jet playing on blades attached to a wheel was the first impulse turbine. Chiefly, however, it was water-wheels which led turbine development to provide power for industry until William Avery, in America in 1832, built a few small but very noisy steam-reaction turbines, one of which was tried in a locomotive.

In 1884 a turbine was used to power a **dynamo**. But it was in England in 1889 that Sir Charles **Parsons** revolutionized impulse steam turbines by fitting several stages of blades to a main shaft in order to improve efficiency and increase the power delivered. In 1897 one of his engines was installed in a 44-tonne boat called *Turbinia* and displayed

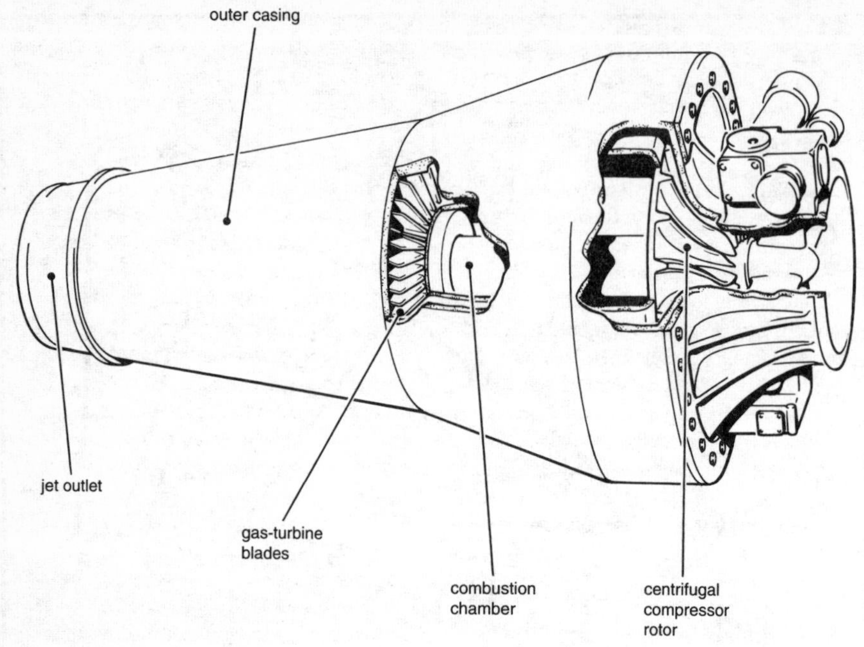

Gas-turbine jet engine

before an astonished world as it sped at 34 knots amongst the British Fleet, assembled at Spithead for Queen Victoria's Diamond Jubilee Review. *Turbinia* was followed in 1899 by HMS *Viper*, a 550-tonne **torpedo-boat**, similarly powered and capable of 30 knots. The scene was set for the widespread introduction of marine turbine engines, infinitely more flexible and reliable than the reciprocating engines they replaced.

The impact of the turbine on the military scene was not confined to ships, dramatic though that was. In the decade preceding **World War II** there was a revived interest (following experimental work in France and Switzerland prior to 1914) in the gas turbine as a means of powering aircraft. The Germans took the lead initially with their **jet**-powered Heinkel HE178 which first flew in 1939. Their subsequent Messerschmitt ME262, though achieving operational service in 1944, was never satisfactory.

In Britain Frank **Whittle** designed turbine blades to withstand temperatures in excess of 3,000°C and the centrifugal effects of rotations at 30,000rpm or more, in an engine which linked an internal combustion turbine with a jet nozzle, the main components comprising a single-stage centrifugal compressor, a single-stage turbine, and a single combustion chamber into which liquid fuel was injected and burnt. Part of the exhaust gases thus generated drove the turbine, the remainder expanding through the jet nozzle. Subsequent refinements yielding improved performance, better fuel economy and quietness in operation have led to jet power being used for most of the world's military and civilian aircraft.

The gas turbine has also found an application as a power source for land vehicles. The American M1 Abrams main battle **tank** and its variants use a Lycoming 1,500bhp gas-turbine engine which can produce more power for a given size than can a diesel; but the much higher fuel consumption requires much larger fuel tanks for an equivalent radius of action. Thus, while the vehicle has an impressive performance, little space or weight has been saved in the overall size of the power pack and its ancillaries. Nevertheless a trend has been set which other designs are likely to follow.

Turing, Dr Alan (1912–54), was arguably the most influential mathematician of the 20th century; he became a Cambridge University (England) don at 22. Following in the intellectual footsteps of Charles **Babbage**, he envisaged a 'thinking' **computer** with a stored memory exploiting binary mathematics. His ideas developed when working during **World War II** to break the German Enigma **codes and ciphers**. While writing brilliant programmes for the Ultra organization, he was crucial in devising Colossus II in 1944, the first electronic computer with a memory. He also made Delilah, a secure speech machine.

Turner, Admiral R. Kelly (1885–1961), was one of the most forceful and versatile American naval officers of the 20th century. A **gunnery** specialist who became an aviator, he served in **destroyers**, **cruisers**, **battleships** and **aircraft-carriers** prior to 1940 when he took charge of Naval War Plans, Organization and Doctrine and Development of **Landing-craft** and **Amphibious Warfare**. In 1942 he was appointed Commander of the Amphibious Force for the landings on Guadalcanal Island as a prelude to the subsequent, decisive **Pacific Ocean War** operations in the **Solomons**, at Tarawa, Makin, the Marshall and the Mariana islands, Iwo Jima and the Okinawa islands.

Typewriter. Attempts to make a machine that could write faster and more legibly than the pen, controlled by a keyboard, date from 1714. Not until 1868, however, was a practical typewriter patented by Christopher Sholes and developed for production in 1873 by the Remington Arms Company. As quickly improved, the Remington machine possessed the essential cylinder, line-spacing, carriage-movement, inked-spool, key-striking and shift-key features, along with the virtually standardized QWERTY keyboard for typewriters, **printing** machines, teleprinters (*see* **Telegraph**) and **computers** (including word processors). Rapidly the typewriter transformed **communications** and commercial and military practices. It was further enhanced by the use of carbon copying paper for duplicating and by such modifications as electrification and improved printing methods to raise efficiency and speed of output. Typewriters also provided many women with employment as typists – whose skills, regardless of sex, remain crucial to full production of text.

U

U-boat. *See* **Submarines.**

UN–Iraq War. On 2 August 1990 Iraq invaded Kuwait for her oil and for control of the head of the Persian Gulf. Organized resistance was quelled in 24 hours. Saudi Arabia was rapidly reinforced by US, Arab and European forces (mainly British and French) and by those of many other nations. A series of United Nations resolutions were passed of which the tenth gave authority to retake Kuwait by force.

When the Iraqis, under Saddam Hussein, fortified Kuwait, UN forces under General Norman Schwarzkopf imposed a **blockade**. **Propaganda** and **psychological warfare** were rife. On 17 January 1991 UN air forces blotted out the Iraqi **C³I** system by attacks on their **communications** and **electronic** installations. Within a few hours the Iraqis were deaf and blind and thus unable to prevent destruction of their economy and war-making capability. The extremely versatile UN attack was made especially deadly by the use of **robots**, **stealth aircraft**, **cruise missiles** and MLRS (*see* **Rockets and guided missiles**) which were very reliable and accurate in finding and destroying vital, pinpoint targets. Iraq's air force was destroyed or grounded, her navy hunted down and her army and **logistic** system hammered. As a result the UN achieved complete surprise for a classic **desert warfare** campaign by deceiving the enemy with the elaborate threat of an **amphibious** attack on Kuwait.

Instead, on 24 February, the UN forces launched a left hook round the Iraqi flank to envelop and overwhelm their army in 100 hours and compel them to obey President George Bush's demand for a cease-fire. Approximately 75 per cent of the 24 Iraqi divisions and their logistic support in southern Iraq and Kuwait were lost along with 4,900 tanks and 2,300 guns. Only 211 UN sailors, soldiers and airmen were killed.

Technology's effects had won a revolutionary operational victory and exposed the obsolescence of the in-service Soviet Russian military systems in Iraqi use. The Russians were forced to change their doctrine to cope with the 'new technology war'; and recognized that the **deterrent** need not necessarily depend upon **nuclear** action.

Uranium is a dense, hard metal discovered in pitchblende by Martin Klapwroth in 1789. Not until 1938 did Lise Meitner, Otto Hahn and Fritz Strassmann show that it could be broken down into radioactive isotopes – thus making feasible the release of **nuclear** energy (the equivalent of three million tonnes of **coal** to one tonne of uranium) and, eventually, the **atom** and **hydrogen bombs**. Strategically significant deposits have been discovered in several parts of the world for use as boiler fuel for nuclear **submarines**, warships and power stations. It has also been alloyed with other metals and used for enriched **armour**-piercing **ammunition**.

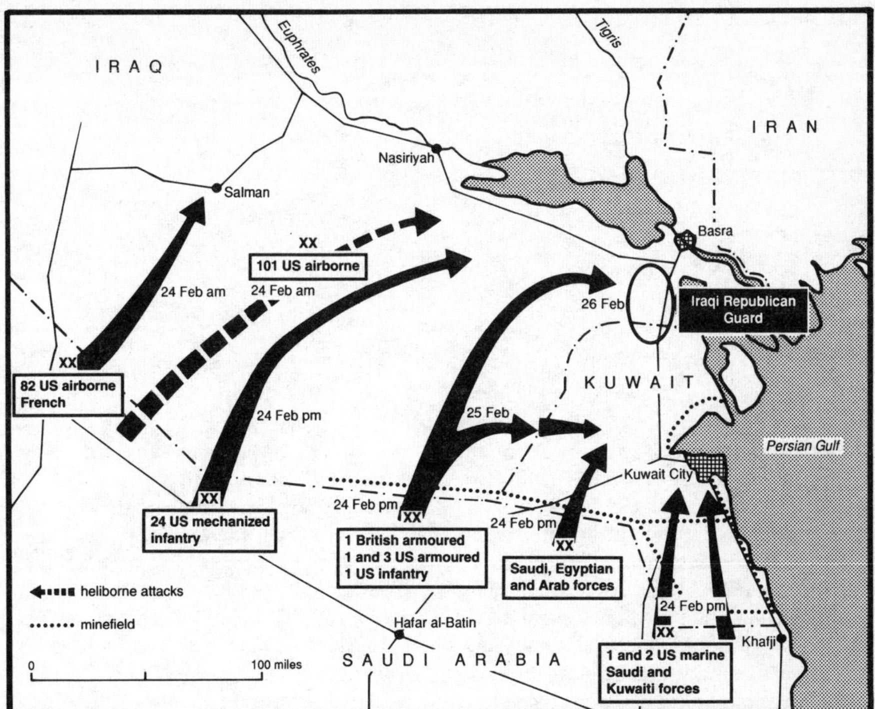

The UN–Iraq War, 1991

V

V1. The 'flying bomb', as it was popularly known, was launched by the Germans against targets in Britain and Belgium from June 1944 until March 1945. It may fairly be described as the world's first **cruise missile**, powered as it was by a pulse jet engine (patented in 1907 by the Frenchman Victor de Karavodine) and flying comparatively slowly (580kph) and at low altitude (600–900m) to its target. With a payload of 900kg and a range of 240km, V1 was guided by a **gyroscopic** automatic pilot monitored by magnetic compass, its dive on to the target at a measured distance being determined by the number of revolutions of a small propeller.

With its simple guidance system (though at least proof against electronic countermeasures), V1 was essentially inaccurate and mostly launched from simple ramps against the general target of London. It could be launched from aircraft, a few northern cities becoming targets to V1s launched from over the North Sea.

Some 10,492 V1s were launched against Britain, of which 3,531 penetrated the defences. Nearly 4,000 were downed by a combination of **balloon** cables, **fighter** aircraft and guns using the new **radar** proximity fuse. The remainder failed either on launch or in flight.

V2. The development of the German V2 stemmed from work by Dr von **Braun** in 1936 on the German rocket A4. Powered by liquid fuel (hydrogen peroxide and liquid oxygen), the **rocket** motor developed 27,180kg of thrust, propelling the missile with its 906kg warhead to an altitude of 97km and giving it a range of 350km. Though V2's autopilot was a wonder of technology, the electric torque motor driving the gyros being the smallest of its kind yet made, it was no more accurate than V1, largely because of launch inaccuracies and motor shut-off variations (arising from current

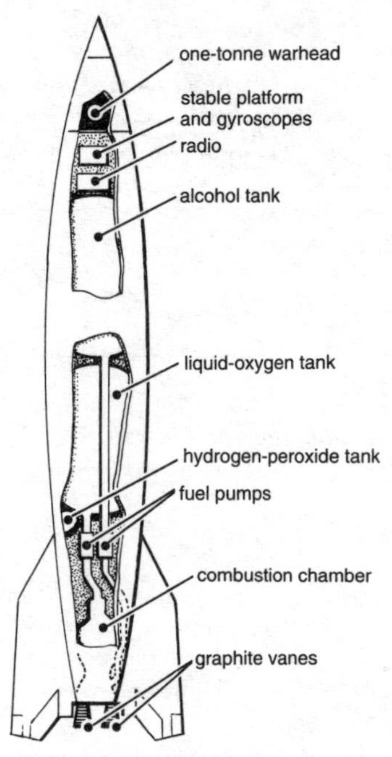

V2 rocket missile, 1944 (German)

supply frequency variations) and poor pick-up from the pitch gyros to the control-surface servos.

V2, like V1, was unaffected by electronic countermeasures and, additionally, could not be intercepted by the defences once it had been launched. Effort was therefore concentrated on attacking the launch sites, once Paris had been attacked on 6 September 1944, soon followed by the launch of 3,195 missiles against Britain and Belgium, attacks ending in March 1945.

Though V2's effectiveness was modest (under two killed per missile, though with some **morale** effect since there was no attack warning), its significance as a forerunner of modern ballistic missiles and **space vehicles** cannot be denied.

Valmy, Battle of. On 20 September 1792 an allied Prussian and Austrian army of 34,000 and 58 guns, under the Duke of Brunswick, which had invaded France to overcome the **French Revolution**, attacked a French army of 52,000 and 56 guns under General Charles Doumouriez. Brunswick lacked enthusiasm because of inadequate **logistics**. Doumouriez's mixture of disaffected regulars and untrained volunteers was unreliable, although the **artillery**, the product of General de **Gribeauval**'s reforms, was good. The allies advanced tentatively across soft ground to the accompaniment of a long-range artillery duel – which caused fewer than 500 casualties all told since many of the 20,000 rounds fired stuck in the mud, but provided a convenient **smoke**screen to cover deployment. Yet, although there was momentary panic among the French, Brunswick lost heart when they appeared to stand firm. So, at the first opportunity, he called off the assault and withdrew. The Revolution was saved, French **morale** bolstered and the Comte de **Guibert**'s theory of an effective citizens' **militia** given credibility.

Vauban, Marshal Sébastien de (1633–1707). Vauban's study of **fortifications** led him to become a master of **siege warfare** and the designer of the sophisticated French star-shaped forts, which were copied by other nations and sufficed for nearly 200 years until made obsolescent by the latest **artillery**. Vauban was engaged continuously in the **French wars** between 1653 and 1706, sometimes in command of **infantry**. He pioneered statistics and wrote authoritatively on river navigation, colonization, economics and peace, besides siege warfare, in *De l'attaque et de la défense des places*, published in 1737.

Vegetius, Flavius (*c.* AD 360–400), was an amateur soldier and military **philosopher** with scanty experience of war, whose *De re militari* (Military institutions of the Romans), based on a study of obsolescent organizations, doctrine and methods, was intended to restore Rome's waning power (*see* **Roman wars**). The book failed because moral and professional decay had gone too far, but survived to be read by leaders of the **Crusades**; to be translated into English, French and Bulgarian; and to be printed in 1473. It was superseded in vogue in the 16th century by **Machiavelli**'s works; although some of its maxims are still quoted, including, 'He who aspires to peace should prepare for war.'

Verdun, Battles of. Standing on the River Meuse and guarding one of the main approaches to Paris, Verdun has been fortified since Roman times and often threatened. In 1792 its **Vauban** star-**fortifications** fell to Austrian invaders prior to the Battle of **Valmy** and during the **Franco-Prussian War** withstood a siege for two months. Starting in 1874 it was modernized by General Brialmont and held firm throughout **World War I**.

In 1916 General von **Falkenhayn** decided on an experiment in attritional warfare at Verdun by limited German **infantry** attacks supported by unlimited **artillery** fire designed to 'bleed the French Army white' in a 'mincing machine'. Correctly he reasoned that the fall of Verdun would deal a shattering blow to French **morale** and perhaps win the war. Unbeknown to him the French no longer had faith in forts and had not only largely disarmed them but also permitted the entrenchments to fall into decay. The offensive started on 21 February on an eight-mile front with support from over 1,200 guns well supplied with **ammunition**. The vital but undefended Fort Douaumont fell on the 25th to cause an almost fatal crisis of confidence. Next day General **Pétain** was placed in command by Marshal **Joffre**, told to hold out and fortuitously was given time to repair the defences before the Germans renewed their main assault.

The key to Pétain's success was supply by **motor vehicles** along the only main road leading into Verdun, the **railway** being out of action from shell-fire. The next German mistake was to change their tactics and expose their infantry to as much attrition as the French, followed by Falkenhayn's decision for political and prestige reasons to continue the assault long after it was recognized as counter-productive. It ground on into July with appalling losses until the British offensive on the **Somme** gave Falkenhayn an excuse to call it off. But the French, now under General Robert Nivelle, counter-attacked, recapturing Douaumont on 24 October and Fort Vaux, which had been lost on 9 June, on 2 November. By then Falkenhayn had been replaced by **Hindenburg**; and the casualty bill amounting to 542,000 French and 434,000 Germans had almost ruined the **morale** of both sides.

Versailles, Treaty of. This post-**World War I** peace treaty, signed on 28 June 1919, redrew many European and colonial frontiers and imposed severe penalties on Germany. These included the disbandment of the Great General Staff, the restriction of her army to 100,000 men and her navy to 16,500 men, and the abolition of **conscription**. She was also forbidden warships in excess of 10,000 tonnes, **submarines**, **fortifications** within 50km of the coast, heavy **artillery**, **tanks**, **chemical warfare** weapons and an **air force**.

Secretly the Germans breached the treaty (*see* **Truppenamt**) and promoted technological **research and development** of forbidden weapons, such as **battleships** and **armoured fighting vehicles** and new weapon systems like nerve gases, **rockets and guided missiles**.

Vickers Company. As flour millers, the Vickers family turned in the 1750s to **iron**-making. Vickers & Sons were formed in 1867 by the brothers Thomas (a brilliant technologist) and Albert and, in the 1880s, started **armaments** manufacture. In 1897 the firm amalgamated with Hiram **Maxim**'s company to make **machine-guns** and **ammunition**; and began **shipbuilding** and the construction of **Holland submarines**. In 1908 it built **aircraft** and **airships**. It prospered during **World War I** and in 1927 amalgamated with the **Armstrong** and **Whitworth** companies (and others) to create a virtual monopoly in British armaments manufacture and exporting, including **armoured fighting vehicles**, prior to **World War II**. Since then its fortunes have fluctuated with the **Cold War** and against considerable competition.

Vicksburg campaign. The essentially decisive moves of the **American Civil War** were made in April 1862 after the Shiloh campaign and the capture of New Orleans (by Commodore David Farragut's squadron) gave the Federals dominance over the Mississippi River. Destruction of Confederate gunboats at New Orleans let Farragut steam up-river against the key city of Vicksburg, with the aim of cutting the Confederacy in half and ruining its war economy. But in June he withdrew after suffering heavy damage against its **forts** and failing to prevent the ironclad *Arkansas* breaking through to the city, which the Confederates, under General John Pemberton, now fortified extensively.

In July command of weakened Federal armies in the West passed to General Grant who won defensive victories at the battles of Iuka (19/20 September) and Corinth (3/4 October). By November, he felt strong enough to move against Vicksburg, sending an **amphibious** force of 40,000 under General **Sherman** and Admiral David Porter down-river from Memphis; and in December himself advancing towards Grand Junction.

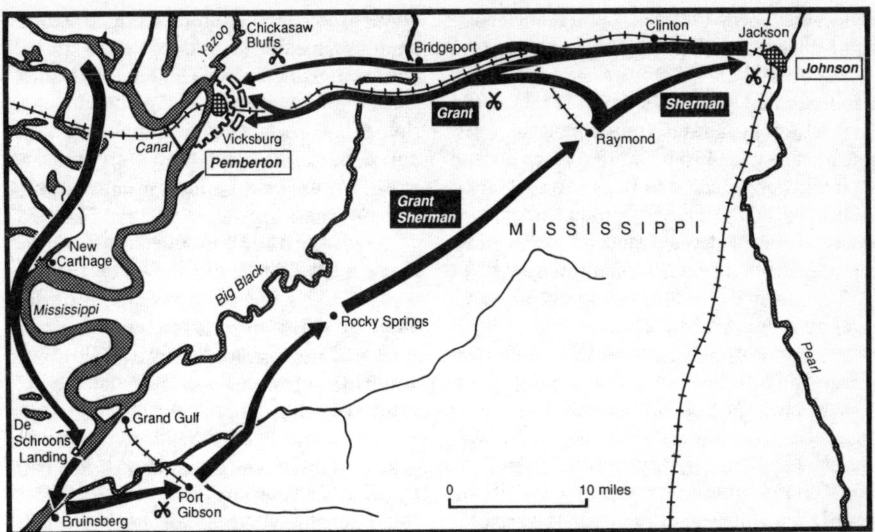

The Battle of Vicksburg, 1863

But he stalled owing to loss of his supplies at Holly Springs to a Confederate raid on the 20th; and Sherman's attempt to storm Vicksburg via Chickasaw Bluffs was repulsed with heavy loss on the 29th.

From January to April 1863, Grant manoeuvred Sherman, Porter and Farragut round Vicksburg via the west bank, dominating the river and progressively shifting the army southwards to cross the Mississippi at Hard Times and strike eastwards towards Jackson. On 1 May Grant cast off from his communications, drove back the thoroughly confused Confederates at Port Gibson, forced General Joseph Johnston out of Jackson on the 14th and then turned west to defeat Pemberton at Champion's Hill on the 16th. Meanwhile Sherman protected Grant's rear at Jackson, prior to using the only available pontoon bridge to cross the Big Black River and complete Vicksburg's encirclement from the north. Two costly assaults on the 19th and 22nd finally convinced Grant that bombardment and raw starvation were the best way to capture the city, leaving Sherman to prevent Johnston interfering from the east. On 4 July Pemberton surrendered with his emaciated force.

Vietnam wars. From the moment the French imposed their rule upon Cambodia, Laos and Vietnam in the 1880s, nationalist resentment was born. There was a minor uprising in 1908 and growing agitation and action in Vietnam during the 1930s by Ho Chi Minh's Indo-Chinese Communist Party. At the end of the Japanese occupation after **World War II**, Ho Chi Minh proclaimed a Vietnamese Republic but his 10,000-man Vietminh **guerrilla** force in the north, under Vo **Giap**, was kept in order by British, Indian, French and Japanese troops, and then tackled head-on by the French in 1946.

The **guerrilla war** which ensued was conducted most skilfully on Chinese Communist lines by Giap against French forces whose equipment, like the Vietminh's, was a mixture obtained from the last war's contestants. Forty thousand Frenchmen under General Henri Leclerc were unable to stamp out 30,000 guerrillas who infested the countryside and the towns, including **Hanoi**. Vietminh **logistics** were tenuous and dependent upon the goodwill of the populace, who were cared for with tact. But things changed decisively after China had fallen to the Communists in 1950. A regular supply route to the north was opened. Attacks upon French communications and bases took place in unison with the taking of control of hinterland communities by the Vietminh. From 1951 onwards, thousands of French-made bicycles, each carrying 450lb, were being pushed through jungle trails to the guerrillas. Such successes as the French sometimes enjoyed in capturing concealed stockpiles were only transitory because replacement was swift.

The crunch came in November 1953 when the French flew 15,000 men with 28 field guns and a few light **tanks** to Dien Bien Phu, which they developed as a jungle base for decisive operations deep into the enemy rear. Giap accepted the challenge by surrounding Dien Bien Phu with numerous **anti-aircraft** guns and 200 field guns. It took nearly four months to assemble these pieces along with the 60,000 infantry (80 per cent of Giap's army) needed for the assault – a concentration neither French troops nor air strikes could check. By 13 March 1954 the French were penned, with air supply strangled by anti-aircraft fire. By holding out until 7 May, however, the French merely delayed the inevitable: the negotiated agreements which put an end to French Indo-China and granted sovereignty to Cambodia, Laos and Communist North and non-Communist South Vietnam.

In January 1959 North Vietnam decided to initiate an 'armed struggle' against South Vietnam and started construction of the **Ho Chi Minh Trail**. In April 1960 they imposed military **conscription** and began the infiltration of the North Vietnamese Army (NVA) cadres into the South to support the indigenous Vietcong (VC) guerrillas. Also in April 1959 President Eisenhower announced that the USA would support the South and began increasing the number of advisers in place.

As the USA intensified its efforts to train, equip and expand the South Vietnamese Armed Forces (SVNAF), Giap increased pressure with widespread attacks. Only slowly did the Americans and SVNAF come to realize they were faced with a well-supplied NVA effort behind the VC's guerrilla warfare. Not until mid-1961, when several army **helicopter** troop-lift squadrons were brought in, were Americans permitted to fire 'in self-defence'. At the end of 1962 11,300 Americans were engaged in a big shooting war, though attacks by land or air upon the North and the Ho Chi Minh Trail were forbidden and always would be hampered by rules of engagement.

In January 1963 the VC inflicted a severe defeat on the Army of the Republic of Vietnam (ARVN) (the old SVNAF) at Ap Bac. Simultaneously their **propaganda** began to undermine resistance within a divided government. Not until June 1964 did the USA really grasp the nettle by appointing General **Westmoreland** to command and permitting a less inhibited war, including **blockade** of the North's coastline and **bombing** of the North in retaliation for an attack on a US destroyer in August. In successive years the intensity of fighting grew at sea, on land and on both sides of the frontier in the air. Hamstrung by politics the frustrated forces of the South fought with one hand behind their backs while Giap, the NVA and VC exploited every opportunity.

Air attacks, opposed by **fighters** and **rocket missiles**, against **logistics** targets in the North and the arrival of Australian, New Zealand, and South Korean troops, were a feature of 1965. As the first American **armoured** forces began to deploy, it began to dawn that this was something more than a guerrilla war and the peace movements sounded louder in the USA. In 1966 the Battles of **Hanoi** began in earnest, there was defoliation by herbicides of the jungle to destroy enemy cover, and Filipino troops joined in, bringing so-called Free World Military Forces in the South to 385,000 Americans, 736,000 AVRN and 52,500 others. The year 1967 brought an overdue change in **tactics** and was also a time for the anti-war movement in the USA to make a significant impact. All this was on the eve of the climactic siege of Khe Sanh and the **Tet Offensive** in January 1968, with the comprehensive defeat of Giap which was rooted very much in the use of **transport aircraft** and helicopters. The latter sometimes flew in flocks of 200 or more

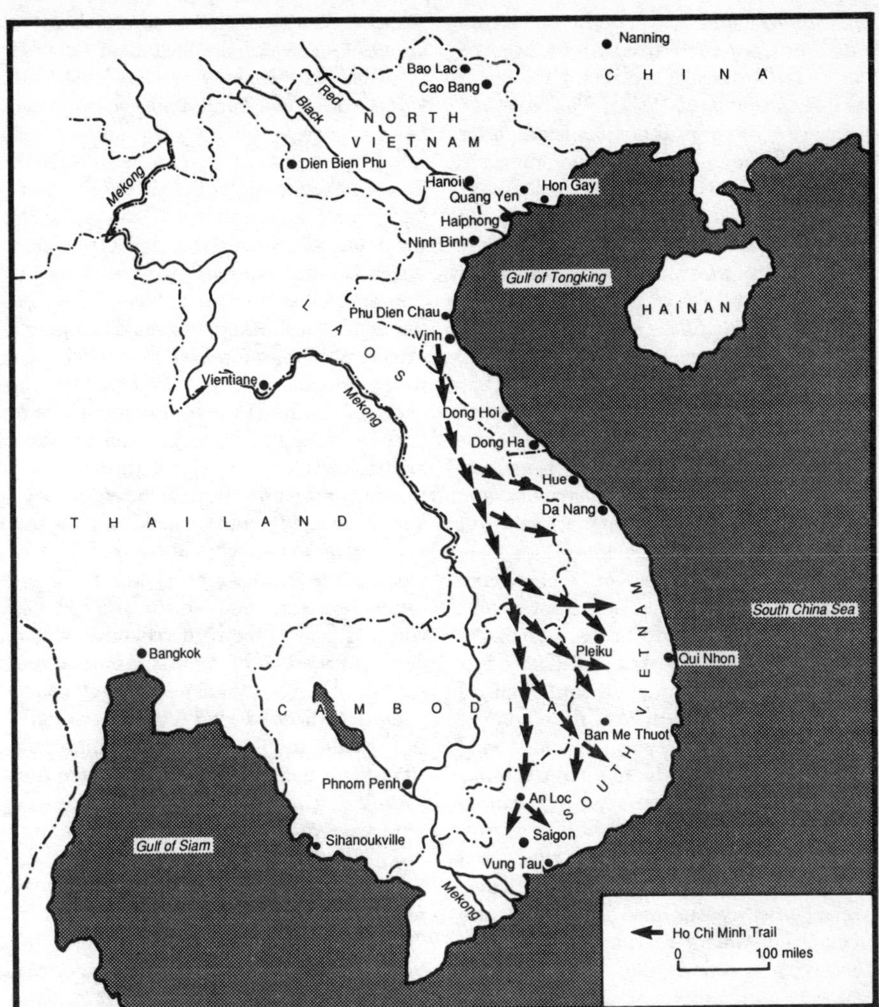

Vietnam and the Ho Chi Minh Trail

machines to break sieges and strike the enemy rear by so-called sky-cavalry tactics.

From then until 1973 successive US governments sought ways out of their dilemma while the fighting went on. The North might have been beaten if General Westmoreland had been allowed to invade Laos and cut the Ho Chi Minh Trail in the same manner as the raids into Cambodia in 1970 eliminated a major NVA base and produced a stinging victory which only had to be followed up. For use of massed fire-power and the latest **technology**, including solid-state **electronics**, surveillance, sea-**mines** and **robotic** weapons, was bringing the North's economy and logistics to the verge of collapse. And defeat of its spring offensive in 1972 seriously undermined the North's confidence as well as Giap's credibility. But as usual a politically scared US government dithered – until, in December 1972, it finally lost patience with the North's procrastination over peace negotiations and ordered unrestricted bombing of all targets connected with the enemy's war effort. This **strategic bombing** brought a cease-fire in January 1973, followed by the progressive withdrawal of US troops. Never before had the USA lost a war. Moreover Giap's successor, General **Dung**, had already topped up the Ho Chi Minh Trail supply line and, on 13 December 1974, launched the attack which swept aside a demoralized ARVN with such rapidity that Saigon was captured with hardly a fight on 30 April 1975 to end the war.

Viking wars of expansion. The Vikings were pagan, outcast Scandinavian warriors, sometimes known as Northmen or Norsemen who, *c.* AD 800, began **amphibious** raiding in their long, 50-man **galleys**. By mid 9th century they had established fortified, feudal settlements on the coasts of Ireland, Scotland, England, Iceland, the Netherlands, France, Spain, Germany, the eastern Baltic states and Russia. They expanded these lodgements into colonies in north-east England and in France where, in 886, their army of 30,000 was thrown back into Normandy after besieging Paris. In the 10th century they also penetrated the Mediterranean to reach southern France.

Lightly **armoured** and armed with **sword** and battleaxe, they were very destructive. Accomplished coastal navigators, they also crossed the Atlantic to America (Vinland). The fiercest resistance they met in Europe, besides that of the French, was organized by King Alfred of Wessex who in 870 and 871 checked their invasion of southern England in nine battles. He beat them many times again through the establishment of Britain's first standing army and strategic **fortifications**, plus a fleet, which stimulated **shipbuilding** and a great **naval warfare** tradition. Measures which paid off when Alfred occupied London in 885 and, between 892 and 896, repelled a Norman invasion at the same time as he raided Viking territory in the north.

Alfred's encouragement of law, order, Christianity and education won time for Viking civilization. For though the Normans eventually defeated the English at Hastings in 1066, they were by then ordered by a feudalism, **chivalry** and a slowly emerging, new technology.

Villers-Bretonneux, Battles of. On 24 November 1870 at a nadir in France's history in the **Franco-Prussian War**, 17,000 soldiers under General Bourbaki gave as good as they got against a superior Prussian force on the ground overlooking this small village, 12 miles to the east of **Amiens** – which fell, notwithstanding.

On 24 April 1917, after the first battle of the Hindenburg Offensive had stopped short on the same ground as on the 4th, British and Australian troops, with three heavy Mark IV and seven light Whippet **tanks**, fought five German divisions, with fifteen heavy A7V tanks, in the first tank-versus-tank battle.

General Georg von der Marwitz's limited aim was to seize the high ground to obtain direct observation of Amiens and then 'act according to the situation'. In foggy conditions, made denser by gas (**Chemical warfare**) and **smoke**, the German tanks entered the village ahead of the **infantry** they were supporting and drove towards the woods and the village of Cachy beyond. Here they encountered the three Mark IV tanks, damaging and driving off two, being shot up by the third, which knocked out one A7V as the German assault lost momentum. To the south of this skirmish the attack made less progress towards Cachy because initially the A7Vs lost direction. They too were unsupported since, in the midst of this fight, the seven British Whippets arrived and scattered the infantry. The attack faded, even though an A7V, unseen by the Whippets, had set fire to one, damaged another and driven off the rest. The demoralized German infantry, however, had suffered enough and declined to go on.

Vimy Ridge, Battles of. This long, northwards-facing, steep escarpment dominating the Plain of Douai and shielding **Arras**, fell into German hands in August

1914 at the start of **World War I**. In October it was a start line during the Race to the Sea to encircle Arras, a struggle which failed to reach its prime objective but nevertheless left the adjacent Ste-Lorette spur also in German hands. There in May, June and September 1915 the French strove with fanatical courage to recapture this vital ground, attacking with little hope of cheap success since they were deficient of adequate heavy **artillery** and **ammunition** to overcome the already very strong enemy **fortifications**. Ste-Lorette fell but not one inch of Vimy Ridge.

In March 1916 the British took over this sector and at once entered into a vicious underground struggle with the Germans, who had tunnelled deeply into the chalk substrata. Mining and counter-mining went on to the accompaniment of raids on the surface, a battle the British won before the Canadian Corps, under General **Currie**, took over in October. In the British offensive at Arras on 9 April 1917, it was Currie's task to secure the left flank by seizing the Ridge with his four divisions on a 7,000 yards frontage. The subjugation of the German defences was carried out by over 1,000 guns (including 400 heavies) which fired 80,000 tonnes of ammunition in three weeks, further to pulverize a quagmire. But the infantry, who had reached their assault positions in the safety of tunnels dug far back behind the lines, went over the top with confidence in a meticulous, well-rehearsed plan. Nearly all objectives were taken – the remainder falling on the 10th and 12th.

For the rest of the war the Ridge remained firmly in Allied hands, easily withstanding a major German assault on 28 March 1918 as part of the Hindenburg Offensive and acting as a springboard for the British and Canadian attack which, on 25 August, pushed the Germans to final defeat.

But in **World War II** the Germans were back again, seizing the Ridge on 24 May 1940 in the aftermath of a stiff fight for Arras and the slopes of the Ste-Lorette spur – a symbolic act of the Battle of France – only to lose it again without a fight to British troops on 31 August 1944.

Vinci, Leonardo da (1452–1519) was a trained military **engineer**, with experience of **siege warfare**, and a clever inventor in addition to being a great artist living at the heart of the Renaissance period. He dabbled in firearms, personal **armour** and **mortars**; and sketched a man-powered **armoured fighting vehicle**, and **aircraft**, **bombs** and a giant crossbow. More practical, however, were his designs of bridges, **siege engines**, **edged weapons**, water-wheels (*see* **Turbines**), gear trains and lifting gear. And most beneficial to mankind was his profound anatomical research, based on carefully recorded dissections of the human body. He was, indeed a true scientist, sometimes known as 'the last great know-all'.

Wagon laager. *See* **Laager, Wagon.**

War games. The origins of games simulating wars, campaigns and battles are almost as obscure as those of chess, which emerged, perhaps, in India in the 7th century AD. For teaching, without troops, they provide an economic advantage in developing **tactics**, **logistics** and **technology**.

Commanders of antiquity probably sketched impending operations on sand tables and debated problems with colleagues, a method improved upon by the arrival of **charts and maps** and made increasingly sophisticated by properly educated **staff** officers after the 17th century. Moreover, in the late 19th century, books of rules (such as an illustrated one by the futurist novelist H. G. Wells) were published to show how to war game at home with model soldiers and equipment. But lack of realism was a serious defect with these games owing to difficulty in simulating terrain; the fact that the enemy, even if 'on the other side of the hill', was in full view; and the use of the throw of dice to inflict, inadequately, casualties.

During **World War II**, **research and development** establishments began gaming to tackle tactical/technical operational **analysis** problems. Realism was improved by imposing control from a central board over the commanders of separate boards displaying only that information which, in reality, would be at their disposal; and by using random numbers up to 100 or more, in lieu of dice, to enhance combat decision options.

In 1957 Mr Scrooby, an American, began formal, home war gaming as a hobby; and was copied in Britain by Donald Featherstone to evolve an international movement. This led in the 1960s to the manufacture, for home use, of **electronic** video war games on visual display units using software in cassettes. The vogue has stimulated elaborate, but cost-effective, **computerized** gaming for operational **training** at all military levels, playable in real time in ships and real headquarters, with relative combat skills, logistic and morale values injected as required.

War, Principles of. The statement of succinct principles of war as a foundation for the conduct of **strategy** and **tactics** is a 20th-century phenomenon. Prior to 1921, when ten principles were first presented in the British **training** manual *Field Service Regulations* (*FSR*) by Colonel **Fuller**, military wisdom had been contained in the often voluminous works of **philosophers** and the maxims of great commanders, works which were often discursive and aimed at a small élite. Fuller wanted an easier way to understanding by the mass of average students, and has been copied by many nations, including those of the **Nato** powers. He argued also in his *Lectures on FSR III* that, regardless of circumstances and new **technology**, strategy and tactics always have been governed, consciously or unconsciously, by his principles of war. They are:

1. *Selection and Maintenance of the Aim*, which speaks for itself and may well be the

product of a commander's original assessment of the situation or simply an adaptation of somebody else's stated *aim*. But it does govern plans, orders and their execution, most of which are affected by the other nine principles.

2. *Maintenance of Morale*. Without sound **morale**, failure is likely no matter how good the plan. The wise commander never ignores its importance and works diligently to enhance it.

3. *Surprise*. An enemy who is left undisturbed and psychologically composed is all the more able to counter the best-conceived plan. The catastrophic consequences of acting without surprise were commonplace in **World War I**, until the advent of the **tank** and registered **artillery** fire at the Battle of **Cambrai** in November 1917 reintroduced the feasibility of attacking without need of prolonged preparations.

4. *Economy of Effort*, designed to make the best use of all resources in time and space, and in order to create and maintain a reserve.

5. *Concentration of Force*. The fairly obvious need to apply maximum effort against the enemy preferably where he is known or thought to be weakest. In other words, to be stronger than the enemy, without necessarily committing larger numbers.

6. *Security*. Meaning the necessity to guard against enemy threats and actions. A principle which calls for first-class **intelligence** and counter-intelligence work. As an element of **C³I** in the **electronics** era this becomes vital.

7. *Offensive Action*, which demands a posture or adoption of attack to win the initiative and throw an enemy off balance. To fail in this aspect through supine inactivity is to court disaster, particularly when faced by resourceful opponents.

8. *Flexibility* is the retention of a capability to react to changing circumstances, especially by mobility, the rapid switching of firepower and arrangement of sufficient **logistic** resources.

9. *Cooperation* is the need to work closely and amicably with all friendly forces, particularly allies. Keys to this are **communications** and the establishment of secure links, as well as sound relationships between commanders, **staffs** and subordinates. This is a never-ending business, best based on long-established personal associations.

10. *Administration*, focusing attention on the vital necessity of foresight in planning and smoothly functioning communications and transport which keep logistics working efficiently.

All principles, notably the briefly expressed kind, are open to debate. There is a tendency to value some among the ten higher than others. For example, after the *aim*, *surprise* and *flexibility* are occasionally rated the most important. Less frequently *administration* is placed high yet, as entries in this Encyclopedia often indicate, in the final analysis operations hinge upon this basic principle.

Warsaw Treaty Organization. This 20-year mutual-defence pact was signed between the USSR, Albania, Bulgaria, Czechoslovakia, East Germany, Hungary, Poland and Romania on 14 May 1955 as a counter to the recently announced admission of West Germany into **Nato**. Mainly it aimed at unified military command under Russian control and legal justification for the location of their troops in Eastern Europe. Like Nato it suffered over the years from recurring outbreaks of anti-Russian nationalism, peaking in Hungary in 1956, Czechoslovakia in 1968 and after 1989, in a widespread manner. It became defunct with the collapse of Soviet Russian power in 1991.

Wars of the 16th, 17th and 18th centuries. After the fall of Constantinople in 1453 and the end of the **Hundred Years War** in 1457, 350 years of slowly accelerating progress occurred before the **French revolutionary wars** and the technological explosion of the 19th century. In this entry, the strategic, tactical and technical changes of this period are summarized, century by century.

16th century. Technically, and therefore tactically, this century was a turning-point in the art of war since it included the invention of wheel-lock and flintlock small arms, which made obsolete the **bow and arrow**. At the same time, although **artillery** failed to keep pace with small arms, it was marginally improved by the development of stronger barrels and better **ammunition**, as well as sturdier carriages. Artillery's chief effect, however, lay in a steady increase in the number and size of pieces produced by a slowly evolving **arms industry**, and their more concentrated use on land and sea.

On land in the 15th century, John **Zizka**'s use of artillery from moving **wagons** as well as static **laagers** won an advantage in mobility and protection during the Hussite Wars. But principally the effect of artillery, especially the heavier pieces of the 16th century, was to make **siege engines** and high, stone-walled **fortifications** obsolete. Henceforward the volume (though not accuracy) of artillery was a potent and, sometimes, decisive factor in battle; and, therefore, increasingly involved in duels with opposing guns as well as being a prime objective for **infantry** and **cavalry** charges. Cavalry, however, became more vulnerable as **armour** declined in protection. These trends were discernible, if not overriding, during the **Ottoman** and **Spanish wars**; and were strongly reflected in **Machiavelli**'s very influential books on military **philosophy** at the start of the century.

At sea, however, shipboard artillery, especially when installed in **carracks** and **galleons**, wrought a revolution. If the virtual demise of the **galley** at the Battle of **Lepanto** in 1571 marked the zenith of Spanish naval superiority and a decline in Ottoman power, the British victory over the Spanish Armada in 1588 was all the more important since the tactical as well as technical superiority of sailing, 'wooden walls' was demonstrated and a new naval power established.

17th century. The **Thirty Years War**, the **French wars**, the Anglo-Dutch wars and the last of the **Ottoman wars** of expansion dominated this century as European technology and techniques became predominant. Close-combat fire-power was increased by mobile **artillery** pieces such as the light, Swedish leather gun, by one-piece **ammunition**, by the introduction of the **musket** and **bayonet** and by the exclusion of the **pike**.

On land the innovative tactics of **Gustavus Adolphus** improved cooperation between **infantry** companies firing volleys, cavalry armed with **pistols** as well as **swords**, and artillery – and suggested the desirability of thinning out formations (to minimize casualties) without fatally reducing volume of fire. Indeed, the high losses from modern weapons often suffered by shoulder-to-shoulder formations frequently deterred commanders from seeking a major battle. The art of manoeuvre and siege grew into a convention designed to minimize loss of lives. The strategy was founded on the latest star-shaped **fortifications** and siege-warfare techniques devised by Marshal **Vauban** – including surrender with honour when a state of calculated indefensibility was reached.

At sea, and especially during the Anglo-Dutch **naval warfare** battles between 'wooden walls', such niceties were at a premium. Once Admiral **Blake**'s *Fighting Instructions* had been issued, lines of broadside-firing ships pounded each other mercilessly at short range, with heavy loss of life, prior to the traditional boarding and hand-to-hand combat. Sinkings were few and the striking of colours mainly at the last gasp. Significantly, by the end of the century it was plainer to the British, Dutch and Spaniards than to the French that navies bestowed a strategic and economic influence at least the equal of armies – a lesson drummed home in the ensuing wars of the 18th century.

18th century. In what was tantamount to a world war, France was at hostilities with her neighbours for more than half this century – and worsted in most of them (until the Battle of **Valmy** in 1792) because of being overstretched and outmatched by superior enemy commanders. This was especially so of Generals **Eugene** and **Marlborough**, who between them, in the Spanish War of Succession (1701–14), defied French **siege warfare** conventions by manoeuvring to seek and win the battles of Chiari (1701), Blenheim (1704), Ramillies and Turin (1706), Oudenarde (1708) and **Malplaquet** (1709). Nevertheless, despite continuous major wars and the founding of the first War College by John of Nassau in 1697, the invention of the Puckle **machine-gun** in 1715 and a breech-loading **rifle** *c.* 1750, the art of war progressed little until the latter half of the century. Marshal de **Saxe**'s *Reveries upon the Art of War*, though technologically aware and highly futuristic, was (perhaps for that reason) rejected.

A turning-point was reached in the **Seven Years War** (1756–63), however, when the victories of **Frederick the Great** and the defeat of the French by the British in India and Canada prompted widespread organizational and technical reform. Frederick's *Instructions to the Generals* began to influence many armies; in France, de **Gribeauval** rationalized the **artillery**, Marshal de Broglie conceived the division as the principal army formation, and **Guibert** promoted conscription for a 'nation in arms' in his *Essai général de tactique* (1772). And in Britain James **Watt** was developing a practical steam engine at the start of the Industrial Revolution which was mightily to stimulate the **armaments industry** during the **American War of Independence** and the **French revolutionary wars**. Of a sudden, inventiveness and the *laissez faire* philosophy of Adam Smith's *The Wealth of Nations* produced new equipment for navies and armies and initiated **air warfare**; including the rifled firearm (*c.* 1750), Cugnot's steam road vehicle in 1769, **Shrapnel**'s air-burst shell in 1786, the **Montgolfier brothers**' hot-air **balloon** in 1783 and the **parachute** in 1797.

On land at the start of the century, Marlborough initiated better **infantry** fire control through volleys by platoons instead of companies, and had been notably thorough with **logistics**. And Frederick's highly disciplined army, with its remarkable strategic triumphs, was recognized as a model of professional excellence. Yet, notwithstanding, Gribeauval's and de Broglie's reforms, doctrinal **tactics** remained almost sacrosanct while it took the genius of **Napoleon** to give a new dimension to **strategy**, significantly helped by the introduction of Chappé's Europe-wide semaphore-signalling system of 1792 (*see* **Communications**). But irregular tactics by skirmishers and **guerrillas**, both in Europe and North America, did indicate a potent doctrine linked to traditional methods.

At sea it was another matter, with the Dutch and British making full use of their strategic insight to dominate world trade at the beginning of the century and, by the end of it, manage usually to **blockade** France and supply her enemies with arms and money. Yet the tactics of line-of-battle remained largely unchanged, notwithstanding the introduction in the 1760s of better flag signalling, which did much to improve command and control in battle. Already, however, the signs of impending technological revolution in the 19th century were in sight, including such primitives as **Bushnell**'s submarine in 1776, the sea-**mine**, the ship-smashing carronade gun and in 1783 d'Abban's steam paddle-boat. Meanwhile, Eli **Whitney**'s and Boulton & Watt's mass-production methods stood ready to benefit from **electromagnetic** discoveries, new **metallurgy** and the wars of the 1850s.

Watson-Watt, Sir Robert (1892–1973). A mathematician and electrical engineer who, during **World War I**, worked as a meteorologist at the Royal Aeronautical Establishment. He experimented with a cathode-ray tube oscilloscope (invented in 1897 by Ferdinand Braun) for weather forecasting. The results of this, in 1934, led to his suggesting Radio Direction Finding (RDF) for locating enemy **aircraft**, instead of trying to destroy them with a high-energy 'death ray', which he dismissed as impractical. As Superintendent of the Bawdsey Research Station and in the Air Ministry he was a leader of **World War II** development of **radar** (as RDF came to be known) for defensive and offensive operations, as well as working on **electronic navigation** aids.

Watt, James (1736–1819). A mathematical-instrument maker who, in 1761, began experimenting with an uneconomic Newcomen steam pumping engine (*see also* **Power-generation equipment**). In 1765 he invented a much more efficient engine with steam piped to the cylinder from a separate condenser. Meanwhile, he planned harbours, canals and bridges but, after 1769, concentrated on developing his condenser steam engine in collaboration with Matthew Boulton. In 1776 John **Wilkinson**'s foundry proved the engine's merits by successfully blowing furnace bellows. By 1796 the Boulton & Watt Company double-action steam engines, designed to convert the reciprocal piston motion into rotation to drive machinery; further modified to deliver greater power for 25 per cent less fuel than a Newcomen engine; and mass-produced in the 'flow-line' Birmingham Soho factory, were a vital force in the Industrial Revolution. And made controllable in speed by Watt's centrifugal governor and more powerful by his so-called compound expansion system. Yet, irrationally, he opposed use of high-pressure steam. Among his many inventions are a steam carriage, with gears for hill-climbing and a spiral ship's propeller. Unlike the Wilkinson brothers he rejected, for patriotic reasons, French offers of employment.

West European campaigns in World War II. At the start of **World War II** Germany possessed on land a pronounced **technological** advantage based on superior **aircraft**, **armoured fighting vehicles** and signal **communications** with a **tactical** doctrine superior to that of other European nations. And, after the conquest of Poland in 1939, proceeded to put it into effect.

1940–43. The German invasions of Denmark and **Norway** started on 9 April and were followed by the invasion of Holland and Belgium on 10 May as the immediate prelude to the Battle of France. These campaigns might have won the war for **Hitler** had he not let the British escape from **Dunkirk**, or if he had been more foresighted and determined in trying to win the Battle of **Britain** before turning against Russia. As it was he gave Britain breathing space to rebuild her forces and mount the **strategic bombing** offensive and the **Commando** raids, neither of which between 1940 and the **Normandy** landing in 1944, brought Germany to her knees, although each contributed strongly to wearing her down and diverting considerable effort from other fronts. Indeed, Commando raids against Norway helped persuade Hitler that the Allies would invade there, while those at Saint-Nazaire and **Dieppe** deeply influenced the 1944 invasion of France since they focused German attention on the vital importance of ports to an invader, causing immense diversion of capital expenditure on static defences. The **fortifications** were hardly tested, because the Allies managed to supply their armies over the beaches and through the prefabricated **Mulberry** harbour.

1944. The massive air offensive and the defeat of the U-boats in the Battle of the **Atlantic**, along with excellent deception measures and extremely well-prepared **amphibious** and **airborne** warfare techniques and technology, ensured the success of the landing in Normandy on 6 June. Thereafter it was mainly a question of who won the reinforcement race between the invaders from the sea and the Germans trying to keep their **rail** and road links open against air attack and sabotage from Special Operations Executive-sponsored **guerrillas**. Because the Germans lost this race and also because Hitler insisted upon unyielding defence of Normandy, the destruction of his forces during the break-out was all the more complete in mid-August. But the war might have ended sooner if Generals Eisenhower and **Montgomery** had coordinated their plans better and taken more care over **logistics**. For whichever was right in the argument over whether a narrow or broad-fronted advance into Germany was best, Eisenhower was at fault (against advice) in setting up his HQ on the west side of the Cotentin peninsula, where its signal **communications** with the rapidly advancing armies were totally inadequate – thus preventing him controlling events. And Montgomery erred, despite warnings from Admiral Ramsay, by his failure to concentrate on the capture and opening up of Antwerp, with inevitably serious supply shortages. As a result the advancing armies were neither properly commanded nor controlled by the Supreme Commander and the bold attempts to win either at **Arnhem** or by breaking through the **Siegfried Line** came to naught with everybody waiting for the deferred Scheldt River battles to be concluded.

This provided Hitler with the opportunity to launch his valedictory and self-consuming Ardennes offensive in December before the Rhineland battles in February and March 1945 announced the closing stages of the Battle of **Germany**.

Westmoreland, General William (b. 1914). An American Gunner who, in **World War II**, fought in **North Africa**, Sicily and North-West Europe, and as an **airborne** regiment commander in **Korea**. On 20 June 1964 he took command of the US Military Assistance Command Vietnam (MACV). On 2 August North Vietnamese **torpedo-boats** attacked an American **destroyer** in the Tonkin Gulf. Almost at once **bombing** of North Vietnam was authorized by Washington and Westmoreland was appointed commander of US forces in Vietnam as the enemy stepped up his offensive, producing an escalation of the **Vietnam War**. Until made Army Chief of Staff in the summer of 1968 he did what was **tactically** necessary, frequently hampered by **strategic** restrictions imposed in election years from Washington and elsewhere. He was forbidden to attack the vital **Ho Chi Minh Trail** upon which the enemy's **logistics** depended and was conscious of the anti-war feeling in the USA which managed to convert his victory in the **Tet Offensive** of 1968 into a defeat. He remains a politically controversial figure in the aftermath of an out-of-court settlement in 1984 when he sued the Columbia Broadcasting System for libel over its implication of suppression or alteration of intelligence data relating to the Tet Offensive.

Wheels with a central hub, made of solid **wood** or planks, probably were invented prior to the 4th millennium, maybe in Mesopotamia, and soon were carrying wagons and **chariots**. Examples of lighter and stronger spoked wheels in Egypt date from 1453 BC. At an early stage metal tyres and hub bearings were fitted. And much later all-metal wheels were manufactured for **railways** and, for road vehicles, solid rubber, followed by pneumatic rubber, tyres, which were patented in 1845 by Robert Thompson and improved and manufactured by John Dunlop in 1888.

Whitney, Eli (1765–1825) is best known for his invention of the cotton gin in 1793. But his greatest innovation for America was the emulation in 1798 of Boulton & Watt's Soho factory 'flow-line' methods for building steam engines, and making use of Henry Maudslay's lathe (1797), to make precision, standardized, interchangeable parts for mass-production of cheap **muskets** – another vital stage in the burgeoning Industrial Revolution during the **wars of the 18th century**.

Whittle, Air Commodore Sir Frank (b. 1907). A test pilot who obtained an engineering degree at Cambridge University and in 1931 began **research and development** into gas-**turbine jet** engines at Power Jets Ltd. Underfunded and frequently frustrated, he managed by 1937 to have working a turbo-jet engine with special nickel-**steel** blades (*see* **Metallurgy** *and* **Propulsion, Means of**). In 1941 one of his axial-flow, compression engines (the forerunner of most types to come) powered the first British jet aircraft. After **World War II** he became a technical adviser to the Ministry of Supply, but in 1948 retired from the RAF and a leading role in jet-engine development.

Whitworth, Sir Joseph (1803–87), began work in a cotton mill but his interest in Henry Maudslay's lathe, the 'flow-line' production methods of the Boulton & Watt Company and Eli **Whitney**'s interchangeability system led him in 1833 to form a machine-tool company. By 1854 his gauges, lathes and standardized nuts and bolts possessed an international reputation. During the **Crimean War** he began manufacturing **rifles**

and **artillery** with hexagonal bores and a slight twist; their performance was superior to rifled pieces but they were unpopular. In the 1860s, however, his breech-loading, homogeneous **steel** barrels (similar to **Krupp**'s) were found superior and more reliable than those of his great rival, Thomas **Armstrong** – especially against **iron armour** with special hollow shells instead of conventional solid shot.

Wilkinson, John (1728–1808). The son of an iron-foundry owner who designed the machine which bored the cylinders of James **Watt**'s steam engine (1776), which he used for manufacturing wrought **iron**. His products were widely used, including parts for the first iron **bridge** and an iron barge. In 1786 he began to make **artillery**, **mortars** and **ammunition**, some of which he smuggled to the French for use against his own countrymen in the **American War of Independence** – when his estranged brother William was teaching the **Schneider** Company at **Le Creusot** how to cast and bore tubes and cannon barrels.

Wilson, Walter (1874–1957), was a naval officer who began designing petrol engines for **gliders** in 1899, and designed **motor vehicles** and an **artillery** tractor for the Armstrong Whitworth Company prior to **World War I**, when he rejoined the Royal Navy and was put to work designing **armoured fighting vehicles** (AFVs). In 1915 he designed for the Foster Company the original rhomboidal-shaped **tank**, the transmissions and tracks of which were crucial elements. He adapted epicyclic gears for the first one-man-driven tank and, after the war, developed self-change gearboxes for motor vehicles and AFVs, leading to the design of a controlled differential box, the prototype for most present-day tracked steering and transmission systems. In **World War II** he contributed to the design of TOG, an anachronistic and faulty heavy tank.

Wingate, General Orde (1903–1944). A Gunner who in 1936 became involved with organizing Israeli **guerrilla** forces against the Arabs in Palestine (*see* **Arab–Israeli wars**). In 1940–41 he performed the same services for Abyssinian guerrillas against the Italians and was next sent to **Burma** to initiate the controversial long-range **jungle** penetration Chindit raids against the Japanese. These culminated in 1944 when he obtained permission to launch a large **airborne** incursion into north-central Burma to disrupt enemy **logistics**. He was killed in an air crash a week after its start when results were still uncertain – and never fully realized.

Wood (xylem). Long before the discovery of metals, wood in its many species, along with **stone**, was the most important material for military purposes. Not only was it made into weapons and missiles, such as clubs, **bows and arrows** and **siege engines**, but was vital for ships, **fortifications**, **wheels** and **chariots**, among many other things. And although it progressively fell into disuse with the discovery of **bronze**, **iron**, **steel** and **plastics**, it continues to have many uses to this day, notably for field **engineering**.

Standing timber has always been a vital strategic resource; and access to forests more important to navies than armies until the mid 19th century. For example, Britain's struggle for control of the Baltic during the **French revolutionary wars** resulted from the need for access to pine for ships' masts and spars. Change, however, was inevitable when insatiable demand in the 18th century depleted the hardwood forests and made essential the conversion to iron and steel for **shipbuilding**. Just as the changes introduced by the invention of gunpowder and firearms in the 14th century made inevitable the redundancy of lightweight, flexible species, such as yew and bamboo, for bows.

By the 20th century demand for wood for ships and weapons had greatly declined – only to be revived for light kinds in **aircraft** construction – providing employment, initially, for furniture-makers. But simultaneously duralumin, **aluminium** alloys and lightweight steels offered superior substitutes. Yet one of the most successful **fighter-**

bombers of **World War II**, the DH Mosquito, was mostly constructed of wood bonded by epoxy glue. Since then, however, **plastics** have joined with metals virtually to exclude wood from weapon systems and vehicles.

World War I. The causes of this war, starting 28 July 1914, were many and varied, ranging through nationalist vengeance to pay off old scores, trade rivalries and the almost uncontrolled militarism which, under bellicose General Staffs, influenced weak royal autocracies and their ineffectual democratic governments. **Propaganda** helped make this a popular war – to begin with. Some states were well prepared for it – notably, on land Germany and France, and at sea Britain and Japan – unlike Austro-Hungary, Russia and Serbia, who were in decline from previous wars and unfit for intense armed conflict. As for Italy, Romania, Portugal and America, who joined later, they became caught up in something they could not easily avoid. The nations that began the war did so in the belief that offensive action would make it a short one and that it would be 'over by Christmas'. But insufficient understanding of **technology's effects** and **logistics** put paid to that gross misconception.

The failure by both the Germans and the French and British to reach a conclusion in 1914 during the operations connected with the Battle of the **Marne** and the subsequent **Race to the Sea**, was their inability to overcome modern defensive systems and solve the **logistic** equation. As was also the case in the Serbo-Austrian battles and, on the east front, the battles of Austria and Germany versus the Russians. The Russians lost at **Tannenberg** but won twice before Warsaw as stalemate set in. **Artillery** and **machine-guns** forced **infantry** underground, behind

World War I in Europe, 1914–18

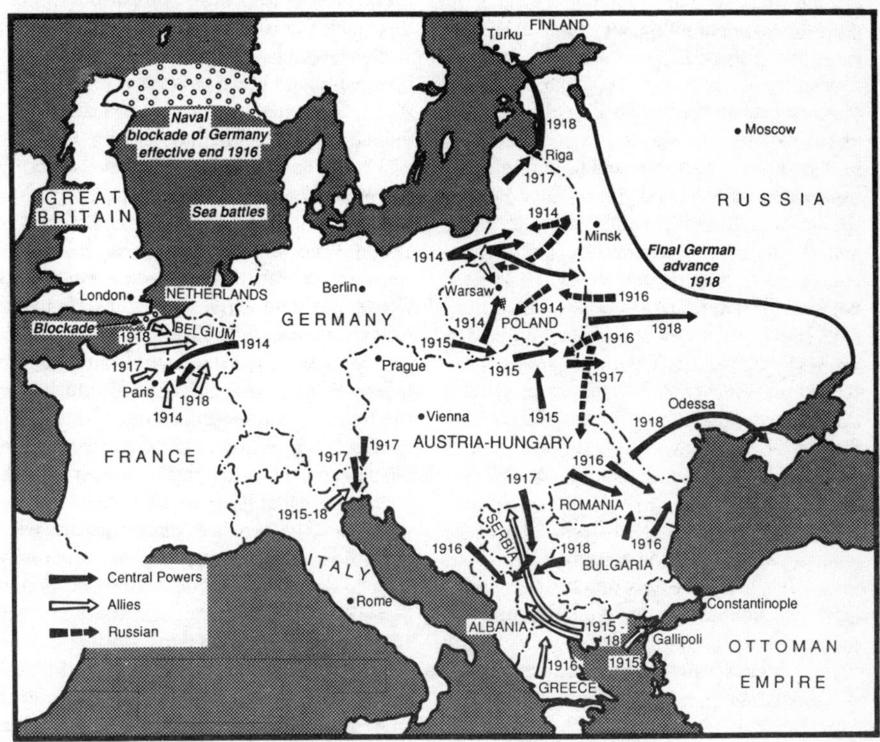

barbed wire, and doomed **cavalry** so that **reconnaissance** became increasingly the function of **air forces**. Likewise at sea the threat of **mines** and **torpedoes** denied admirals the decisive battles they had envisaged. **Blockade** and warfare against commerce predominated; **submarines** and aircraft played an ever larger part. Nothing had gone to plan. Nobody could think of a way out of the impasse through diplomacy, because each country was led to believe in victory. And the military leaders, faced by a sort of **siege warfare** against unturnable **fortifications**, were barren of ways to win without vast expenditure of lives and material – thus fostering the philosophy of attritional warfare. Indeed, at this stage, only Japan profited by siding with the Allies. For a mere 1,207 casualties, she drove the Germans out of Tsing-tao in 1914 and in due course acquired ex-German islands to strengthen her position in the Pacific Ocean.

Only the Gorlice Offensive, in 1915, had a decisive look as the Russians fell back in disorder and demoralization. Elsewhere that year, and also in 1916 and much of 1917, stalemate predominated; on **Vimy Ridge**, in Champagne, at **Verdun** and on the **Somme**, as well as in Italy, the Balkans at **Gallipoli**, in Egypt, Palestine, **Mesopotamia** and, to some extent, in East and West Africa. Neither the defeats of Serbia in 1915, nor of Romania and the Brusilov Offensive in 1916, were conclusive. And the sea battles of **Dogger Bank** and **Jutland** did little more than consolidate the Royal Navy's obvious supremacy yet without making it possible to enter the Baltic and help Russia or land armies behind the German lines. Indeed the paucity and ineptitude of **amphibious** operations were as remarkable as the exaggerations of **strategic bombing**'s effects.

Not until 1917 did attrition break the impasse when blockade began to hurt both sides. The Russian Revolution, Germany's unlimited submarine campaign against commerce, America's entry into the war and the mutiny of the French Army in the aftermath of the Battle of Chemin des Dames, conjointly disturbed the military balance. Initially it went Germany's way in Flanders and at **Caporetto** in October. But the Battle of **Cambrai**, with the first use of massed **tanks** and unregistered artillery fire to restore the principle of surprise (*see* **War, Principles of**) to land battles, hinted at radical military change, one that became doctrine after the Hindenburg Offensive in 1918 had been snuffed out by successive surprise counterattacks which preceded the definitive tank counterstroke Battle of **Amiens** on 8 August. This all-arms battle cracked the German **morale** and broke her leaders' resolve as Allied armies everywhere else overran the disastrously weakened Central Powers; in Palestine, the Balkans and Italy. The unexpected collapse washed away monarchies and governments in waves of revolution by the masses, who were stimulated by hopes of a new dawn in Russia, clouded by the Russian revolutionary wars. As it reshaped the postwar world, the **Versailles Peace Treaty** took note of these events in Russia, but other than extract the Central Powers' forces from their midst, did little to curb them.

The total cost in lives from combat has been estimated at a little over 8 million, and there were also, very approximately, $6\frac{1}{2}$ million civilian deaths attributable to a war which some were deluded into thinking would end war.

World War II. This war, though sharing some of the causes of its predecessor, was affected by two other important elements. Nobody was ready for it and it was unpopular. Arguably the initiation of the Chinese–Japanese War on 7 July 1937, which was the next stage of Japan's wars of expansion, started World War II. The Japanese flaunted a policy of greed and aggrandisement, stimulated by a belief that the Occidental powers might bend to their bellicosity. The sins were like those that corrupted the criminally minded minions of Adolf **Hitler** and Benito **Mussolini** as, in the guise of revolution, they imposed Fascist rule on Germany and Italy.

Militarily, too, this war was different from **World War I** since Hitler, as the motivating aggressive force in Europe, was a political

World War II in Europe: Axis expansion, 1939–42

opportunist whose shrewd grand **strategy**, which unopposed conquered Austria and Czechoslovakia, was not complemented by a properly thought-out military strategy. The invasion of Poland on 1 September 1939 was a gamble with a Wehrmacht that was ready only for a short war. The decision to attack France, via Holland and Belgium in May 1940, was as unexpected and unprepared a notion as the subsequent spur-of-the-moment decision to precede it by the invasions of Denmark and **Norway** in April. Hitler got away with all three by a combination of surprise, the Wehrmacht's prowess, the introduction of sophisticated **tactics** and **propaganda**, and the latest **technology**, plus the inescapable fact that none of his opponents (like his ally Italy when she joined forces in June 1940 in the hope of easy pickings after the Battle of France), was ready for modern war. (*See* **West European campaigns in World War II**.)

Hitler's superficial understanding of military strategy was revealed the moment he failed to concentrate on winning the Battle of **Britain** with an immediate (and feasible) invasion. He suicidally overstretched Axis capability by preparing not only for the gigantic Russo-German War in 1941 but also, simultaneously, being drawn into Italy's **North Africa battles**, the invasions of Yugoslavia, Greece and **Crete** and peripheral adventures in Syria and Iraq – besides an attempt to bring Britain to her knees by night **bombing** and in the Battle of the **Atlantic**. It was another symptom of Axis capriciousness that as Japan (deprived by American embargoes of vital war materials) indicated a clear intention to spread the war into the Pacific Ocean and South-East Asia, no joint strategic plan and little political coordination was arranged. Indeed, Hitler did not share his plan to invade Russia with Japan and was kept in the dark about the timing of the attack on **Pearl Harbor** on 7 December and Japan's subsequent plans.

Yet despite these defects and the extraordinary overconfidence of the Axis in challenging the rest of the world, it is remarkable that it came so close to winning in 1941 and 1942. Undeniably this was more a manifestation of Allied unreadiness and outmoded thinking than of moral or material Axis superiority – claim the opposite as Axis leaders would. Pre-war sins of omission lay at the heart of defeats which rolled the Russians back to the Caucasus and Stalingrad in August 1942, drove the British into Egypt, and forfeited to the Axis China's entire coastline (including Hong Kong), the Philippines, **Malaya**, the Dutch East Indies, **Burma** and many key islands in the **Pacific Ocean War**. So too did the unwillingness of the Allied leaders to recognize the horror of the threat and create firm policies with modern military forces as a **deterrent** to Axis ambitions.

It was the Allied tragedy that, having allowed the enemy to seize such vast territories at such relatively low cost in less than three years, it would take more than three years at very high cost, starting with the Battle of **Midway** on 5 June 1942, to win them back. But the length of time needed to raise, train, equip and supply the Allied counter-offensives might have been a lot longer had they not, by use of superior **communications** and assisted by a great willingness to cooperate among themselves, devised a coherent, joint, aggressive strategy in January 1942. This was the most vital achievement of Winston **Churchill** and President **Roosevelt** when they pooled their ideas and resources for offensives which would concentrate more against Germany than Japan, yet providing strength enough to throw the latter on the defensive.

The unveiling of this strategy was slow and controlled by **logistic** restraints. It was also conditioned by political factors such as the thorny problems of collaboration with Russia and China, to keep them in the war, and plans to reshape the world when peace returned. It was accompanied by disagreements over priorities related to prestige and practicalities, for example, the merits of **strategic bombing** as a war-winning concept as well as insurance of the survival of independent air forces. And the American readiness, for prestige and politics, to risk disaster with an invasion of France in 1942 which gave way to the prudence of British peripheral attacks upon southern Europe via Algeria and Tunisia and into Sicily and Italy in 1943, thus postponing the invasion of France until 1944. Or the high-level American debate between those like Admiral **Nimitz**, who wanted to head more directly against Japan, and General **MacArthur**, who persuaded Roosevelt of the political and military advantages of tackling the Philippines first in 1944. And so many other crucial decisions like the deliberate introduction of the cult of **guerrilla warfare** into politically sensitive lands (with untold postwar consequences); the development of the **atom bomb** to pre-empt any possibility of the enemy building one; and dropping them on Hiroshima and Nagasaki to present Japan with a face-saving excuse to admit defeat and save an entire nation from hara-kiri.

Whatever the short-term consequences on the Axis powers in terms of destruction and on the world in loss of life from the controversial and unpremeditated announcement of the policy of unconditional surrender in January 1943, no Versailles Peace Treaty was

World War II in Europe: Allied advances, 1942–5

needed to reshape the old order this time. It was reshaped instead by *force majeure* of the victorious powers and, shortly, by instinctive nationalist reactions to correct the depredations of the colonial expansions of the past and restore racial and religious independence, by protracted wars if necessary. To attain that state of freedom, the roughly estimated price in military dead was some 15 million, plus attributable civilian deaths anything between 26 and 34 million.

Wright brothers, Orville (1871–1948) and **Wilbur** (1867–1912). Inventors, magazine publishers and bicycle manufacturers who, in 1896, began experimenting with large kites and **gliders**. They designed an original rigid biplane controlled by rudder, elevator and wing 'warping' (later replaced by ailerons). Powered by a 12hp engine and driven by a propeller of their own design, the Flyer contained all the essentials of future **aircraft**; carefully thought out and meticulously tested in workshops and wind tunnel before making the first cautious, manned flight on 17 December 1903. The Wright Company began manufacturing aircraft for governments, which used them mostly for military **research and development**. But after Wilbur's death, Orville tended to concentrate more on scientific research than the distractions of commerce.

Y

Yamamoto, Admiral Isoruku (1884–1943). As a young officer Yamamoto was seriously wounded at the Battle of **Tsushima**, but survived to become a **gunnery** specialist prior to taking a lead in the development of **aircraft-carrier** operations and, in the Aeronautics Department, direction of its **technology**. From being Navy vice-minister in 1936 he was given command of the Japanese Combined Fleet in 1940. With personal knowledge of the USA and Britain he was better aware than most Japanese of the danger of war against them. But he was also a gambler who, though warning of the considerable risks of war, staked all upon the **Pearl Harbor** raid to win a decisive victory. It was his misfortune to be served by leading subordinates who did not thoroughly understand his concepts. As a result neither at Pearl Harbor nor with any of the ensuing aircraft-carrier raids were comprehensive results achieved. Defeat at **Midway** in June 1942 marked the start of a decline which he recognized prior to being intercepted, shot down and killed by American fighters on 18 April 1943.

Yi Sung Sin, Admiral (d. 1598). During the Korean–Japanese War (1592–98), the Japanese fleet covering army reinforcement **convoys** in the Yellow Sea was intercepted piecemeal and wiped out in a series of actions by a Korean fleet brilliantly commanded by Yi Sung Sin. In Yi's fleet were one or two low-silhouette **galleys** of his own design, armed with **artillery** (unusual in the Far East at that time) and with overhead **iron armour**, said to be the first of its kind in the world. Chiefly, Yi's **tactics** were conventional **ram** and boarding, supported by archers firing flaming arrows. Chinese intervention then pushed the overstretched Japanese back to Pusan in the south. In 1597 the Japanese drove north again and, at a moment when Yi was out of favour, also won a naval battle. Recalled, he harassed Japanese convoys but, when inflicting another decisive defeat at Chinhae Bay, was killed. The Japanese, who lost 200 out of 400 ships, then felt compelled to negotiate peace – and entered a period of nearly 300 years of isolation.

Z

Zeppelin, Brigadier Count Ferdinand von (1838–1917). A technically educated German army officer whose interest in flight began during the **American Civil War** (in which he fought) when he went up in a **balloon**. He fought in the Austro-Prussian War and the **Franco-Prussian War** but retired in 1890 to promote **airships** for passenger, cargo and military purposes. With Theodor Kober a design for a rigid airship driven by petrol engines and propellers was evolved. The German military authorities were sceptical until 1900, when LZ1 flew successfully. Thereupon both the Navy and the Army adopted zeppelins, mainly for **reconnaissance**. But it was by strategic **bombing** during **World War I**, that they achieved most success. When, by the end of 1916, their extreme vulnerability proved fatal, the Zeppelin Company turned to building **bombers**.

Zhukov, Marshal Georgi (1896–1974). A peasant who fought in **World War I** and commanded **cavalry** in the Red Army in the Russian revolutionary war. He first won fame in command of the victorious Russian forces at the Battle of the Khalkin River in August 1939 and in 1941 was appointed Army Chief of Staff by Josef **Stalin**. In that capacity he failed to convince Stalin of the imminence of the German invasion on 22 June 1941. But for nearly every major battle of the Russo-German War he was employed to plan and/or command at the front. For example, he fought the Battle of **Moscow** in December 1941, and planned the victories of Stalingrad in 1942 and **Kursk** in 1943. He dominated the re-entry into White Russia and Poland in 1944, and personally commanded the capture of Berlin in 1945 during the Battle of **Germany**.

In 1946 he was welcomed home as a hero and almost immediately exiled to the provinces by the ever-paranoiac Stalin. But he was brought back after Stalin's death in 1953 as First Deputy Minister of Defence, becoming Minister of Defence in 1955 and, shortly after, a member of the Praesidium of the Communist Party Central Committee. He was dismissed in 1957.

Zizka, John (Jan) (*c.* 1376–1424), led the Bohemian contingent at the Battle of **Tannenberg** (Grunwald) in 1410 on the victorious Polish side. In 1419 he joined the Protestant Hussites in their war (*see* **Religion, Wars of**) against Catholicism and won many battles and sieges by innovation. Realizing that the latest plate **armour** was defeating the **bow**, he taught **artillery** and **infantry** to shoot to kill instead of only for noisy effect; and told the infantry to fire from the shoulder. And he formed wagon **laagers** not only as **fortifications** to hold commanding ground, but occasionally, as triumphantly at the Battle of Kutna Hora in 1421, having them fight in formation on the move. That same year, at the Siege of Rabi, he lost the sight of his remaining eye. Yet until his death from plague, he continued to command with success against the Catholics and also stopped a Hussite civil war.

Bibliography

The bibliography of war and technology is so vast as to be beyond the capacity of any book, let alone one the size of this Encyclopedia. Therefore this bibliography has deliberately been kept brief in order to provide the readers with a digestible list of the most useful reference works. The emphasis is upon the best encyclopedias, bibliographies and specialized works. The list covering specific wars is very select indeed and includes only those few official histories which are almost unique in themselves.

See also references to books (printed in italics) in the Index.

Aircraft and Air Warfare

Brown, D., Shores, C., and Macksey, K., *The Guinness History of Air Warfare*, Guinness, 1976

Frank, N., *Aircraft versus Aircraft* (4 vols.), Bantam, 1986

Jablonski, E., *Air War* (4 vols.), Doubleday, 1971

Jane's All the World Aircraft, Jane's Publishing

MacBean, A., and Hogben, A., *Bombs Gone*, Stephens, 1990

Robinson, D., *The Zeppelin in Combat*, Foulis, 1961

Armour and Weapons

Bailey, J. B., *Field Artillery and Firepower*, Military Press, 1987

Crow, D., and Icks, R., *Encyclopedia of Tanks*, Barrie & Jenkins, 1975

Jane's Armour and Artillery, Jane's Publishing

Jane's Infantry Weapons, Jane's Publishing

Jane's Weapon Systems, Jane's Publishing

Ogorkiewicz, R., *Armour*, Stevens, 1960

Padfield, P., *Guns at Sea*, Evelyn, 1973

Wilkinson, F., *Antique Firearms*, Guinness, 1969

Wilkinson, F., *Battle Dress*, Guinness, 1970

Wilkinson, F., *Edged Weapons*, Guinness, 1970

Bibliographies

Albion, D., *Naval and Maritime History, An Annotated Bibliography*, Manson Institute, 1972

Eggenberger, D., *Dictionary of Battles*, 1967

Ensor, A. G., *Subject Bibliography of the Second World War*, Deutsch, 1977

Ensor, A. G., *Subject Bibliography of the First World War*, Deutsch, 1979

Higham, R., *Official Histories*, Kansas State University, 1970

Higham, R., *A Guide to the Sources of British Military History*, California University Press, 1971

Higham, R., *A Guide to the Sources of US Military History*, Archon, 1975

Keegan, J., and Wheatcroft, A., *Who's Who in Military History*, Hutchinson, 1976

Pinard, M., *Dictionnaire de biographie française*, Libraire Letouzy, 1956

Scribner's Dictionary of Scientific Biography, Scribner, 1980

Tunney, C., *Biographical Dictionary of World War I*, Dent, 1972

Communications and Code Systems

Baker, W. J., *A History of the Marconi Company*, Methuen, 1970

Jane's Military Communications, Jane's Publishing

Kahn, D., *The Code-Breakers*, Weidenfeld & Nicolson, 1966

Encyclopedias and Works of General Interest

Clarke, I. F., *Voices Prophesying War*, OUP, 1966

Collins Modern Thought, Collins, 1988

Dupuy, R. E. and T. N., *The Encyclopedia of Military History*, Macdonald, 1970

Encyclopedia Britannica, Benton

Fuller, J. F. C., *The Decisive Battles of the Western World* (3 vols.), Eyre & Spottiswoode, 1956

Ladd, J., *Commandos and Rangers*, Macdonald, 1970

Ruffner, F. G., and Thomas, R. C., *Code Name Dictionary*, Gale, 1963 (includes military slang)

The Times Atlas of World History, Times Books, 1978

United States Government, *Soviet Military Power 1981–*

Fortifications

Toy, S., *A History of Fortification 300 BC – AD 1700*, 1955

Intelligence (*see also* **Communications and Code Systems** and **Wars**)

Hinsley, F. H. *et al.*, *British Military Intelligence in the Second World War* (5 vols.), HMSO, 1979–88

Lewin, R., *The Other Ultra*, Hutchinson, 1982

Logistics

Jane's Combat Support, Jane's Publishing

Macksey, K., *For Want of a Nail*, Brassey's, 1989

Thompson, J., *The Lifeblood of War*, Brassey's, 1991

Philosophers of War

Earle, E. M., *Makers of Modern Strategy*, Princeton, UP, 1952

Phillips, T. (ed.), *Roots of Strategy*, Lane, 1943 (includes: *The Art of War* by Sun Tzu; *Military Institutions of the Romans* by Vegetius; *Reveries Upon the Art of War* by de Saxe; *Military Instructions* by Frederick the Great; *Military Maxims* by Napoleon)

Quotations

Heinl, R. D., *The Dictionary of Military and Naval Quotations*, US Naval Institute, 1966

Ships and Naval Warfare

Archibald, E., *The Wooden Fighting Ship*, 1968

Archibald, E., *The Metal Fighting Ship*, 1971

Dyer, G. C., *The Amphibians Came to Conquer*, US Government, 1969

Frere Cook, G., and Macksey, K., *The Guinness History of Sea Warfare*, Guinness, 1975

Jane's Fighting Ships, Jane's Publishing

Kemp, P. (ed.), *The Oxford Companion to Ships and the Sea*, OUP, 1976

Macintyre, D., and Bathe, B., *The Man of War*, Methuen, 1968

Padfield, P., *Guns at Sea*, Evelyn, 1973

Space

Pebbels, C., *Battle for Space*, Beaufort, 1983

Technology

McGraw-Hill Encyclopedia of Science and Technology (20 vols.)

Macksey, K., *Technology in War*, Arms & Armour, 1987

A History of Technology (8 vols.), Oxford

Wars (listed chronologically)

Alexander the Great by R. Lane Fox, Penguin, 1987

The Roman Legions by H. Parker, 1959

The Art of War in the Middle Ages (2 vols.) by C. Oman, Greenhill, 1990

The History of the Mongol Conquerors by J. Saunders, 1971

The Crécy War by A. Burns, Greenhill, 1990

The Agincourt War by A. Burns, Greenhill, 1991

The Art of War in the Middle Ages by C. Oman, Greenhill, 1990

A History of the Art of War in the 16th Century by C. Oman, Greenhill, 1990

The Decisive Battles of the Western World, vols. 1 & 2, by J. F. C. Fuller, Eyre & Spottiswoode, 1954

John Zizka and the Hussite Revolution, by F. Heymann

The Thirty Years War by C. V. Wedgewood, 1938

Marlborough, His Life and Times by W. S. Churchill, Cassell, 1933

The Battles of the Crimean War by W. Baring Pemberton, 1962

A Centennial History of the (US) Civil War (2 vols.) by B. Catton, 1965

Sadowa by H. Bonnal, 1907

The Franco-Prussian War by M. Howard, Hart-Davis, 1962

The Spanish War 1898 by G. J. A. O'Toole, Norton, 1982

The Great Anglo-Boer War by B. Farwell, Harper & Row, 1976

The British Official History of the Russo-Japanese War (5 vols.), HMSO, 1909

Purnell's History of the First World War (8 vols.), ed. B. Pitt, BPC, 1969

Purnell's History of the Second World War (8 vols.), ed. B. Pitt, BPC, 1966

The Decisive Battles of History, vol. 3, by J. F. C. Fuller, Eyre & Spottiswoode, 1956

The Arab–Israeli Wars by C. Herzog, Arms & Armour, 1982

The Communist Insurrection in Malaya 1948–60 by A. Short, Muller, 1975

Vietnam War Almanac by H. G. Summers, Fact on File, 1985

The British Army in Ulster by D. Barzilay, Century

The Royal Navy and the Falklands War by D. Brown, Cooper, 1987

Military Lessons of the Gulf War by B. W. Watson *et al.*, Greenhill, 1991

Index